Date Due

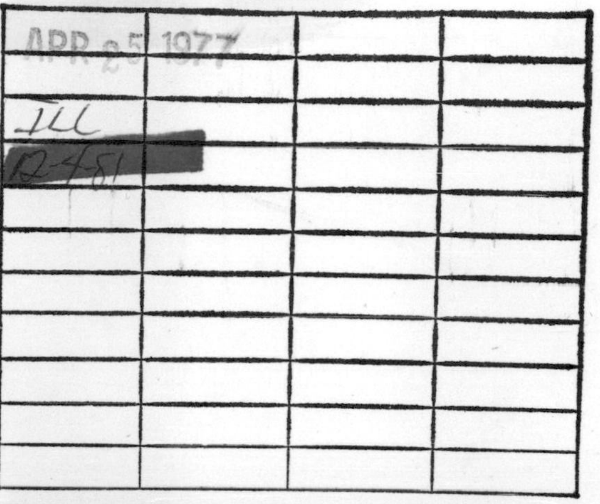
APR 25 1977
ILL
12-4-81

NEW MEXICO
EASTERN
UNIVERSITY
E
N
M
U
LIBRARY

Introduction to

DYNAMIC MORPHOLOGY

Introduction to
DYNAMIC MORPHOLOGY

EDMUND MAYER

Department of Experimental Pathology
Lederle Laboratories
Pearl River, New York

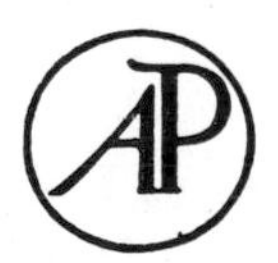

1963

Academic Press • New York and London

ACADEMIC PRESS INC.

111 Fifth Avenue
New York 3, N. Y.

United Kingdom Edition
Published by
ACADEMIC PRESS INC. (London) Ltd.
Berkeley Square House, London W. 1

Library of Congress Catalog Card Number 62-21934

PRINTED IN THE UNITED STATES OF AMERICA

Preface

Morphology, the study of visible structures of organisms, occupies places of variable importance in different fields of biology. The role of morphology is obvious in surgery, embryology, and taxonomy. It is less obvious in chemotherapy, genetics, and enzyme biochemistry. *Dynamic morphology* includes static structures as well as development and functional changes of structures.

Morphological procedures range from Roentgen-ray analysis to histological staining, from mechanical dissection to electron microscopy, from perfusion of an isolated heart to centrifugation of living amebas. In the face of ever-increasing specialization, this book offers (1) a survey of areas or problems in which morphological procedures are needed, and (2) a review of morphological procedures in their relation to various types of projects.

The material is organized in such a fashion that continuity is maintained between macroscopic, light microscopic, and electron microscopic levels of analysis. Interaction between morphological and biochemical approaches is stressed throughout. Pathological phenomena are classified from both the biological and the medicopathological points of view.

The book is intended primarily for pharmacologists, microbiologists, biochemists, chemists, physicists, and statisticians who need an understanding of morphology but who have had no training in this discipline. Therefore, detailed discussions are devoted to matters that most morphologists take for granted. The principles governing morphological studies are identical with those of natural science in general, but certain rules of procedures in morphology are stated explicitly. The term "introduction" in the title indicates that this book does not aim at completeness; it is not supposed to replace textbooks or laboratory manuals.

Readers trained in normal and pathological morphology may use this book as a guide for traveling unusual routes through familiar scenes. They also may enjoy occasional excursions into unfamiliar areas at the borders of dynamic morphology.

Pearl River, New York
November, 1962

EDMUND MAYER

Acknowledgments

The writing of this book was encouraged by the management of the American Cyanamid Company and by my colleagues at the Stamford Research Laboratories and the Lederle Laboratories. Drs. Earl H. Dearborn and J. T. Litchfield, Jr., read the entire manuscript. Their critical and stimulating comments were most helpful. Large portions of the manuscript were read by Dr. C. W. Dunnett, who made valuable suggestions on statistical and other matters, and by Dr. T. G. Rochow, who corrected some errors in my presentation of microscopic procedures.

I wish to acknowledge the help of other colleagues at the Lederle laboratories who advised and instructed me in their special fields of study: Drs. D. A. Buyske, J. H. Clark, S. B. Davis, J. P. English, M. Gertrude Howell, and R. G. Shepherd (chemistry and biochemistry); Drs. M. Forbes and H. J. White (bacteriology); Drs. R. I. Hewitt, S. Kantor, and R. E. Thorson (parasitology and zoology; Dr. C. N. Latimer (neuroanatomy and neurophysiology); Dr. W. J. Sullivan (renal physiology); Dr. J. R. Cummings (blood pressure determination). The members of the Department of Experimental Pathology were helpful in various ways. Dr. W. M. Layton, Jr., patiently discussed items of the book with me almost every day for a period of five years.

Other colleagues who also advised me generously are Dr. D. W. Fawcett (Department of Anatomy, Harvard Medical School), electron microscopy of sections; Dr. R. V. Grieco (Methodist Hospital of Brooklyn, New York), Roentgen-ray procedures; Dr. F. Jacoby (Department of Histology, University College, Cardiff, Wales), tissue culture problems; and Dr. O. Krayer (Department of Pharmacology, Harvard Medical School), isolated heart preparations. It is impossible to mention all the scientists who have been consulted with respect to special problems.

Members of the Photographic Department of Lederle Laboratories cooperated effectively. All India-ink drawings were made by Mr. W. Hearn, either from my pencil sketches or from published originals. I appreciated the constant assistance of our librarians, in particular Mrs. J. M. Fantini and Miss G. A. Irby.

During the later stages of the manuscript, secretarial and editorial work were done by Mrs. Erla Pratt with great devotion and meticulous care. My wife, Hildegard W. Mayer, typed all first drafts and joined me in the preparation of the subject index. Without her encouragement and help through the years, this book could not have been written.

Finally, I wish to thank the publishers and scientists who permitted reproduction of illustrations. Personal communications are acknowledged in the text.

Contents

PART I

Introduction

PART II

Morphology, The Study of Visible Structures

PART III

Procedures, Interpretations, and the Problems of Presentation in Dynamic Morphology

PART IV

Elementary General Structures

PART V

Classification and Identification of Biological Structures

Part I

INTRODUCTION

Purpose of the Book

The systematic study of visible structures of organisms, briefly biological morphology, is part of the curricula for students of medicine, veterinary medicine, zoology, and botany. A number of useful textbooks and reference books are available for students of these sciences. The present book is not intended to compete with these books. It is meant for scientists who need an understanding of morphology, but who have had no satisfactory training in this discipline.

I have had the opportunity to collaborate over a period of forty years with physicists, chemists, biochemists, pharmacologists, and bacteriologists. This collaborative work was done in universities, in institutes for fundamental research, in hospitals, and, during the last eighteen years, in industrial research laboratories. In all these places peculiar difficulty was encountered when morphological information was needed by scientists experienced in other areas: they could rarely understand morphological publications or textbooks.

These difficulties proved to be more formidable on the microscopic than on the naked-eye level and appeared to be almost insurmountable when pathological conditions were involved. Before discussing the probable causes of this problem, I will give a few examples to illustrate how physicists and other nonmorphologists get involved in morphological matters. Physicists who are occupied with ionizing radiation find themselves faced with biological effects, such as leukemia and mutations, which are characterized by important morphological aspects. Bacteriologists recognize the necessity of analyzing morphological changes in infected experimental animals. Biochemists interested in the enzymes of the kidney have to determine the localization of enzymatic activity in the intricate structures there. Pharmacologists who study the distribution of drugs in the animal body are forced to consider the morphological as well as the chemical composition of the different organs. Why is it that these highly trained scientists are not able to use the existing textbooks in order to obtain morphological information?

The first complaint concerns terminology. Conventional morphological terms which seem to be satisfactory for communication among insiders, present puzzles to the outsider. Who can guess that the term cuboidal epithelium refers not to cubes but to hexagonal prisms in which the height does not exceed the width? Who can guess that the term fibroblasts refers in ordinary histology to fiber-forming cells, but in tissue culture language to a cytoplasmic network pattern with or without fibers? Who can guess that a malignant tumor characterized by an aggregation of oval nuclei and the absence of any recognizable pattern of cell bodies is called carcinoma when it occurs in the lung and sarcoma when it occurs in the leg?

The second complaint is that photographs and drawings of microscopic structures seem to show what they are supposed to show only to the eyes of the initiated. Besides uncertainty in identifying the structures, three-dimensional interpretation of two-dimensional sections seems to be particularly difficult for persons not trained in morphology.

The third complaint is that there is a lack of continuity between the presentations of normal and pathological morphology. Illustrations of pathological changes of microscopic structures are rarely accompanied by a picture of the normal condition. The M.D. is supposed to remember the normal condition, or to refresh his memory with a textbook of normal histology. This raises too many barriers, however, for a physicist who wants to make a reasonable effort to understand bone marrow injury by radiation.

One may advise students of any branch of natural science to acquire a sound training in mathematics, but I do not feel that every scientist should study normal and pathological morphology even though he may need this knowledge on special occasions.

In my association with other scientists, I have made numerous attempts to explain morphological points by using a minimum of technical language, by selecting exceptionally clear drawings and photographs on the microscopic level, and by making extensive use of diagrams including three-dimensional ones. This proved helpful in many cases. Yet there remained a mysterious veil which seemed to cover morphology. Finally it became clear to me what prevented my colleagues from understanding morphology. In modern science it seems to be obvious that *results should be presented as functions of procedures.* Physical, chemical, physiological, genetical, and bacteriological results are not given without stating the particular procedures by which particular results were obtained. This is true not only of research publications, but also of textbooks. Students and technicians are taught in this way. Morphology textbooks are the only exception: as in textbooks of history the emphasis is on results, and procedures or documentation may or may not be mentioned. What is missing in both the textbooks and the technical manuals of morphology are the *principles of morphological procedures,* or, in other words, *explicit statements of the rules of the morphological game.* This generalization needs some supplementary comments. Morphological papers contain descriptions of material and methods. However, these publications take for granted a certain amount of morphological background which is presumably to be found in textbooks. Unfortunately, too much is taken for granted. The situation varies in different areas of morphology. Most laboratory manuals of macroscopic dissection serve their practical purpose in schools of human and veterinary medicine. Yet there is rarely a discussion of the differences between the appearance of organs in a fresh and a preserved cadaver, of the principles of dissection, of the different ways of dissecting a brain, of the role of injection techniques, or of the use of a magnifying glass or dissecting microscope. By and large, the modern branches of morphology such as histochemistry and electron microscopy have cultivated the habit of detailed presentation of procedures and careful derivation of results. In traditional morphology, the special textbooks of anatomy and histology of the nervous system show a laudable tendency to coordinate presentations of procedures and results. Many general textbooks of normal or pathological morphology contain paragraphs, or even whole chapters, in which results are properly backed by procedures. As an example I mention W. E. Le Gros Clark's book (1958), in which anatomical methods are given for the study of nervous tissues, lymphatic vessels, and postembryonic growth of bone (increase in size), but not for blood, bone marrow, skin, or other organs. I do not know of any textbook on morphology in which the relation between procedures and results is established as a principle and followed throughout. Similarly, most books on microscopic techniques give satisfactory descriptions of procedures, but do not include the principles of interpretation of morphological phenomena. The *third dimension* of biological structures presents special problems of technique and interpretation that are treated in various ways in existing textbooks. In the present book substantial space will be given to topographic and three-dimensional analysis on macro- and microscopic levels.

One of the most important tasks of the modern morphologist is the *dynamic interpretation of static pictures.* How does one know whether certain structural variations seen in fixed and stained sections represent a chronological sequence? It seems that the rules for handling such problems have never been published. They will be discussed in detail since time-associated changes of structures are of particular interest to pharmacologists, biochemists, and bacteriologists.

Special efforts will be made to *connect results with procedures* in the same way in which it is done in physics, chemistry, bacteriology, physiology, or pharmacology, and to present pathological conditions in close connection with normal ones. The fact has to be faced that the purpose and efficiency of morphological procedures are obscure to scientists other than morphologists. This uncertainty produces a peculiar paradox. In the mind of some scientists, morphology still enjoys the authority it had during the 19th century, whereas in the opinion of others the study of visible structures is superseded by modern biochemistry and biophysics. Physiologists, biochemists, and pharmacologists may, at unpredictable occasions, place great confidence in the morphological verdicts of the pathologist, or they may adorn their publications with histological data and pictures because it is done traditionally. The present book will point out what to expect and what not to expect from morphological analysis.

Organization and Scope of the Book

Macroscopic and microscopic structures, normal and pathological, are the subject of this book. The relations between structural and functional aspects will be discussed in detail. However, the organization of the book is based on *procedures* rather than on structural or functional classifications. The term procedure is meant here to include technical manipulations as well as intellectual planning and interpretations. Seemingly unrelated biological items will frequently be tied together by a common procedure. Macroscopic and microscopic examples can illustrate the same principle, if no scaling problems are involved. The banding of birds and the vital marking of embryonic cells will be treated together as tagging procedures. Potential spaces will be discussed first on a macroscopic level, with the pleura cavity as an example, and then on the microscopic level of tissues and cells. The subject of axial polarity of biological structures will be introduced by the planes of orientation in the mammalian body and concluded by comments on polarity of cells. Under the common heading of "Natural and Artificial Units" the reader will find the isolated heart-lung-kidney preparation of a dog, and also the question of applicability of the cell concept to protozoa.

In contrast to tradition, the present book does not start with the cell, for a number of reasons. What is simpler, a cell or a dog? Claude Bernard (1866) stated that a highly organized animal offered simpler conditions for experimentation than so-called lower organisms did, since the latter seemed to perform the same basic functions—metabolism, motility, reproduction—without visible specialized structures. At the present time the tissue cell has become complicated for the opposite reason: electron microscopy and cytochemistry have revealed a wealth of functional structures within each cell, and new ones are being discovered at a high rate. To emphasize this change, Novikoff (1960) published an impressive set of illustrations: liver cells as presented by Rudolf Virchow in 1858; the generalized cell as conceived by Edmund B. Wilson in 1896; and a diagram of a rat liver cell based on recent cytochemical and electron microscopic information. The last picture not only is very involved, but some of its features are better established than others. Sjöstrand (1956) pointed out difficulties in judging the *average picture* of a cell because of the limited number of cells that can be examined under the electron microscope and because of the variability of structure in different functional stages which are not synchronized from one cell to another. Moreover, cells of different tissues and species now show so many differences that one should be reluctant to start a textbook of histology with the description of "the generalized cell." In organizing the present book along procedural lines I was guided, to some extent, by Eugen Albrecht's (1907, p. 247) comment that Virchow's cellular pathology produced the side effect of "cellular myopia." Now there is also the danger of subcellular myopia. Probably the best prophylaxis is a balanced consideration of macro-, micro-, and submicroscopic levels. The importance of maintaining procedural continuity between different orders of magnitude will be stressed at many occasions. It is well to remember P. W. Bridgman's (1927, p. 51) statement that "the large may not always be analyzed into the smaller."

The present book is not a systematic treatise on morphology. No attempt will be made to give complete descriptions of all morphological structures of each system and organ. Neither is it within the scope of this book to discuss, or even mention, all the harmless and pathological variations which may occur in the different organs. The normal and pathological morphology of systems and organs is described with great completeness in the existing textbooks. Guided by the present book, scientists without morphological training should be able to utilize the material in these textbooks.

It is obvious that the choice of procedures should depend on the purpose and nature of each study. The selection of the best technique is not always simple, and an investigator may not know what techniques are available to tackle his problem. In the present book a large variety of procedures will be mentioned and their application will be illustrated by examples. Some procedures will be described briefly, others more elaborately. The reader will be introduced to the principles of histological staining by a detailed discussion of one of the simplest stains, hematoxylin. However, this description will hardly enable anyone to stain a section in a satisfactory way. The present book cannot replace laboratory experience and the use of technical manuals, but it should give an insight into the workshop of the morphologist.

The examples which illustrate procedures are taken mostly from the morphology of man and laboratory animals, but references to comparative morphology and physiology will be frequent. This means that not only vertebrates, but also invertebrates, protozoa, and plants may be discussed. Examples from bacteriology will be used extensively.

Embryological material is found in all parts of the book. Modern experimental embryology is particularly suitable for demonstrating the relation between morphological and physicochemical procedures. An understanding of visible structures and their variability is hardly possible without a study of their production during embryonic and postembryonic life. Most of the pharmacologists, bacteriologists, chemists, and physicists with whom I have been associated showed no desire to be informed on problems of experimental embryology. If static morphology was hardly accessible to them for the reasons stated before, the mere idea of unstable, changing structures was certainly forbidding. Moreover, these scientists may have been so fascinated by genetics and the evolution of species that no enthusiasm was left for the problems encountered in the study of embryos. It is hoped that a greater interest in experimental embryology will be kindled by the present book, in which embryology is presented within the framework of physiology.

This book is devoted primarily to the study of visible structures of organisms, or biological morphology. Certain activities of the organism produce visible structures, and, in turn, visible structures are necessary for particular activities. The activities of organisms are called functions. As a rule, functions and visible structures are related in some way, but there is a great variety of relations, many of which will be discussed in the book. These relations will be compared to similar relations in man-made machines. The study of biological functions is known as physiology, and the study of visible structures in their relation to functions is known as *functional or dynamic morphology.* The first parts of this book deal with fundamental aspects of physiology in order to supply the background for the subsequent parts, which are devoted to functional morphology. Therefore the first parts will contain physiological items which seem remote from the study of visible structures. There will be analyses of muscular contraction and of the different phases of sunburn; there will be discussions of genetics and of immunology; and there will be dose-response curves. Attempts will be made to consider the various areas of physiology from the common view of stimulus, response, and reactivity. All this is necessary in order to place morphological procedures in their proper relation to the problems of physiology.

The same item may be discussed in different contexts. Thus, the contractility of the dog's spleen will be mentioned in several places of the book: illustrating procedures for observing internal organs during life; requiring special postmortem techniques; playing an important role in the distribution of blood in the body; and finally resembling other erectile organs. This type of repetition had to be faced as inherent to the plan of the present book. I was encouraged by a statement in Albert Einstein's 1918 semipopular presentation of the theory of Relativity: "For the sake of clarity, frequent repetitions were deemed necessary while no attention was given to elegance of presentation: I followed conscientiously the advice of the great theoretician L. Boltzmann, that elegance should be left to tailors and shoemakers" (translated by E. M.).

The last part of the book will be devoted to the classification and identification of biological structures under normal and pathological conditions. In classifying organs, tissues, and cells, not only structural and functional characteristics are used but also origin, potentialities, and immunological relationships. Pathological phenomena may be classified according to abnormal stimuli, such as bacterial infections, or according to abnormal responses, such as neoplasms (cancer).

Therefore, many items, which were treated in earlier chapters from the procedural point of view, will appear again in the discussion of alternative classifications.

Terminology

A term expresses a concept or a definition. It is hardly worth while to argue over a term if the concept is sound and the definition is clear. The rules applied in the present book can be characterized as follows.

(1) Clarity is given preference over traditional use, and plain English over unnecessary technical terms. The open space inside a tube will be called the bore, instead of the lumen, which is the customary term in biological morphology. In my experience, nonmorphologists expect the word lumen to mean unit of illumination. Certain cells which have the shape of hexagonal prisms will be called prismatic rather than cuboidal, although the latter term is generally used. At many occasions the self-explanatory descriptive terms will be supplemented with the conventional terminology.

(2) Controversial terms will be avoided, especially if they are not needed for the objectives of this book. The word *protoplasm* will not be used. Some authors apply the term protoplasm both to cell bodies and to nuclei; others exclude the nuclei and, therefore, consider cell body, cytoplasm, and protoplasm as synonymous. Moreover, the traditional definition of protoplasm as living substance cannot be easily applied to intercellular substances.

As a rule, the term *degeneration* will not be applied to microscopic structures. Microscopic structures should be interpreted either as living or dead, and, if living, it should be stated whether the structures are in good or poor condition; criteria of this appraisal should be given. The term degeneration is always properly specified with respect to the nervous tissue. For instance, "degeneration of a tract" in the spinal cord means the loss of myelin sheaths of the nerve fibers in that particular tract.

(3) Ill-defined general terms will be replaced by specific terms or by qualifications of the general term. Instead of using the term *growth,* reference will be made to increase in length, surface, volume, cell number, intercellular substances, wet weight, dry weight, total protein, nucleic acids, or other components. If the term growth is intended to summarize the effect of several factors, the factors will be listed, and the result will be called the balance sheet of growth. The term *norm* will be qualified by distinguishing the statistical from the teleological norm. The statistical norm refers to an arbitrary area on both sides of the mode of a frequency distribution curve. The teleological norm, synonymous with the evaluating, desirable, or idealistic norm, refers to the optimum of health. Besides these two useful concepts of norm there are also bizarre norms such as the textbook norm, which offers nonexisting or exceptional conditions as standards; and the

hospital norm, which interprets observations in hospital patients as being representative of the population.

(4) Terms that imply different criteria of classification will be explained. The term *epithelium* may refer to the visible pattern of a cell aggregate (mosaic pattern), or to the origin of mature cells from specific precursors (cytogenesis). According to a widespread but unsubstantiated opinion, the two criteria of epithelium are *always* correlated. To avoid this problem, the term epithelium will be replaced as follows: liver cord cells for liver epithelium; lining of thyroid follicles instead of thyroid epithelium; cells of renal tubules instead of kidney epithelium. *Problems of nomenclature are inseparable from problems of classification.* Organic chemists as well as botanists and zoologists will agree to this statement. In a handbook used for the registry of diseases, entitled "Standard Nomenclature of Diseases," the editors are constantly faced with problems of classification.

(5) Some conventional terms are not used in the same sense by all investigators. To be on the safe side, the terms reproducibility, precision, and accuracy will be defined operationally in the present book, in agreement with those authors who follow the same principle. Preference will be given to the term reactivity over its synonyms irritability, excitability, responsiveness, and susceptibility. Although the latter four terms are found more frequently in the literature, reactivity has the advantage of not being associated with special branches of biology (see Table 2).

In the present book organisms and man-made machines are compared on various occasions. Neither machines nor biological objects can be described without evaluating teleological concepts such as order and randomness, or fuel materials and waste products. The philosophical implications of teleology have been the subject of a number of treatises. I mention Immanuel Kant's "Kritik der Urteilskraft," which appeared in 1790, and Morton Beckner's book "The Biological Way of Thought," which appeared in 1959. These books represent *philosophies of biology.* I do not pretend to offer more than a *philosophy of morphological procedures* or, perhaps, a philosophy of biological procedures with special emphasis on morphological techniques.

Comments on the Literature Selected

It is hoped that the scientists for whom this book is written will benefit from it in two ways: they should be able to understand morphological matters, and also be able to judge the merits of morphological data and interpretations. References to textbooks of physiology, histology, and pathology are given at many occasions. Morphological and histochemical laboratory manuals are mentioned for those who wish to acquire familiarity with special procedures.

References to original publications were selected with the following purpose in mind. If my condensed presentation of a subject arouses the desire for more detail, the quotation of review articles or of an individual paper on this subject will put the reader on the track to the literature in the particular area. My choice of papers in each field was arbitrary, to some degree.

The historical development of important procedures or factual discoveries is given in many instances, usually with more emphasis on the logical steps than on the chronological sequence. However, efforts have been made to credit major contributions to the proper authors.

A discussion of biophysics and biochemistry as related to dynamic morphology serves as an introduction to the bewildering flow of cytophysical and cytochemical publications.

Part I: References

Albrecht, E. (1907). *Frankfurt. Z. Pathol.* **1**, 221-247.

Beckner, M. (1959). "The Biological Way of Thought." Columbia Univ. Press, New York.

Bernard, C. (1866). "Leçons sur les propriétés des tissues vivants." Baillière, Paris.

Bridgman, P. W. (1927). "The Logic of Modern Physics." Macmillan, New York.

Clark, W. E. Le Gros (1958). "The Tissues of the Body. An Introduction to the Study of Anatomy," 4th ed. Oxford Univ. Press, (Clarendon) London and New York.

Einstein, A. (1918). "Über die spezielle und die allgemeine Relativitätstheorie," 3rd ed. Vieweg, Braunschweig.

Kant, I. (1790). "Kritik der Urteilskraft" (K. Vorländer, ed.) 1924 ed. Meiner, Leipzig.

Novikoff, A. B. (1960). *In* "Developing Cell Systems and Their Control" (D. Rudnick, ed.), p. 167. Ronald, New York.

Sjöstrand, F. S. (1956). *Intern. Rev. Cytol.* **5**, 455.

"Standard Nomenclature of Diseases and Operations." (1961), 5th ed. (E. T. Thompson and A. C. Hayden, eds.). McGraw-Hill (Blakiston), New York.

Virchow, R. (1858). "Die Cellularpathologie in ihrer Begründung auf physiologische und pathologische Gewebelehre." Hirschwald, Berlin.

Wilson, Edmund B. (1896). "The Cell in Development and Inheritance," 1st ed. Macmillan, New York.

Part II

MORPHOLOGY, THE STUDY OF VISIBLE STRUCTURES

Chapter 1

Morphology Characterized by Procedures

A. Morphology as a Method of Physiology

Structure, form, and pattern are synonymous terms indicating the orderly arrangements of parts of a whole, in contrast to chaos or randomness. This is a very general definition and therefore is not related to particular procedures. Structure in this wide sense has been studied by optical, tactile, magnetic, electrical, and chemical procedures or even by procedures outside natural sciences. For the purposes of the present book, however, biological structures are characterized by their accessibility to optical procedures. Then, biological morphology can be defined operationally as that method of physiology which deals with the visible structures of organisms. This definition refers to the study of whole organisms as well as to the study of its parts such as systems, organs, and their components.

Warburg (1924) included morphology in the field of physiology when he stated that of all problems of physiology the problem of form is the most inaccessible. Fischer and Mayer (1931) went one step further by declaring *morphology to be a method of physiology.* I was responsible for that statement and am still convinced that this is a useful point of view. Fruitless arguments on the relation of structure and function are avoided. It is possible to analyze the production and variability of visible structures of organisms within the general physiological framework of stimulus, response, and reactivity. Finally, a reasonable basis is provided for comparisons between organisms and man-made machines.

In order to understand the functioning of a man-made machine, one has to know its blueprint on the one hand, and its energy distribution and transformations on the other hand. Similarly, the functioning of a living organism, or any part of it, can be understood only if its structures are studied as well as its conversions and distributions of energy. The main difference between man-made machines and organisms seemed to be that organisms reproduce themselves whereas the known man-made machines cannot reproduce their own kind. This difference has sometimes led to the exaggerated conclusion that the production of structures of organisms is altogether inaccessible to rational study. The production of structures, or morphogenesis, has been studied successfully by modern experimental embryology and will be discussed extensively in this book. At this point we limit the machine comparison to the *finished structures.* Then we may say that morphology deals with the blueprint, whereas the studies of metabolism and of transmission of stimuli deal with the conversion and distribution of energy. All these lines of investigation are summarized under the name of physiology. It is in this sense that I consider morphology a method of physiology. *Dynamic* morphology is the study of time-associated changes of biological structures. Any study of changing structures includes, of course, the study of static structures. Figuratively one may say that in static morphology still photographs are studied, whereas motion pictures are the subject of dynamic morphology.

This differs from Frey-Wyssling's (1953, p. 7, footnote) view that morphology deals with spatial arrangements in the organism as opposed to physiology, which deals with processes, i.e., changes related to time. It seems to me that the trabeculae in bones have a static function, while their chemical composition changes with time through metabolic replacement. To Frey-Wyssling, dynamic morphology is a contradiction in terms.[1] One must conclude that there is no room for experimental morphology in Frey-Wyssling's concept of morphology, since, according to him, changes of structure belong to physiology. In our machine analogy, the apparently static aspect of a blueprint should not be overrated. Arrows on blueprints indicate the direction of movement of parts, electric current, etc. Moreover, animated cartoons could be made of blueprints showing actual movements of switches, wheels, valves, and fluids.

A special discussion (Part II, Chapter 2, H) will be devoted to comparisons between organisms and machines, including the problems of their production. One of these problems should be mentioned here. Since blueprints are needed for manufacturing a machine, the central question of biological morphogenesis is: where is the equivalent to a blueprint? The plan for the final organism must be present in the unfertilized ovum, but it is a long way from this general postu-

[1] Goethe, who introduced the concept of morphology into biology in 1795, defined morphology as "the discipline which deals with the shape (Gestalt), formation, and transformation of organisms" (translated by E. M. from Beutler's edition, 1949, p. 115).

late to the demonstration of the earliest patterns. The origin of basic patterns, such as axes and poles of the ovum, is not easily determined.

In his famous book "On Growth and Form," D'Arcy Wentworth Thompson stated that "the form of an organism is determined by its rate of growth in various directions" (2nd ed., 1942, p. 79). This statement covers not only the organism as a whole, but also any parts of it.[2] Thompson's term growth refers to increase in volume irrespective of chemical composition. His book is devoted mainly to the mechanical engineering of nature. Discussions of the origin of forms are more or less restricted to the mathematical transformation of one form into another. By and large, Thompson is not concerned with those physicochemical or biological factors which could be responsible for the *production* of forms in organisms.

Problems of the physiology of finished organisms and problems of embryonic development overlap: embryonic cells carry out general metabolic functions, and in the adult organism embryonic mechanisms are maintained, for instance, in regenerating lost structures. In spite of this overlap, a separation of the two sets of problems is indispensable for an orderly discussion.

The title of this book, "Dynamic Morphology," refers not only to time-associated changes of structures, but also to the relations between structures and functions. This dual use of the term dynamic is unavoidable here since changes in structure can be produced or conditioned by functional activities. The shape of the body depends on nutritional factors, the volume of a muscle increases after a period of sustained exercise, and the structures of a fractured bone are repaired according to mechanical demands. On the other hand, many functions are known to depend on the presence of special structures. This will be discussed in detail in later chapters.

Dynamic morphology then deals with visible biological structures, their time-associated changes, and their relations to functions. A somewhat different concept of dynamic morphology was proposed by L. von Bertalanffy in an article which appeared in 1941.

B. Operational Definition of Visible Structures

I have defined biological morphology as the discipline dealing with visible structures of organisms. By the adjective *biological* the field is separated from the morphology of inanimate objects, such as crystals, and also from morphology

[2] Unequal rates of increase in size in adjacent areas led Wilhelm His (1874) to the "principle of differential growth," which was supposed to cause folding of cell sheets in the early embryo. However, unequal rates of cell multiplication or increase in cell volume are neither the only nor the primary mechanisms in the development of form (see Part III, Chapter 3, B; Fig. 32).

in social sciences, such as Spengler's "Morphology of History."[3] Yet, the remaining terms of the definition, namely, *visible* and *structures* can cause, and have caused, enough misunderstandings to warrant a discussion of both.

The term *visible* refers not only to perception by the unaided eye, but also by the eye aided by magnifying glasses and microscopes. Contrast is increased by differential staining or phase contrast microscopy. Fluorescent screens and photographic emulsions make it possible to use ultraviolet light, Roentgen rays, and electron microscopy; the latter has extended the power of resolution far beyond the limit of the light microscope. All phenomena demonstrable by any of the direct and indirect methods listed above will be termed *visible* in the present book. In many cases, the visual impressions can be supported or supplemented by tactile impressions. In the definition of visible structures the accessibility to optical procedures is obligatory, whereas the accessibility to tactile procedures is optional. It is obvious then that the term *visible phenomena,* or *structures,* covers properties which can be seen but not felt, such as color, transparency, or glossiness. One should distinguish between extensions of optical procedures, such as electron microscopy, and translations into visible records, such as temperature charts. There are hardly any phenomena that cannot be translated into visible records. Such recordings will not be considered as visible phenomena in the present book.

The definition of morphology as a method of physiology implies that some aspects or phases of physiological problems lend themselves to this method but others do not. In any special problem the applicability of optical procedures to biological structures may be connected with important features of that problem, or it may be accidental. This will be illustrated by a physical model (Fig. 1) in which visible particles are related to magnetic forces. Chemical or other forces could be substituted in constructing similar models. As shown in Fig. 1a, a hollow sphere with thin rubber walls, resting in a rigid basket with wide meshes, is completely filled with water.[4] The water contains little rods, some of magnetic iron and the others of nonmagnetic brass. Each rod is coated with an amount of paraffin so that its specific gravity equals 1. Thus all these rods will be distributed at random in the water. Because of the paraffin coating, there is no visible difference between the iron and the brass rods. If a pole of a strong electromagnet is brought near the surface of the sphere, the magnetic iron rods will aggregate at the corresponding inside area. If the magnet has enough mechanical force, the pull on the aggregated iron rods will cause a bulge in the rubber wall. This involves a certain deformation of the rest of the sphere, since other diameters must shorten to compensate for the elongation of that diameter which goes through the bulge. The volume of the system remained unchanged, while its form changed from spherical to pear-shaped.

[3] This is the subtitle of his "Decline of the West," in the German original.

[4] Deviation from the exact spherical shape caused by gravity is neglected.

Let us assume that an observer watches these phenomena without knowing anything about either magnetism or the actions of the experimenter. The observer sees a sphere in a basket, and by touching it discovers that it can be deformed. If the wall and the interior of the sphere are transparent, he sees little rods which move at random when he knocks at the sphere. He notices that, at certain times, a number of little rods aggregate near the wall opposite a conic piece of steel. If the thin wall is for some reason not transparent, the

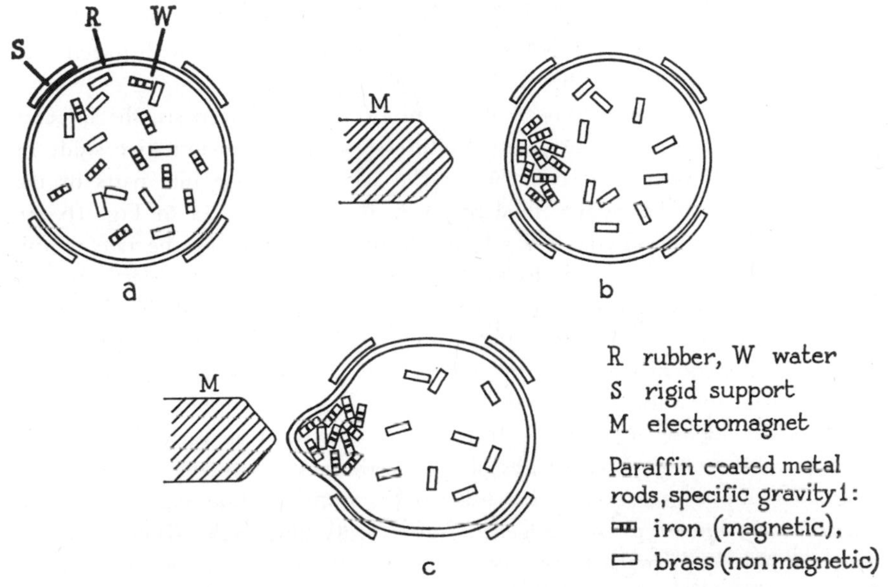

FIG. 1. Model illustrating two different things: the usefulness and limitations of optical procedures, and the change from randomness to order (explanation in text). Hollow rubber sphere filled with water in which paraffin-coated magnetic and nonmagnetic rods float. (a) Random distribution of magnetic iron rods and nonmagnetic brass rods; (b) moderate action of an electromagnet resulting in aggregation of magnetic iron rods in area near magnetic pole; (c) stronger action of electromagnet resulting in bulging of rubber wall where magnetic iron rods are aggregated; note a nonmagnetic brass rod trapped among the iron rods.

observer will not see the little rods, and, therefore, will not be aware of any change inside when the electromagnet is switched on. If, on the other hand, so much force is applied that a bulge is produced, the observer will notice this, irrespective of transparency or opacity of the rubber. If the sphere is transparent, the observer will try to find out why a certain proportion of little rods is at times concentrated in one area. Suppose black marks were placed on the magnetic iron rods. Then the observer will conclude that the factor which causes the local aggregation of marked rods has selective power. He will be right in correlating the black marks with the local aggregation of rods at certain periods, but he would be wise to leave the question open whether the black marks play a causal role in the phenomenon of selective aggregation, or have indicator value only.

In the model certain properties of the rubber wall, such as transparency or opacity, are unrelated to magnetism, yet these properties control the *optical accessibility* of the main phenomenon, which is the selective aggregation of rods under the influence of the magnet.

Besides the technical aspects, the model demonstrates how a system can change from randomness to order if an external directive factor acts on the system. Figure 1b shows the establishment of a pole without change in the exterior form of the system, but in Fig. 1c, the exterior form is distorted into a bulge which accentuates the polarity.

In the structural development of organisms, or morphogenesis, the increase in visible patterns is known as morphodifferentiation. A quantitative study of such changes is possible if one focuses the attention on definite parts of the system. In the rubber sphere model the change from Fig. 1a to Fig. 1b can be observed and measured. It would be possible to determine the rate of migration of the magnetic rods to the area near the pole of the electromagnet and the final concentration of magnetic rods in that area. Similarly, the transformation of Fig. 1b into 1c could be expressed by indicating the change of ratios of different diameters of the rubber sphere from the constant ratio 1:1 in Fig. 1b to the variable ratios in Fig. 1c.

In a developing organism, such as a frog embryo at an early stage, it has been possible to tag special groups of cells with dyes and thereby observe their migration and aggregation which lead to the differentiation into organs and therefore are known as morphogenetic movements (Fig. 32). There is a certain resemblance between these movements and the migration and aggregation of magnetic rods in the model leading to the condition in Fig. 1b. However, the forces which cause this change are inside the system of the frog's egg, but they are outside the rubber sphere system in the model. Similarly, the bulging of the model, seen in Fig. 1c, has its counterpart in the change in surface of the frog embryo which, of course, is produced by forces inside the embryo. Besides migration of cell groups, local differences in the rates of cell enlargement, cell multiplication, and cell death are instrumental in molding the external form of the whole embryo as well as its internal structures.

It is hoped that now the concept of visible structures as applied to organisms has been sufficiently clarified. However, it might be worth while to remember that, outside of biological morphology, the concept of structure is used with reference to energy configurations. Physicists talk of the structure of the field around the poles of a magnet as demonstrated by iron filings on a piece of cardboard. In chemistry the structural formulas of compounds start as tentative pictures or symbols but are gradually verified by continued experimentation and calculation; finally, not only the spatial arrangement of all components but also the energy configurations of the compound are established. In his "Submicroscopic Morphology of Protoplasm and Its Derivatives" (1953), Frey-Wyssling devotes many pages to structural chemistry, which he considers a morphological science. In the present book, structures on or below the molecular level will be mentioned on various occasions.

The term *organism* is not controversial, but the terms *organization* and *organized* can be ambiguous. Different elementary structures of an organism show unequal degrees of organization: cells have the highest degree, certain products of cells such as fibers have a lower degree, and some cell products known as unorganized ground substance show no visible organization. According to Hess (1955), the unorganized ground substance of the brain can be identified histochemically (see Part IV, Chapter 3, B). Gersh and Catchpole (1950) published a paper with the title "The Organization of Ground Substance and Basement Membrane and Its Significance in Tissue Injury." This paper dealt with what I would call the chemical composition of the unorganized ground substance. In the present book the concept organization will be restricted to visible organization, visible being used in the same operational sense as before.

C. Functional, Vestigial, and Ornamental Structures

As stated above, Warburg in 1924 called the problem of form the most inaccessible of physiological problems. I agreed at that time, but now I believe that a considerable part of the inaccessibility stems from a vagueness of concepts. With respect to physiology three concepts of form or structure should be distinguished. *Functional structures,* such as the valves of the heart, are clearly related to known functions. *Vestigial structures,* such as the umbilicus and the cranial sutures, are obvious relics from embryonic development and have no function in the mature organism; examples of microscopic vestigial structures are embryonic glomeruli found commonly in the kidneys of adult dogs. *Ornamental structures* cannot be classified either as functional or vestigial. Examples are the crest on the heads of male and female cardinal birds, and the black, brown and white color patterns of a beagle. I give preference to the term "ornamental structures" over "structures without known function" to avoid the impression that functions will doubtless be discovered in the future.

The vast majority of visible structures in organisms are functional structures. The same structure may play different roles in different types of animals. Darwin pointed out that the sutures in the skulls of young mammals have a temporary function insofar as they facilitate parturition, but no such function can be ascribed to the cranial sutures in birds and reptiles since their young merely escape from a broken egg ("Origin of Species," Modern Library edition, 1936,

p. 145). It seems to me that the difference between a functional and an ornamental interpretation of an organ is illustrated clearly by the example of mammalian tails. They vary all the way from the functional prehensile tail of the spider monkey to the ornamental coil of the pig, the dog's wagging tail being intermediary, since expression of emotions may be within either the functional or ornamental range. Ornamental structures can be correlated with important functions. For instance, some strains of mice with hereditary differences in fur color differ also in their susceptibility to infection with mouse typhoid bacilli (see Part II, Chapter 2, E; Fig. 6a–c). The possibility of using a visible characteristic as an indicator of an invisible property can be very advantageous in research.

It is worth mentioning that the triple concept of structure applies to man-made machines as well as to organisms. In an automobile most structures such as motor, wheels, frame, etc., serve the function of moving. Vestiges of the manufacturing process are the marks left at the parting lines of the molds in cylinder blocks; other vestiges are plugged holes that were cored during the casting process in order to allow access for machining such items as the camshaft mountings and oil galleries (Trowbridge,[5] personal communication, 1957). A two-tone finish on the car's body is ornamental since the use of more than one color has nothing to do with the function of moving or the manufacturing process; yet, as in organisms, different ornaments may indicate differences in the functioning parts. There is a tendency to mix functional and esthetic values, e.g., use of the term streamlined. This tendency has developed in recent architecture: a building is supposed to satisfy esthetic standards when its functional structures stare the onlooker in the face (functional architecture).

The present book is devoted to the functional structures. However, consideration must be given to vestigial and ornamental structures. This differentiation will be necessary in discussing the variations of structures, particularly the harmful (pathological) variations. Sometimes, it will be difficult or impossible to classify a certain structure as functional, vestigial, or ornamental, as illustrated above by the example of mammalian tails.

D. Statistical and Teleological Norm. Morphological and Functional Abnormalities

The statistical and the teleological norm have already been mentioned briefly in Part I. Two additional examples may help to illustrate the dual norm. There are strains of mice in which the occurrence of cancer is statistically normal. Although the experimenters may be pleased with this situation, cancer is

[5] I am indebted to Mr. Roy P. Trowbridge, GM Engineering Standards, General Motors Corporation for a list of vestigial structures in automobiles.

not desirable from the viewpoint of the mice. In the United States the most of the adult population have one or more cavities in their teeth: this is statistically normal. The small minority without any cavity represent the teleological or desirable norm.

It may be well to emphasize that both norm concepts are fictive but at the same time indispensable. This is a frequent situation in the exact sciences. The one-dimensional straight line of geometry is fictive: it is needed in physics though it does not exist in any physical sense. In Mainland's excellent book "Elementary Medical Statistics" (1952), one finds a strange condemnation of the teleological or idealistic norm. According to him, that concept of norm which refers to the ideal or perfect state "ceases to be applicable to actual beings and things." It seems to me that the so-called normal frequency or distribution curve, without which Mainland's book could not have been written, is equally remote from "actual beings and things." In his subsequent discussion of health and disease Mainland cannot avoid the use of evaluating terms which are all based on the tacit assumption of a teleological norm.

I defined the teleological norm as the optimum of health, and the statistical norm as an arbitrary area on both sides of the mode of a distribution curve. Distribution curves resulting from measurements are rarely available in the field of pathological morphology. The arbitrary area left and right of the mode is identical with the majority of items, and the minus and plus deviators are small minorities by definition.

For every special problem different groups should be compared with respect to frequency of one criterion. In Table 3 several groups of scouts drenched for different periods will be compared and the occurrence of pneumonia used for calculation of the range. Frequently the range of 95% is used which covers 19 out of 20 samples of the population; then, each small area under the ends of a symmetrical distribution curve represents 2.5%. Although it is not difficult to apply the norm concepts to one criterion, either structural or functional, things become very difficult if one tries to consider several criteria simultaneously.[6]

A person may be normal with respect to ten criteria and abnormal with respect to ten other criteria. This holds for either norm. Concerning the statistical norm, Roger J. Williams (1956) pointed out that a population in which 95% are normal with respect to any one criterion would contain only 60% who are normal with respect to ten independent criteria, since $0.95^{10} = 0.60$.

Teleological concepts are indispensable in describing man-made machines as well as organisms. Terms such as wheel, valve, or handle imply a purpose. The very term *function* is obviously teleological, meaning action for a purpose. The sentence "the heart functions as a pump" expresses the usefulness of the

[6] In genetics the statistical handling of combinations of criteria is an important technique.

heart for the whole organism. In the study of biological variations, the teleological as well as the statistical norm has its place. It is the muddled use of the two norms which is objectionable.

Pathologists, when evaluating morphological deviations from the statistical norm, used to take the stand that all deviations are to be considered harmful (lesions) unless the contrary is proved. It seems to me that the investigator should be prepared to find that some morphological variations are harmful, some indifferent, and some advantageous. Surprisingly in studies on the safety of drugs it was observed that the drug-treated experimental animals might differ from the controls not only in a detrimental, but also in a beneficial, direction (Part V, Chapter 4, Introduction).

To be fair to traditional pathology, not all deviations from the statistical norm have been considered harmful. Suppose one kidney has been destroyed or removed, and, after some time, the remaining kidney is found to be much larger than a normal one. This condition is called compensatory enlargement, which means a useful deviation from the statistical norm. A similarly reasonable interpretation is current with respect to the dilation of chambers of the heart accompanied by thickening of their wall. When such conditions are found in the presence of damaged valves, the abnormal dimensions of the chambers are referred to not as lesions, but as attempts of the heart muscle to compensate for the inefficiency of the valvular mechanism.

During the 19th century, partly as the result of the overwhelming authority of Virchow's "Cellular Pathology" (1858), the idea developed that for every functional disturbance the pathologist should discover some morphological equivalent. This equivalent became known as "the underlying pathology." Diseases with visible morphological abnormalities were called organic, and those without such abnormalities were called functional, diseases. The functional diseases were under a double cloud of suspicion: when the autopsy did not show morphological abnormalities, either the pathologist was not competent enough to discover them, or the patient had faked his disease and death. Harvard's late physiologist, Walter Cannon (1929, p. 113) remarked ironically that fears, worries, and states of rage and resentment which leave no clear traces in the brain are obviously "unreal or of minor importance" just as other functional disturbances for which no accompanying morphological disturbances could be found. During the 20th century the increasing knowledge of metabolic functions dispelled, to some extent, the expectation that every chemical or enzymatic process in the organism should be accompanied by visible morphological phenomena. In addition, increased emphasis is being placed on the distinction between reversible and irreversible phenomena. Obviously not only metabolic, enzymatic, and other functional processes, but also morphological changes, may be reversible or irreversible. These are the reasons why the separation into organic and functional diseases should be abandoned.

Irrespective of the presence or absence of morphological abnormalities, the *classification of diseases according to sites* is still the best for purposes of registry in hospitals and statistics (see Part V, Chapter 4, B, 2). Epilepsy is a good example. Some forms of epilepsy are connected with visible morphological injury whereas others are not, but the site of the disturbance is always in the brain.

Chapter 2

Stimulus, Response, and Reactivity, a General Frame of Reference

A. The Concept of Function and the Scope of Physiology. An Analysis of Sunburn and Tanning

In the preceding chapters, functions have been discussed in relation to visible structures. The concept of function needs some clarification. In physiology, a function is the activity of a part of an organism. The part can be a system, an organ, or a component of an organ such as a group of cells in a definite arrangement, or a group of chemically defined substances which are distributed and reacting in a special pattern. To some extent, functions can be described in physical or chemical terms. It is the function of the kidney to eliminate certain substances from the blood and to discharge them in the urine: various phases of this process can be stated in terms of chemistry and physical chemistry. It is the function of a nerve to conduct impulses: electrical changes can be observed while the nerve is conducting an impulse, and there is evidence that movements of sodium, potassium, and chloride ions through the outer layers of the nerve fiber are associated with this activity. Because of the possibility of describing functions in physical or chemical terms, physiology has been defined as the study of energy transformations and distributions in the living organism. This definition expresses the conviction of many physiologists that, eventually, all activities of the living organism will be understood as physicochemical[1] processes: "Physiology has for its final goal, far away and dim as yet, but steadily held in view . . . a physicochemical explanation of life" (Mitchell, 1932, p. XV). Let us not argue here whether such expectation has any clear meaning, but rather ask the question: what proportion of physiological activities can, at the present time, be described in physicochemical terms? An ex-

[1] In biological discussions, physical and chemical processes are conveniently summarized as physicochemical processes. This term should not be confused with the subject of physical chemistry, i.e., physical mechanisms of chemical processes.

ample will illustrate the meaning of this question. A white-skinned[2] man with brown eyes and light brown hair exposes himself to sunshine at the beach. Several hours after the exposure, his skin becomes red and hot, and stays so for several days: usually the discomfort increases toward evening and sleep at night is disturbed by burning pain. After some days, the painful irritation of the skin subsides; it starts to itch and to peel, i.e., lose its horny layer. Finally, a brown pigmentation of the skin develops: the desired tan has taken place. At the same time the horny layer of the epidermis not only regenerates but becomes thicker than it had been before the sunburn. In order to translate this familiar chain of events into the language of physiology, Table 1 has been prepared.

In column A, the sunburn story is divided into eight phases stated in non-technical terms. In columns B, C, and D such processes are entered which are known, or assumed, to be connected with the respective phases in A. The eight phases of column A do not occur in every individual. Some people tan without experiencing any marked reddening and discomfort, whereas others suffer the burn but do not acquire the tan (see Blum, 1944, pp. 1148–1150).

Column B shows that certain physical or chemical processes are known to occur during the phases 1, 3, 8, and, possibly, 4, of the sunburn story. The most complete physicochemical information is available with respect to phase 8. A brown pigment called melanin is formed from a colorless precursor, an amino acid, through the action of an oxidizing enzyme. The physicochemical factors involved in this process have been studied extensively. In Table 1 the action of sex hormones has been placed in column C rather than in B for the following reason. Although the substances involved are chemically defined (steroids), the chemical processes by which they act are unknown. It had been known for some time that white-skinned eunuchs are subject to sunburn but do not tan. However, the most remarkable phenomenon was reported by Hamilton (1948). A white male castrate was exposed to repeated sun radiation in August. As was expected, he developed redness (erythema) but did not tan. Five months after the exposure, testosterone, a sex hormone, was administered to him, with the result that all those areas which had been exposed to sun radiation five months earlier turned brown. This tanning occurred in January; the castrate had not been exposed to any source of ultraviolet rays in the meantime. It would be beyond the scope of this book to discuss in detail the fascinating relations between sex and skin.

[2] In the present book, the term white-skinned is used as an abbreviation for the yellowish-pink complexion as analyzed by Davenport and Davenport (1910), and by Edwards and Duntley (1939). It is contrasted to other colors, e.g., brown and black. Social labels, such as white in contrast to colored, have no place in biological studies of pigmentation. An adequate use of social and biological labels is found in a study by Glass and Li, "Dynamics of Racial Intermixture — Analysis Based on the American Negro" (1953).

TABLE 1

Phases of Sunburn and the Processes Involved

A. Phases of sunburn	B. Physicochemical processes	C. Neurovascular and hormonal processes	D. Morphological processes on microscopic level
1. Exposure to sunshine	Ultraviolet rays are absorbed in the skin	Original condition	Original condition
2. Period of latency between phases 1 and 3	?	Decrease in blood pressure	In biopsies of pigs 15 min. after exposure, capillaries in lower layer of skin seem to show dilation[a]
3. Reddening of skin	Release of histamine (β-iminazolylethylamine)	Dilation of small blood vessels (capillaries) in skin	In biopsies of pigs, capillaries of skin appear dilated and congested; plasma is found leaking from these vessels[a]
4. Pain	Increased accumulation of histamine	Increase in permeability of capillary walls; leaking of fluid from blood stream (edema), possibly stimulating pain receptors	Leaked fluid fills spaces in skin, probably compresses and distorts cells and nerve endings
5. Increased discomfort at night	?	Nocturnal redistribution of blood? Changed reactivity of nervous system during sleep?	Increased compression of cells?

6. Subsiding of reddening and discomfort	?	?	?
7. Peeling and itching	?	?	Nutritional disturbance resulting from edema causes accelerated shedding of horny top layer of skin
8. Tanning	In the skin, a colorless amino acid, *l*-3,4-dihydroxyphenylalanine (DOPA) is oxidized to brown melanin by an enzyme (DOPA oxidase)	Presence of sex hormones is necessary for tanning	(a) Colorless granules in cells of skin turn to brown melanin. Pre-existing brown melanin granules migrate from deep to superficial layers of skin (b) Regeneration of horny layer with increase in thickness beyond original

[a] Ham (1957).

The purpose of the sunburn story is to demonstrate three points. (1) Not all phases of the sunburn story can be expressed in physicochemical terms; in other words, column B of Table 1 cannot replace columns C and D. (2) Not all phases are accompanied by *known* physicochemical processes. (3) Those physicochemical processes which are known proved to be closely interwoven with neurovascular, hormonal, and morphogenetic processes. This leads to the following conclusion. Since it is neither possible, nor satisfactory, to express all physiological activities in physicochemical terms, a more general frame of concepts is needed.

B. Sunburn and Tanning in Terms of Stimulus, Response, and Reactivity. Self-regulating Mechanisms

A useful frame of reference for any orderly discussion of physiological problems is available in the conceptual triad of stimulus, response, and reactivity. Synonyms are: environmental factor for stimulus; reaction for response; and irritability, excitability, or responsiveness for reactivity. The author who first used the concepts stimulus, reaction, and irritability seems to have been the physiologist and anthropologist Johann Friedrich Blumenbach in "De Generis Humani Varietate Naturale" (3rd ed., 1795; pp. 193–197 of Bendyshe's translation, 1865).

The concept of stimulus may be introduced best by the following reasoning. From a continuum of phenomena a living system is selected for study; this may be a whole animal, an organ, or a cell of an organ. With reference to a defined living system, all other phenomena are summarized as environment. Air and food are part of the external environment of an animal, whereas the circulating blood is part of the internal environment of its organs. The frontiers between living systems and their environment vary greatly with respect to definiteness and stability. The more obvious stimuli are changing factors in the environment, such as a sudden noise within hearing distance or the appearance of potential food in the field of vision. Examples of constant factors acting as stimuli are the oxygen and water needed for the maintenance of vital functions of animals, or the concentration of thyroid hormone in the blood necessary to keep the basal metabolism at the proper level. The differences between constant factors, steady states, and changing factors are quantitative, depending on sensitivity of measurements.

A *stimulus* is a changing or constant factor of the environment which produces certain responses in a particular living system of a given reactivity. A *response* is a phenomenon observed in a living system of a given reactivity and produced by a certain factor in the environment, the stimulus. *Reactivity* expresses the relation of stimulus and response.

Stimuli and responses are observed. Reactivities are inferred. At any given moment the reactivity of a living system depends on its genetic, embryological, and environmental history.

The triad has been applied successfully in physiological studies of muscles and nerves, in genetics after the rediscovery of Mendel's laws in 1900, and in experimental embryology. In other areas of normal and pathological biology, the triad either has not been recognized or has been hidden under various terminologies, as will be shown in the next chapter (Table 2).

The example of the sunburn as presented in Table 1 can be used to illustrate the relative position of stimulus and response, and the flexible way in which these concepts should be handled. Any chain of events made the subject of an analysis has necessarily been carved out from a continuum of phenomena. The first and the last link of the chain should be stated in order to limit the problem on hand reasonably. At another time, the study may extend beyond the limits of the original problem. The sunburn story started with exposure of a white-skinned man to sunshine at the beach. This was considered to be "the stimulus." Suppose the heat of a summer day stimulated the man to go to the beach. Then the sunbath with exposure to ultraviolet radiation was an indirect response to the original stimulus. The absorbed ultraviolet rays acted as a stimulus when they caused release of histamine as a response. This in turn played the role of a stimulus in producing dilatation of capillaries and leakage of their walls. The chain of neurovascular stimuli and responses ends with the disappearance of the erythema (phase 6) and some minor sequelae (phase 7). In the case of eunuchs or some red-haired persons who cannot tan, this would be the only chain of events. However, in the brown-haired man of Table 1, a second chain, the process of pigmentation, was set off by the absorbed ultraviolet rays. Here the early responses are unknown. The final responses are listed in phase 8. To what extent the tanning chain depends on the erythema chain is difficult to tell since tanning can be produced without, or with a very mild, erythema by gradual exposure or by application of suntan lotion. In Table 1 the last link in the pigmentation chain is the tanning response. This again is arbitrary. For instance, the bleaching of the man in winter could encourage a study of the stimuli which cause disappearance of the pigment. This may suffice to illustrate the principle that response R_1 to stimulus S_1 can become stimulus S_2 in causing response R_2 and so forth. For each phase the relation between stimulus and response is defined as reactivity. In other words, in every special investigation reactivity has meaning only with reference to a specified stimulus and specified response. Change of reactivity by stimuli will be the subject of special discussion (Part II, Chapter 2, F).

Some change in reactivity is involved in the sunburn story since increasing tanning and thickening of the horny layer of the skin produce an increasing protection against sunburn. From this point of view the eight-phase story of

Table 1 can be interpreted as one of the self-regulating mechanisms which are encountered everywhere in physiology. A simpler self-regulating mechanism is the pupillary reflex of the eye (Fig. 2). It is a typical example of negative feedback in living organisms.

An individual steps from a dark room into bright light. At the first moment his pupils are dilated since the iris muscles were adapted to the dark. In strong illumination an excessive amount of light passes through the pupils and reaches the retina (stimulus S_1). The retina responds by sending impulses through the optic nerve (response R_1), which reach nerve cells in the brain (S_2). These cells respond by sending impulses[3] through a chain of motor nerves (R_2), which ends in the iris. The iris muscles are stimulated (S_3) and respond by contracting the pupil (R_3). Thus the amount of light which im-

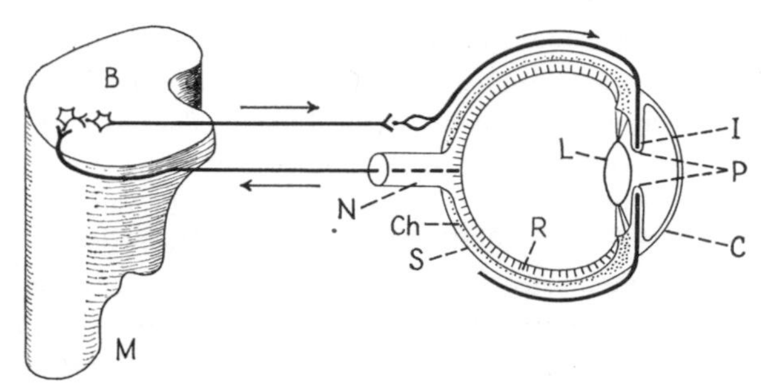

FIG. 2. Diagram of a pupillary light reflex arc to illustrate feedback. Arrows indicate direction of impulses. *C*, cornea; *I*, iris; *P*, pupil; *L*, lens; *R*, retina; *Ch*, choroid; *S*, sclera; *N*, optic nerve; *B*, brain; *M*, medulla oblongata. The anterior chamber, bordered by cornea and iris (or lens at the pupil), contains aqueous humor; the space enclosed by retina and lens contains the vitreous body.

pinges on the retina is reduced to the desirable level. If the illumination decreases so much that recognition of objects becomes difficult, the retina will send weaker or rarer impulses through the optic nerve. At the end of this stimulus-response chain the pupil will dilate. Incidentally, the pupils respond mainly to *change* of illumination. If moderate illumination follows darkness, they will contract, but if the same moderate illumination follows bright light, they will dilate.

The use of the triad, stimulus, response, and reactivity, may be demonstrated by an additional example, which, in some respects, overlaps the sunburn story. Four men are examined at the end of the summer and again during the following winter. At the end of the summer they all show a light-brown skin of the same hue. When biopsies of their skin are taken, brown melanin pigment in similar amounts is found, with identical distribution in the different layers

[3] The transmission of stimuli at the synapse will be described later (Part III, Chapter 1, A).

of the skin. Let us assume that each of these four men is 30 years old, has straight black hair, brown eyes, lips of medium thickness, and a nose which is neither aquiline nor flat. Why are these four men brown? On re-examination during the following winter, No. 1 shows a white skin, and Nos. 2, 3, and 4 are as brown as before. We conclude that No. 1 is a white-skinned man whose skin had acquired a reversible tan after exposure to sun irradiation. Continued analysis shows low blood pressure in No. 2. This suggests Addison's disease, a destruction of the adrenal cortex, which, among other phenomena, causes pigmentation of the skin and mucous membranes. The history of No. 2 reveals that originally his skin was white and that it had turned brown gradually without exposure to ultraviolet irradiation. Pigmentation of his oral mucous membranes and signs of general weakness confirm the diagnosis of Addison's disease in No. 2. Concerning Nos. 3 and 4, who proved to be irreversibly brown but have no abnormally low blood pressure, one can be sure that they are neither tanned nor suffering from Addison's disease. By elimination, Nos. 3 and 4 can only be either the product of two brown parents with similar ancestors, e.g., from the Mediterranean area; or the product of cross-breeding — such as a mulatto with white-skinned and brown parents. In order to decide which alternative is correct, one must study their pedigrees. Suppose that No. 3 is a Mediterranean, and No. 4 is a mulatto. Then the four men are brown for different reasons. The stimuli to which their skins had responded by being brown were part either of the external or the internal environment. The external environment of the skin contains stimuli such as sun radiation, humidity of the air, and the nature of the clothes. The internal environment of the skin is composed of stimuli such as the supply of oxygen, nutritional material, and hormones. Products of metabolic transformation of these substances, e.g., CO_2, also belong to the internal environment.

The way in which the skin responds to these stimuli, in other words, its reactivity, is determined genetically but is modified throughout life by environmental factors. Subjects Nos. 1 and 2 were born with a white skin which was endowed with the following genetic reactivity: to the ordinary stimuli of life their skin responds by remaining white, but to certain changes in the external or internal environment their skin responds by pigmentation. Their reactivity was modified by hormonal factors, since the tanning response would not have taken place without the presence of adequate sex hormones. The skin of No. 2, the man with Addison's disease, probably had the same genetic reactivity as that of No. 1. When his adrenal cortex was destroyed, the blood no longer carried the proper proportion of hormones.[4] This was a change in the inner environment of the skin, or, in other words, a new stimulus to which

[4] The hormones involved are those of the adrenal cortex and the glandular lobe of the hypophysis. I do not know whether sex hormones play a role here; would a white-skinned castrate turn brown when his adrenal cortex is destroyed?

the skin responded by pigmentation: No. 3, the Mediterranean, and No. 4, the mulatto, were born with unpigmented skin, but after birth acquired their brown color gradually, in response to ordinary environment.

Further comments are necessary concerning the reversibility and irreversibility of pigmentation. Although this difference is important, it should not be overrated. With respect to No. 1, it is well to remember that many white-skinned people who lead an outdoor life for many years, such as sailors, acquire an irreversible pigmentation of the exposed parts of their skin (so-called weather-beaten skin). This is why the four brown men were characterized as being 30 years old. Concerning No. 2, the pigmentation of a person with Addison's disease may, eventually, prove to be reversible if the proper hormone substitutes should be discovered. Finally, No. 3, the Mediterranean, and No. 4, the mulatto, may suffer a partial depigmentation of their skin: small or large areas may change from brown to white, a condition called vitiligo. Some cases are caused by infection of the skin with fungi; in other cases the cause is unknown. There seems to be no record of complete bleaching by vitiligo in genetically brown people. Albino children of brown or black Negro parents occur, but they are born and remain without pigment; this has nothing to do with reversibility of pigmentation. Reversibility and irreversibility of genetic and environmental characteristics will be discussed again in Part II, Chapter 2, D.

C. Stimulus, Response, and Reactivity in Various Areas of Normal and Pathological Biology

Stimulus and response are observed, whereas reactivity is the deduced relation between the two. Therefore stimulus, response, and reactivity define each other quantitatively. If two of these magnitudes can be measured, the third one can be calculated. This is the principle of mutual definitions which has proved indispensable in all areas of natural science. The most general expression of such mutual definition of three variables is the equation $z = f(x,y)$. In physics, examples of mutual definitions are force, acceleration, and mass; amount of heat (calories), temperature, and specific heat; electromotive force (volt), intensity of current (ampere), and conductance. In the last group, one measures either conductance (mho) or its reciprocal, resistance (ohm).

Table 2 shows a variety of terminological triads which are traditionally used in different fields of normal and pathological biology. As mentioned above, it has not always been recognized that these terms are equivalent to stimulus, response, and reactivity. At the top of the table physiology has been placed. Physiology can now be defined as that branch of biology which deals with the responses of organisms to physical, chemical, or psychological stimuli. The relation between stimulus, response, and reactivity has been discussed in the story of the sunburn and the story of the four brown men. The second item, pharmacology, is a special branch of physiology characterized by the emphasis on chemical stimuli such as extracts of herbs, mineral waters of spas, and well-defined compounds of natural or synthetic origin, usually summarized as drugs; vitamins and hormones are also included in the domain of pharmacology. An example may illustrate the usefulness of the stimulus-response-reactivity triad

TABLE 2

Stimulus, Response, and Reactivity in Various Areas of Normal and Pathological Biology

Area of Biology	Observable		Inferred
1. Physiology	Stimulus[a]	Response[b]	Reactivity[c]
2. Pharmacology	Chemical stimuli (drugs, hormones, etc.)	Response	Susceptibility[d]
3. Toxicology	Harmful chemical stimuli (poisons)	Effects of poisoning	Susceptibility[d]
4. Immunology	Antigens, i.e., chemical stimuli causing antibody formation	Antibodies	Capacity of forming antibodies
5. Genetics	Environmental stimuli[e]	Phenotypic responses	Genotypic reactivity
6. Experimental embryology	Evocators, inductors, organizers	Chemical and morphological differentiation	Competence, potentiality
7. Medicine	Injury	Disease[f]	Disposition, constitution
8. Pathological morphology	Injury	Lesion[f]	Susceptibility[d]

[a] Synonym: environmental factor (changing or constant).

[b] Synonym: reaction.

[c] Synonyms: responsiveness, excitability, irritability.

[d] Reciprocals of susceptibility are tolerance and resistance.

[e] In genetics, the term environmental stimuli refers to all factors except genes; it is used in contrast to the action of genes on genes.

[f] In this table, conventional terms of medicine and pathological morphology are included although they are based on a mixture of the statistical and teleological norms.

as a frame of reference for handling pharmacological problems. J. T. Litchfield, Jr. (unpublished data, 1958) analyzed the role of tranquilizers in veterinary medicine. The problem was how to make excited domestic animals tractable. He considered the following possibilities. One may eliminate environmental stimuli, for example, by putting blinders on a horse to prevent shying; or one may make the response physically impossible, for instance, by muzzling a dog to prevent it from biting. The purpose of tranquilizers is not to eliminate stimuli or responses, but to modify the reactivity of the animal. On this basis, different types of tranquilizers were considered: (1) those which prevent noticing the stimuli (anesthetics); (2) those which prevent integration of stimuli in the central nervous system (hypnotics, analgesics); and (3) those which block the emission of signals from the central nervous system.

Toxicology, the third item, is a subclass of pharmacology. Both disciplines are concerned with chemical stimuli, but in toxicology the emphasis is on harmful effects. Substances known mainly by their harmful effects are called poisons or toxins. The concept of toxin (or poison) has *no absolute meaning*. Whether a chemical agent is useful, indifferent, or harmful depends on the following factors: dose, rate of administration, route of application; environmental conditions such as temperature and humidity; and finally the reactivity of the animal, organ, or tissue at the time of exposure to the agent. Some bacterial toxins, such as those of food poisoning, are dealt with in toxicology, and others are left to pathology, immunology, and other fields of medicine. Usually, in pharmacology and toxicology the reactivity is called susceptibility, frequently expressed as tolerance or resistance, the reciprocals of susceptibility.

The fourth item, immunology, deals with special chemical stimuli and, therefore, is related to pharmacology and toxicology. The chemical stimuli which are the subject of immunology are known as antigens, or potential antigens, and are characterized by the peculiar chemical response of the stimulated animal or man. This response consists of the production of antibodies, which occur in the globulin portion of blood and lymph plasma and are able to neutralize the effect of the antigens. The relation of antibody to antigen shows an amazing specificity. This specificity has remained one of the basic problems of immunology ever since Ehrlich expressed the problem graphically by the famous lock and key symbols.[5] The capacity of the stimulated animal to form specific antibodies has been entered in the column for reactivity (Table 2). Foreign proteins which reach the organs of an animal parenterally, i.e., by any route except the digestive tract, act as antigens.

The interaction of antigens and their specific antibodies may be either demonstrated *in vitro* (e.g., by precipitation of proteins) or inferred from animal experiments. If a guinea pig, No. 1, has been given an injection of a minute dose of horse serum, a larger dose injected after a proper interval will cause death by suffocation. At this time a sample of the guinea pig's serum mixed with horse serum shows precipitation of protein. An excised part of the uterus from a guinea pig, No. 2, which received one injection of horse serum, will show contractions when placed in a saline bath containing horse serum, provided a certain interval since the injection has elapsed; the necessary interval is similar to that between the two injections in guinea pig No. 1. A specific antigen-antibody reaction is

[5] The comparison was introduced originally by Emil Fischer (1894) to illustrate the specificity of enzyme-substrate relations; it was adapted by Ehrlich to an antigen which combines with a sessile antibody (Ehrlich, 1900, Fig. 1).

considered to be the basis of the contraction of the uterus because of two facts. First of all, the uterus of the guinea pig sensitized against horse serum will not respond to the serum of animals other than horses. Secondly, the quantity of serum which is required to elicit a marked contraction may be as small as 1 ml. of a 10^{-7} dilution of horse serum in a 150-ml. bath of saline. Other aspects of immunology will be discussed in Part II, Chapter 2, F.

The fifth item in the table is genetics. The units of hereditary transmission are called genes, and their sum total in one individual organism is called the genotype. The phenotype represents the sum total of observable properties at any stage of an organism, from the fertilized ovum through embryonic life, childhood, adult life, and old age. In other words, the relation of the phenotypic observable responses to observable environmental stimuli is obtained by inference: it is the genotypic reactivity. As Woltereck (1932) expressed it, the individuals of each species inherit the "Reaktionsnorm," their standard mode of reaction. During embryonic life the hereditary reactivity determines how the embryo responds to environmental factors such as temperature, egg yolk in birds, and the maternal metabolism in mammals. The sex is decided genetically, but as soon as the male or female gonads are established in the embryo, they produce hormones which, from then on, modify the genetic reactivity to stimuli of the internal and external environment. Genetic reactivity modified by hormones is illustrated by the failure of tanning in white castrates (Part II, Chapter 2, A).

In 1909, Johannsen introduced the concept of gene as an abstract unit of heredity. In 1926 he still referred to the gene as "eine Art Rechnungseinheit", a kind of *mathematical* unit, although, in the meantime, studies in the fruit fly (*Drosophila*) by Morgan and his associates had transformed the gene into a *material* unit. Linear arrangement of genes on chromosomes was established, relative distances of genes were determined, and elaborate maps of gene locations in chromosomes were constructed. This new phase of genetics was expressed in the title of Morgan's 1919 book: "The Physical Basis of Heredity." Subsequently, spectrophotomicrography of giant chromosomes in ultraviolet light demonstrated relations between bands of high concentrations of nucleic acids and the pattern of gene distribution (work of Caspersson and Schultz, starting 1939: see Caspersson's 1950 book). Bacterial genetics confirmed the role of deoxyribonucleic acid in the biochemistry of genes. "The gene, once a formal abstraction, has begun to condense, to assume form and structure and defined reactivity" (Sinsheimer, 1957). This transformation of the gene concept is reminiscent of the change of the atom concept. At the beginning of the 19th century, the atomic weight of an element was introduced as a relative value; it was defined as a multiple of the unknown weight of a hydrogen atom to which the value 1 was assigned. When the periodic table of elements was published in 1869 and 1870, atomic weights were still relative values. Finally, the oil drop experiment of Millikan (1911) allowed the determination of the electric charge of one electron and, thus, the calculation of the absolute weight of a single hydrogen atom. This does not mean that the present concept of the gene approximates the quantitative meaning and definiteness of the atom. As Swanson (1957, p. 420ff.) pointed out, there is still competition among four definitions of the gene. It may be considered as an ultimate unit of recombination, or of mutation, or of physiological activity, or of cell reproduction.

The sixth item of Table 2, experimental embryology, centers mainly around the problem of differentiation, that is the development from something homogeneous and chaotic into something heterogeneous and patterned. The something may be morphological or chemical, but in both cases, the terms homogeneous and heterogeneous are meant operationally. In other words, conditions or phenomena in which, at a given time, no heterogeneity can be observed, change in such a way that heterogeneity becomes observable. A change from randomness to order is illustrated by our inanimated model: an electromagnet and a hollow, water-filled rubber sphere containing iron and brass rods (Fig. 1). The differentiation of a frog's egg and early embryo will be described now (Fig. 3). The unfertilized egg is one cell. It is stimulated by the entry of a spermatozoon, or

---

FIG. 3. Diagram of development of frog's egg and early embryo. (a) to (f) and also (g) show external appearance of whole egg or embryo; (e′) is a vertical section through (e); (f′) is a section cut through (f) between the two crosses; (h) to (l) are cross sections through the presumptive head region.

(a) Fertilized ovum. (b) First cleavage (cell division) starting. (c) Second cleavage completed, 4-cell stage; *AP* animal pole on top of animal hemisphere which contains little yolk and is marked by pigmentation; *VP* vegetal pole at bottom of vegetal hemisphere, which is heavily laden with yolk and is unpigmented. (d) Eight-cell stage with large yolk cells near vegetal pole. (e) and (e′) Late blastula stage: upper part consists of two or more layers of small cells which form the roof of a cavity (blastocoel = segmentation cavity SC in Fig. e′). (f) and (f′) Early gastrula stage: the layer of small animal cells has overgrown the vegetal yolk cells *Y*. Halfway between the equator and the vegetal pole a notch forms. This is the dorsal lip of the blastopore marked by crosses in (f) and (f′). Here the cell sheet turns inward, forming a second layer which lines a new cavity, the archenteron or primordial intestinal cavity *AE*. As more animal cells cover the yolk, the ventral and lateral parts of the blastopore *(BP)* are formed. The animal cells are so small now that their outlines cannot be shown at the scale of these pictures. This leads to a superficial resemblance of (a) and (f). As drawn in (f′) the animal cells are too few in number and too large in size; this distortion was necessary for the sake of clarity in the diagram.

Note that up to (f) the volume of the system does not change. The continuous increase in cell number is associated with a decrease in volume of individual cells.

(g) Late gastrula with neural groove *(NG)*. Change from spherical to oval shape of gastrula is not shown in the picture. (h) Section through early neurula; *NF*, neural folds. (i) Section through late neurula; *NG*, neural groove; circle ventral to it indicates notochord (primordium of vertebral column). (j) Section through head region of embryo with neural tube *(NT)*. *(k)* Left side: outpocketing of eye vesicle *(EV)* from brain part of neural tube. Right side: head ectoderm responds to proximity of eye vesicle by thickening *(L)* which is the start of lens formation; eye vesicle changes to eye cup. *(l)* Left side: *EC*, more advanced eye cup of which the concave side is the presumptive retina; *L*, advanced stage of lens in close contact with eye cup. Right side: final stage of lens, no longer connected with head ectoderm; secondary (final) cavity of eyeball is established.

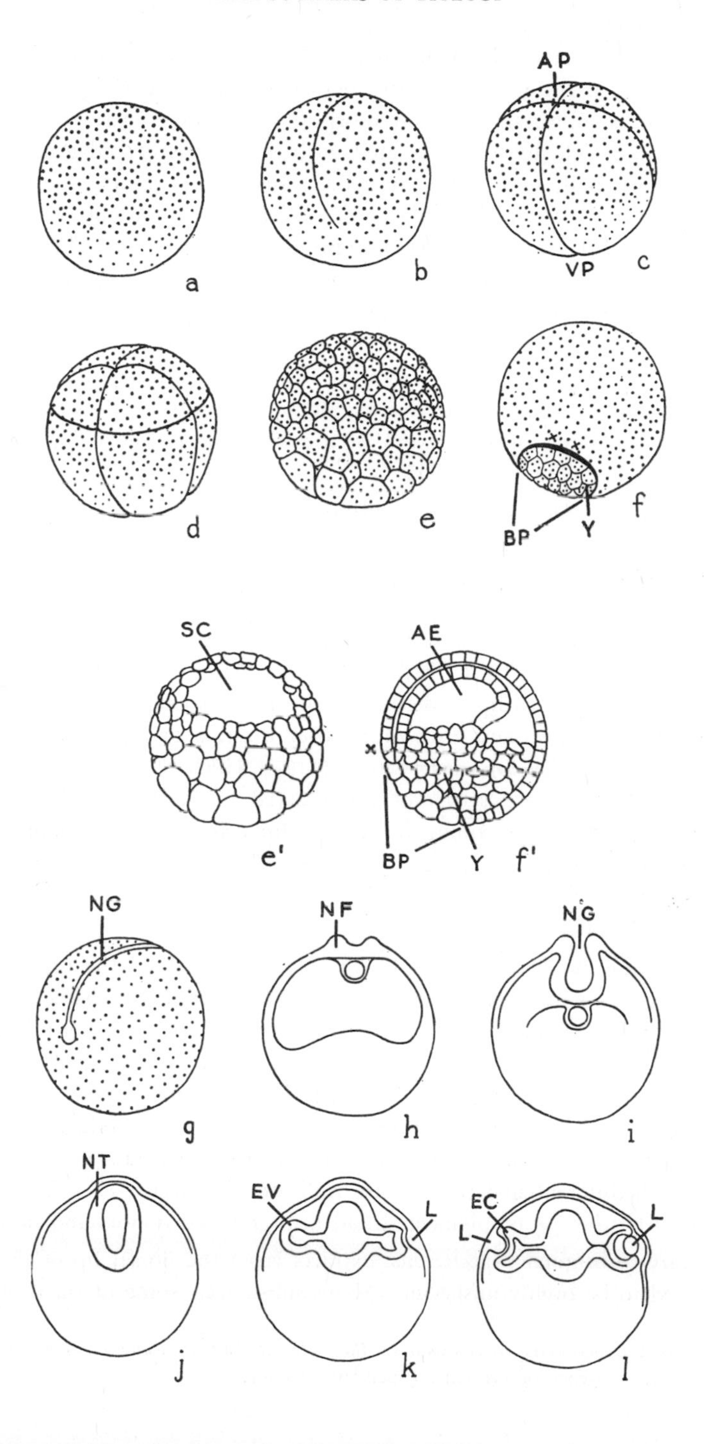
AP
VP
a
b
c
d
e
f
BP
Y
SC
AE
e'
f'
NG
NF
NT
EV
L
EC
g
h
i
j
k
l

by the prick of an experimental needle, to divide into two daughter cells (first cleavage). This is the start of a chain of cell divisions in the course of which local differences in metabolism and morphology become increasingly distinct. Unequal rates of cell division associated with migrations of cell groups lead to a folding of originally simple layers of cells. This folding results first in a double layer of cells (Fig. 3f′ and Fig. 32), and subsequent in- and outpocketings produce structures with a higher degree of complexity (Fig. 3g–l). These structures are indirect responses to the cleavage-producing stimulus, and represent primordia of organs. The primordia, in turn, act as stimuli in causing undifferentiated adjacent structures to differentiate. Part of the brain, in the process of forming the primordium of the eye, moves from the inside of the head toward its surface. When this brain material reaches the stage of an eye cup it acts as a stimulus on the adjacent ectoderm (surface layer) of the head and causes it to fold in, thus forming the lens[6] (Figs. 3k and l).

However, this reactivity (potentiality or competence) is a property not only of the ectoderm of the head but also of the ectoderm of other regions. Figure 4 shows that an eye cup which is transplanted to an unusual area stimulates lens production in the adjacent ectoderm. A special area of the early embryo, the dorsal lip of the blastopore, marked *x* in Fig. 3f and f′, has been given the name of organizer (Spemann, 1921) because of two characteristics. (1) If material which includes the dorsal lip of the blastopore is transplanted from a blastula or early gastrula (*A*) into any place of another blastula or early gastrula (*B*), a second embryo will develop in that area of (*B*) which received the graft. This indicates strong inductive power of the grafted material. (2) In its natural situation, the material containing the dorsal lip differentiates into presumptive chorda mesoderm and neural plate. This also takes place if the material is isolated from its natural environment. In other words, the organizer material is capable of self-differentiation.

It is because of characteristic (1) that I have classified the organizer as a stimulus, together with the evocators and inductors. Because of characteristic (2), one might also have entered the organizer in the last column, as a substrate endowed with reactivity (competence, potentiality). In an excellent illustration, Holtfreter and Hamburger (1955, Fig. 82) show an embryo with an organizer-induced second embryo. The structures which are derived from the host (induced tissues) are indicated in white, and those derived from the transplanted organizer material (self-differentiated tissues) are indicated in black. This difference can be observed by the use of pigmented and unpigmented donors and hosts.

Contrary to original expectations, extracts from the dorsal lip of the blastopore proved to be highly unspecific. Many substances, some of them of simple

---

[6] This is the necessary mechanism in *Rana fusca,* but in *Rana esculenta* lenses can originate in the absence of eye cups (Spemann, 1938).

chemical composition, can act as inductors. Similar results are obtained with inert foreign bodies whose primary effect is merely mechanical. For some time the organizer seemed to play the peculiar role of a stimulus whose action did not depend on the reactivity of the substrate. This is historically paralleled by early phases of other research areas. There was a time when the action of enzymes did not depend on variable substrates, and there was a period when pathogenic

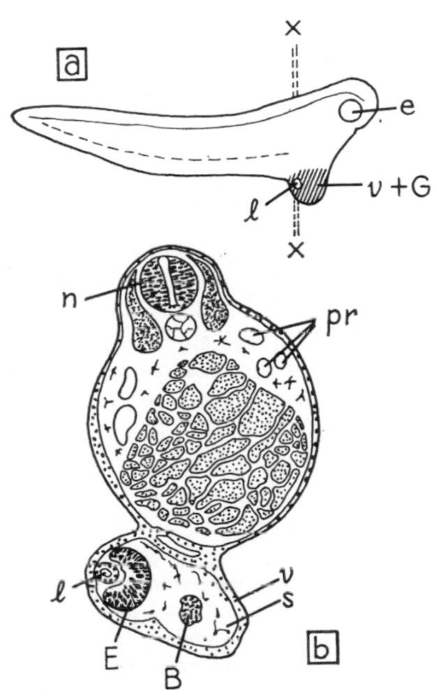

FIG. 4. Induction of lens in unusual place of ectoderm illustrating extent of reactivity (potentialities). From an early gastrula of *Triton alpestris* (donor), primordia of brain and eye were implanted into a blastocoel of *Triton taeniatus* (host). Later stage of the host (larva in side view) is shown in (a); section through the plane x—x is shown in (b). Lower case letters indicate structures and products of the host; capital letters indicate grafted material and its products. Diagram (a): e, eye of the host larva; v + G, vesicle formed from host tissue enclosing the graft; *l*, area of lens formed from the ventral trunk epidermis of host. *Diagram (b): v,* vesicle wall formed by host, continuous with the epidermis of the host; *s,* space of vesicle; *B,* part of the grafted brain; *E,* grafted tissue differentiated into eye cup; *l,* lens formed from *v* in response to proximity of *E; n,* spinal cord of host; *pr,* pronephric tubules of host. Redrawn from Mangold (1929, Fig. 35).

bacteria produced diseases undisturbed by the variable reactivity of hosts. The search for a definite substance as an organizer had to be abandoned. However, as Shen pointed out (1958), the phenomena of embryonic induction and the idea of a chemically defined organizer have been a major source of inspiration for chemical embryology.

Medicine and pathological morphology in medicine, items 7 and 8 of Table 2, are characterized by the emphasis on harmful stimuli called injury. In medicine any response to injury is referred to as disease. The response may be chemical or psychological, simple or complex, localized or generalized. Both in medicine and in pathological morphology the reactivity is expressed either as susceptibility, or as its reciprocal, resistance. The reactivity of the organism as a whole is usually called constitution. Pathological morphology is concerned with changes of visible structures; the changes are traditionally called lesions.

D. Statistically Normal and Abnormal Stimuli, Responses, and Reactivities

Statistically normal and abnormal stimuli, responses, and reactivities may be illustrated by a fictive example. Let us assume that three groups of ten scouts were drenched by rain for different periods of time and responded to this abnormal stimulus with common colds, pneumonia, or no ill effects. Table 3 shows

TABLE 3

Fictive Example of Three Groups of 10 Scouts, Exposed to Three Degrees of the Same Stimulus (Heavy Rain), with the Result of Three Different Responses (No Ill Effect, Common Cold, Pneumonia)

Group designation	Stimuli: duration of drenching (hours)	Responses (number of scouts)			
		No ill effect	Common cold	Pneumonia	Total
A	2	4	6	0[a]	10
B	4	1	8	1	10
C	6	0	6	4[a]	10

[a] Different at a 10% significance level (Mainland *et al.*, 1956).

different distributions of responses in the three groups. By definition, all these responses were a result of the interaction of different stimuli with different reactivities. As far as statistical significance is concerned, Group C with 4 cases of pneumonia and Group A with no cases were different at a 10% significance level for a sample of 10 (Mainland *et al.,* 1956).

Which had the greater effect in determining the responses, the stimulus or the reactivity? If in Group C, which was drenched for 6 hours, each of the ten scouts had contracted pneumonia, one would say that the stimulus was so strong that differences in reactivity became negligible. Suppose, on the other hand, the five scouts of Groups B and C who contracted pneumonia after 4 or 6 hours of drenching, are subsequently exposed to a few minutes of drenching. If one of them responds with pneumonia again, his reactivity (or susceptibility in medical language) at the time of the second exposure would be considered extremely high relative to the very weak stimulus. There is the possibility that his original reactivity had been less and was increased by the first attack of pneumonia. The effects of repeated stimulation will be discussed in Part II, Chapter 2, F.

Stimuli in the external and internal environment, genetic reactivity, and the modification of genetic reactivity by hormones are illustrated by the example (Part II, Chapter 2, A) of four brown men: the white-skinned man with a suntan, the white-skinned man with Addison's disease, the Mediterranean, and the mulatto. Evidently, the interaction of genetic and environmental factors cannot be analyzed without the concepts of stimulus, response, and reactivity.

The *relative importance of genetic and environmental factors* can be studied in a constructive way, provided definite phenotypic criteria are the target of the study and the environmental and genetic factors involved are not too complex. Yet, there is no sense in the familiar question whether *in general* the genotype or the environment is more important in determining the phenotype. "Most of this discussion has been due to faulty thinking" (Dobzhansky, 1937, p. 15). It is pointless to ask whether *cancer in general* is caused more by hereditary or by environmental factors. Both in humans and in animals certain forms of tumors occur in which hereditary factors play a great role. There are strains of mice in which more than 75% of the animals over the age of one year develop tumors of the lung, in the ordinary laboratory environment[7] (Grüneberg, 1952, p. 450). However, by restriction of caloric intake, the incidence of spontaneous tumors in mice can be reduced markedly (Tannenbaum, 1947). Details will be given in connection with the biological classification of malignant neoplasms (Part V, Chapter 4, C, 1). Those cancers which develop from a well-defined local injury demonstrate the strong influence of an environmental factor. Pipe-smokers' cancer develops in that place of the lower lip on which the hot pipestem rested: the location of the cancer is clearly determined by the environmental stimulus. However, only a small proportion of pipe smokers develop cancer of the lip. This shows that the reactivity of the lips of individuals varies, owing either to genetic or environmental factors, or to both. Finally, there are cancers which originate from malformations of tissues, in other words, from disturbances which happened during embryonic development. These disturbances, in turn, may be ascribed to genetic or to environmental factors.

In concluding the discussion on heredity and environment, a few words may be useful concerning reversibility and irreversibility of characteristics. Most genetically determined characteristics are irreversible, but this rule is not without exceptions. The occasional bleaching of genetically brown individuals known as vitiligo was mentioned before. A second example may be helpful. An adult woman will grow a beard and acquire a masculine voice when her adrenal cortex develops a certain type of tumor. Her sex was determined genetically, and she remains a female with respect to the genetic units (genes, located in chromosomes) which are present in all of her cells. But certain hormone-controlled

[7] This is probably pure chromosomal heredity, whereas mammary tumors of mice are transmitted from generation to generation by a factor in the milk of the mothers plus some chromosomal factor.

expressions of her sex, such as a beardless face and a feminine voice, proved reversible. On the other hand, some characteristics which develop as responses to environmental factors may become irreversible. As stated above, a sailor who has genetically a white skin may acquire irreversible pigmentation by long exposure to outdoor life (weatherbeaten skin, more or less identical with prematurely senile skin). Obviously, mechanical factors can produce irreversible changes, e.g., the loss of a limb by an accident. Minor destructions are followed either by complete restoration of the original condition or by a scar which is a mechanical repair with or without decreased functional value. Scars are irreversible structural changes which represent the last link in a chain of responses to environmental stimuli. The process of aging involves an interesting combination of irreversible changes which are partly controlled genetically and partly a result of environmental wear and tear.

E. Quantitative Relations of Stimulus, Response, and Reactivity

Quantitative relations between stimuli, responses, and reactivities have been studied in various areas of biology, but there are no absolute units of stimulus, response, or reactivity. It is necessary to define the units for each area of study. Neuromuscular activities are very suitable for introducing the reader to quantitative stimulus-response relations. In skeletal muscle electric stimuli of moderate frequency elicit one contraction per shock. Such single contraction is called a twitch. If the frequency of shocks per second is increased, the contractions decrease in amplitude, since the muscle cannot return to the level of complete relaxation after each contraction. If the frequency is further increased, the contractions fuse into a sustained tetanic contraction.[8] An external eye muscle of the cat will show complete tetanic fusion without a trace of individual contractions, if stimulated at a frequency of 100 or more shocks per second. The most rapidly repeated muscular contractions yet measured occur in those muscles of small insects (Diptera) which drive the wings. These muscles respond at frequencies up to 1000 per second as concluded from the frequency of their wing beat (Chadwick, 1953).

The following example illustrates how the choice of different units of reactivity can produce qualitatively different responses to seemingly similar stimuli. When an isolated muscle fiber is stimulated electrically, there will be a maximal contraction at a certain intensity of stimulus, but there will be no contraction at all below this intensity. This phenomenon is known as *all or none response.* On the other hand, a whole muscle, which is composed of many fibers, responds

[8] A classic chart illustrating the gradual transformation of single contractions into tetanus is that of Howell (1926, p. 43). It is reprinted by Winton and Bayliss (1955, p. 331).

with increasing degrees of contraction to increased stimulation. This behavior of the whole muscle results from the fact that each muscle contains some layers with predominantly slow fibers and other layers with predominantly fast fibers. The sequence of responses, starting with the fastest and ending with the slowest units in the muscle, is known as *recruitment.*

A motor neuron (nerve cell and nerve fiber) and the group of muscle fibers controlled by the neuron are called a motor unit. The number of muscle fibers which belong to one motor unit vary from two to six in an external eye muscle to about one hundred in a leg muscle. Eccles and Sherrington (1930) calculated the average tension developed by a single motor unit in various muscles of the cat's leg as follows. The tension of the whole muscle produced by maximal stimulation of the motor nerve was measured in grams. This value was divided by the total number of motor units in the whole muscle as obtained by counting the fibers of the motor nerve before branching.

Depending on the purpose of the particular study the unit can be one muscle, a motor unit, one muscle fiber, or a molecular thread of actomyosin which is the contractile constituent of muscle.[9] While it is being studied, the muscle may still be in the animal or it may be isolated and surviving for the duration of the experiment. Similarly, a single muscle fiber may be observed either surviving as an isolated preparation or still in its natural situation in the muscle. It may be useful to give a more detailed description of a study of whole muscles in which graded relations between stimuli and responses were observed.

Cooper and Eccles (1930) determined the effect of stimulation by induction shocks on the tension of muscles. Two of their graphs are reproduced in Fig. 5a and b. In the example of Fig. 5a, one of the muscles which moves the eyeball was stimulated by a bipolar electrode placed on the muscle's nerve. This was done in cats of which the brains were removed, while the muscles remained in their natural situation. Clamps were fixed to the two ends of the muscle so that its contraction produced a negligible amount of shortening.[10] A lever recorded the force of contraction, expressed as tension in grams, at each stimulation of the muscle. When 100 to 300 shocks per second were applied, fusion of contractions took place. Such tetanic contraction was obtained in each experiment of Fig. 5a. The resulting tension (force of contraction) increased with the number of stimuli per second until a plateau was reached at approximately 200 shocks per second. With a shock frequency of 100 to 300 per second, it took a minimum of 20 to 30 shocks (about 1/10 second) until tension developed fully. A similar experiment carried out with a leg muscle is shown in Fig. 5b. To produce tetanic

[9] See Hanson and Huxley (1955) on the structural basis of contraction in striated muscle.

[10] The fact that contraction and shortening are not identical can be observed in any person who tries to lift an object which is too heavy for him: his muscles contract, but do not shorten. (This example was suggested to me by Dr. E. H. Dearborn, 1958).

contraction the leg muscle required only 20 to 100 shocks per second, and the increase in tension with increased number of stimuli per second reached a plateau at approximately 30 shocks per second. At this rate, about 10 shocks proved the minimum required to obtain fully developed tension. In other words, the two muscles showed a considerable difference in reactivity. Of course, the leg muscle, being much larger than the eye muscle, produced tensions of a higher order of absolute magnitude. By plotting changes in tension against frequency of shocks,

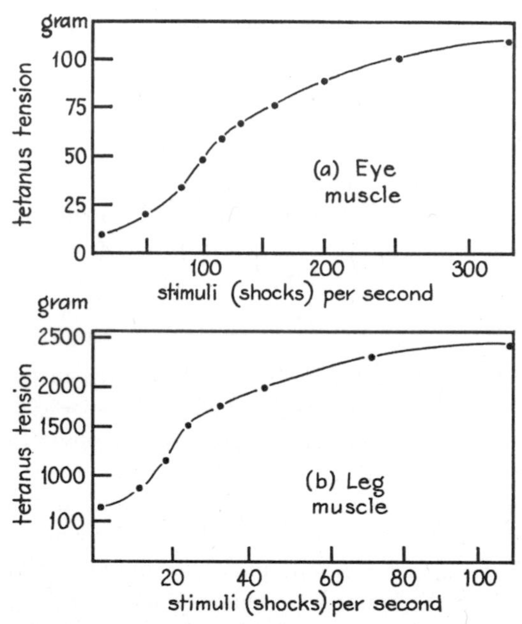

FIG. 5. Differences in reactivity of muscles. Muscular tension as response to electric stimuli of varying frequency (after Cooper and Eccles, 1930). Muscles of decerebrated cat stimulated electrically to tetanic contraction; length of muscles kept constant. Tension (force of contraction) in grams plotted against frequency of stimuli (shocks) per second. (a) One of the muscles that move the eyeball (internal rectus); (b) one of the leg muscles (extensor digitorum longus).

one obtains some sort of S-shaped curve with either muscle. This is not surprising since S-shaped curves are the most common cumulative curves of increments. What is the relation of these experiments to the function of muscles in the organism? One learns from the experiments described here that the eye muscle, if compared to the leg muscle, can tolerate stimuli of both greater frequency and greater absolute number without being forced into tetanic fusion of contractions. This is understandable from a teleological point of view since the movements of the eyeball need faster and more precise controls than those of the leg. The difference between eye and leg muscle is also accentuated by the fact mentioned before that the number of muscle fibers controlled by one nerve fiber is about 100 in a leg muscle and two to six in an eye muscle.

Curves like those shown in Fig. 5a and b are known in pharmacology as dose-response curves. A large collection of such curves is found in Clark's 1933 book, with mainly pharmacological examples. Most dose-response curves in pharmacology deal with studies of groups. Suppose a group of ten mice was used to determine the dose of a certain agent which kills 50% of the mice within a fixed period of observation (LD_{50}). The next step would be a study of several groups, to determine the variability of responses to the same dose of one agent within each group of animals. When reproducible results have been obtained, responses to different doses of the same agent can be compared. Finally, different doses of different agents can be applied to a number of groups. In such studies, the individuals are not labeled separately, but are considered as interchangeable components of the group. Obviously, information on the variation of reactivity of individual animals is very limited when groups are used as units. Readers interested in the relation between statistics of individuals versus statistics of groups will find Figs. 13 and 14 of Clark's book very useful; they show the effects of digitalis on 537 cats, after a paper of van Wijngaarden (1942). Group studies in humans are illustrated in my Table 3, which shows a fictive example of three groups of scouts exposed to rain for different periods. Wells (1958; Figs. 1-2 and 1-3) gives a brief but very lucid presentation of the relation between the S-shaped cumulative curve of the mortality of groups and the bell-shaped distribution curve of the mortality of individuals (intraperitoneal injection of pentamethylenetetrazol, Metrazol, in rats).

Suitable illustrations of group studies in laboratory animals are found not only in pharmacological experiments but also in investigations of bacterial infections. In bacterial infections, the stimulus-response-reactivity relation is peculiar in that the host responds to the bacteria, but also the bacteria respond to the host.

Two studies on the effect of bacterial infections in mice will supply the material for our discussion. The schemes of these studies are shown in Table 4, and the graphs of the results are presented in Fig. 6a–c, and Fig. 7a and b. In both examples, the dose of injected bacilli and the period of observation are kept constant. In the first example (Gowen, 1952) the infection is allowed to take its own course, and in the second example (Redin and McCoy, 1957, unpublished data) an antibacterial drug is applied. In the first example, the intensities of bacillary action represent the stimuli and are entered on the abscissa. In the second example, the action of the bacilli is inhibited by different doses of an antibacterial drug. These doses are entered on the abscissa and represent the depressed bacterial stimuli. In both cases the survival of the mice serves as the response and is entered on the ordinate. In the first example there are three variates: different strains of one species of bacilli, different strains of one species of host (mice), and the percentage of surviving mice during the period of observation. In the second example, only one strain of bacilli and one strain of mice are used. There are two variates in this example: different doses of an antibacterial drug and the per cent of mice which survive the period of observation.

TABLE 4
Two Schemes of Quantitative Studies of Interaction between Disease-Producing Bacteria and Mice

Factors of experiment	Example 1[a]		Example 2[b]	
	Constant	Variable	Constant	Variable
Bacterial strains	—	Several strains	One strain	—
Bacterial doses	One dose	—	One dose	—
Antibacterial drugs	No drugs	—	—	Different doses of one drug
Mouse strains	—	Several strains	One strain	—
Criteria	—	% surviving mice	—	% surviving mice
Periods of observation	Same for all experiments	—	Same for all experiments	—
Total constants	3	—	4	—
Total variables	—	3	—	2

[a] Gowen (1952).
[b] Redin and McCoy, unpublished data (1957).

The results of the first example are illustrated in Fig. 6a–c. A three-dimensional graph was published by Gowen (1952) as a summary of his extensive studies on the interaction of genetically different strains of mice and genetically different strains of bacteria. His diagram is reproduced here (Fig. 6a). Some strains of mice were recognizable by different fur colors, others by the presence or absence of waltzing; the hereditary nature of these properties was verified by standard techniques of cross-breeding. The different strains of bacteria were mutations, i.e., hereditary variations from a known strain of *Salmonella typhimurium* which causes mouse typhoid. The bacterial mutants were identified by their varying need for adenine among their nutritional requirements. The visible properties of the mouse strains were correlated with different degrees of susceptibility to infection with mouse typhoid bacilli.[11] The nutritional differences of the bacterial strains were correlated with their ability to multiply in the host, which, in disease-producing bacilli, is known as virulence.

In order to construct Fig. 6a, one could start either with one strain of mice

[11] In Part II, Chapter 1, C, these mice with different fur colors and different susceptibility to infection were mentioned as an example of ornamental structures which are correlated with an important function.

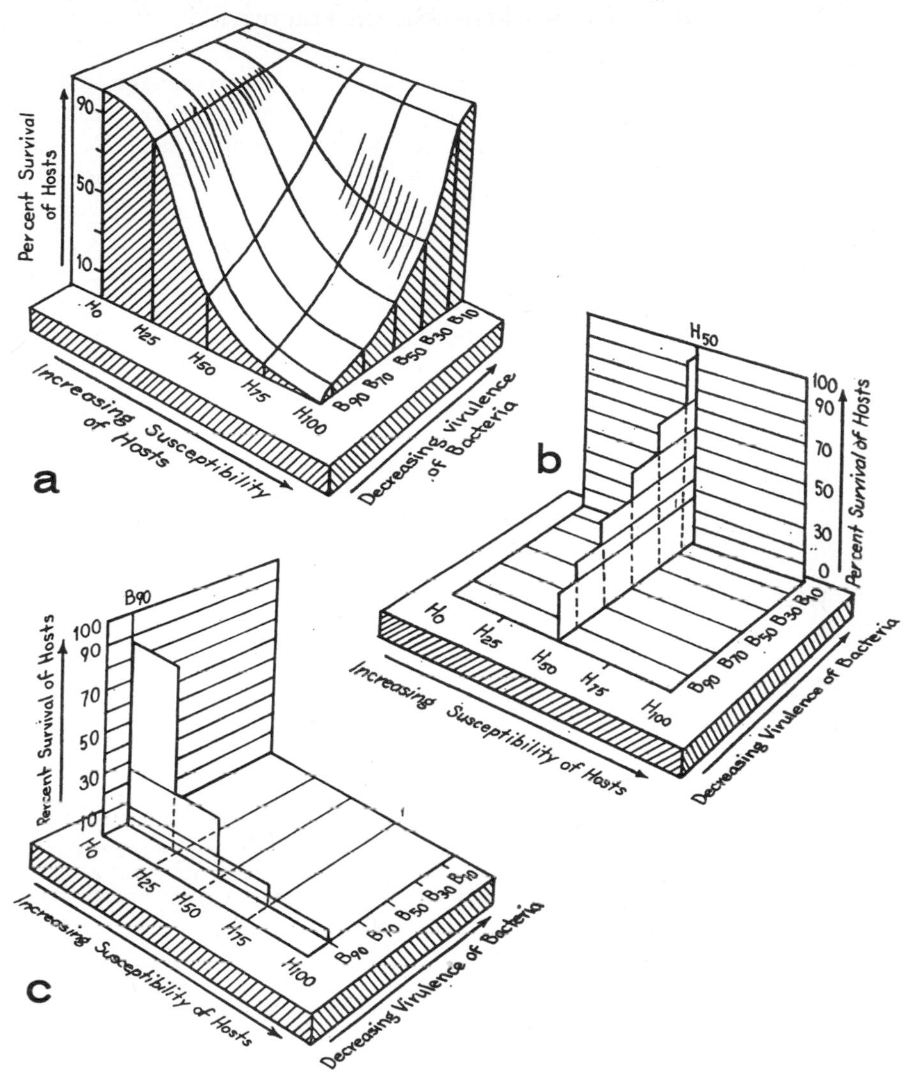

FIG. 6a. Interaction of seven strains of *Salmonella typhimurium* with five strains of mice. Diagrammatic three-dimensional graph compiled by Gowen (1952) from his observations. The designations $H_0 - H_{100}$ for five different hosts and $B_0 - B_{100}$ for seven strains of bacilli indicate merely the ranking of susceptibility and virulence, respectively; i.e., they do not represent absolute values. The lettering for B_0 and B_{100} was omitted to avoid crowding.

Construction of Fig. 6a is shown in Figs. 6b and 6c (graphs made by the present author). Either add six histograms to 6c to cover the remaining *B*'s, or add four histograms to 6b to cover the remaining *H*'s. Fill spaces of completed 6a or 6b, and draw a rounded-out surface through all the points of the steps to produce the solid 6a.

FIG. 6b. Effect of seven bacterial strains on one host strain, H_{50}. By traveling from each corner of the histogram to the back wall, per cent survival for each *B* strain is obtained. For instance, B_{70} caused approximately 43% survival.

FIG. 6c. Effect of one bacterial strain, B_{90}, on the five *H* strains. Reading of histograms as in (b). For instance, H_{50} had approximately 25% survival.

or with one strain of bacilli. To illustrate these procedures Fig. 6b and 6c have been prepared by the present author. Virulence is measured by the death rate of the hosts: The higher the per cent survival of mice, the lower the virulence of the bacilli. Since it is easier to observe the death of mice in cages than the death of bacilli in mice, the survival of mice is the more convenient criterion. This measures the reactivity of mice[12] when the bacilli are considered as the stimulus, and it measures the reactivity of bacilli, if the mice are considered as the stimulus. It is obvious that the virulence of bacteria cannot be determined in test tubes, but only in hosts. However, biochemical or morphological characteristics of bacteria *in vitro* may be associated with different degrees of virulence in a host. Different nutritional requirements paralleled by differences in virulence were mentioned above. In many species of bacteria a change from a smooth to a rough appearance of their colonies (on solid media) is connected with a decrease in virulence (Wilson and Miles, 1955, p. 355). In some laboratories, tubercle bacilli which form cords when cultivated in the proper media appeared to be more virulent than those which multiply without producing this characteristic pattern (Middlebrook *et al.*, 1947); in other laboratories this relation between virulence and morphological aggregation in cords was not observed (Conalty and Gaffney, 1955).

The second example of a dose-response curve represents unpublished studies of G. Redin and E. McCoy (1957). Experiments were made to determine the effect of different doses of chlortetracycline in mice infected with pneumococci (*Diplococcus pneumoniae*) by intraperitoneal injection. The strain of mice used is known as Carworth Farm 1. A single subcutaneous dose of the drug was administered immediately after the infection. The mice were observed for a period of 14 days. All deaths occurred within the first 6 days after infection and were reasonably ascribed to the infection. Five different doses were applied, and each dose was given to a total of 40 mice; the per cent survivals are shown in Table 5. Simultaneously with the drug experiment, 60 mice were infected with the same dose of pneumococci, but without subsequent drug administration. Of these 60 untreated infected mice, 60 died within 2 days. Of the 40 infected mice which received the lowest dose of drug, 40 died within 4 days, 29 of them within 2 days. As Fig. 7a shows, the dose-response curve was S-shaped.

For various reasons it may be preferable to present data as a straight line rather than as an S-shaped curve. In order to obtain a linear expression of their data Redin and McCoy applied the rapid graphic method of Litchfield and Wilcoxon (1949). The data were plotted on logarithmic probability paper

[12] An excellent survey of genetic reactivity of mice to bacterial infections is found in Chapter XX of Grüneberg's "Genetics of the Mouse" (2nd ed., 1952), covering the work of Gowen and other investigators in this field. Many unpublished data of Gowen are used in Snedecor's book (4th ed., 1946) for illustration of statistical procedures.

(Fig. 7b). This paper has a logarithmic abscissa and an ordinate which is divided in such a way that a symmetrical S-shaped distribution curve appears as a straight line. The doses were entered on the abscissa and the per cent survival on the ordinate. Then the best-fitting straight line was drawn through the points of measurement. This was easy in the present case since the points for the three middle doses happened to be exactly in a straight line.[13] The intercepts of the drawn line with the horizontal lines at the 16%, 50% and 84% levels supply the values which permit calculation of the slope function S and the 95% confidence limits both for S and for ED_{50}, the effective dose for 50% of the animals. This procedure is indispensable for estimating relative potencies of two or more drugs.

TABLE 5

Effect of Different Doses of Chlortetracycline on Survival of Mice Infected with Same Dosage of Pneumococci[a]

Group no.	Drug dose (mg./kg.)	Mice	
		Alive/total	% Survival
1	54	40/40	100
2	38	38/40	95
3	27	23/40	57.5
4	19	4/40	10
5	13	0/40	0

[a] Experimental plan: Each of 200 mice received a single intraperitoneal injection of the same amount of a suspension of pneumococcal broth cultures. Pool of 4 experiments with 50 mice in each. The 50 mice in each experiment were divided into 5 groups of 10 mice each. Immediately after infection each mouse received a single dose of chlortetracycline and each group received a different dose of chlortetracycline. (Redin and McCoy, 1957, unpublished data.) For graphic presentation of same data, see Fig. 7a and b.

Quantitative studies of stimulus, response, and reactivity in connection with morphogenesis need special discussion. The production of visible structures, or morphogenesis, has been discussed previously. A hollow, water-filled rubber sphere with magnetic and unmagnetic particles (Fig. 1) served as an analogy to the developing frog's egg and as a model for the illustration of morphology as a procedure. The concept of differentiation was discussed at that occasion and then again in applying the triad, stimulus, response, and reactivity, to embryology, (Part II, Chapter 2, C). Both the external form of the whole embryo and its internal structures are molded by factors which are accessible to quantification, namely, the migration of cell groups and local differences in the enlargement, multiplication, and death of cells (cf. Part II, Chapter 1, B). The rates of these processes as a function of time have been determined, and

[13] If the fitting had been less obvious, most weight would have been given to the points within or near the 16–84% range.

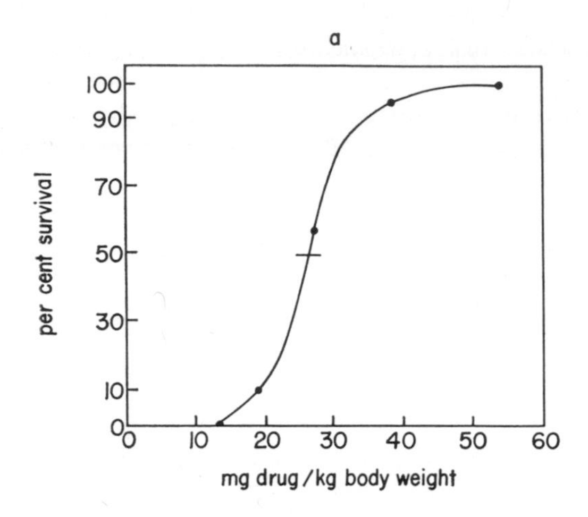

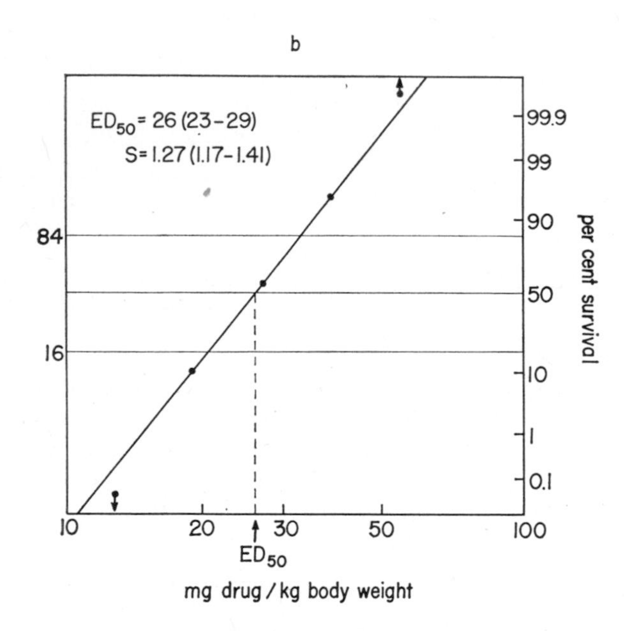

FIG. 7. Effect of different doses of chlortetracycline on survival of 200 mice infected with the same dose of pneumococci (Redin and McCoy, 1957, unpublished data). For experimental data see Table 5. (a) Arithmetic plotting of per cent survival of mice against doses of drug. (b) Same experiment and data as in (a). Plotting on logarithmic probability paper. Abscissa, doses on logarithmic scale; ordinate, per cent survival on probability scale (Litchfield and Wilcoxon, 1949). ED_{50}, dose producing 50% survival; *S*, slope function; figures in parentheses, 95% confidence limits.

they have been correlated with other measurable variables such as environmental temperature, oxygen pressure, or radiation effects. Two examples may illustrate the way in which quantitative biochemistry can be related to morphological differentiation.

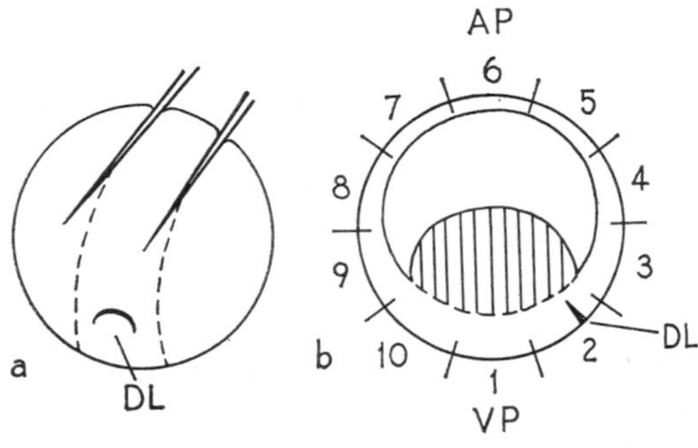

FIG. 8. Determination of oxygen uptake in different areas of early axolotl gastrula (Gregg and Løvtrup, 1950). Technique of isolating pieces from gastrula. (a) Cutting a slice with glass needles; *DL*, dorsal lip of blastopore (picture modified by the present author). (b) Slice or ring obtained in (a) with indication of ten isolates used for respiration *in vitro*; *AP*, animal pole; *VP*, vegetal pole; *DL*, dorsal lip.

The early development of amphibian embryos is illustrated in Fig. 3a–l. By the early gastrula stage (3f) a certain amount of morphological differentiation has taken place although no definite primordia of organs can be recognized. This stage was used by Gregg and Løvtrup (1950) for a biochemical study of different areas (Fig. 8a and b). Ten pieces were excised, kept surviving, and examined for their respiration. Significant differences in their oxygen uptake were measured (Fig. 9, after Gregg and Løvtrup, 1950). The respiration of

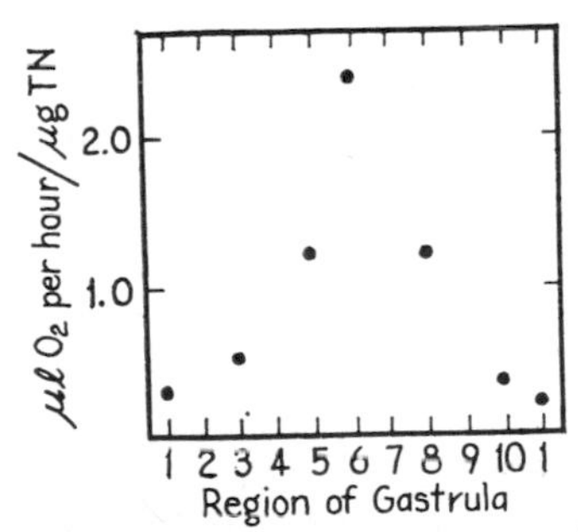

FIG. 9. Variation of oxygen uptake in the different areas of axolotl gastrula shown in Fig. 8b (Gregg and Løvtrup, 1950). *TN* on ordinate means total nitrogen.

the animal pole explant (area 6) was approximately 10 to 12 times greater than that of the vegetal pole explant (area 1) if the total nitrogen of each piece was taken as a base of reference. In similar experiments, not only the oxygen uptake, but also the activities of dipeptidase and beta-glycerophosphatase showed higher values in the animal pole fragments. This difference was attributed, in part, to the fact that the ratio of cytoplasm to yolk is high in the cells of the animal pole and low in those of the vegetal pole.

The second example refers to a well-coordinated qualitative and quantitative study of morphological and biochemical differentiation in the *brains of mammalian embryos.* Flexner and his associates observed critical periods in the development of the cerebral cortex of guinea pigs. Between the 42nd and the 45th day of embryonic life, the nuclei of the nerve cells reached mature volumes, cytoplasmic inclusions (Nissl bodies) increased in number, the ramifications of nerve cells became more pronounced, and nerve fibers appeared more mature. On the biochemical side there was a marked increase in the activities of adenylpyrophosphatase and succinic dehydrogenase. Study of the composition of intra- and extracellular fluid showed a change in the ratio sodium distribution : chloride distribution. According to Flexner, this change suggests relations to the appearance of cortical electrical potentials on the 46th day. The most remarkable feature of these phenomena is the simultaneous progress of chemical and morphological differentiation along several lines (for summary, see Flexner, 1950; extended studies were reported by Flexner in 1955). The relations between morphological differentiation, cholinesterase activity, and functional developments in the brain and retina (Boell and Shen, 1950; Shen, 1958) will be reported in connection with histochemistry of enzymes (Part III, Chapter 7, D, 4). In his investigations of the morphological and chemical development of the retina Coulombre (1955) included determinations of changes in cell number, cell size, cell volume, and changes in size of the whole eye; his studies will be discussed in connection with the variability in composition of organs (Part V, Chapter 3).

F. Continuous and Repeated Stimulation. Change of Reactivity. Some Aspects of Immunology

Continuous stimulation as contrasted to repeated stimulation can mean two different things: (a) with the most sensitive methods available, no discontinuity in time or change in intensity can be discerned when stimulation is recorded over a period of time; or (b) for the purpose of a particular investigation, the stimulation applied over a certain period is considered as a continuous single stimulation or compound stimulation rather than a series of discrete stimuli.

In the discussion of tetanic contractions of muscles produced by electric stimulation of the nerve, the frequency of shocks per second was used as the unit of stimulation (Fig. 5a and b). Fusion into a tetanic contraction was obtained at certain frequencies and after a certain number of shocks. Continuous stimulation may change the reactivity of a muscle or a sense organ, depending on both intensity and duration of the stimuli. To a maintained stimulation of constant intensity, muscle receptors or sense organs may respond with electric discharges which show high initial frequency but gradually increase in intervals until a plateau rate of discharge is reached and maintained for some time. Such response decrements can be interpreted either as fatigue or as adaptation. For a discussion of this subject see Ruch (1949).

Change in reactivity may result not only from continuous, but also from repeated, stimulation. Whether the response to the second stimulus will be stronger or weaker than that to the first stimulus, or unchanged, depends on the following factors: the time interval between first and second stimulation, the intensity of first and second stimulus, and the basic reactivity of the organ (or other unit) subjected to repeated stimulation.

Both in skeletal and in cardiac muscle, the twitch after an electrical stimulus is followed by a period during which a second stimulus produces no effect. This so-called absolute refractory period is very short in skeletal muscle, but in cardiac muscle it lasts at least as long as the phase of contraction. Consequently in the skeletal muscle, repeated stimuli at the proper intervals produce either summation of contractions or the fusion of contractions known as tetanus, whereas neither summation nor tetanus can be produced in heart muscle. Because of pre-existing differences in reactivity between skeletal and cardiac muscle, repeated stimuli produce different changes in reactivity.

Many changes in reactivity resulting from continuous or repeated stimulation are associated with modifications of visible structures. The sunburn story can be used again for illustration. Repeated exposure to sunlight causes gradual tanning and also a slight thickening of the horny layer of the skin. Consequently there is a decrease in penetration of ultraviolet rays, and this results in decreased reactivity.[14] Another well-known example is the response of the skin of hands and feet in places of repeated mechanical injury by pressure or friction. The first response consists of the production of blisters, but after they have healed, the new epidermis is thicker and more keratinized than the original one. In this way the reactivity to further mechanical irritation is decreased. In these two examples it was possible to follow the steps which led to changes in reactivity. This is not possible, as a rule, in those cases where repeated *chemical* stimuli lead to a decrease in reactivity.

One of the best-established cases of repeated chemical stimulation with decreasing reactivity is the habitual intake of morphine. Larger and larger doses are requested by the addict, who may eventually tolerate two hundred

[14] In contrast with ultraviolet irradiation, a single dose of long, soft Roentgen rays can produce a series of erythema responses, at intervals of several weeks and of decreasing intensity. Like ultraviolet, repeated doses of Roentgen rays eventually cause tanning in conjunction with thickening of the horny layer of the skin. Whether sex hormones are necessary for the Roentgen-ray tanning effect is unknown to me.

times the therapeutic dose, or twenty times the dose which is fatal to other people. Yet nothing is known concerning the biochemical processes which lead to decreased reactivity in the addict. Under certain social conditions, the decrease in reactivity to tobacco smoking or alcohol may be desirable; this decrease in reactivity certainly becomes harmful when no limit is set to increasing dosage. The present interpretations of changing tolerance and habit formation are well presented in the textbooks of physiology and pharmacology. For readers interested in these problems I suggest that they begin with the study of conditioned reflexes.

Finally, let us discuss immunology, that branch of physiology which is entirely devoted to repeated stimulation and changes of reactivity. The clearest part of immunology concerns antigen-antibody relations, which were described previously (Part II, Chapter 2, C). Besides these chemical relations, there is a confusing variety of observable responses which one may call symptomatologic immunology.

The symptomatologic side of immunology is burdened with historical terms which have been retained along with the modern development of chemical immunology. Terms such as immunity, hypersensitivity, atopy, idiosyncrasy, allergy, and anaphylaxis are used with different connotations by different authors; Cooke (1947) expressed the opinion that all these terms should be included in the term allergy, with anaphylaxis a subdivision of allergy. No attempt will be made here to explain the terminology. A certain knowledge of symptomatologic immunology is necessary in order to understand the variability of morphological responses to infectious agents (Part V, Chapter 4, A, 3). In symptomatologic immunology, particular forms of response are hives (urticaria) or the disappearance of white cells from the blood (agranulocytosis). Variations in reactivity can be: abnormally high or low reactivity; desirable or undesirable changes in reactivity; changes in reactivity depending on the number, intensity, and timing of the same type of stimuli; and change in reactivity by qualitatively different stimuli. These possibilities are illustrated best by the effects of three types of repeated stimuli: (1) parenteral administration of foreign protein, (2) food, and (3) bacteria.

It was mentioned earlier (Part II, Chapter 2, C) that injection of a guinea pig with a small dose of horse serum followed by a larger dose after the proper interval will cause death by suffocation. A more detailed study shows that the circular smooth muscles of the guinea pig's bronchi contract strongly and remain contracted so that the bore of the bronchi is practically closed. As a consequence the passage of air to and from the lung is made impossible, and the animal dies of respiratory failure. In the rabbit, the smooth muscles of the small pulmonary arteries contract so that the right ventricle of the heart has to pump against greatly increased resistance. The animal dies of failure of the heart muscle. Finally, in the dog, the smooth muscles in the vessels of the liver and other abdominal viscera are affected in such a way that a tremendous congestion of these organs is produced. This abnormal redistribution of blood causes a fatal lack of blood supply in vital organs such as the brain and the heart. Evidently, guinea pig, rabbit, and dog have this in common that the reactivity of smooth muscles, but not of striated muscles, is changed by repeated injections of horse serum. However, the organs in which the most sensitive smooth muscles are located are different.

Suppose each of ten individuals eats a large quantity of strawberries. Eight respond with a moderate indigestion, one with a severe indigestion, and one with no ill effects. With still larger quantities of strawberries the distribution would probably shift in the direction of the severest response, whereas with smaller quantities it would shift in the direction of the mildest response. This stimulus-response relation is similar to that shown in Table 3. Although the majority of people respond to strawberries in the way described, a minority show a qualitatively different response, namely transient hives (urticaria). It seems that one strawberry can be enough to produce this response. In the terms of the present book those individuals who respond with hives to strawberries possess a reactivity which is qualitatively different from that of the majority of the population. Quantitatively this is a very high reactivity since the stimulus needed to elicit the exceptional response (hives) is of a much smaller order of magnitude than that which produces the majority response (indigestion). It is not known whether quantitative dose-response relations can be demonstrated in individuals who are subject to strawberry rash.

Suppose an individual A could originally eat strawberries without unpleasant effects, but now responds with hives. Evidently his reactivity has changed from a desirable zero level, representing the teleological norm, to an undesirable positive level. Besides becoming teleologically abnormal, he also became statistically abnormal, since most people do not respond with hives to strawberries. In the case of individual A, it is assumed that the repeated intake of strawberries is in some way responsible for the increase in reactivity. If such a condition is produced experimentally by applying the same stimulus repeatedly, the first stimulus is called the sensitizing dose while a later stimulus which is expected to produce the altered response is called the challenging dose.

Suppose an individual B who had no previous contact with strawberries is given such berries for the first time and responds with hives. Then his peculiar reactivity was either this way since birth, or developed unnoticed during some period of his postnatal life. If it can be demonstrated that the peculiar reactivity has been present since birth, the next question should be whether genetic or intrauterine factors were responsible. In most cases it is difficult to exclude the possibility of prior contact with sensitizing substances when a person with unusually high reactivity comes to the attention of the immunologist.

A desirable change in reactivity is observed in many bacterial infections.

The primary response to infection with typhoid bacilli consists of increase in body temperature and of other signs and symptoms of typhoid fever. This is the normal reactivity of people whose bodies are invaded by typhoid bacilli for the first time. Those who survive the disease will not usually respond with another attack of typhoid fever if exposed again to an infection with typhoid bacilli. A person who had typhoid fever may even harbor living and moderately multiplying typhoid bacilli for the rest of his life. Such a person, a carrier of virulent typhoid bacilli, is a menace to others, but his own reactivity to the presence of the bacilli has dropped to zero. In the case described the abnormally low reactivity was acquired, but a person who never contracted typhoid fever, no matter how much exposed to infection, possesses either genetically or congenitally low reactivity to typhoid bacilli.

In strawberry rash, antigens and antibodies are not demonstrable. By contrast, the infection with typhoid bacilli as well as many other bacterial infections leads to the production of demonstrable antibodies by the host in response to the antigens of the bacilli. It is the acquisition and maintenance of the antibodies which protect the individual against another attack of typhoid fever. If one uses the clinical manifestations as a standard, the reactivity of the infected person started at a high level and ended at zero level. Using the defense mechanisms of the host as a standard, one may say that the reactivity of the infected person increased when more and more antibodies became available to neutralize the antigens.

Abnormal reactivity to drugs after repeated administration deserves a special discussion. A thousand patients with excessive activity of the thyroid gland will respond to the administration of thiouracil by a decrease in thyroid activity. Although their dose-response relations differ quantitatively, all show decreasing thyroid activity with increasing dosage. However, among these thousand patients a small minority produce, in addition to the antithyroid effect, an entirely different response, namely, a sudden decrease in granulated white cells in the circulating blood (agranulocytosis). A similar decrease in white cells is observed after administration of amidopyrine. Again, this occurs in a small minority of the people who take amidopyrine, and no relation to dosage or frequency of administration has been demonstrated. The usual effect of amidopyrine is that of an antipyretic or fever-reducing agent. The remarkable point is that agranulocytosis is produced in a minority of people by two drugs of different chemical composition and with different principal activity, one being an antipyretic and the other an antithyroid drug. Because of the relative rarity of agranulocytosis, it is not known whether the same individuals respond in this way both to amidopyrine and thiouracil. However, there is evidence that the same person may respond with hives, asthma, or equivalent disturbances when exposed to feathers, horse dandruff, wool, and other stimuli of the ordinary environment. Abnormal reactivity to drugs can have serious consequences: agranulocytosis is fatal in some cases. The study of these abnormal reactivities to drugs has been hampered by the fact that, as yet, they have not been reproduced in experimental animals. Moreover, no antibodies have been found in humans with such abnormal reactivity (Thomas, 1955; p. 482). If the search for antibodies remains unsuccessful, one should look for chemical mechanisms different from antigen-antibody formation. Then, of course, abnormal reactivity to drugs would lose its relationship to chemical immunology.

Quantitative variations of the stimulating dose or of reactivity can produce qualitatively different responses in the same species of animals. These are not quantitative dose-response relations in the sense defined previously. As an illustration an experiment may serve which is basically identical with the sensitization of guinea pigs or rabbits by horse serum described above. The only additional feature is the size and number of challenging doses. The first stimulus is produced by parenteral administration of a minute quantity of foreign protein to a rabbit, i.e., administration through any route other than the digestive tract. The second stimulus consists of an entry of the same protein into the blood circulation, also by any of the parenteral routes, but the amount of protein used for the second stimulation and the degree of reactivity of the animal at this time decide what type of response will occur. (A) Large amounts of protein or high reactivity lead to shock. (B) Small amounts of protein, or low reactivity, may have two different effects: (1) if the protein deposit remains localized, i.e., by subcutaneous injection, a local, transient wheal forms; (2) if the protein is distributed in the body, transient hives or swelling of the skin will occur in many areas. Both in (A) and in (B) the first as well as the second stimulus consisted of a single injection of protein. The reactivity of the rabbit can be modified in still another way by injection of small amounts of the same protein several times. Then two entirely different responses may be observed. Either attenuated (not fatal) shocks occur after each of the repeated injections, or there will be a local response of the skin as in (B, 1); however, whereas a few injections produce a transient wheal, continuous injections cause a lasting redness and swelling which leads to local death (necrosis) of the skin. All these different responses to repeated parenteral application of foreign protein are accompanied by the formation of specific antibodies.

Unspecific changes in reactivity to infectious agents play a great role in empirical medicine. Changes in frequency of various infections have been related to changes in nutrition or climate. Experimental studies of unspecific changes in reactivity became possible with the use of corticosteroids or Roentgen irradiation. Particularly impressive is the fact that not only certain bacteria and parasitic worms, but also transplanted tissues, may gain a foothold in animals which were treated with corticosteroids but they would not gain a foothold in untreated animals.

The area of immunological research is much larger and much more complex than one might conclude from my highly simplified presentation. In host-parasite relations it is not only the reactivity of the host that changes, but also the reactivity of the parasite. Extensive observations are available on the development of resistance to drugs by microorganisms (see Schnitzer and Grunberg, 1957). Although, historically, bacterial infections took the lead in the study of immunological phenomena, there is now an increasing interest in the immunology of virus infections on the one hand, and helminthic infections on the other hand. Problems of embryological differentiation can be studied with immunological methods (Part II, Chapter 2, H, 2). Fluorescent antibodies are used as histochemical stains (Part III, Chapter 7, B, 2). Antigen-antibody interactions of cells, cellular aggregates and organs represent important criteria of classification (Part V, Chapter 2, C).

G. Various Relations between Functions and Visible Structures in Organisms

The model presented in Fig. 1 shows the action of an electromagnet on paraffin-coated metal rods which float in a water-filled rubber sphere. Depending on various conditions, including quantitative factors, the effects of the magnetic force are, or are not, accessible to direct visual observation. In its application to organisms this model proved helpful by establishing a flexible approach to the various possible relations between functions and visible structures. As stated before, demonstrable relations between a certain structure and a certain function may be correlative, with or without time as a variate. If there is a correlation with time as a variate, this may, or may not, represent a casual relation. In a casual relationship the visible structure may be the cause, and the functional phenomenon the effect, or the reverse. Some examples will illustrate the variety of relations between functions and visible structures. In discussing the examples, logical classifications, such as correlation or causation, will remain in the background. The variety of available observations should convince everyone that more than one kind of relationship exists between the functions and visible structures of organisms.

1. *One Function Related to One Structure*

Obvious relations between functions and visible structures in organisms have been described in innumerable instances. Mechanical engineering techniques of nature have been compared to those of man. A well-known example is the trabeculae of bones, which have been found to satisfy the rules of human engineering. When a fractured bone heals it frequently acquires a shape which differs from the original shape, e.g., a straight bone may become curved. Then the trabeculae are rearranged in compliance with the new shape. This is included in "the law of transformation in bones" as proposed by Julius Wolff (1892). For recent illustrations see Weinman and Sicher (1955, pp. 128–135). A complicated configuration in which different organs are adjusted to a common function is shown in my Fig. 10 (after Hesse, 1924). In three different aquatic animals, frog, crocodile, and hippopotamus, the shape of the head

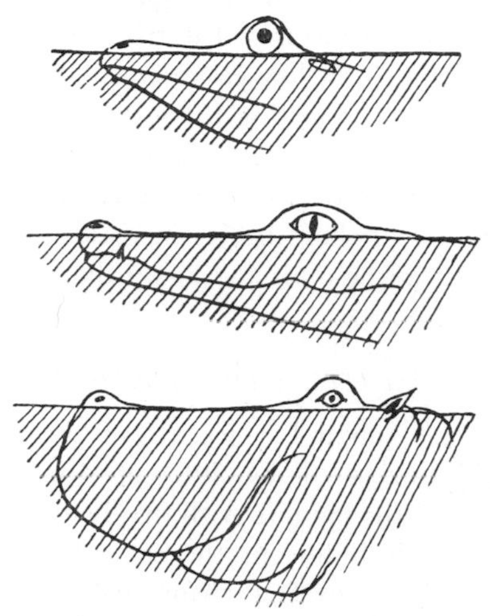

FIG. 10. Animals of different classes and different orders of magnitude using a similar structural pattern for the same function: frog, crocodile, and hippopotamus, each with eyes and nostrils above water level when the remainder of the head is submerged. (After R. Hesse, 1924, Fig. 81.)

allows the nostrils and eyes to stay just above the surface of the water when the remainder of the head is submerged. It is certainly remarkable that such a similarity of functional structures is found in animals which belong to different classes of vertebrates and, in addition, differ tremendously in size.

2. *Multiple Functions of One Structure*

The bones of vertebrates have three main functions, two of them mechanical and one chemical. The chemical function is to serve as a calcium storehouse. Whenever the concentration of calcium ions in the blood threatens to decrease temporarily, e.g., because of a calcium-deficient diet, the bones discharge calcium into the blood. Sufficient concentration of calcium in the blood is vital. If the concentration decreases below certain critical values, the heart muscle cannot work and the blood cannot coagulate. With less extreme deficiencies, convulsions (tetany) occur. One mechanical function of bones is to supply a supporting framework for the body and levers for the contractions of skeletal muscles. Another mechanical function is that of a protective shell for vital soft organs, such as the central nervous system and the bone marrow. Some highly specialized mechanical functions of bones may be mentioned: dogs have a bone to support the penis, and kangaroos have a supporting bone for the pouch. In most birds, the bones incase air sacs which are extensions of the bronchi outside the lungs; the function of the air sacs is not clearly understood (Sturkie, 1954).

Masses of aggregated fat cells, known as fat tissue, are another illustration of multiple functions of one structure. Fat tissue stores combustible energy, but also serves as insulation against low environmental temperature in the polar zones. In whales and seals the insulating masses of fat add to the buoyancy of the animals in water. Finally, fat masses have various mechanical functions. Layers of fat form pads under the feet of human beings and elephants. A plug of fat in the cheeks of newborn babies increases the efficiency of sucking. Probably the masses of fat in the orbit and along the seminal cord have special mechanical functions.

3. *Similar Functions Carried Out by Different Structures*

Locomotion of animals supplies a wealth of examples illustrating how similar functions can be carried out by different structures. Mammalian feet seem to be equally equipped for running with two hoofs, one hoof, or no hoofs, if one compares animals which live in the same habitat, such as antelopes, zebras, and cheetahs. Without any feet, many snakes move very fast on land. Flying is done with feathered wings by birds, whereas bats and insects use membrane wings. A variety of structures and techniques are used for swimming.

One of the most intriguing structural differences associated with similar function is revealed by a comparison of the eyes of vertebrates and the equally elaborate eye of an invertebrate, the octopus. Both types possess similar refractive media (cornea, lens, and vitreous body) and a retina which contains the visual cells. Each visual cell is bipolar. One pole carries a process which serves as a receptor to photic stimulation, and the other pole tapers into a nerve fiber which transmits the impulse to a chain of neurons. The receptors are rods and cones in some mammals, such as man, and only rods in other mammals, such as guinea pigs. Only rods are present in many classes of vertebrates and in the octopus. The striking structural difference between the eyes of vertebrates and the octopus eye is found in the position of the receptors in the retina (diagrams Figs. 11 and 12). In the octopus eye the receptors point toward the light, while in the vertebrate eye they point away from the light. A comparison of sense organs shows a similarity of orientation of their receptors. With the exception of the acoustic receptors in the inner ear, which present special conditions,[15] the receptor ends of the polarized sensory cells point toward the source

[15] Sound waves are changed into fluid waves; the fluid waves stimulate the sensory cells of Corti's organ.

of their respective stimulation, as seen in the organs of touch, taste, and smell. Since this is considered the standard orientation of receptors, the retina of vertebrates is described as inverted and that of the octopus as noninverted. It is

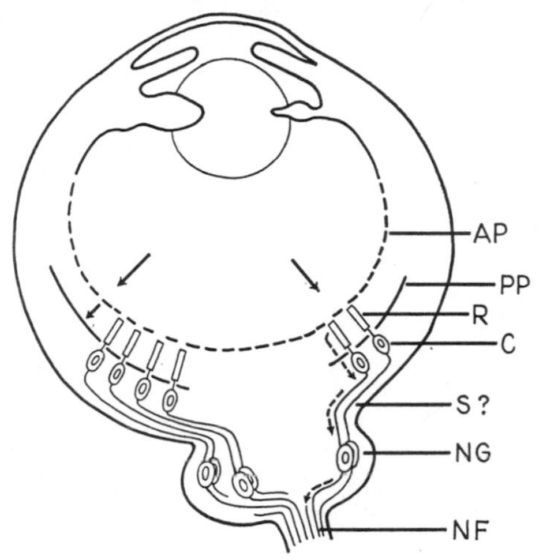

FIG. 11. Diagram of noninverted retina of octopus eye. *C* body and *R* receptor of visual cell (first neuron). The octopus has rod cells only; 6 of them are represented. Location of next neuron and synapse *(S?)* is uncertain. *NG*, nerve cells in optic ganglion; *NF*, nerve fibers assembled in optic nerve; *AP*, anterior pigment layer; *PP* posterior pigment layer. Solid arrows show direction of light; interrupted arrows show direction of impulse transmission. Cornea and iris similar to human eye (Fig. 12); lens spherical as in some mammalian species. Thickness of retina relative to other parts of the eye is exaggerated in the diagram.

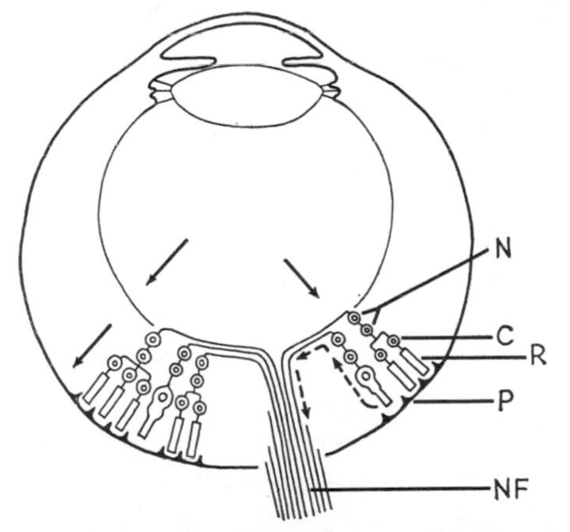

FIG. 12. Diagram of inverted retina of human eye. *C* body and *R* receptor of visual cell (first neuron); 7 rod cells and 2 cone cells are represented. *N*, second and third retinal neurons; *NF*, nerve fibers assembled in optic nerve; *P*, pigment layer of retina. Solid arrows show direction of light; interrupted arrows show direction of impulse transmission. Thickness of retina relative to other parts of the eye is greatly exaggerated; for more natural proportions see Fig. 2.

obvious that both types of retina function satisfactorily. There is hardly any difference between the environment of marine vertebrates, especially fishes, and that of the octopus.

The puzzle of the inverted retina of vertebrates has been approached from the embryological as well as from the phylogenetic angle. I follow here the presentation of Parker (1908). In vertebrates the retina is the only sense organ which does not develop directly from the surface layer of the embryo, the so-called ectoderm. The indirect development of the vertebrate retina is shown in my Fig. 3g–l. It starts with an inpocketing of the medullary plate,

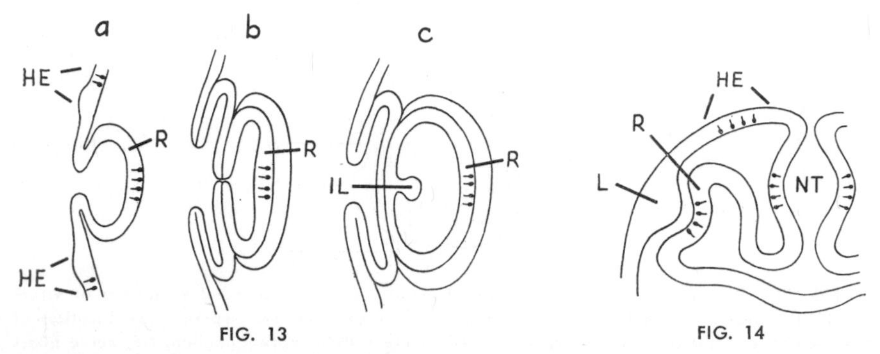

FIGS. 13 and 14. Diagrams illustrating Balfour's (1881) embryological theory of inverted retinas in vertebrates, and noninverted retinas in invertebrates. Group of four imaginary primordia of polarized sensory cells are drawn with their receptor ends originally pointing to the surface of the embryonic ectoderm. *HE*, head ectoderm; *R*, presumptive retina; *NT*, neural tube of vertebrate (presumptive brain); *IL*, inner primordium of lens of octopus; *L*, primordium of vertebrate lens.

FIG. 13. Three stages of octopus eye (after Korschelt, 1936). (a) Direct inpocketing from ectoderm. (b) Multiple folding to form iris and cornea. (c) Inner lens primordium has formed. Unchanged orientation of sensory cells, now visual cells of the retina.

FIG. 14. Development of vertebrate retina condensed in one diagram (after G. H. Parker, 1908); for separate stages see Fig. 3, i to k. Multiple in- and outpocketing carries the primordia of sensory cells from the head ectoderm to the neural tube, the optic vesicle, and finally to the optic cup, where they become the visual cells of the presumptive retina. Inverted orientation of visual cells is the result.

an area of thickened ectoderm; subsequent outpocketing produces the eye vesicle, and a final inpocketing produces the eyecup, of which the concave inner lining is the presumptive retina. All other sense organs, including the ear vesicle, are established as a direct inpocketing of the ectoderm, although secondary folding processes may follow. In the octopus and other invertebrates the embryology of the eye resembles that of other sense organs. As shown in Fig. 13a–c, the retina of the octopus eye forms by direct inpocketing from the head ectoderm, without any relation to the development of the nervous system; connections with the nervous system are made later. In contrast, the vertebrate retina has been termed properly a part of the brain, although it is indirectly derived from the ectoderm. These circumstances became the basis of an embryological theory of the inversion of vertebrate retinas proposed by Balfour (1881). It is illustrated in my Fig. 14. According to this theory the ectoderm of the gastrula stage of the early embryo contains the primordia of all sensory units. The primordia are polarized in such a way that their potential receptor-ends point to the external environment from which sensory stimuli are expected to come. In the developing vertebrate eye the primordia of visual cells are originally located in the head ectoderm, and are polarized and oriented in the same way as the primordia of other sensory cells. In the process of inpocketing and outpocketing the primordia of visual cells are carried from their location in the head ectoderm into the wall of the neural tube, the eye vesicle, and finally the eyecup, where they represent the presumptive retina. The orientation of the sensory units with respect to the surface of the embryo has necessarily been reversed in the course of these transformations. Since the retina of the octopus eye is the result of a single inpocketing of the head ectoderm, the orientation of sensory primordia is not reversed. The main objection to Balfour's theory stems from the fact that the eyes of several invertebrate species possess inverted retinas, although their eyecups develop directly from the ectoderm. In some spiders two eyes are found side by side, one of them with an inverted retina and the other with a noninverted retina (Korschelt, 1936, Fig. 711). The inverted and noninverted retinas illustrate how similar functions can be performed by surprisingly different structures. Some additional examples follow.

As a rule stimuli are conducted by nervous fibers, but in the heart cross-striated muscle fibers conduct the stimuli by which the contractions of the different parts of the heart are coordinated.

Hormones can be produced by a variety of structures. The structures may be permanent or transient. Permanent structures are endocrine glands, such as the thyroid, the hypophysis, and the adrenals. Transient structures with hormone production are the placenta and the corpora lutea in the ovary. The architecture of most endocrine glands resembles that of glands with external secretion except for the absence of ducts in the endocrines (Fig. 87). Both types of glands are characterized by a mosaic of closely packed cells whose surfaces are in intimate contact with blood capillaries. Closely packed cells may form solid blocks, as in the parathyroid, or they may form a wall which surrounds a large central space as in the follicles of the thyroid (Figs. 25, 26, and 37). Certain hormones can be extracted from the adenohypophysis, others from the neurohypophysis. The adenohypophysis has a mosaic pattern with or without small spaces inside the cell blocks. The neurohypophysis shows the typical pattern of nervous tissue, namely, a network of specialized cells and fibers. It is controversial whether the neurohypophysis produces the hormones or stores hormones which are imported from the adjacent part of the brain (hypothalamus). Either alternative proves the point which matters here: structures with the pattern of nervous tissue can produce hormones. In spite of differences in their arrangement, individual cells may exhibit characteristics of secretory activity. It is on the basis of cellular criteria that E. and B. Scharrer arrived at the concept of neurosecretion (1945). Bodian (1951) described a distinct organization in the neurohypophysis of the opossum which makes this organ particularly suitable for the study of neurosecretion. For recent developments see the transactions of the Second International Symposium on Neurosecretion (Bargmann *et al.*, 1958).

Unequal structures can be the source of similar, or possibly identical, hormones. Thus adrenaline is formed in the trabecular chromaffin tissue of the adrenal medulla, but a similar substance is released from sympathetic nerve endings throughout the body (see Part V, Chapter 2, C, 3). The highly variable interstitial cells of the testis are an example of a site of endocrine function without any definite arrangement of the specialized cells; these interstitial cells probably produce androgen. Finally, a specific hormone may be discovered which for a time cannot be associated with any special structures. Renin is an example of this type. The kidney consists of a system of blood vessels which is interlaced with the renal tubules (Fig. 86). This arrangement serves for excretion of waste products with preservation of useful chemical material. Besides having this function which is clearly correlated with the renal structures, the kidney produces a special enzyme, known as renin, which plays a definite, though indirect, role in the control of the contraction of small arteries. Although it was observed that renin occurred only in the cortical zone of the kidney, it remained unknown in which histological component it formed. Subsequently a correlation was found between the renin content of the renal cortex and the amount of granules in the juxtaglomerular body, which is a small aggregation of cells adjacent to the glomeruli (in the interstitium). Finally, Hartroft and Edelman (1960) reported that they had succeeded in localizing fluorescent antirenin antibodies in juxtaglomerular cells of rabbits. These positive findings were limited to juxtaglomerular cells particularly rich in granules as a result of a sodium-deficient diet. (For a description of the fluorescent antibody technique of Coons *et al.*, see Part III, Chapter 7, B, 2.) A discussion of the relations between high blood pressure, renin, and juxtaglomerular cells is included in Tobian's review (1960). In summarizing, one may state that the function of hormone production can be associated with the structure of the classic endocrine glands, with the structure of nervous tissue, or with groups of cells without any definite pattern.

Changes in interpretation of structure-function relations are to be expected in biological research. An interesting example is the myelinated nerve, in which the neurofibrils lost their role as conducting structures and the nodes of Ranvier became part of the conducting mechanism. This will be described in connection with fibers (Part IV, Chapter 3, A).

4. *Stability and Variability in Function-Structure Relations*

The variability of functional structures will be discussed at various occasions. Within the frame of the present chapter, two examples will suffice to illustrate how structural and functional variability are interwoven.

The cyclic variations of the mammalian uterus are both structural and functional. During pregnancy the organ enlarges with the increasing size of the fetus, serving the dual function of protection and nutrition of the fetus. The termination of pregnancy is controlled by hormonal and mechanical factors. After delivery the uterus contracts until it is approximately as small as it had been before pregnancy. From a functional point of view the small nonpregnant uterus with its narrow bore facilitates the travel of spermatozoa toward the Fallopian tubes (which are the usual place of fertilization) and the implantation of the fertilized ovum after its discharge from the Fallopian tubes. The estrous cycle entails periodic changes in the architecture of the inner layers of the uterus (endometrium) and the lining of the vagina. The cycle is governed by hormonal controls and shows different features in different species.

Before the advent of the antithyroid drugs, such as sulfonamides or thiouracil and its derivatives, morphological activation of the thyroid was considered to be inseparable from functional activation. This was justified in the light of experiments with thyrotropic hormones, observations on the effect of low environmental temperature, and experiences with human patients suffering from hyperthyroidism (Graves' disease). In all these cases the signs of increased functional activity of the thyroid (increase in basal metabolic rate, pulse rate, etc.) were associated with conspicuous morphological changes, such as increased height of the follicular cells, and dilution and accelerated discharge of colloid (see diagrams Figs. 25 and 26). The same picture of morphological activation was observed after administration of antithyroid drugs, but it was associated with functional inactivation. This dissociation of structure and function is the result of a three-phase mechanism which has been demonstrated in rats: (1) drug-induced inhibition of thyroxine production, (2) increased discharge of thyrotropic hormone from the adenohypophysis as a response to the thyroxine deficiency in the circulating blood, and (3) morphological activation of the thyroid as a response to the increased stimulation by thyrotropic hormone (Mackenzie and Mackenzie, 1943; Astwood, Sullivan, *et al.*, 1943). There is little doubt that, in this mechanism, the morphological activation of the thyroid is futile since the functional inactivation of the gland persists.

Structure-function relations play an important role in the classification of organisms (taxonomy), organs, and tissues (Part V, Chapters 1 and 2).

H. Tabular Comparison of Organisms and Man-Made Machines

Analogies between organisms and machines were mentioned earlier. At this point a systematic comparison of organisms and machines seems useful. In Table 6 several criteria are listed. The first seven criteria refer to the structures and

TABLE 6

Comparison of Structure and Function in Organisms and Man-Made Machines

Criteria	Organism	Machine
1. Functional structures	Valves of the heart	Valves of a pump
2. Vestigial structures (relics of manufacturing process)	Umbilicus of adult	Mold marks on cylinder block in automobile
3. Ornamental structures	Color pattern of beagles	Two-tone finish of automobile
4. Planned transformation and distribution of energy	Skeletal muscle	Steam engine
5. Necessity of blueprints (plans of organization)	Development of embryo	Manufacture of machines
6. Self-regulating controls (feedback) built into structures of organism or machine	Pupillary reflex	Photoelectric control of exposure in photographic camera
7. Adaptation of structures to function by use	Acquired callosities of skin	Breaking-in of new automobile
8. Repair of functions by inside forces	Wound healing with scar	Self-sealing of punctured tires
9. Replacement of functional parts by inside forces	Continuous regeneration of skin and mucous membranes	Continuous self-baking electrodes
10. Reproduction of a whole organism by the organism, or of a whole new machine by the machine	Sexual and asexual reproduction	—

functions of the finished products. All of them are shared by organisms and machines. Criteria Nos. 1, 2, and 3, namely functional, vestigial, and ornamental structures, were described previously (Part II, Chapter 1, C). Comments are hardly needed on criteria Nos. 4, 5, and 6, namely planned transformation and distribution of energy, necessity of blueprints (plans of organization), and self-regulating controls (feedback) built into structures of organisms or machines. Criterion No. 4 is part of the definition of any organ or any machine. The comparison between the eye and a photographic camera used as the example for criterion No. 6 will be discussed later in greater detail. Criterion No. 7 is adaptation of structures to function by use. Well-known examples in the organism are the increase in mass of muscles that are used for hard work over a long period of time, and the formation of callosities of the skin in places subject to mechanical pressure or friction.[16] The analogy in machines would be the breaking-in of an automobile or of a piano.

Criteria Nos. 8 to 10 need some discussion. Repair of a damaged function may or may not involve structural replacement. The two alternatives are listed as criteria Nos. 8 and 9 in the table. Criterion No. 8, restitution of functions by inside forces without any structural replacement, may be illustrated by two biological examples. In vertebrates, destruction of a motor center in the cerebral cortex or of a skeletal muscle cannot be repaired anatomically, but other nervous centers or muscles may take over the function. Among man-made machines, maintenance of function in spite of continued loss of structure is found in automatic grinders. After a certain number of actions the grinding wheel is sharpened automatically, and then the resulting loss in diameter is compensated by adjusting the position of the wheel relative to the object; both steps are performed automatically. Thus, function is restored although the material of the wheel is not replaced.

Criterion No. 9, replacement of functional parts by inside forces, may have two different meanings. Either the original part lost is replaced by an identical structure carrying out the function, or makeshift material replaces the original part with a satisfactory functional result. Such replacements by inside forces are tied to a colloidal state of material, both in organisms and in man-made machines. Healing of wounds (skin, muscles, inner organs) resulting in a scar illustrates such process in the organism. In machines, examples of colloidal repair of automatic parts are the self-sealing after puncturing of tires or of the lining of gasoline tanks. Replacement of loss by identical functional structures in organisms is known as regeneration. It involves restoration of material and shape, and, at the same time, restoration of function. It should be noticed that the terms "parts" and "inside forces" can be ambiguous. Both in organisms and in machines there are transitions between a part and a product. Staples may count as parts of the stapler as long as they are in the container, but they are considered products after delivery. The shape of a staple changes when it hits the anvil; it remains unchanged when ejected without meeting the anvil. There are many examples of this relation in organisms. Secreted milk is a product, but the fat droplets were parts of glandular cells before secretion. Spermatozoa or ova are parts of the testis or ovary until they are discharged as products. "Inside forces" is a concept requiring a definition of the system which is to

[16] Callosities develop in anticipation of future use on the wrists of warthog embryos (Cuénot, 1941, Fig. 21).

be discussed. Since all systems of interest here exchange matter and energy with their environment, there are minimum requirements in the environment on which each system depends for its functioning, repair, or reproduction. With respect to a man-made machine, *Q*, let us define as "outside forces" any activities of men, and any activities of machines other than *Q*. Suppose *Q* is a typewriter. If it could repair itself, replace its damaged parts, and produce new typewriters, these would be activities of inside forces. Evidently, all these activities would need a blueprint which directs the inside forces and is built into structures of the machine *Q*. Criterion No. 9 refers to self-replacement of parts, and criterion No. 10 to self-reproduction of a whole machine.

Can any existing machine utilize amorphous material for automatic regeneration of functional parts? Mr. A. S. Taylor of the Mechanical Research and Development Department of the Lederle Laboratories attracted my attention to a device which fulfills these requirements, namely, the continuous self-baking electrodes of Söderberg. I follow here the description of these electrodes in Mantell's "Electrochemical Engineering" (1960, p. 634). Large electric furnaces used in the melting of materials such as calcium carbide contain carbon electrodes of cylindrical shape, each up to 180 inches long and 20 inches in diameter. They are placed vertically, and their bottom tips, which form the arc, continuously melt away. Replacement is supplied from a funnel-like reservoir, containing a carbon paste, mounted on top of each electrode. By gravity the crude paste moves into the cylindrical holder of the electrode. The holder serves as a mold in which, at an appropriate temperature, the amorphous masses of carbon are pressed and baked into solid cylinders. At the same time the grip of the holder is loosened so that the electrode can slide into the proper position by its own weight. Thus, the continuous regeneration of material and shape of the electrode allows uninterrupted functioning.

In mammals the continuous wear and tear of the skin and mucous membranes is compensated by multiplication and migration of cells, as illustrated in Fig. 33. Examples of periodic replacement of surface structures (molting) will be given later. Healing of injured parts is also a form of regeneration.

Self-reproduction of *whole machines* by inside forces is, as yet, in a theoretical stage. Therefore, a minus sign has been entered in Table 6, criterion No. 10. This point will be discussed at the end of the chapter.

In biology many transitions are found between replacement of a part, known as regeneration, and the production of whole new organisms, known as reproduction. A tree can regenerate a twig, but an isolated twig of a poplar can also regenerate a tree. In the latter instance the twig is called a cutting or scion. If invertebrate animals such as hydras, planarias, or earthworms, are cut into pieces, some of the pieces produce whole individuals. This is regeneration as well as reproduction. In such plants and animals the blueprint for the production of a whole organism must be contained in the isolated cuttings or fragments. If the organism possesses special cells which serve the reproduction of the species, one speaks of sexual reproduction. Then the blueprint for a whole organism must be contained in the reproductive cells. In some species of plants and animals the potentiality of reproducing a whole organism is present both in the reproductive cells and in the other cells of the body, known as somatic cells.

In the following discussion, first the finished organisms and machines will be compared from different points of view, and then the processes of manufacturing organisms and machines will be reviewed briefly.[17]

1. *Comparison of Finished Products: Nature's Engineering* ***versus*** *Human Engineering*

J. Z. Young (1951, p. 24) rightly emphasized that the habit of comparing living bodies with machines "is at the basis of the whole modern development of biology and medicine," irrespective of unavoidable crudities and limitations of such procedure. Leonardo da Vinci (see Richter and Richter, 1939) and Galileo (1638) made remarkable analyses of mechanical properties of biological structures. D'Arcy Wentworth Thompson's book "On Growth and Form" (1942) contains fascinating comparisons between the mechanical engineering of nature and that of man. During the 19th and 20th centuries the knowledge of chemistry and electricity has developed at an astonishing rate, and in parallel, the chemical and electrical properties of living organisms have moved into the foreground of interest. During the last decade comparisons between organisms and man-made machines have become more and more operational. The reader is referred to Wiener's book "Cybernetics, or Control and Communication in the Animal and the Machine" (1948). Examples of feedback in organisms were mentioned previously in the present book. As Wiener pointed out (p. 157 of his book), organisms are able to compress elaborate communication systems into much smaller spaces than those required by man-made machines, but modern electrical communication techniques exceed in speed anything that can be achieved by organisms. The highest known frequency of muscular responses, synchronous with the stimuli, seems to be of the order of 1000 per second, as observed in the wing beat of insects (see Part II, Chapter 2, E). The nervous apparatus of the ear can respond with a frequency equal to that of acoustic waves up to 3000 per second, whereas the responses of the retina and optic nerve cannot possibly attain the frequency of light waves, which is 10^{14} per second.

The greater speed of man-made communication systems as compared to those

[17] In a much quoted statement, Immanuel Kant (1790, p. 338 of original edition) claimed that no Newton will ever arise to explain the production of a blade of grass. It seems to me that Kant failed to separate problems of a finished system from those of its production. Newton's laws of gravity revealed the basic mechanics of the existing solar system, but did not lead him to any satisfactory theory of its origin (see A. Wolf, 1953).

of animals is, to some extent, a result of the fact that in electric machines conduction is mainly electronic, whereas in organisms electric conductions are usually ionic. This was pointed out by Rosenblueth, Wiener, and Bigelow (1943). These authors also contrasted the general energy economy of machines with that of organisms. Most machines exhibit large differences in potential which allow rapid mobilization of energy; in organisms, the energy is distributed more evenly and is not very mobile. It seems to me that the absence of large differences in energy potential in organisms is related to their inability to tolerate wide ranges of temperature. Temperature ranges of thousands of degrees as used in machines have no counterpart in living organisms. Obviously both the differences in speed of communication systems and range of power energetics are connected with fundamental differences in raw material. Colloidal material such as proteins plays a dominant role in the living machine, whereas rigid materials such as metals prevail in the man-made machine. For early discussions of the organism as a colloidal machine, see J. Loeb (1906).

Rigid components, such as bone or chitin, occur in organisms; and colloidal components, such as rubber, in man-made machines. However, many animals are built without any rigid supports (jellyfish, slugs), and the vital organs of all animals consist of colloidal material. Among the man-made machines, squirting rubber balls and similar simple devices are the only ones which consist entirely of colloidal material. Hard-rubber structures, flexible plastic tubes, and metal wires, including springs, occupy an intermediary role between soft rubber and rigid steel in machines.

In vertebrates, the organs of posture and locomotion consist of bone, cartilage, ligaments, and skeletal muscles. Mechanically the bones are comparable to the rigid parts of man-made machines. In most cases, the shapes of articulating bones indicate the function of the particular joint. For instance, in hoofed animals the talo-crural joint consists of a groove with which a tight-fitting segment of a disc articulates. As one would expect, the only movement possible is rotation around the common horizontal axle of the disc and the groove. In other instances the shapes of the bones are less indicative of their mechanical functions.[18] In the transverse arch of the human foot there are three wedge-shaped bones (cuneiforms), but only two of them have their broad sides facing upward toward the convexity of the arch (Fig. 15A). The third bone, which is nearest the instep, has its broad side facing downward toward the concavity of the arch. This appears paradoxical by the standards of human engineering, as illustrated by the Roman arch (Fig. 15B). However, the human foot depends for its mechanical efficiency on a combination of rigid bones and colloidal parts with different mechanical properties, such as cartilage, ligaments, tendons, muscles,

[18] Some controversies in paleontology center around the functional interpretation of fossil bones and articulations. An entertaining discussion of such matters is found in Holland's 1910 paper, which deals with the mounting of the dinosaur *Diplodocus carnegiei*.

and the plantar pads of fat tissue. The tendons of three muscles of the calf of the leg insert into bones of the foot so that they form a slinglike support for both the transversal and longitudinal arches of the foot; this supporting effect varies, of course, with the state of contraction of the three muscles (see Jones, 1941; also Woodburne 1957, p. 600). Good descriptions of the locomotor apparatus are available in textbooks of anatomy and physiology, and in Thompson's "On Growth and Form" (1942). I also found Fenn's article "Mechanics of Muscular Contraction in Man" (1938) particularly useful.

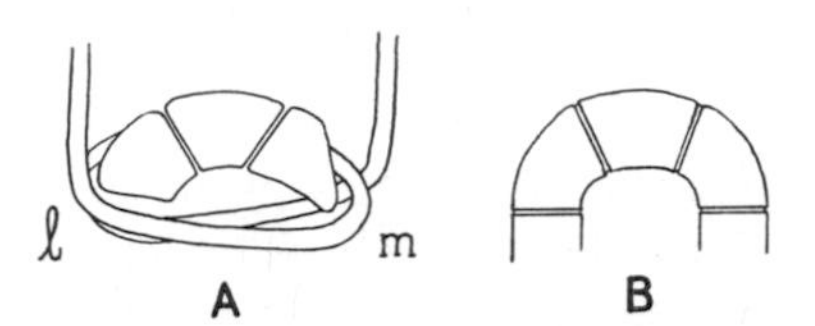

FIG. 15. Comparison of arch in the human foot with engineer's Roman arch. (A) Simplified diagram of transverse arch of foot showing the three cuneiform bones. The bone on the lateral side (l) and the bone in the central position point with their broader ends upward. The bone on the medial side (*m*) points in the opposite direction, which seems to decrease static efficiency. However, a large part of the mechanical load is carried by a sling of tendons (indicated in the picture) and especially by ligaments (not shown in the picture). (B) Roman arch in which all building stones point upward with their broad end, in keeping with static requirements.

From the point of view of human engineering it is difficult to understand the movements of those organs which have no rigid support at all. Most of the organs in this class consist of unstriated muscle, such as the stomach and the intestines. A remarkable example of a colloidal machine is the striated muscle of the heart. Up to the present, human imagination has concentrated on rigid machines. Therefore one can hardly predict whether or not an automatic pump consisting entirely of soft material will be constructed by human engineers.

There is one instance of a striking resemblance between a living apparatus of colloidal material and a man-made apparatus of rigid material: the eye and the photographic camera. The main points in common are the projection of an image on a screen by means of a lens and the control of the amount of entering light by a diaphragm. The focusing methods in animals, called accommodation, may or may not resemble those of the photographic camera. In fishes, amphibians, and snakes, the position of the lens is shifted so that the distance between lens and screen is variable, as in the photographic camera. In other animals, such as birds and mammals, focusing is achieved by varying the shape of the lens, thereby adjusting its focal length. This change is possible only because of the colloidal consistency of the lens. The amount of entering light in the eye is controlled by a negative feedback mechanism, the pupillary reflex (Fig. 2). In the photographic camera, the time of exposure can be controlled by a photoelectric cell which acts on the shutter. If the photoelectric cell is mounted inside the camera in the plane of the film, as in the Zeiss Ultraphot II, the device represents a negative feedback loop. If the photoelectric cell is located on the external surface of the camera, it also controls the time of exposure, but not as a feedback mechanism.

In all comparisons between organisms and man-made machines certain restrictions are necessary. When man's standards of mechanical, electrical, or chemical engineering are used to understand a certain mechanism of nature, let us remember that all aspects of the natural mechanism may not be known and that usefulness should not be the only yardstick. In an earlier chapter I pointed out that besides functional structures, vestigial and ornamental structures must be recognized both in organisms and in machines. A different approach is found in Cuénot's book "Invention et Finalité en Biologie" (1941), in which nature's inventions are presented as useful, useless, harmful, or exaggeratedly useful. In his book "On Growth and Form" (1942) Thompson has been carried away, on some occasions, by his enthusiasm for mathematics and human engineering. Calculations led him to the conclusion that birds larger than hummingbirds "must fly fast or not at all"; and that the hen's egg must travel through the oviduct with the blunt end foremost to comply with the requirements of streamlining. Nonetheless, large birds, such as kestrels and ospreys, do hover using light headwind; and the hen's egg travels with its pointed end foremost.

The efficiency of organisms as well as of man-made machines is to be appraised in terms of compromises between different requirements. Two machines with the same purpose will be constructed differently if one is to last much longer than the other. Taking the broadest view, one may say that in the organism the maintenance of the individual competes with the maintenance of the race. For the sake of the individual, a steady state is maintained, during which losses by wear and tear are replaced and injuries are repaired. For the sake of the race, the individuals have to go through a life cycle of youth, maturity, and old age, whereby reproduction is mainly a function of the mature phase. The way in which these processes are interwoven will be discussed more fully in Part V, Chapters 2 and 3.

In organisms, damaged rigid structures pass through phases of colloidal transformation for structural repair. A bone fracture heals in the following way. The external membrane of the bone (periosteum) and certain components of the marrow send out blood capillaries and migratory cells which invade the clotted blood and debris between the bone fragments. Gradually, a bridge of soft connective tissue joins the fragments. Finally, this bridge is solidified by calcification. In the meantime, the irregular ends of the broken bones and all scattered fragments are decalcified, and dead tissue is absorbed. The processes described are representative of the techniques of self-repair which are based on the colloidal properties of organisms.

In a man-made machine, the replacement of functional parts is, as a rule, carried out by some action from outside, namely, either by a human being or by another machine. However, automatic replacement of damaged parts by the same machine is possible. Examples of self-repair, both of colloidal and of rigid parts, were given earlier (Table 6, criterion No. 9). Techniques of structural repair or replacement are closely related to manufacturing processes, our next subject.

2. *The Manufacturing of Organisms by Nature*

It has been pointed out repeatedly in this book that the blueprint for the production of an embryo is built into the ovum. There is no equivalent to this in the manufacturing of man-made machines: the engineer who plans the machine, the draftsman who makes the blueprint, and the sheet of paper which carries the blueprint are not parts of the machine in any phase of its existence.

In manufacturing and assembling the parts of a machine, scaffolds and molds are used which are not material components of the machine at any time. There is no counterpart to such extraneous aids in the production of an organism, although transient structures frequently occur in embryonic and postembryonic development. Extensive postembryonic replacements of transient structures by permanent ones are conspicuous in the metamorphosis of amphibia and insects. Larval organs such as tail and gills of a tadpole, or the biting mouthparts of the caterpillar, disappear in the course of the metamorphosis. The gills of the tadpole do not help in forming the lung, and the tail disappears without any successor organ. The caterpillar's biting mouthparts are not the structural forerunners of the sucking tube of the butterfly. In these examples the absence of scaffolds or molds is obvious. A very different process is the transformation of cartilage into bone which occurs in vertebrates during embryonic and postembryonic development. This transformation happens in two ways: the cartilage cells change their shape and pattern of association, and calcium salts are deposited in the substance between the cells. In loose parlance, one may say that the cartilage serves as a mold or scaffold for the bone. This is not correct from the engineering point of view, since the material of the cartilage cannot be recovered from the finished bone structure, and, therefore, cannot be used again.

Many manufacturing processes go on in the adult organism. The replacement of losses by wear and tear, and the healing of wounds, including the healing of fractured bones, were mentioned previously. Molting is, under certain conditions, a periodic replacement of surface structures in animals which have reached their final body size (feathers of birds), but under other conditions it is connected with the increase in size of animals (shell of lobsters). Periodic *increase* of organized material in one organ occurs in the antlers of deer when a new prong forms every year. Remarkably rapid replacement has been observed in *Oikopleura,* a tunicate. This marine organism secretes a jelly house, equipped with funnels to allow rapid passage of water through the house when the animal is in motion. The funnels are traversed by screens which allow only such particles to pass as are small enough to serve as food. After a few hours the screens are clogged. Then the animal leaves the house through a kind of trapdoor and soon secretes another expendable jelly house (Claus, Grobben, and Kühn, 1932).

The main factors of embryonic morphogenesis were mentioned earlier, namely, the migration of cell groups and local differences in the increase in size, multiplication, and death of cells. These factors lead to changes in thickness of cellular layers, to outpocketing and to inpocketing, as illustrated by the development of the vertebrate eye (Fig. 3, g to l). The interplay of inductors and substrates was also discussed.

Since morphological and chemical differentiation are equally important aspects of embryology,[19] time relations between the two chains of phenomena have been the object of interesting investigations. These time relations are involved in the studies of Flexner (1950, 1955) and of Gregg and Løvtrup (1950), which were recounted previously in a different context (Part II, Chapter 2, E). Flexner showed that in the embryonic development of the guinea pig brain there is a period of a few days during which a number of chemical and morphological characteristics appear almost simultaneously. Gregg and Løvtrup (1950) isolated fragments of the gastrula of axolotls and observed pronounced regional differences in oxygen uptake between the areas of the animal and vegetal poles (Fig. 8a and b, and Fig. 9). Since a certain amount of morphological differentiation is present before the gastrula stage, it would be worth while to investigate *how early* the regional difference in oxygen uptake can be detected. Comparisons of morphological and chemical timetables in the embryo depend on the relative sensitivity of morphological and chemical methods available. In order to substantiate this point, let us discuss again the embryonic development of the lens.

In the vertebrate embryo an outpocketing of the brain produces the optic vesicle. As shown in Fig. 3k and l, the optic vesicle approaches a certain area of the head ectoderm which responds to the new neighbor by thickening, bulging, and finally forming a lens. The optic vesicle cooperates in this process by changing into a cup, thus preparing the necessary receptacle for the lens. Morphogenetic analysis reveals two facts. First of all, the ectoderm does not respond to the approaching optic vesicle by forming a lens, if contact between the two partners is prevented either by intervening connective tissue cells (Lewis, 1907) or by a cellophane strip (McKeehan, 1951). Secondly, within a critical period of 12 hours, the nuclei and cell bodies in the presumptive lens area of the ectoderm become elongated and assume a parallel position (palisading), but no pattern of any sort is visible earlier. At the same time the cytoplasm changes from a vacuolated sponge-like condition to a more solid one. These studies were made by McKeehan (1951) in chick and duck embryos. Chemical investigations of the development of the lens have as yet been limited to immunological procedures. Ten Cate and van Doorenmaalen (1950) injected chicken lens into rabbits.

[19] For morphological and chemical *growth* in the embryo see Part V, chapter 3, B.

The rabbit serum produced specific antibodies in response to the antigenic foreign protein. By adding the lens antiserum of rabbits to saline extracts of embryonic chick lens, specific precipitin reactions were observed. Chick embryos from 48 to 192 hours were examined. The earliest definite reaction was found when the lens extract came from 60-hour-old embryos. With the more mature embryos higher dilutions of lens extract could be used to produce the precipitin reaction. This work was expanded by Langman (1958) using the agar diffusion techniques of Oudin (1948). Extracts of embryonic chick eyes near hatching gave at least seven separate precipitin bands indicating at least seven different antigens in the eye. Going back to younger and younger ages, the number of identifiable antigens decreased until in the ten-somite embryo[20] no lens-specific protein could be found.

By combining these morphological and chemical data of chick embryos on a time scale, one arrives at the following relations. At approximately 30 hours of incubation the head ectoderm shows neither chemical nor morphological characteristics different from body ectoderm. A few hours later the head ectoderm shows at least one specific lens protein, if examined with rabbit antiserum and the agar diffusion technique. The first morphological characteristics appear also shortly after the 30th hour of incubation, namely, consolidation of cytoplasmic structures and starting orientation of nuclei. With an older immunological technique, the precipitin reaction in saline, specific lens proteins could not be discovered before 60 hours of incubation. My attempt to place the morphological and chemical data on a time scale needs several reservations. Since some investigators indicated the hours of incubation, whereas others characterized the phase of development by somite numbers, I converted somite numbers into hours of incubation. The degree of specificity of the immunological reactions used is not clearly defined.[21] With the agar diffusion technique Langman (1958) found that his antiserum reacted against presumptive lens ectoderm, iris, and retina, but not against body ectoderm. It is not known how the antiserum reacted with head ectoderm adjacent to the presumptive lens area (W. M. Layton, 1958, personal communication). On the morphological side, analysis has been limited to light microscopy. It seems perfectly possible that electron microscopy would detect differences between presumptive lens ectoderm and adjacent head ectoderm before 30 hours of incubation. However, electron microscopy of early embryonic material is hampered by particular difficulties of interpretation. Whether cytochemical procedures would or would not reveal early characteristics of the presumptive lens ectoderm is difficult to predict. As stated above, the

[20] Embryos beyond the stages illustrated in Fig. 3 show a gradual elongation of their bodies. Their dorsal mesoderm divides into increasing numbers of paired units, so-called somites, producing a pattern which resembles two parallel rows of beads (see Rugh, 1948, pp. 96 and 97). The number of somites is an index of morphological stages. If environmental conditions are kept constant, stages and ages are correlated.

[21] See Tyler's article "Ontogeny of Immunological Properties" (1955).

detection of the earliest signs of morphological or chemical differentiation depends on the sensitivity of the methods available.

Besides chemical and morphological analysis, a study of properties such as mechanical adhesion proved important. McKeehan (1951) found that the lens ectoderm cannot be separated from the eyecup by dissection until approximately the 50th hour of incubation (26-somite stage), but that such separation is readily possible in later stages.

Other techniques have been applied in order to push back the frontiers of demonstrable differentiation. Harrison, Astbury, and Rudall (1940) made an attempt to detect orientation in early primordia which showed no visible differentation as yet and were also known to be isotropic with reference to the axes of the embryo. Primordial material is considered isotropic, as long as its development is not inhibited by transplantation, irrespective of the orientation of the graft relative to the host. In an isotropic primordium neither anteroposterior nor dorsoventral polarity has been established in an irreversible way (see Part III, Chapter 6, A, 1). What Harrison *et al.* tried to find was *molecular orientation* as shown by Roentgen-ray diffraction techniques. Although their results were not conclusive, their approach seems remarkable.

It is reasonable to postulate that, in the very first stages of embryonic development, differentiation on the molecular level precedes differentiation on the cellular level, since the ovum with its poles and axes consists of many protein and nucleic acid molecules, but only of one cell. This postulate does not extend to later stages. Probably the morphological differentiation of the intestines is fairly advanced before its digestive enzymes can develop. Some enzymes are detectable only at late embryonic stages. There are remarkable instances of a *sudden* enzyme development in a particular organ. Morgan (1930) observed that xanthine oxidase is not present in the chick embryo liver before the 21st day of incubation. On that day the enzyme appears in the liver after the chick has put its head through the shell. As long as the egg is not chipped, the enzyme cannot be detected in the liver. In other organs, xanthine oxidase appears earlier, e.g., on the 15th day in the kidney.

In the manufacturing techniques of nature, the production of new cells occupies a key position. When cells divide, each daughter cell receives half of the genic material that was present in the mother cell. To avoid dilution of genic material each daughter cell before dividing has to regenerate the full amount of genic material. In chemical terms, this regeneration requires synthesis of deoxyribonucleic acid from other material available in the cell. Mazia and Prescott (1954) pointed out how the regeneration and distribution of genic material is achieved by the nuclear cycle, the alternation of mitotic and intermitotic phase (see my Fig. 61). The dispersed state of the chromosome in the intermitotic phase, with the presence of a nuclear membrane and of nucleoli, represents a condition favorable for metabolic processes which result in the synthesis of deoxyribonucleic acid. During the mitotic phase the chromosomes are in a condensed state in which they can be moved around passively by the mitotic mechanism, as necessary for an orderly spatial redistribution of genic material.

In my opinion the nuclear cycle is characteristic of mechanisms which leave no room for the question whether chemical or morphological processes take the lead. The borderline between the two phases of the mitotic cycle became even less distinct in view of some recent observations. It seems that the condensed chromosomes, when prevented from entering interphase, may be capable of synthesis of deoxyribonucleic acid, though this synthesis normally takes place during interphase (personal communication from Professor Mazia, 1958).

3. *Interaction of Function and Development in Organisms*

The fact that special chemical compounds are synthesized by the living machine in preparation for the production of a new identical machine is one of the profound differences between the manufacture of machines by man and the manufacture of organisms by nature. Another remarkable aspect in the developing organism is the maintenance of respiration and important metabolic processes by the changing living structures. The maintenance of general functions in all embryonic parts is necessary both for the sake of the parts and for the sake of the whole embryo (Streeter, 1938; p. 412). To use a striking (unpublished) expression of Professor Streeter: "Embryonic tissues are open for business during alterations." In the chick embryo, for instance, glycogen is handled by the yolk sac membrane until the embryonic liver has developed the morphologic structure suitable to take over this important part of the carbohydrate metabolism (Willier, 1955; p. 609, footnote).

The maintenance of a balanced internal environment (homeostasis) requires functioning communication systems not only in the finished organism, but also in all stages of its development. Function and development are inextricably interwoven in the nervous system and the endocrine system. It is fascinating to study the ways in which these systems develop morphologically, start to function in the embryo, and reach mature stages of functioning while serving the changing needs of the embryo. The reader will find valuable reviews of these subjects in Willier, "Ontogeny of Endocrine Correlation" (1955) and in Hamburger and Levi-Montalcini, "Some Aspects of Neuroembryology" (1950).

4. *Concluding Remarks on Organisms and Machines*

Organisms and machines have become suprisingly comparable, since communication systems of both have moved into the center of interest. A remarkable result of these developments is the fact that "the whole mechanist-vitalist controversy has been relegated to the limbo of badly posed questions" (Wiener, 1948, p. 56). It seems to me that, from any philosophical or theological point of view, it is no longer objectionable to ask: Has a particular problem in organisms been solved by nature in a way similar to the approach a human engineer would have chosen? Questions of this type have been asked and discussed repeatedly in the present book. Both in organisms and machines many solutions of engineering problems are compromises between competing functional requirements. In order to produce a special structure in an organism or machine, either a variety of manufacturing processes may be available or rigid restrictions to one process may be imposed. As mentioned in an earlier chapter, not all structures of organisms and machines have functional significance: some structures are ornamental and others are vestiges left over from manufacturing processes. With respect to organisms it is sometimes difficult to tell whether the variability of structures has functional significance. In the eyes of amphibians, the pupillary patterns show an amazing variability from species to species: rhomboid, pear-shaped, heart-shaped and other forms occur. Evidently all of them are compatible with the environment in which their owners live, but there is no evidence of special advantages of any particular pupillary shape. I quote from the "Comparative Anatomy of the Eye" by Prince (1956, p. 201) the following statement concerning the pupillary pattern of amphibians: "It is almost as though they have decided to experiment with everything that will achieve any degree of contraction." It seems to me that similar motives may induce human engineers to vary their designs of machines.

Differences between organisms and machines were discussed in previous chapters. The extensive use of colloidal protein material in organisms was contrasted to the predominance of rigid steel material in machines. Consequences of this difference in building material were pointed out, such as the wider range of power engineering and the greater speed of communication in machines. However, many differences between finished machines and finished organisms are quantitative rather than fundamental. The decisive difference between organisms and machines reside with manufacturing processes. Self-reproduction is, as yet, a prerogative of organisms. The fact that the ovum contains the blueprint for the whole organism is without parallel in man-made machines.

The presentation given here refers to the state of affairs as of 1961. I mention some future possibilities concerning the relation of organism and machine. Von Neumann (1951) has demonstrated the theoretical possibility of designing self-reproducing machines which select the necessary raw materials from their environment. However, the structural components of such machines would necessarily be rigid rather than colloidal. Conducting, insulating, and ferromagnetic material would be necessary to operate the machinery by electrical and mechanical methods. Soft components would be accidental, for instance, rubber and lubricating oil. As Moore (1956) pointed out, our knowledge of biochemistry is still so meager that we would not know how to design a system operated by hormones and enzymes.

Jacobson (1958) described the construction of a self-reproducing toy train. His model is closer to regeneration and asexual reproduction than to sexual reproduction of organisms. The toy train model seems to indicate that self-reproduction does not depend on colloidal material. According to Kemeny (1955), von Neumann's theoretical machine carries its blueprint for self-reproduction in a "tail" which is several hundred times larger in volume than the "box" which represents the functional part of the machine. In living organisms the blueprint carriers are condensed in very small spaces such as cells or even chromosomes. This ability of condensation may still be linked with the properties of colloidal material.

My discussion of self-repair and self-reproduction has been limited to comparisons between organisms and man-made machines. It is beyond the scope of this book to extend the comparison to other nonbiological systems. A few references may suffice. Analogous phenomena in crystals were the subject of Przibram's 1926 book on the inorganic areas bordering biology. Studies of dynamic morphology in crystals and in organisms share a peculiar dualism of approach (*"Ursache* vs. *Urbild"*), as pointed out by the crystallographer Niggli (1954). Hypotheses on possible self-duplication of protein and nucleic acid molecules have been proposed by Astbury (1945). A special possibility of self-reproduction in double-stranded helical molecules of deoxyribonucleic acid was contemplated by Watson and Crick (1953).

PART II: REFERENCES

Astbury, W. T. (1945). *In* "Essay on Growth and Form, presented to D'Arcy Wentworth Thompson" (W. E. Le Gros Clark and P. B. Medawar, eds.), p. 352. Oxford Univ. Press (Clarendon), London and New York.

Astwood, E. B., Sullivan, J., Bissell, A., and Tyslowitz, R. (1943). *Endocrinology* **32**, 210.

Balfour, F. M. (1880, 1881). "A Treatise on Comparative Embryology," Vols. 1 and 2. Macmillan, London.

Bargmann, W., Hanström, B., Scharrer, B., and Scharrer, E., eds. (1958). *Trans. 2nd Intern. Symposium Neurosecretion.* Springer, Berlin.

Blum, Harold F. (1944). *In* "Medical Physics" (O. Glasser, ed.), Vol. I, p. 1148. Yearbook Publ., Chicago, Illinois.

Blumenbach, J. F. (1776, 1795). *In* "De Generis Humani Varietate Naturale," Section II, 1st ed. 1776; 3rd ed. 1795; translation by Thomas Bendyshe (1865). Rosenbusch, Göttingen.

Bodian, D. (1951). Bull. Johns Hopkins Hosp. **89**, 354.

Boell, E. J., and Shen, S. C. (1950). *J. Exptl. Zool.* **113**, 583.

Cannon, W. B. (1929). "Bodily Changes in Pain, Hunger, Fear and Rage," 2nd ed. Appleton, New York and London.

Caspersson, T. O. (1950). "Cell Growth and Cell Function." Norton, New York.

Chadwick, L. E. (1953). *In* "Insect Physiology" (K. D. Roeder, ed.), p. 637. Wiley, New York.

Clark, A. J. (1933). "The Mode of Action of Drugs on Cells." Arnold, London.

Claus, C., Grobben, K., and Kühn, A. (1932). "Lehrbuch der Zoologie." Springer, Berlin and Vienna.
Conalty, M., and Gaffney, E. (1955), *Am. Rev. Tuberc.* **71**, 799.
Cooke, R. A. (1947). "Allergy in Theory and Practice." W. B. Saunders, Philadelphia, Pennsylvania.
Cooper, S., and Eccles, J. C. (1930). *J. Physiol.* (*London*) **69**, 377.
Coulombre, Alfred J. (1955). *Am. J. Anat.* **96**, 153.
Cuénot, Lucien (1941). "Invention et Finalité en Biologie." Flammarion, Paris.
Darwin, C. (1859). "The Origin of Species by Means of Natural Selection." Modern Library Edition, New York, 1936.
Davenport, C. B., and Davenport, G. C. (1910). *Am. Naturalist* **44**, 641.
Dobzhanski, T. (1937). "Genetics and the Origin of Species," 1st ed. Columbia Univ. Press, New York.
Eccles, J. C. (1930). *Proc. Roy. Soc.* **B106**, 326.
Edwards, E. A., and Duntley, S. Q. (1939). *Am. J. Anat.* **65**, 1.
Ehrlich, P. (1900). *Proc. Roy. Soc.* **B66**, 424.
Fenn, W. O. (1938). *J. Appl. Phys.* **9**, 165.
Fischer, A., and Mayer, E. (1931). *Naturwissenschaften* **19**, 849.
Fischer, E. (1894). *Ber. deut. chem. Ges.* **27**, 2985.
Flexner, L. B. (1950). *In* "Genetic Neurology" (P. Weiss, ed.) p. 194, Univ. of Chicago Press, Chicago, Illinois.
Flexner, L. B. (1955). *In* "Biochemistry of the Developing Nervous System" (H. Waelsch, ed.) p. 281, Academic Press, New York.
Frey-Wyssling, A. (1953). "Submicroscopic Morphology of Protoplasm and Its Derivatives," 2nd Engl. ed. Elsevier, New York.
Galileo, G. (1638). "Unterredungen und mathematische Demonstrationen über zwei neue Wissenszweige," Ger. transl. by A. von Oettingen (1917). Engelmann, Leipzig.
Gersh, I., and Catchpole, H. R. (1950). *Am. J. Anat.* **85**, 457.
Glass, H. B., and Li, C. C. (1953). *Am. J. Human Genet.* **5**, 1.
Goethe, J. W. (1795). *In* "Naturwissenschaftliche Schriften" (E. Beutler, ed.), Part II, p. 111. Artemis, Zürich, 1952.
Gowen, J. W. (1952). *Am. J. Human Genet.* **4**, 285.
Gregg, J. R., and Løvtrup, S. (1950). *Compt. rend. trav. lab. Carlsberg, Sér. chim.* **27**, 307.
Grüneberg, H. (1952). "Genetics of the Mouse," 2nd ed. Martinus Nijhoff, The Hague.
Ham, A. W. (1957). "Histology," 3rd ed. Lippincott, Philadelphia.
Hamburger, V., and Levi-Montalcini, R. (1950). *In* "Genetic Neurology" (P. Weiss, ed.), p. 128. Univ. of Chicago Press, Chicago, Illinois.
Hamilton, J. B. (1948). *In* "Biology of Melanomas" (R. W. Miner, ed.), p. 341. *N. Y. Acad. Sci. Spec. Publ. No.* **4**.
Hanson, J., and Huxley, H. E. (1955). *Symposia Soc. Exptl. Biol. No.* **9**, 228.
Harrison, R. G., Astbury, W. T., and Rudall, K. M. (1940). *J. Exptl. Zool.* **85**, 339.
Hartroft, Phyllis M., and Edelman, R. (1960). *In* "Edema" (Moyer and Fuchs, eds.), p. 63. W. B. Saunders, Philadelphia, Pennsylvania.
Hess, A. (1955). *A.M.A. Arch. Neurol. Psychiat.* **73**, 380.
Hesse, Richard (1924). "Tiergeographie auf ökologischer Grundlage." Fischer, Jena.
His, Wilhelm (1874). "Unsere Körperform und das Physiologische Problem ihrer Entstehung." Vogel, Leipzig.
Holland, W. J. (1910). *Am. Naturalist* **44**, 259.
Holtfreter, J., and Hamburger, V. (1955). *In* "Analysis of Development" (B. H. Willier, P. Weiss, and V. Hamburger, eds.), p. 230. Saunders, Philadelphia, Pennsylvania.

Howell, W. H. (1926). "A Textbook of Physiology," 9th ed. Saunders, Philadelphia, Pennsylvania.

Jacobson, Homer (1958). *Am. Scientist* **46**, 255.

Johannsen, W. (1909, 1926). "Elemente der exacten Erblichkeitslehre" 1st ed. 1909, 3rd. ed. 1926. Fischer, Jena.

Jones, R. L. (1941). *Am. J. Anat.* **68**, 1.

Kant, I. (1790). "Kritik der Urteilskraft" (K. Vorländer, ed.). Meiner, Leipzig, 1924.

Kemeny, John G. (1955). Man viewed as a machine. *Sci. Am.* **192** (4), 58.

Korschelt, E. (1936). "Korschelt and Heider's Vergleichende Entwicklungsgeschichte der Tiere," new ed., 2 vols. Fischer, Jena.

Langman, J. (1958). *Anat. Record* **130**, 329.

Lewis, W. H. (1907). *Am. J. Anat.* **6**, 473.

Litchfield, J. T., Jr., and Wilcoxon, F. (1949). *J. Pharmacol. Exptl. Therap.* **95**, 99.

Loeb, Jacques (1906). "The Dynamics of Living Matter." Columbia Univ. Press, New York.

McKeehan, M. S. (1951). *J. Exptl. Zool.* **117**, 31.

Mackenzie, C. G., and Mackenzie, J. B. (1943). *Endocrinology* **32**, 185.

Mainland, D. (1952). "Elementary Medical Statistics. Principles of Quantitative Medicine." W. B. Saunders, Philadelphia, Pennsylvania.

Mainland, D., Herrera, L., and Sutcliffe, M. (1956). "Tables for Use with Binomial Samples — Contingency Tests, Confidence Limits, and Sample Size Estimates." Dept. of Medical Statistics, New York University College of Medicine.

Mangold, O. (1929). *Wilhelm Roux' Arch. Entwicklungsmech. Organ.* **117**, 586.

Mantell, C. (1960). "Electrochemical Engineering," 4th ed. McGraw-Hill, New York.

Mazia, D., and Prescott, D. M. (1954). *Science* **120**, 120.

Middlebrook, G., Dubos, R. J., and Pierce, C. (1947). *J. Exptl. Med.* **86**, 175.

Millikan, R. A. (1911). *Phys. Rev.* **32**, 349.

Mitchell, P. H. (1932). "A Textbook of General Physiology for Colleges," 2nd ed. McGraw-Hill, New York.

Moore, E. F. (1956). *Sci. Am.* **195** (4), 118.

Morgan, E. J. (1930). *Biochem. J.* **24**, 410.

Morgan, T. H. (1919). "The Physical Basis of Heredity." Lippincott, Philadelphia, Pennsylvania.

Niggli, Paul (1954). *Experientia* **10**, 193.

Oudin, J. (1948). *Ann. inst. Pasteur* **75**, 30, 109.

Parker, G. H. (1908). *Am. Naturalist* **42**, 601.

Prince, J. H. (1956). "Comparative Anatomy of the Eye." C. C Thomas, Springfield, Illinois.

Przibram, H. (1926). "Die anorganischen Grenzgebiete der Biologie (insbesondere der Kristallvergleich)." Borntraeger, Berlin.

Richter, J. P., and Richter, I. A. (1939). "The Literary Works of Leonardo da Vinci," 2nd ed. Oxford Univ. Press, London and New York.

Rosenbluceth, A., Wiener, N., and Bigelow, J. (1943). *Phil. Sci.* **10**, 18.

Ruch, T. C. (1949). *In* "A Textbook of Physiology" (John F. Fulton, ed.), 16th ed., p. 292. Saunders, Philadelphia, Pennsylvania.

Rugh, R. (1948). "Experimental Embryology." Burgess, Minneapolis, Minnesota.

Scharrer, E., and Scharrer, B. (1945). *Physiol. Revs.* **25**, 171.

Schnitzer, R. J., and Grunberg, E. (1957). "Drug Resistance of Microorganisms." Academic Press, New York.

Shen, S. C. (1958). *In* "Symposium on Chemical Basis of Development" (W. D. McElroy and B. Glass, eds.), p. 416. Johns Hopkins Press, Baltimore, Maryland.

Sinsheimer, R. L. (1957). *Science* **125**, 1123.
Snedecor, G. W. (1946). "Statistical Methods Applied to Experiments in Agriculture and Biology," 4th ed. Iowa State College Press, Ames.
Spemann, H. (1921). *Wilhelm Roux' Arch. Entwicklungsmech. Organ.* **48**, 533.
Spemann, H. (1938). "Embryonic Development and Induction." Yale Univ. Press, New Haven, Connecticut.
Spengler, O. (1920–1922). "The Decline of the West: Form and Actuality," English translation by C. F. Atkinson. Knopf, New York, 1926.
Streeter, G. L. (1938). Cooperation in Research *Carnegie Inst. Wash. Publ. No.* **501**, 397.
Sturkie, P. D. (1954). "Avian Physiology." Cornell Univ. Press (Comstock), Ithaca, New York.
Swanson, C. P. (1957). "Cytology and Cytogenetics," Prentice-Hall, Englewood Cliffs, New Jersey.
Tannenbaum, A. (1947). *Ann. N. Y. Acad. Sci.* **49**, 5.
Ten Cate, G., and van Doorenmaalen, W. (1950). *Koninkl. Ned. Akad. Wetenschap. Proc.* **53**, 894.
Thomas, L. (1955). *In* "Textbook of Medicine" (R. Cecil and R. Loeb, eds.), 9th ed., p. 482. Saunders, Philadelphia, Pennsylvania.
Thompson, D'Arcy W. (1942). "On Growth and Form," 2nd ed. Cambridge Univ. Press, London and New York.
Tobian, L. (1960). *Physiol.* Rev. **40**, 280.
Tyler, A. (1955). *In* "Analysis of Development" (B. H. Willier, P. Weiss, and V. Hamburger, eds.), p. 556. Saunders, Philadelphia, Pennsylvania.
van Wijngaarden, C. de L. (1924). *Arch. exptl. Pathol. Pharmakol. Naunyn-Schmiedeberg's* **113**, 40.
Virchow, R. (1858). "Die Cellularpathologie in ihrer Begründung auf physiologische und pathologische Gewebelehre". Hirschwald, Berlin.
von Bertalanffy, L. (1941). *Biol. Generalis* **15**, 1.
Warburg, O. (1924). *Naturwissenschaften* **12**, 1131.
Watson, J. D., and Crick, F. H. C. (1953). *Nature* **171**, 964.
Weinmann, J. P., and Sicher, H. (1955). "Bone and Bones. Fundamentals of Bone Biology," 2nd ed. Mosby, St. Louis, Missouri.
Wells, J. A. (1958). *In* "Pharmacology in Medicine" (V. A. Drill, ed.), 2nd ed., p. 1. McGraw-Hill (Blakiston), New York.
Wiener, N. (1948). "Cybernetics." Wiley, New York.
Williams, R. J. (1956). "Biochemical Individuality." Wiley, New York.
Willier, B. H. (1955). *In* "Analysis of Development" (B. H. Willier, P. Weiss, and V. Hamburger, eds.), p. 574. Saunders, Philadelphia, Pennsylvania.
Wilson, G. S., and Miles, A. A. (1955). "Topley and Wilson's Principles of Bacteriology and Immunity," 4th ed. Williams & Wilkins, Baltimore, Maryland.
Winton, F. R., and Bayliss, L. E. (1955). "Human Physiology," 4th ed. Little, Brown, Boston, Massachusetts.
Wolf, A. (1953). Article "Cosmogeny," Encyclopedia Britannica, Vol. 6, p. 488.
Wolff, J. (1892). "Das Gesetz der Transformation der Knochen." Hirschwald, Berlin.
Woltereck, R. (1932). "Grundzüge einer allgemeinen Biologie." Enke, Stuttgart.
Woodburne, R. T. (1957). "Essentials of Human Anatomy." Oxford Univ. Press, London and New York.
Young, J. Z. (1951). "Doubt and Certainty in Science." Oxford Univ. Press, London and New York.

Part III

PROCEDURES, INTERPRETATIONS, AND THE PROBLEMS OF PRESENTATION IN DYNAMIC MORPHOLOGY

Since procedures and interpretations are interdependent, the subdivisions of the present part are arbitrary to some extent. Similar morphological techniques may be used for different purposes. The isolated, perfused heart and lung of a dog may serve for the study of cardiac output, or may be used to investigate the arrest of circulating leucocytes in the lung.

A survey of morphological techniques can be organized on the basis of different criteria. The following four ways of classification suggest themselves: (1) different orders of magnitude of animals, organs, and microscopic samples; (2) species differences; (3) procedures for living material and procedures for dead material; and (4) natural and artificial conditions of the material to be studied.

I have decided to devote special discussions to (3) and (4), and to mention (1) and (2) here and in other parts of the book whenever necessary. A few comments will clarify the meaning of the classifications (3) and (4).

In dividing morphological techniques into those for living material and those for dead material, one should be aware of the fact that many morphological procedures can be applied to studies of living as well as of dead material. Roentgen rays show differences between structures no matter whether they are alive or dead. Microscopes can be used with biological structures during life or after death. My classification of a procedure will depend on its *predominant* use in living or in dead material. Therefore, Roentgen-ray procedures will be discussed in connection with living material, and microscopic procedures mainly in connection with dead material.

Living material may be observed with a minimum of interference, or under highly artificial experimental conditions. An organ may be studied in its natural position in the animal or removed from the animal and kept surviving by perfusion with saline. Dead material may be studied in a natural or in a preserved condition. These procedural alternatives form the basis of interpretation and presentation of morphological data.

Chapter 1

Combined Study of Live and Dead Material

A. Examples of Successful Combinations of Studies of Live and Dead Material

There is an understandable tendency to consider biological studies of living material more valuable than studies of dead biological objects. It has been stated that "the study of a physiological process cannot be carried out with the dead organ" (Ellinger, 1940, p. 332). This unassailable statement needs a supplement: physiology cannot be studied without dead organs. To a great extent the success of biological investigations depends on the skillful combination of studies of living material with studies of dead material. In the history of biology this combination has proved extremely useful no matter whether morphological, biochemical, or other procedures were applied. Claude Bernard was a master at combining observations in the living and in the dead animal; see, for example, his "Introduction to the Study of Experimental Medicine" (1865).

Some of the examples which were discussed in the preceding parts of this book represented combinations of studies in living and dead material. In the sunburn story (Table 1), observations of the skin in the living organism were supplemented by investigations of dead samples which had been processed for microscopic analysis. The role of motor units in the contraction of skeletal muscles was demonstrated by stimulation of the living nerve and muscle and by a count of nerve fibers in the dead preparation (Part II, Chapter 2, E). Some additional examples may illustrate how studies of live and dead material are interwoven.

The discovery of the *blood circulation* was a result of observations in living and dead organisms. Long before Harvey's time (1578–1657) anatomical knowledge was available of the valves of the heart, of the existence of two types of vessels, namely, arteries and veins, and of the presence of valves in the veins. Without this knowledge it would have been impossible for Harvey to ask the proper questions, to conceive the return of blood to the heart, and to prove it by his compression experiments on the cutaneous veins of the human arm.[1] Finally,

[1] For a reproduction of Harvey's pictures see Winton and Bayliss (1955, Fig. II. 1.).

Malpighi's (1661) discovery of capillaries in the living frog's lung established the connection between arteries and veins, and thus the existence of a complete circuit.

A modification of the classic scheme became necessary when direct connections between arteries and veins were observed two centuries after Malpighi. Arteriovenous anastomoses which circumvent the capillaries were discovered by injection of dyes. Sucquet (1862) injected alcoholic solutions of resins mixed with soot into arteries of human corpses. Although this material was not able to pass through the capillaries, it appeared in the veins. Sucquet claimed that he saw the places of transition, consisting of vessels approximately 1 mm. in diameter; according to his estimate, the average diameter of capillaries was less then 0.01 mm. For a detailed survey of the historical developments, see Clara's 1956 book from which I abstracted the work of Sucquet. Injections with dyes and observations in the ear chamber of the rabbit (Part III, Chapter 2, A, 2) confirmed the interpretation of the structures seen, but did not prove their functional significance. What was needed was evidence that particles much larger than the diameter of capillaries could travel from the arterial to the venous system in the living organism. As an example of such a demonstration, I mention the work of Prinzmetal *et al.* (1948). Glass spheres ranging in diameter from 10 to 440 μ were suspended in saline and injected into the right ventricle or the pulmonary artery of anesthetized rabbits, and dogs (see my diagram of circulation, Fig. 84). The animals were killed within 30 seconds after injection. Not all glass spheres were arrested in the pulmonary capillaries, which are 10–20 mm. in diameter. Glass spheres as large as 320 μ in diameter were recovered from the pulmonary vein and liver, evidently having passed through arteriovenous shunts of the pulmonary circulation.

In the field of endocrinology the discovery of *insulin* offers a perfect illustration of successful alternation of studies during life with studies after death. The first step was the observation of severe diabetes in dogs from which the pancreas had been removed (von Mehring and Minkowski, 1889); this disturbance of carbohydrate metabolism in the depancreatized dog was cured by implanting a piece of living pancreas into any part of the animal, for example, under the skin. The next step was the recognition that the maintenance of a proper carbohydrate metabolism was related in some way to a special component of the pancreas, the islands of Langerhans. Of course, these islands were seen in dead preparations, scattered between the acini which form the bulk of the pancreas and which secrete digestive enzymes into the intestine (see diagrams Fig. 57 A–G). Relations between pancreatic islands and diabetes were suggested by microscopic findings in human patients and in experimental animals. Patients who died of diabetes frequently showed destruction of the islands of Langerhans in an otherwise intact pancreas. In animals, the experimental obstruction of the pancreatic duct led to a conspicuous shrinkage of the pancreas, but not to diabetes. It turned out that ligature of the duct caused the destruction of the acini, but not of the islands of Langerhans (Ssobolew, 1900; Schulze, 1900). The third and decisive step was taken by Banting and Best (1922). They ligated the pancreatic duct in dogs, waited until the acini had disappeared, and extracted the remainder of the pancreas, which consisted of islands and connective tissue. The extract contained insulin.

Since all branches of physiology depend on combined studies of live and dead material, more examples will be found in later parts of the present book. One basic procedure proved to be remarkably fruitful, namely, *the determination of functions by elimination of morphological structures.* Particularly in the study of the nervous system, answers to functional questions were obtained when a certain part of the system was eliminated anatomically. Such eliminations resulted either from an accidental injury, from a planned experimental destruction, or from a congenital malformation. A few examples follow. The brains of human patients who had suffered the loss of well-defined motor functions were examined after death and showed destructions of special areas by hemorrhages. Areas of the cerebral cortex containing different motor centers are shown in Fig. 49. Destruction of the middle part of area 4 was associated with paralysis of the right hand, arm, and shoulder whereas destruction of the lower part of area 6 resulted in a form of language disturbance. Subsequently, these localizations were confirmed by electric stimulation in living persons. Observations in human beings were supplemented successfully by experimental ablation of cortical areas in apes (Fulton's textbook, 1949, pp. 267-274). A remarkable analysis of localization of hearing receptors in the inner ear resulted from combined observations of living persons with microscopic study after death. Boilermakers who have been exposed to loud clanging sounds for many years sometimes become deaf to tones of the same pitch as that of the noise although retaining their hearing over other parts of the scale. In a number of such individuals postmortem analyses of the cochlea were carried out. Microscopic study showed destruction of Corti's organ in the basal turn of the cochlea, or the nerves supplying this part (Best and Taylor, 1955, p. 1207).

Occasionally, even the study of dead structures on the subcellular level has been used in conjunction with studies in living animals. One good example is the successful coordination of cytological and statistical genetics: chromosomes and the loci of genes in chromosomes were studied in fixed and stained cells while the role of these structures in genetics was determined by breeding experiments in plants and animals (for details see discussion of Table 2, item 5). Another example refers to the determination of the synaptic cleft. The synapse is the site of transmission of stimuli from one neuron to another. When the impulse in the first neuron (the response to the original stimulus) reaches the synapse, a stimulus is generated for the second neuron which, in turn, responds with an impulse. The mechanism of transmission at the synapse consists of the release of a transmitter substance, such as acetylcholine, and an increased permeability to ions in the membrane of the receptor. Eccles (1957, p. 217) pointed out that the transmitter substance would be efficient if it crossed a synaptic cleft of 200 Å. or less in about 1 μsec., and similarly the requirement for the ionic flow would be satisfied by a cleft of 200 Å. The values for the rate of ionic migration were obtained by inserting microelectrodes into motor neurons of the spinal cord of a living cat (Fig. 2 of Eccles). The width of the cleft postulated experimentally was confirmed by electron micrographs. The surface of an end bulb *(bouton terminal)* and the surface of the next nerve cell[2] proved to be separated by two more or less parallel lines which enclosed an area of increased permeability for electrons. The distance between the two lines measured approximately 200 Å. (Fig. 1E of Eccles, electron micrograph supplied by Palade and Palay, 1956; for complete electron microscopic documentation see Palay, 1958).

In the training of students of medicine an important place is occupied by the coordinated use of live and dead material. This training is continued in hospitals with high standards by clinical-pathological conferences which serve for the joint evaluation of postmortem findings and data obtained during the life of the patient.

B. Planning Studies Combining Morphological and Other Procedures

Our examples showed how problems could be solved by observations during life followed by appropriate postmortem studies. If different procedures are needed to study a certain problem, some procedures may interfere with others. Special planning is necessary to satisfy the competing requirements. Compromises must be made between morphological techniques on the one hand, and bacteriological or pharmacological procedures on the other hand.

Routine autopsies of human beings are performed according to procedures derived from the standards which were established by Virchow and von Rokitanski in the 19th century. The sequence of dissection is usually such that the abdominal, thoracic, and cranial cavities are explored and the organs dissected either in their natural situation or after removal from the body. Extremities are dissected at the end of the autopsy. For bacteriological studies after death the following considerations are important. The opening of blood vessels may dislodge the clotted or liquid blood locally as well as over some distance. As the autopsy progresses, the chances increase that blood, pus, and the contents of the stomach and intestines will not only contaminate the surfaces of organs, but also

[2] A brief description of the structures involved in synapses will be given in our discussion of interfaces on the cellular and subcellular level (Part IV, Chapter 1, C).

their interior. This cannot be remedied by sterilization of surfaces before sampling, although a hot spatula or the flame of a Bunsen burner can be applied freely in postmortem bacteriology. Therefore the following technique was developed by von Gutfeld and Mayer (1932). Before making any other incisions into the corpse, the axillary and the femoral veins were exposed in a sterile way, whereupon a long needle mounted on a syringe was inserted into the vessel. Usually at least one cubic centimeter of liquid blood could be aspired. A sterile sample of femoral marrow was obtained by exposing the middle of the femur, cutting it transversally, and dislodging the end nearer the body. The cut surface and at least one centimeter of the marrow were sterilized by carbonization with a Bunsen burner. A sterile tube was pushed through the carbonized segment and a sample from the upper part of the marrow cavity was aspired. After this, the heart was exposed, and a sterile sample of blood obtained from the right ventricle. Depending on the case, sterile samples were obtained from the spleen, the gallbladder, and other organs while the organs were in their natural site and none of their blood vessels had been opened. After the bacteriological sampling, the dissection proceeded in the usual order.

In *experimental studies of infectious agents in small laboratory animals,* either whole organs are ground in a sterile way, or samples are obtained by thrusting a metal loop into the organ and pushing it in different directions before withdrawing it. Since these procedures prohibit morphological study in organ sections, separate groups of animals have to be used for bacteriological and morphological examination.[3] In the postmortem studies of a *human patient* the same organs must be used for bacteriological and morphological analysis. Therefore, the searing of surfaces and the insertion of needles or metal loops has to be limited so that enough undisturbed material remains for morphological purposes. By this restriction the bacteriological sampling loses part of its efficiency. Sometimes one has to sacrifice a morphological point, and sometimes a bacteriological one. Bacteriological examinations of human corpses are made for two reasons: (1) if an infection has been diagnosed during the life of a patient, it may be important to determine the distribution of the infectious agent in different parts of the body; (2) if a patient dies before a definite diagnosis was possible, the autopsy is expected to reveal the nature of the fatal disease.

In pharmacological and toxicological experiments with animals, postmortem studies are very useful provided they are planned as part of the whole project. The animals most extensively used in such studies are dogs and rats. The difference in size between dogs and rats imposes differences in postmortem procedures.

Autopsies of experimental animals may be handled in different ways. This

[3] An instructive example of the use of separate groups of mice for virus study and for histologic study is found in a paper by Melnick and Godman (1951, Table I).

will be discussed in the next chapter. Organs or systems which showed disturbances during the life of the animal are of particular interest. If a drug of the sulfonamide type had caused bloody urine or cessation of urination, special attention would be given at autopsy to the possible deposits of the drug in the urinary system. In the bladder, ureter, and renal pelvis precipitates can be seen with the naked eye, but the substance of the kidney has to be searched for crystals under a dissecting microscope. For this purpose a kidney is split lengthwise and placed under the dissecting microscope. By scratching the fresh cut surface with a needle or pointed knife, deposits of crystals can be discovered. The isolated crystals can be examined microscopically in ordinary and in polarized light. If necessary, crystallographic analysis can be applied to identify the drug. For this purpose, the study of the fresh kidney is by far the best method. Besides the determination of drug precipitates one might wish to ascertain the morphological condition of the renal structures by histological methods. A dog's kidney is so large that some areas can be examined in the fresh condition, as described, while others remain intact for fixation and histological study. A rat's kidney is too small to be divided for investigation of both fresh and preserved material. Moreover, transversal sections through the whole kidney of a rat are most favorable for histological study after fixation, whereas longitudinal splitting is preferable for inspection of the renal pelvis in the fresh kidney. As a consequence, one may use one kidney of a rat for determination of drug precipitates, and the other for histology, or one may use both kidneys of some rats for one procedure and both kidneys of other rats for the other procedure. These alternatives are feasible in drug experiments, since, as a rule, the rats are divided into groups in which each animal receives the same dose.

Chapter 2

Morphological Techniques

A. Morphological Techniques for the Study of Live Material

1. *Surface Structures and the Eye*

Besides the simple inspection of superficial structures, measurements of their visible properties are possible. Human skins with different degrees of pigmentation have been compared by colorimetric scales (Davenport and Davenport, 1910). As discussed in connection with the sunburn story (Part II, Chapter 2, A) the so-called white skin is subject to great variations in color, partly caused by changes in pigment and partly by changes in vascularization. These variations have been recorded photoelectrically by means of the Hardy spectrophotometer (Edwards and Duntley, 1939). Microscopic studies of the living skin have been made mainly for the purpose of observing blood capillaries. A very suitable object is the web of a frog's foot. To some extent, the webs between the fingers of man can be used. In this type of study, adequate illumination is the main problem. The wall of the human finger nail can be made transparent by oil so that the capillaries can be seen with the help of a microscope. Pictures of the apparatus and of the capillaries are included in Houssay's "Human Physiology" (1955, p. 210). The capillary patterns vary among different healthy individuals and among fingers of one individual, but the pattern remains constant in a given finger as verified in observations covering a year (Walls and Buchanan, 1956, with numerous illustrations).

The mucous membrane of the mouth, the eyelids, and other openings of the body lend themselves to observation during life. The upper eyelid of man can be readily inverted for examination of the so-called fornix. This area is particularly important in the early diagnosis of trachoma. In the frog, the tongue can be pulled out, exposed outside the mouth, and flattened gently on a board so that the capillaries can be studied microscopically. A picture illustrating this classic experiment of Cohnheim (1872) is reproduced in Florey's "General Pathology" (1958). Funnels and short tubes (specula) may be inserted into openings such as the external ear, the nose, or the vagina, in order to observe deeper locations. With small mirrors, mounted on handles, the oral cavity, the pharynx, and part of the larynx can be examined.

The eye offers special opportunities for the study of superficial and interior parts. The presence of a pupil and the transparency of cornea, lens, and vitreous body make the use of relatively simple mirrors possible. The ophthalmoscope not only permits the observation of the retina in the background of the eye, but also permits microsurgical manipulations, such as the insertion of a tiny pipette into the retinal artery of an experimental animal for direct determination of blood pressure in this vessel (Duke-Elder, 1938, Vol. 1, Fig. 479). Observation of choroid vessels require scleral windows (see later).

2. *Structures below the Surface*

Structures hidden in the depths of the body can be made accessible to direct observation by various techniques: Roentgen rays, insertion of illuminated tubes into natural openings, surgical exposure, and transparent windows. In addition, single organs or groups of several organs can be isolated from the remainder of the animal and observed in a surviving condition. Isolation preparations will be described later.

a. *Roentgen-ray techniques.* By means of Roentgen rays inner structures of the body are observed either on a fluorescent screen or photographed on a film. Adjacent structures can be distinguished from each other provided they differ in their capacity to absorb Roentgen rays. This difference in Roentgen-ray density or opacity results in black-white pictures showing outlines which may or may not coincide with optical outlines. In addition, areas of various distinctness are seen in black, white, or different shades of gray. These areas may have optical equivalents, or may represent chemical differences. For instance, the difference between the calcified cortex of the femur and the marrow cavity shows equally in a Roentgen-ray picture and in a bone that has been split lengthwise for inspection. On the other hand, different concentrations of calcium in the cortex of the femur can be detected by Roentgen ray, but not by inspection of the split bone.

Many soft structures of the organism have similar radiopacity. One cannot expect Roentgen rays to distinguish between muscles, fat, nerves, or blood vessels if these structures are mixed or adjacent to each other. However, air is much more permeable to Roentgen rays, or, briefly, more radiolucent, than the tissues of the body. Therefore the lung, which is normally filled with air, can be clearly distinguished from the heart, the aorta, and other structures. Figure 16 (from Bartone and Grieco, 1955) shows a diagram of human lungs studded with small radiopaque foci. This picture may indicate very different pathological conditions, such as miliary[4] tuberculosis, histoplasmosis, or a par-

[4] Miliary means similar to millet seeds (Latin, *milium*).

ticular deposition of inhaled dust. The descriptive name given to such pictures is miliary radiopacities. Which of the pathological conditions mentioned produced this picture in an individual can be concluded only from other criteria such as the presence or absence of fever, history of body weight changes, and the frequency of the respective diseases in the area in which the individual lives or lived. For instance, histoplasmosis is prevalent in the Mississippi River

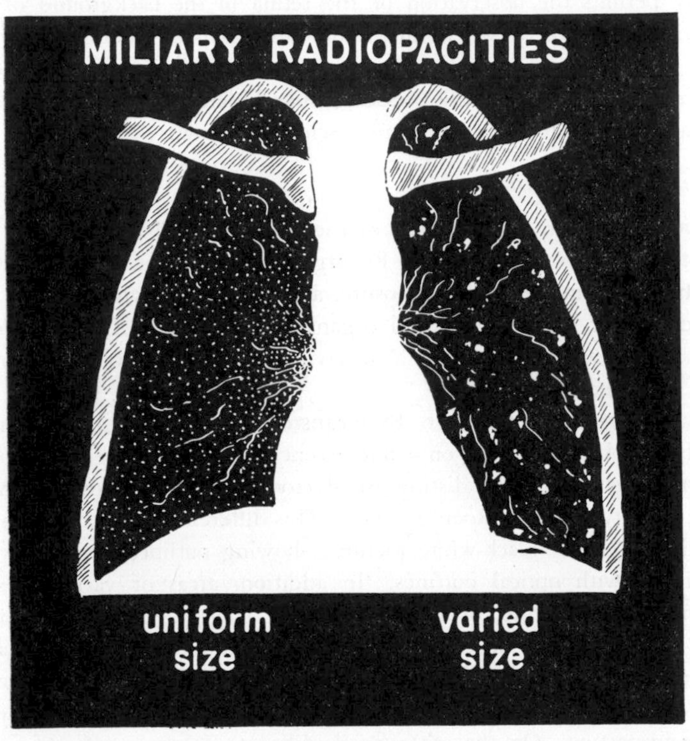

FIG. 16. Diagram of Roentgen ray picture of human lung with abnormal radiopacities. Radiopaque structures, white; radiolucent structures (e.g., air), black. Note small, so-called miliary radiopacities of uniform size in the right lung, of varied size in the left lung. For possible interpretations see text. (After Bartone and Grieco, 1955.)

basin. In routine radiological examinations Chaves and Abeles (1952) discovered disseminated small nodules in the lungs of 75 persons. Four years of continued observations revealed in 56 of the cases complete or incomplete disappearance of the radiopacities within several months or years. Six of the 75 individuals developed interferences in the pulmonary function, but the remainder did not show any respiratory disturbances. These examples illustrate the importance of classifying morphological data in a descriptive manner prior to any attempt to study causes and mechanisms.

Hollow structures can be accentuated artificially by injection or insertion of either radiopaque or radiolucent material. When air or other innocuous gases are injected into the ventricles of the brain or into the joints, these cavities will stand out by radiolucency. On the other hand, filling with radiopaque material will enhance the visibility of blood vessels, the intestines, or the ureters. In order to visualize the blood vessels of the human brain, Diodrast, an iodine-containing compound, is injected into the carotid artery. The first film which is exposed during the injection will show the arteries, and a second film, exposed 3 to 5 seconds later, will show the veins (Gross, 1944, p. 1262). A meal that contains radiopaque barium, administered to a human being or animal, travels through the whole digestive tube. Depending on the time which has elapsed since the intake of the meal and on certain technical modifications, the radiopaque barium is shown by Roentgen rays in the esophagus, the stomach, the duodenum, or the other parts of the intestinal tube, including the appendix. Peristaltic movements can be observed and abnormalities of the inner surface of the digestive tube may be discovered. A list of contrast media and their administration can be found in an article by Newell (1944).

Radiopaque flexible tubes have been equipped with a thickening at one end so that they could be readily swallowed. The thickened end passed through the pylorus and finally through the whole intestine. With Roentgen rays the position and movements of the tube in all parts of the digestive tract could be observed (Van der Reis and Schembra, 1924; Miller and Abbott, 1934; Underhill, 1955). This procedure will be described in comparing the length of the human intestine during life and after death (Part III, Chapter 3, C, 1). Radiopaque tubes can also be inserted into the heart and blood vessels. The position of a tube in the heart (intracardiac catheter) is shown by Best and Taylor (1955, p. 434).

Finally I mention a combination of Roentgen-ray techniques and experiment surgery which opened a new avenue to the physiology of the spleen. Barcroft *et al.* (1925) exposed the spleens of dogs and cats surgically and inserted a number of metal clips into the capsule, whereupon the abdominal cavity was closed again. This made it possible to take Roentgen-ray pictures of the outline of the spleen. Such pictures were taken from two different directions. The volume of each spleen was computed from the two projected areas. By this procedure it was demonstrated that, in dogs and cats, the spleen can change its volume to an astonishing extent.

b. *Inspection through natural openings with the use of illuminated tubes.* For the purpose of exploring the urinary bladder, the bronchi, the esophagus, the stomach, or the rectum, special tubes have been constructed which carry near their tip an electric bulb and a mirror (cystoscope, bronchoscope, etc.). Through an opening on one side of the tube the bulb illuminates a small area of the inner organ, and the mirror projects the picture of this area into the eye of the observer. These techniques are known as *endoscopy.* For illustrations and descriptions, see Jackson and Jackson (1944).

c. *Surgical exposure of inner organs.* Surgical procedures play a prominent role in experimental physiology. Many of these procedures, with their respective purposes, are described in the book by Markowitz *et al.* (1954). After surgical exposure, tubes have been inserted into large blood vessels, or the heart, in order to determine the blood pressure manometrically. Artificial openings, so-called fistulas, have been constructed which lead from various parts of the digestive tract through the body wall to the surface of the animal so that the secretions of the different parts could be collected. A classic example of experimental surgery is Pavlov's gastric pouch. A part of the dog's stomach is transformed into a pouch which opens to the outside. The pouch is separated from the cavity of the main stomach, but shares with it the blood vessels and nerves. Under these conditions food offered to the dog stimulates not only secretion in the main stomach but also in the pouch. This made it possible to determine various factors that control gastric secretion.

My discussion will be limited to those procedures of experimental surgery which make inner organs accessible to direct visual observation. Some of the procedures require not more than an incision in the skin, whereas others involve opening the abdominal cavity or similar major surgery. The optical devices which are needed in conjunction with the surgical procedures vary from simple illumination by daylight or a good lamp to ingenious devices such as insertion of a Lucite rod into the trachea of a mouse for study of the thyroid in transmitted light (Williams, 1944). In the mouse a small incision into the ventral skin of the neck exposes a bridge between the right and left thyroid lobes, called the isthmus. Since in the mouse this part may not be thicker than one layer of thyroid follicles, morphological detail was observed with high magnifications and microsurgical experiments were performed on single follicles, and even on cells (see my Fig. 22, after Williams). Different types of transluminated quartz rods and their applications are described by Hoerr (1944) and Knisely (1950).

Fluorescence microscopy has been applied to inner organs after surgical exposure. Innocuous fluorescent dyes were injected intravenously into frogs by Ellinger and Hirt (1929). Subsequently, studies of the exposed liver, thyroid, and kidney were also made in mammals (Hartoch, 1933; Grafflin, 1947). Of course, only structures near the surface of organs can be oserved with fluorescence microscopy in ultraviolet light.

The examples mentioned referred to observations of short duration, usually a few hours. Organs remained either in their natural position or were moved to the surface by a gentle pull on their vascular stalks or on other connections nearby, without ligatures of blood vessels or other surgical alterations. Florey

(1958, p. 51; Fig. 5) shows an interesting picture of the apparatus used by Thoma in 1878 for short-term microscopic studies of mesenteric capillaries in small warm-blooded animals.

Let us consider procedures which allow visual observation of inner organs over long periods, ranging from a few days to several years. These procedures fall into two different classes and may be labeled *surgical rearrangement methods* or *transparent window methods.* Although minor surgical rearrangements of skin and muscles are necessary for the installation of transparent windows, this technique will be described separately.

One of the most important types of surgical rearrangement is known as *exteriorization,* which means the transfer of an inner organ into a position outside the body wall. This can be done mainly with organs which are readily mobilized because of their stalklike connection with the rest of the body. Through the stalk, their vascular and nervous connections remain intact although the organ lives outside the skin. Protection from loss of heat, from desiccation, and from mechanical injury can be achieved by movable bandages. In dogs, loops of the intestine have been exteriorized with their mesenteric attachments intact. The frequency of peristaltic contraction could be studied under the effect of various stimuli, such as food, sleep, sectioning of vagal or splanchnic nerves (for illustrations of exteriorized intestines, see Markowitz *et al.,* 1954, Figs. 296–298).

In their extensive studies on the volume of the spleen in dogs and cats, Barcroft and his associates arrived at exteriorization as the most satisfactory technique. Their first attempt was to make the outline of the spleen visible by Roentgen rays. When it became desirable to observe color and general appearance as well as the size of the spleen, abdominal windows were installed (Barcroft and Stephens, 1927). They were useful for relatively short periods, until they lost their transparency by deposition of fibrin on the inside. This was the reason for the third step, exteriorization (Barcroft and Stephens, 1927). In all cases the volumes of the spleens were computed from outline drawings. The outline drawings were projections of the visible margins of each spleen on cellophane sheets and paper. With the Roentgen-ray technique, projections in two different planes were available for each spleen whereas the window technique allowed only one projection. On the other hand, the outlines were more precise in the window projections than in the Roentgen-ray pictures. To follow changes in size of exteriorized spleens, Barcroft and collaborators used again projection on a cellophane sheet, similar to the window technique. From the outline drawings, volumes were computed; and from the volumes, estimates of weights were derived. I find it difficult to understand why the volume of exteriorized spleens was not determined directly, either by caliper measurements in different directions or by making molds. However, the main results of Barcroft and his associates seem to be well established.

Observations of changes in size and color in exteriorized spleens have helped confirm the interpretation of the function of this organ as a blood reservoir (see Part III, Chapter 6, A, 3, and Part V, Chapter 2, C, 1). Dogs with exteriorized spleens have been maintained in good condition for several years. A picture of such a dog is given by Wagoner and Custer (1932; Fig. 16). A fourth technique is used for measuring changing volumes of the spleen in its natural site. The abdominal cavity is opened and an air-tight sterilized chamber (plethysmograph) is placed around the spleen. The large wound is closed, but through a small hole a rubber tube leads from the chamber to the outside so that changes in volume of the spleen can be recorded (Markowitz *et al.*, 1954, p. 602).

d. *Transparent windows.* Transparent windows in the abdominal wall, like those mentioned above, were first made by Katsch and Borchers in 1913. An opening is made in the skin and muscular wall of the abdomen and a piece of cellophane is fixed to the edge of the opening by threads or metal clips. The presence of the foreign body is tolerated remarkably well by the adjacent tissues. Loss of transparency by fibrin deposits limits the useful period of such windows.

Blood vessels of skin and mucous membranes can be observed microscopically in the living organism with special illuminating devices. However, observations of this type, such as capillary microscopy in the nail wall, are limited to short periods. In order to permit long-term microscopic observation of small blood vessels of the skin, two different window procedures have been invented. The first one is known as the rabbit ear chamber of Sandison (1924) and Clark *et al.* (1930). A hole is punched in the ear of a rabbit and covered by means of two small glass plates, one on the inside and one on the outside of the ear. The resulting transparent chamber is gradually invaded by small blood vessels which sprout from the tissues around the hole. Good illustrations of the ear chamber are found in Florey's "General Pathology" (1958). This preparation proved very useful in pharmacological studies. For instance, Clark and Clark (1943) observed that newly formed muscular arteries responded to epinephrine before they became innervated. In addition to small blood vessels, cells and fibers can be seen in the rabbit's ear chamber. I mention an observation of Stearns (1940): the formation of new connective tissue fibers in the chamber proceeded at a speed which had not been expected from the study of fixed and stained preparations; a dense network developed within 48 hours.

The rabbit ear chamber technique was adapted to the skin of mice by Algire in 1943. I follow here the description given by Algire and Legallais (1949). The skin on the back of the mouse is very loosely attached to the underlying muscle. Therefore it was possible to elevate a large fold of the skin and to keep it permanent by a supporting frame of plastic. A hole was cut in one side of the skin fold. The hole was covered by a plate of mica fastened to the plastic frame (our diagram, Fig. 17). In the resulting shallow chamber, with the skin on one side and the mica plate on the other side, microscopy in transmitted light could be used for the study of blood vessels, cells, and fibers.

To observe the blood vessels in the choroid of the eye, A. W. Vogel made windows in the sclera of albino rabbits (reported by Leopold, 1952). The opening was covered with a piece of plastic. These windows could be used for two weeks; after this, scleral tissue started covering the window.

Hens' eggs have been supplied with transparent windows covering a hole in the shell for observation of the developing embryo (for illustrations, see Mayer, 1942). Eggs with transparent windows have also been prepared for

FIG. 17. Transparent window in elevated fold of skin on the back of a mouse. This permits microscopic observation of living structures. Blood vessels in window were drawn too large for sake of distinctness. (Diagram after photograph of Algire and Legallais, 1949, Plate 45, Fig. 1.)

the study of viruses which were inoculated on the chorioallantonic membrane (Himmelweit, 1938). Such windows are particularly useful for observing fleeting reactions to newly discovered viruses (personal communication by Dr. V. Cabasso, 1959).

The cornea of the eye represents a natural window in the surface of the body. Markee (1929) took advantage of this fact by developing a technique which proved extremely fruitful. He transplanted the endometrium (inner lining of the uterus) of a guinea pig into its anterior eye chamber (see my Fig. 2). A short time after transplantation, the endometrial fragment was supplied adequately with new blood vessels originating from the eye. The structure and reactivity of the endometrium did not seem to be altered by transplantation. It became possible to observe directly the periodic changes of the endometrium under the influence of the estrous cycle.[5] In other words, the ovulation clockwork could be watched in the eye of the guinea pig. Factors which modify these mechanisms could be studied readily. The principle of Markee's technique has since been applied for various problems. For instance, fragments of human tumors have been transplanted successfully into the anterior chamber of guinea pig eyes (H. S. N. Greene, 1941).

Also of interest is a method of observing events inside the shell of a snail which is inhabited by a hermit crab *(Eupagurus bernhardus)* and a worm (an annelid, *Nereis fucata*). The opaque shell was replaced by a transparent glass model so that one could observe how the worm obtained some of the food of the crab (see pictures in R. and M. Buchsbaum's "Basic Ecology," (1957, p. 49).

3. *Isolation Preparations*

Let us assume that the kidneys have been removed from a dog and the dog is kept alive for observation: in this way the functions which are lost can be determined. On the other hand, the removed kidneys may be the subject of study. They can be kept surviving by perfusion and observed *in vitro:* this represents an isolation experiment. In many cases such isolation of an organ permits direct observation of some of its functions. Isolation preparations may also consist of part of an organ or a group of two or more organs.

a. *Perfusion of single organs and several connected organs; organs and organ slices kept surviving without vascular perfusion.* The study of *isolated single organs* has yielded many important discoveries of which I mention the following example. At the beginning of this century the idea developed that nerve endings may release special substances which act upon the innervated organ. This chemical theory of the transmission of nervous stimuli was suggested by the similarity of the effect of adrenaline and sympathetic nerve stimulation, and between the effect of pilocarpine and parasympathetic nerve stimulation. In 1921 Loewi studied the well-known inhibition of the heart muscle by the vagus nerve. He used the isolated frog heart technique which had been developed by Straub (1901), with the modification that the left vagus nerve was dissected out, and kept in its connection with the heart. After thorough rinsing, a certain amount of saline was kept in the beating heart for some time without changing it, and then removed and stored in a flask (sample A). The heart was refilled with fresh saline, and repeated electrical stimulations of the vagus were applied for a period equal to that during which sample A remained in the heart. The saline which had been present during vagal stimulation was removed and stored (sample B). As usual, the heartbeat had been inhibited by vagal stimulation. When the heart had recovered from this effect, it received alternately portions

[5] A similar experiment is sometimes made by nature. In rare instances, nodules of displaced endometrium are found in the umbilici of women. Bleeding of these nodules, in a way, makes menstruation directly visible.

of the two saline samples. Sample A had no effect, but sample B produced an inhibition of the heartbeat similar to that which results from stimulation of the vagus. Evidently, an inhibiting chemical stimulus was present in sample B. The agent proved to be identical with acetylcholine.

Single organs have been isolated and studied for various physiological and pharmacological purposes. For most experiments it was necessary to perfuse the isolated organ with oxygenated saline or with blood to which anticoagulants had been added. In the process of dissecting organs for isolation, assembling the glassware, and preparing the perfusion fluids, bacterial contamination is difficult to avoid. Since most of these experiments last not more than a few hours, the bacteria in the organs and perfusion fluids do not multiply to a degree which might interfere with the study. The situation is different when perfusions are intended to continue through days and weeks. The perfusion apparatus of Lindbergh solved the problem of sterile perfusion. A description of the apparatus and of some experiments performed with it are to be found in the 1938 book by Carrel and Lindbergh. The Lindbergh apparatus has not been used extensively, probably because there are not enough problems that require continuing perfusions beyond a few hours. The experiments of Carrel and Lindbergh have shown that perfused organs do not die as fast as some investigators had stated (see Part III, Chapter 3, C, 2).

A perfused, isolated organ may survive while several of its functions deteriorate. In order to keep an isolated heart in a condition of good performance, Sarnoff *et al.* (1958) combined perfusion and cross-circulation techniques; for a description of cross circulation see Part III, Chapter 4, B. The heart of one dog was isolated and perfused so that the left ventricle and atrium formed a closed system with the perfusion reservoir and tubes. Some blood, however, escaped through the coronary circulation into the right ventricle. This blood was led to the veins of a second dog, the support dog. From the arteries of the support dog, blood was channeled to the reservoir. The authors found that the isolated heart did not deteriorate for 6 hours. During this period aortic pressure, cardiac output, and heart rate could be altered experimentally and returned to near the same levels; diastolic pressure in the left ventricle did not change markedly.

Let us now turn to procedures in which *two or more organs are kept in their functional vascular connection,* but isolated from the rest of the body. The heart-lung preparation of Knowlton and Starling (1912) opened new avenues of experimentation. A diagram of the experimental design is included in Starling's textbook (5th ed., 1930) and reprinted by Winton and Bayliss (1955, p. 29). It is beyond the scope of this book to discuss the results of experiments in which the heart-lung preparation was used. It was possible to vary independently the venous input into the heart. Parallel with these variations the output of the heart varied. This is very different from a rigid mechanical pump. If, on the other hand, arterial pressure is varied within moderate ranges, the output of the heart does not change. In this respect the heart resembles a rigid mechanical pump. These items, including the comparisons with a man-made pump, are abstracted from Winton and Bayliss (1955). Problems other than those of hemodynamics have also been tackled with the heart-lung preparation. For instance, the arrest of circulating leucocytes in the capillaries of the lung was verified by this preparation (Ambrus *et al.,* 1954); this will be described in connection with the methods for following the fate of cell populations (Part III, Chapter 3, B).

The heart-lung preparation has been extended in various ways. Heart-lung-kidney perfusion and heart-lung-hindleg perfusion have proved particularly useful. The dependence of renal function on the presence of a living lung (Starling and Verney, 1925) will be discussed later (Part III, Chapter 4, A).

An astonishing achievement along these lines is a preparation called the visceral organism which was described by Carrel in 1913 but became practical only in 1930 when Markowitz and Essex improved the technique. Figure 18, taken from their book, shows the thoracic and abdominal viscera of a dog detached from their natural environment and kept alive in an incubator. The vascular system is perfused every one and a half hours with heparinized blood. The lungs are ventilated by compressed air or by oxygen containing 4% CO_2. Blood pressure and many other functions are maintained in visceral organisms for several hours. However, it is difficult to tell from the book of Markowitz *et al.* (1954) what problems have been tackled and solved with the use of the visceral organism.

An interesting type of incomplete isolation is used for the study of the cerebral cortex. The brain of the animal is exposed, and small slabs of a definite cortical area are isolated from adjacent nervous tissue by means of razor blade fragments and fine wire; the clefts are kept open by mineral oil. Since the blood vessels of the pia mater are preserved, they constitute the only connection of the cortical slab with the rest of the animal. This preparation is particularly suitable for the study of transmission of electric impulses (Burns, 1950).

A whole uterus of a guinea pig, an excised segment of the intestine, and rings or spirals carved from an artery can be maintained for some time in a warm, oxygenated saline bath. Such preparations play an important role in the study of hormones and drugs. For technical details see Sollmann and Hanzlik (1939). Thin surviving slices of various organs are used in Warburg's (1923) method for short-term metabolic studies *in vitro*. This technique proved very valuable in spite of the destructions on the cut surfaces of slices which are obtained from organs such as the liver or brain. A minimum of injury, if any, is involved in careful chemical or thermic isolation of the epidermis from the underlying corium (Medawar, 1941; Baumberger *et al.,* 1942).

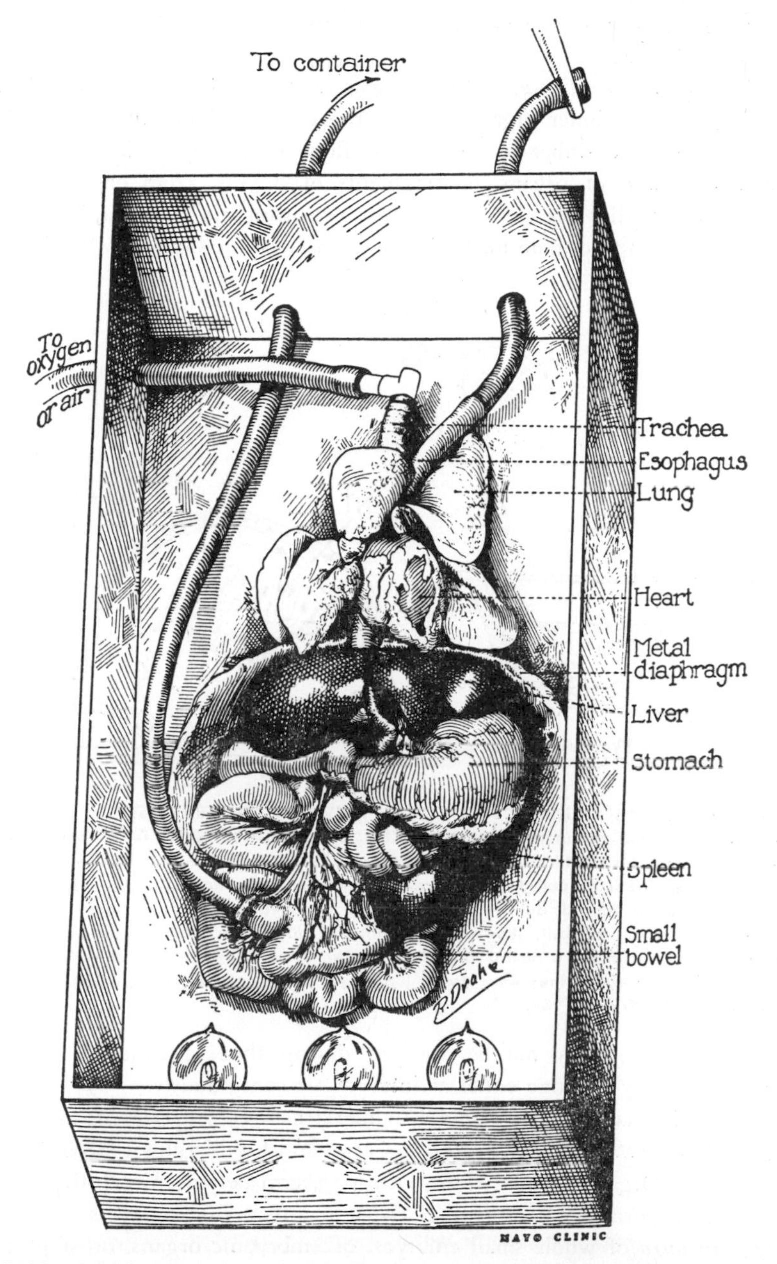

FIG. 18. Surviving thoracic and abdominal organs removed from a dog but kept structurally and functionally connected with each other. The preparation is maintained in an incubator saturated with water vapor at 39° C. The lungs are ventilated and the vascular system is perfused with heparinized blood. (From Markowitz and Essex, 1930.)

The isolated epidermal sheets are suitable for metabolic as well as for morphological studies.

b. *Explantation* (*tissue culture, organ culture*). It was mentioned before that fragments of inner organs such as the endometrium can be transplanted into the anterior chamber of the eye to allow direct observations. In contrast to transplantation of organs, or fragments of organs, from one place of an organism to another place, or to another organism, the transfer into artificial containers with appropriate media is known as explantation. The media in the

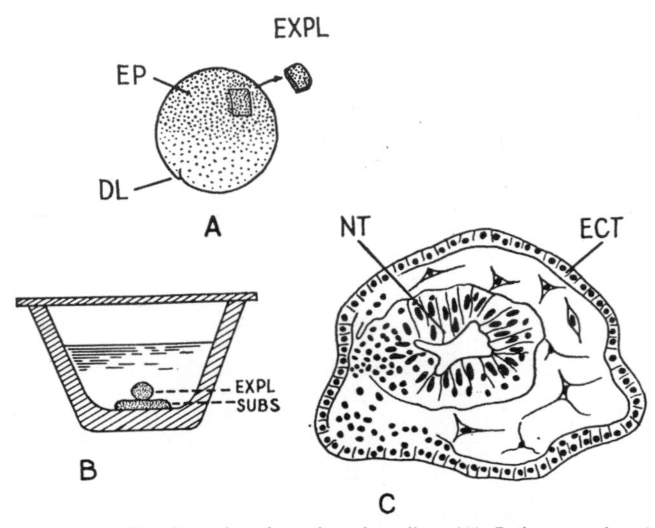

FIG. 19. *In vitro* cultivation of embryonic primordia. (A) Early gastrula of *Triton*, surface view. From the presumptive epidermis (*EP*), a small piece (*EXPL*) is cut out for explantation. For dorsal lip of blastopore (*DL*), compare Fig. 3f. (B) Jar with saline containing a dead substrate (*SUBS*) prepared from marginal zone of *Triton* gastrula. The explanted piece of ectoderm (*EXPL*) placed on the substrate has assumed a spherical shape. (C) Section through the spherical explant, showing differentiation into a cover of ectoderm (*ECT*) and an interior vesicle resembling a neural tube (*NT*). In the space between *ECT* and *NT*, the stellate mesenchymal cells formed; they were drawn too large for the sake of clarity. Semidiagrammatic drawings after Holtfreter, 1934.

containers may or may not be changed during the experiment. If they are changed this may be done either at intervals or continuously.

The maintenance of explanted living structures is referred to as *organ culture, tissue culture, and cell culture*. Carrel and Lindbergh's 1938 book, which was entitled "Organ Culture," dealt with long-term perfusion of fully developed organs. In experimental embryology the term organ culture refers to the maintenance *in vitro* of whole small embryos, of embryonic organs, or of primordia of organs. Such studies require a regular change of media, but vascular perfusion is neither feasible nor necessary, since culture fluids readily diffuse into small explants. Pictures showing the cultivation of whole embryos are found in Rugh's manual (1948, p. 432). I have illustrated the cultivation of primordia of organs in Fig. 19 A–C adapted from Holtfreter (1934). In this experiment the explant formed an external vesicle of ectoderm and an interior vesicle with the characteristics of the neural tube, as it develops in the late gastrula stage (Fig. 3j). The neural tube in Fig. 19C is surrounded by mesenchymal cells as it would be in the embryo.

Depending on the technique of cultivation, an explanted embryonic organ may continue its differentiation as if it were still part of the whole organism, or it may give rise to undifferentiated cell colonies. These alternatives are illus-

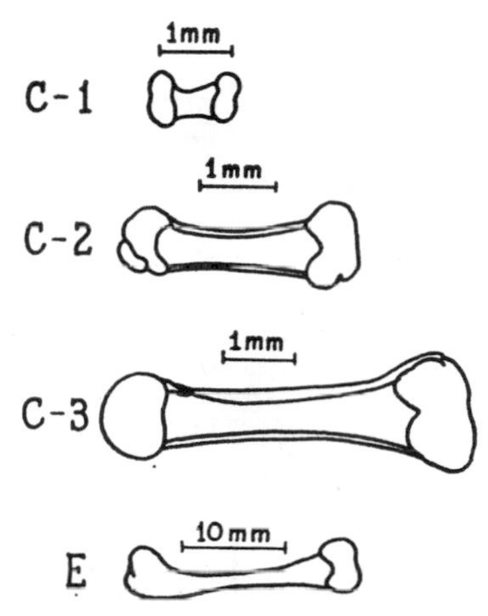

FIG. 20. *In vitro* cultivation of whole embryonic organs. C-1, femur explanted from 5½-day-old chick embryo. C-2, after 9 days of cultivation. C-3, after 21 days of cultivation (age *in situ* + *in vitro* 26½ days); total length of femur approximately 5 mm. E, a femur which developed in the embryo for 21 days; total length approximately 22 mm. Note that histo- and chemodifferentiation of the femur progressed similarly in the culture and in the embryo though the rate was slower in the culture. (Tracings made from originals of Fell and Robison, 1929.)

trated in Figs. 20 and 21. In both cases the starting material is a femur bone from a 6-day-old chick embryo. At this stage the femur consists of cartilage only. In order to obtain continued development, Fell and Robison (1929) placed the little femur on the surface of a plasma clot containing embryo extract (so-called watchglass culture; see Willmer, 1954, Fig. 2C). Every 3 days the femur was washed in embryo extract, transferred to the surface of a new clot, and returned to the incubator. Under these conditions the cartilaginous femur increased in size, approached the shape of the mature bone (Fig. 20, C-3 and E), and also differentiated histologically into calcified bone. Parallel with the morphological changes, phosphatase, which is necessary for bone calcification, increased from nondetectable to detectable quantities. In other words, morphological and chemical differentiation proceeded in the cultured embryonic bone much as it would have in the embryo. However, there were quantitative differences. Length, weight, and phosphatase content of femurs after 12 days of development in the embryo exceeded the maximum values attained in explanted femurs after 27 days of cultivation.

A strain of undifferentiated cell colonies can be obtained from the femur of a chick embryo in the following way. The ends of the femur are cut into small pieces and placed on a coverglass or other solid surface, covered with clotted chicken plasma and embryo juice (hanging drop or Carrel flask cul-

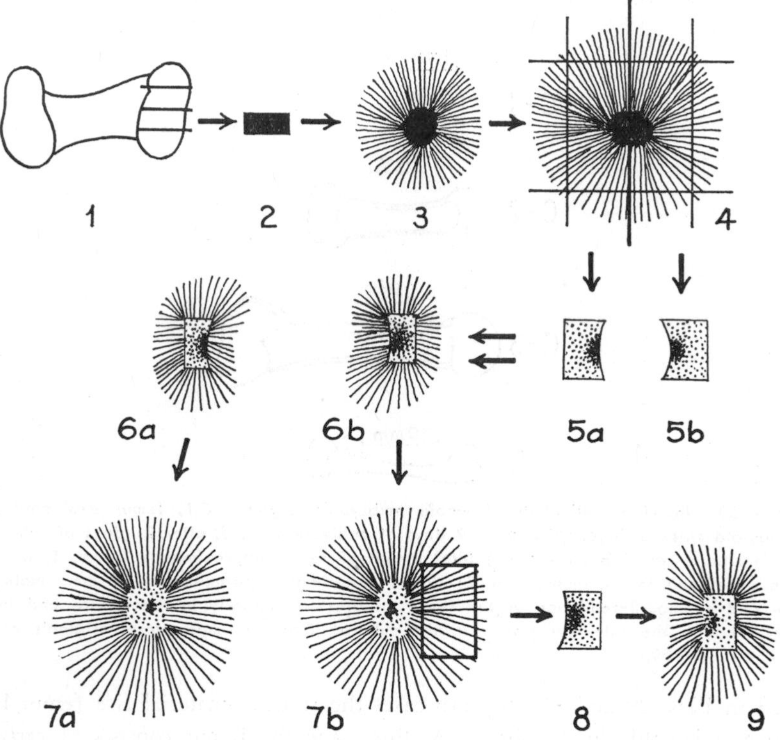

FIG. 21. *In vitro* cultivation of tissue cell colonies. Transfer (passage) technique. (1) Embryonic chick femur. (2) One of the cartilaginous fragments cut from (1) to be used as explant. (3) Small zone of cells produced by emigration from explant and subsequent mitotic divisions. (4) Larger zone resulting from continued migration and multiplication of cells; rectangles indicate cutting of colony into halves prior to transfer into new media. (5a and 5b) Halves retracted to small areas due to release of elastic tension. (6a and 6b) New zones of cells formed around each transferred fragment. (7a and 7b) Each daughter colony has attained a circular shape and the size of the mother colony shown in (4). (7b, 8, and 9) Technique for promoting homogeneity of cell population: transfer of a piece of the newly formed zone, without admixture of central part, which contains remainders of original heterogeneous explant. Note that the lines radiating from centers of colonies are diagrams of cell chains with cross connections (network pattern). The proportions of the bone (1), the explanted fragment (2), and the colonies (3–9) are not drawn to scale. Diameters of colonies (4), (7a) and (7b), 48 hours after explantation or transfer, are approximately 5 mm.

tures; see Willmer, 1954, Figs. 2A and 2B). After incubation, a zone of radially arranged cells appears around each explanted fragment, as shown diagrammatically in Fig. 21. As the zone increases in width, the cells form a network with narrow meshes. The radial extension of the zone is due to centrifugal migration of cells, whereas the density of the network is maintained by cell multiplication. Instead of a differentiated organ, a more or less undifferentiated cell colony has been produced. Although the original explant contained various components, subsequent passages, with division of colonies (Fig. 21, diagram 4) will increase the uniformity of the cell colony. Since, in our example, the cells were cultivated from cartilage, they would be labeled chondroblasts by most tissue culturists. Their appearance is shown diagrammatically in Fig. 76. The classification of cells in tissue cultures will be considered later in connection with the classification of cells in their natural situation in the organism (Part V, Chapter 2, B, 1). Cell colonies *in vitro* will be discussed again in connection with living models, regeneration, and tumor formation.

4. *Micrugical Techniques*

Mechanical operations on living biological material on the microscopic level are known as micrurgy (short for microsurgery) or micromanipulation. The special activities are referred to as microdissection, microinjection, etc. If low magnifications are used, freehand manipulations are possible, but special attachments to the microscope are needed for operations at high magnifications. In experimental embryology freehand manipulation plays a great role. Some of the instruments used are small knives and scissors which had been designed for eye surgery during the 19th century, but a new era of experimental embryology began when Spemann (1906) introduced fine glass needles and hairs for operations on amphibian and echinoderm eggs. A plate showing the simple but ingenious instruments of this type is found in Hamburger's "Manual of Experimental Embryology" (1942). Instructive pictures are also contained in the short article by Hörstadius on freehand manipulations in McClung's Handbook (1950); the techniques described refer to magnifications not exceeding 100 times. The use of glass needles for the isolation of pieces from an axolotl gastrula is illustrated in my Fig. 8a.

For operations under high magnification the delicate instruments must be guided by devices that reduce the crude movements of the fingers to movements of microscopic orders of magnitude. These devices resemble either the racks, pinions, and screws used in focusing the microscope and moving the mechanical stage, or are combinations of levers with ball and socket joints. A variety of micromanipulators for high magnifications are described and illustrated in the chapter by Chambers and Kopac in McClung's Handbook (1950). Another chapter of the same book, written by Knower, reviews dissecting apparatus with a wide range of control.

Micrurgical techniques can be used for the study of morphological structures as such, but their main object is the isolation or handling of minute morphological structures for special experimental purposes. Protozoa and ova have been cut in such a way that surviving fragments with and without nuclear material were obtained. Individual bacteria and tissue cells have been picked in order to cultivate populations which are entirely derived from one known cell. Such one-cell cultures are biologically termed clones. Pictures of single tissue cells which were isolated by micropipettes and gave rise to cell colonies *in vitro* are included in the paper by Sanford *et al.* (1948) mentioned previously.

Transplantation of nuclei from one cell to another was achieved by various authors during recent decades. However, the improved micropipette technique of Briggs and King (1953) was necessary for the transplantation of nuclei of frog blastulae and gastrulae into enucleated eggs. These studies shed light on the role of the nucleus in differentiation; they will be discussed in later chapters.

Fluids from microscopic cavities were aspired with tiny pipettes held by micromanipulators. Wearn and Richards (1924) succeeded in obtaining separate glomerular and tubular urine from the kidney of a living frog; the quartz pipettes used in these studies had points of 10–20 μ inner diameter. It was mentioned before that a certain area of the thyroid of the mouse lends itself to microdissection experiments. Figure 22, modified from Williams (1944), shows the puncturing of a blood capillary adjacent to a thyroid follicle. Red blood corpuscles leaked into the proteinic content of the follicle, the so-called thyroid colloid, and did not change in appearance during 12 hours of observation. It was concluded that the osmotic pressure of the proteins in the thyroid colloid is not markedly different from that of the blood plasma proteins.

Instead of mechanical devices, such as needles and knives, ultraviolet radiation has been used for microdissection. Schleip (1923) directed a beam of monochromatic ultraviolet (280mμ) at fertilized ova of *Ascaris* before the first cleavage. If a 14 μ^2 area of one of the two pronuclei[6] was irradiated for 1 minute, the development was disturbed to the extent that gastrulation did not take place in the majority of experiments. Similarly, gastrulation was suppressed when a 14 μ^2 area of cytoplasm was irradiated for 8 minutes. Thus, a quantitative difference in reactivity between nuclear and cytoplasmic material was demonstrated. Improved techniques by Uretz *et al.* (1954) permitted the use of ultraviolet microbeams for the irradiation of parts of a chromosome. Two examples of localized ultraviolet irradiation under low magnifications may be mentioned. Geigy (1931) irradiated the pole cell region of *Drosophila* eggs in order to inhibit the formation of germ cells. Mayer (1933) irradiated various areas of chick fibroblast colonies cultivated *in vitro* in a study of regeneration.

[6] Nuclei of ovum and spermatozoon before their fusion.

Whereas Fischer (1930) and Ephrussi (1933) had observed healing of mechanical wounds in tissue cell colonies, Mayer produced similar local defects by lethal doses of ultraviolet, without tearing the plasma clot in which the cell colonies live. Some results of these experiments will be reported in the discussion on living models (Fig. 46).

The elimination of minute areas inside the brain of living animals has become possible by the development of stereotaxic apparatus (see Part III, Chapter 6, A, 4). Physical and chemical procedures as applied to individual cells require microelectrodes, micropipettes, and microburettes which have to be guided by micromanipulators.

Separation of cellular components can be achieved not only by mechanical dissection, but also by chemical dissection or by microcentrifugation. Nuclei have been isolated from the cytoplasm by various means. The discovery of nu-

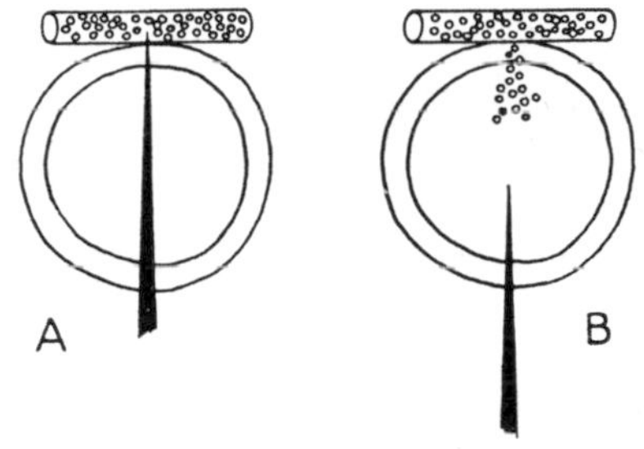

FIG. 22. Diagram showing puncturing of capillary adjacent to thyroid follicle in the living mouse. (A) The microdissection needle, indicated in black, pierces follicle and capillary. (B) Needle withdrawn; red blood corpuscles have escaped from the hole in the capillary into the colloid which fills the follicle. (From R. G. Williams, 1944, Plate 1, Fig. 6.)

cleic acids by Miescher (1871) resulted from a joint application of chemical, mechanical, and microscopic procedures. Pus cells were treated by prolonged digestion with dilute HCl and subsequent shaking in a separatory funnel with ether. A solid layer which formed in the bottom of the separatory funnel consisted of nuclei without cytoplasm, as verified microscopically. Chemical analysis of the pure nuclear material showed the presence of 2.5% phosphorus and 14% nitrogen. This ratio had not been found in any known compound. Miescher gave the name nuclein to the new substance. In a later phase of cytochemistry it became desirable to obtain isolated nuclei in an undamaged condition. This was achieved by Dounce (1943) by grinding a sample of frozen rat liver in a Waring blendor containing citric acid and ice water. Most enzyme activities of the isolated nuclei proved to be undisturbed as compared to the activity of the whole tissue.

Separation of cellular components by microcentrifugation produced important results. Andresen (1942) obtained separate layers in centrifuged amebas. One of the layers contained fat globules; another, the nuclei together with food and crystal vacuoles; structureless cytoplasm (hyaloplasm) was found in three different layers. Subsequently the various layers were studied chemically. The centrifuge microscope designed by Harvey and Loomis (1930) allows the observation of living cells while they are rotating. By means of two prisms, the image of the rotating preparation is projected into a stationary microscope mounted above the axis of the centrifuge. For illustrations see Harvey and Loomis (1930) and Harvey's article in McClung's Handbook (1950).

5. *Vital Staining*

Vital staining refers to the uptake of dyes by living cells without noticeable injury to the cells. The classic example is the staining of the so-called reticuloendothelial system by trypan blue or carmine after injection of these dyes into living animals. A description of the reticuloendothelial system is found in Part V, Chapter 2, C, 1.

Zeiger (1938), Doan and Ralph (1950), and other authors restrict the concept of vital staining to the intake of dyes by cells in the living organism (*in situ*) after oral or parenteral administration of the dye. If surviving cells *in vitro,* e.g., blood cells in a saline suspension, are stained, these authors prefer the term supravital staining. Evidently, the authors wish to indicate that the cells *in vitro* are deteriorating and will die soon.[7] It seems to me that this concept of supravital staining makes no provision for staining of healthy cells *in vitro,* such as the cells in organ cultures and in cultivated populations of multiplying cells. On the other hand, one cannot always establish a hard and fast borderline between the staining of living and the staining of dead material; there are no safe criteria of life with respect to individual cells in histologic sections (cf. Raczkowski, Kloos, and Opitz, 1953). Attempts have been made to correlate particle size, electric charge, and other properties of dyes with their capacity for being ingested and stored by living cells. It is beyond the scope of the present book to discuss this subject. Survey articles are found in McClung's Handbook (3rd ed., 1950) and in Cowdry's "Laboratory Technique" (1952). Among older presentations Zeiger's 1938 book on the physicochemical foundations of histological methods offers a thorough analysis of these problems.

Some dyes have a different effect in living and in preserved material. Methylene blue is a vital stain for nerve fibrils, but stains both nuclei and cytoplasm in fixed material. When a suspension of carmine is injected into a living animal the dye particles are taken in by the cell bodies (cytoplasm) of the reticuloendothelial system. If applied to fixed material, carmine acts as a nuclear stain in all cells reached by the dye. As described before, $AgNO_3$ stains different structures when applied to fresh or fixed material (Fig. 51a and b). Under

[7] Roulet (1948) applies the terms of postvital or supravital staining to the uptake of dyes by fresh dead tissues.

both conditions this black stain is produced by reduction of $AgNO_3$ to silver oxide or metallic silver.

Some vital stains are not simply an intake of dyes by cells, but rather a result of metabolic exchanges. If, for a period of time, an animal is given food containing fat which has been colored with Sudan III, gradually the unstained fat depots of the animal are replaced by red-stained fat. Another example of metabolic staining is the response of bones to the feeding of madder: only those portions of the bones stain red which form after madder became available in the circulation of the animal.

The use of fluorescent dyes in the microscopic study of living organs was mentioned before. Although most vital stains leave the nuclei unstained, certain fluorescent dyes, diaminoacridines, prove to have an affinity for living nuclei without harming them (DeBruyn, Robertson, and Farr, 1950). In a later chapter vital staining of embryonic cell groups will be described when tagging procedures for following the fate of living structures are discussed.

B. Morphological Techniques for the Study of Dead Material

1. *General Comments on Macro- and Microscopic Procedures for the Study of Dead Material*

The present chapter deals with the study of dead organisms and their parts, and includes material killed by removal from a living organism. Studies of dead material require a wide range of procedures depending on the different purposes and circumstances which are peculiar to zoology and human anatomy, experimental physiology and pathology, hospital pathology covering autopsies and biopsies, veterinary pathology, and finally, medicolegal pathology.

There is a striking contrast between the laboratory work of a zoologist and the routine activity of a hospital pathologist. In most cases the zoologist chooses his own problems, and no external pressure prevents him from using the best techniques, irrespective of time. If the hospital pathologist is given a biopsy for microscopic diagnosis, he is frequently expected to arrive at a decision within 10 minutes or less while the patient remains on the operating table. The time available for an autopsy of a hospital patient cannot be extended beyond a few hours. The family of the deceased may impose restrictions on the extent of dissections. Otherwise the conditions for postmortem studies are quite favorable in a modern hospital, since the corpses are kept refrigerated until autopsy. In contrast, the medicolegal pathologist has an unlimited amount of time for a human autopsy and its extent is unrestricted, but, as a rule, the material is in a bad condition. Frequently he has to obtain all information from a decomposed body or from mutilated parts. To compensate for these handicaps ingenious methods have been designed to use circumstantial evidence with seemingly hopeless material (see Gonzales *et al.,* 1954). Microscopy of hair has been developed to a remarkable degree for medicolegal purposes.

The medicolegal pathologist and the hospital pathologist are sometimes faced with unique cases. Some diseases are so rare that a pathologist may autopsy only one case during his whole career. As an example I mention a form of leukemia with green coloration of the abnormal tissue (chloroma). The zoologist and experimental pathologist give preference to studies which can be reproduced. This does not exclude the possibility that a unique observation may be of great value (cf. Part III, Chapter 3, C, 2, discussion of precision and reproducibility). It will be mentioned later that in experimental pathology, the procedures for handling large numbers of small animals (mice, rats) have to be different from those for small numbers of large animals (dogs).

Although there are considerable differences between the usual tasks of a zoologist on the one hand and those of a hospital pathologist on the other hand, there is also overlapping of their interests and techniques. Many hospital pathologists do histological and cytological research, and some zoologists have developed interest in tumors. Parasitology is a science which combines zoology and pathology. Most laboratory work in zoology is of a delicate type, but collecting of hypophyses from whales on a whaling vessel (J. T. Litchfield, Jr., 1935, personal communication) requires procedures comparable to those used by the veterinarian in autopsying a dead horse on a farm.

Manuals of morphological procedures have been written from different points of view. Some books deal with the dissection of one species, such as the rabbit or the frog. The demands of human and veterinary medicine have produced the necessary manuals of normal and pathological anatomy. Laboratory handbooks of normal histology are frequently written for the needs of zoologists, whereas those on pathological histology are adjusted to human pathology.

Manuals for applied and fundamental experimental pathology do not seem to exist except in the form of introductions for students. I mention three books of this type, one by Salomonsen (1919) in Danish, the second by Wagoner and Custer (1932) in English, and the third by Meessen (1952) in German. None of them gives much space to postmortem techniques in experimental animals. Meessen gives numerous photomicrographs of histopathological conditions. He describes how these conditions are produced experimentally, but does not mention techniques of postmortem dissection and histology. The book by Markowitz *et al.* on experimental surgery (1954), to which I have referred repeatedly, deals with surgical procedures as used in experimental pathology. It does not cover postmortem procedures. In the comprehensive book by Cohrs *et al.* (1958) on the pathology of laboratory animals, the first volume gives detailed descriptions of postmortem findings in diseased laboratory animals. Brief technical comments are scattered throughout the book, but special paragraphs devoted to technical procedures are found only in the chapters on blood, bone marrow, and the nervous system.

Applied experimental pathology comprises macroscopic and microscopic studies of laboratory animals for practical purposes, such as the development of new vaccines, drugs, or food additives. Fundamental experimental pathology is the study of macroscopic and microscopic abnormalities from the standpoint of biological research. While it is not the purpose of the present book to supply a laboratory manual for experimental pathology, certain general technical principles of experimental pathology will be discussed here.

Efficient planning of postmortem studies requires selection of procedures adjusted to the different orders of magnitude of the respective objects. Animals of different size, such as a dog and a mouse, obviously require different procedures for dissection. The heart of a dog can be opened during autopsy by separate incisions into each atrium and ventricle; the thickness of the muscular walls of each part is estimated at this occasion. The heart of a mouse is too small for such a procedure. A cut across both ventricles is almost the only incision feasible in the fresh condition, and systematic inspection must be left to microscopy after fixation. Cross sections through the fresh kidney of a mouse can be made with the naked eye or under a dissecting microscope, but the adrenals of this animal are too small and delicate to handle in the fresh condition; preservation of the intact adrenals for microscopic study is preferable. In a fairly large dog, on the other hand, sections through the fresh adrenals are feasible.

The rules for macroscopic procedures are not uniform. In one laboratory a dog is autopsied by one person who also handles the sampling for microscopy, the weighing of organs, and the bookkeeping. In another laboratory a team of trained workers collaborates in the autopsy of each dog. Ideas of the necessary scope of autopsies vary greatly. Some laboratories seem to be satisfied with information obtained by examination of thoracic and abdominal organs, and therefore ignore the cranial cavity and bones. The attention given to topographic relations is not the same in all laboratories. Some pathologists leave the macroscopic autopsy to technicians, require microscopic samples from every organ, and concentrate their professional effort on the microscopic diagnosis. I feel that the pathologist is responsible for the macroscopic autopsy, since the microscopic diagnosis cannot correct mistakes that were made in preceding macroscopic procedures. In order to safeguard continuity between macro- and microscopic procedures the pathologist who selected the samples for fixation should also handle them after fixation. He should cut them into the proper shape (so-called trimming) and decide the planes of histologic sectioning. At this occasion the necessary instructions are given to the technician who carries out the microscopic techniques. With this system, the same dog or rat will occupy some of the pathologist's time on at least two different days. As a consequence, one pathologist cannot give his full time, day after day, to the performance of autopsies.

A laboratory with a small staff and relatively low standards may be willing to autopsy eight or more dogs a day, whereas another laboratory with a similar number of workers but with higher standards will probably refuse to do more than two or three dog autopsies a day. Analogous differences prevail in the handling of rat autopsies. Under these circumstances it will be understandable to the readers of the present book that morphological results of different laboratories should be compared only if the procedures are stated explicitly, no matter how trivial the detail may appear. Usually, more is taken for granted on the macroscopic than on the microscopic level.

Macroscopic and microscopic postmortem studies can be made on fresh or on preserved material. As a rule preservation of large objects requires rinsing of the blood vessels with saline followed by injection of a fixing fluid, whereas smaller objects can either be injected with, or immersed in, the respective fluids. Hollow organs such as the stomach may be filled by the fixative and subsequently immersed so that the fixative can penetrate into the wall of the organ both from inside and outside. Fresh material can be frozen and stored for a long period and finally thawed for study.

It was discussed previously that postmortem studies may involve biochemical or bacteriological questions besides morphological problems. Examples were described that illustrate compromises between the different requirements. Similarly, there may be competition of problems within a morphological study. It was just mentioned that the best fixation of large organs is obtained by injection of the fixatives into the vascular system. This, however, necessitates removal of the blood from the vessels. Therefore, fixation by injection should be avoided if the distribution of blood might be important in the study of microscopic sections from the respective organ.

Special injection techniques have been developed for three-dimensional presentations of hollow structures. Small blood vessels and spaces can be injected with India ink, which remains fluid after injection. Subsequent fixation of the organ will keep the ink in its place during embedding and sectioning (Fig. 43a,b). By means of serial sections the injected structures can be reconstructed in three dimensions. Large hollow structures are represented best by means of casts. Figure 23 shows the bronchial casts of a human lung. Such preparations are made by injection of stained vinylite into the bronchi of the fresh lung, and the same method is also applicable to the presentation of blood vessels.[8] When the injected masses have solidified after approximately 24 hours, the surrounding

[8] Bronchi, arteries, and veins may be injected with differently stained vinylite. This led to the discovery of bronchovascular segments (see discussion of units, Part III, Chapter 4, C).

tissues are removed by corrosion with acid or alkali (Liebow *et al.*, 1947). As a rule, sections for microscopic study cannot be made from vinylite-injected organs: before hardening of the vinylite, manipulation must be avoided, and when the hardening is complete the tissues have deteriorated by postmortem autolysis. This conflict can be resolved in paired organs, such as lungs, kidneys, or thyroids, if one can reasonably assume that the left and right organ are in the same condition. Then one of the pair may be used for injection and the other for microscopy. If

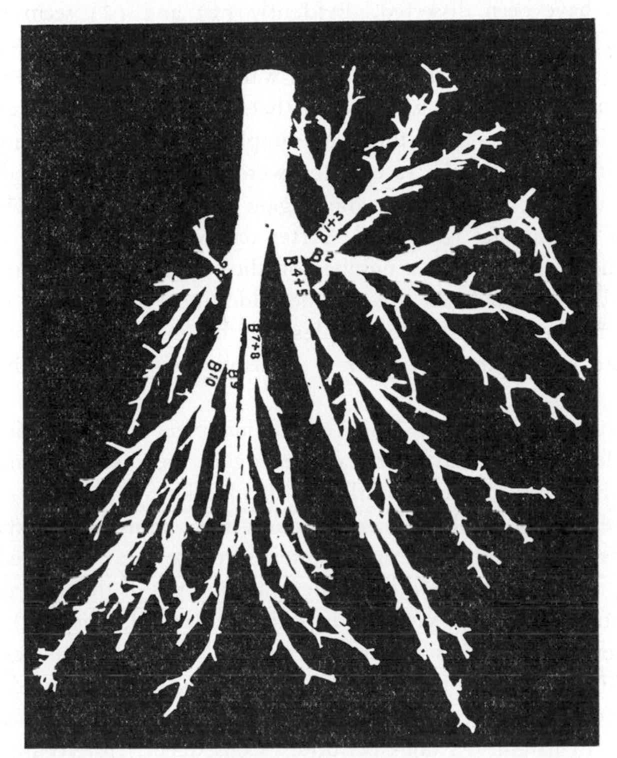

FIG. 23. Bronchial cast of human left lung, viewed from the median plane. The cast was obtained by vinylite injection of fresh lung and subsequent removal of surrounding tissue by corrosion. Labeling of bronchi is according to Boyden (see monograph, 1955). Silhouette was made from a photograph supplied by Drs. A. A. Liebow and R. A. Vidone (Laboratory of Pathology, Yale University School of Medicine).

hollow structures less than 50 μ in diameter are to be injected, neoprene (synthetic rubber) is preferable to plastics.

Macroscopic dissections for the verification of topographic relations may require so much time that the sampling for fixation is retarded. As a consequence, the microscopic study of the organs concerned may lose in efficiency.

An example may illustrate how, with a proper autopsy technique, competing requirements can be reconciled. Let us assume that three aspects in the postmortem study of the liver of an experimental dog are equally important: (1) dissection of the portal vein and bile ducts which connect the liver with neighbor organs, (2) determination of the weight of the liver, and (3) obtaining of samples for histological and histochemical analysis as soon as possible after the animal has been killed. Procedure (1) is time consuming and hardly possible before the intestines have been removed. Procedure (2) requires removal of the whole liver; this should not be done before delicate adjacent organs such as the adrenals have been dissected. Evidently (1) and (2) seem incompatible with (3). The conflict can be resolved as follows. As soon as the liver has been exposed by an incision into the abdominal wall, samples are removed as required for (3). These samples are weighed so that their weight can be added to that of the remaining liver. Since the liver samples are to be placed in receptacles (with or without fluid), the receptacles were labeled and weighed before the start of the autopsy, and then weighed again with the samples. Thus delay of treatment of the samples is avoided. After removal and weighing of the liver, the organ should be cut into a number of slices as usual so that numerous cut surfaces can be inspected. At this occasion additional samples can be taken for microscopy. Although these late samples may be unsatisfactory with respect to some delicate points, they represent a valuable supplement to those samples which were obtained earlier.

Compromises between the different requirements are sometimes difficult or impossible.[9] This is particularly disturbing in the case of human autopsies and rare pathological findings in animals. It is easier to apply a variety of techniques, if certain morphological conditions can be reproduced experimentally. Then groups of animals can be used: one procedure can be applied to half of the animals, and another procedure to the other animals of the same experimental group (cf. study of drug precipitates in kidneys, Part III, Chapter 1, B). However, in group experiments certain advantages and disadvantages are to be balanced. In the study of a new drug, with the use of different doses, groups of ten rats for each dose can easily be handled, but few laboratories would be in a position to handle ten dogs per dose. Consequently, the chances of reproducing certain morphological conditions experimentally are greater in rats than in dogs. The main advantage in using larger species, such as dogs, is the greater attention given to the individual animal. Observations in a living dog are, as a rule, much more complete than those in a rat. In postmortem studies of the larger animal, macroscopic dissection can be very thorough, resembling that of a human being. In smaller animals, such as rats and mice, microscopic study either covers the whole organ, or covers samples which are more representative of the whole organ than in the case of a large animal. Quantitative aspects of sampling will be treated in a later chapter (Part III, Chapter 8, C, 1).

[9] This is a frequent situation in archeology and paleontology. In order to study the natural cast of a fossil brain the skull must be sacrificed (Edinger, 1948; p. 11).

The following basic rules for routine autopsies apply to both human patients and experimental animals. (1) Topographic relations should be verified before any organ is displaced. (2) The presence of abnormal fluid (blood, pus) or of pathologic adhesions between neighboring organs in the thoracic and abdominal cavities is to be ascertained. (3) Organs which are apt to change their volume substantially at removal need special treatment (e.g., lungs and spleen). (4) Possible pathological structures should not be dislodged. For instance, the pulmonary artery should be opened to ascertain the presence or absence of an embolus, and the ureters should be inspected *in situ* (possibly also palpated) so that ureter stones will not be missed or displaced.

In the routine described above, the abdominal cavity is opened first and, after proper inspection, the intestines are removed with ligatures in the places of cross-sectioning so that no content of stomach or intestines contaminates the other organs. This sequence has the advantage of making other organs readily accessible. However, there are other points of view which may overrule technical convenience. Some organs deteriorate faster by autolysis than others. Therefore, in an experimental animal, one may give preference to the removal of such organs which, in the light of the experiment, seem particularly important and, at the same time, are perishable.

Continuity between macro- and microscopic procedures is desirable in postmortem studies. Continuity is assured best if the same investigator who performed the autopsy and selected the samples for microscopy also attends the samples after fixation. This individualistic technique is particularly necessary in studies of large animals such as dogs. The situation is different if the nature of the experiment requires large numbers of animals. Mass experiments with mice and rats as used for screening and toxicity studies necessitate abbreviated selective techniques which may approach assembly line methods. Under these conditions the workers who carry out the autopsies may not be the same ones who trim samples for microscopy or examine histological sections.

The transition from macro- to microscopic procedures is as precarious as it is important. In a previous chapter it was mentioned that some structures are seen best in the fresh condition of the organs, for instance, drug precipitates in the kidneys of dogs or rats. On the other hand, organs of delicate consistency are easily damaged by dissection in the fresh condition. Sometimes a useful compromise is made by fixing the delicate organ for a few hours before making any incision. In the pathology of the human brain, local softening is a characteristic consequence of circulatory disturbances. Obviously changes in consistency are ascertained best by palpation of the fresh material. However, the detection of softened areas inside the brain requires the cutting of the brain into thin slices. This cannot be done in the fresh brain without crushing it. Its normal consistency allows, at best, making a few thick slices. Under these circumstances it is preferable to fix the brain as a whole and to cut thin slices when the brain has attained the proper consistency. Fortunately, softened areas can still be recognized after fixation.

Sometimes, no satisfactory technique of fixation is available. In the adult rat the membrane which includes the testis (tunica albuginea) is too thick to allow the diffusion of fixatives. If one makes an incision into the membrane the soft testicular tissue protrudes from the opening, thereby causing structural distortions. It is only by cutting off one polar cap that tolerable fixation of a central slice may be obtained (Fig. 27). This difficulty is not present in smaller animals such as mice and very young rats, in which the membrane is delicate enough to permit diffusion of the fixative without incisions. Neither are the conditions so unfavorable in larger animals such as dogs. Here incisions are necessary to allow the entry of fixatives. However, the testicular tissue is firmer than in the rat and therefore is not distorted by the incisions. When the tissue is reached by the fixative, its natural architecture is preserved.

2. *Mechanical, Chemical, and Metabolic Dissection*

Mechanical dissection is used to separate adjacent parts of different consistency. In this way, loose connective tissue is teased off from a muscle, a blood vessel of rubberlike consistency is isolated from the soft substance of the central nervous system, and cartilage is scraped from hard bone. Mechanical dissection can be carried out with the naked eye, a hand lens, a dissecting microscope for low magnifications, or special microdissecting devices for high magnifications. Different types of micromanipulators were described previously.

A special variety of mechanical dissection is done by shaking an organ in a container with fluid. The structure of lymph nodes has been elucidated in this way. A lymph node consists of a three-dimensional network of connective tissue cells and fibers; this network is obscured by densely packed lymphocytes which fill the meshes. If a slice of a fresh lymph node is shaken in saline for a short time, the lymphocytes are removed and the reticular scaffold can be clearly recognized.

Chemical dissection is based on different reactions of adjacent structures to treatment with solvents, swelling agents, or enzymes. The epidermis can be loosened from the underlying corium by various agents such as ammonium hydroxide or heating to 50° C. (Baumberger *et al.*, 1942), or by tryptic digestion (Medawar, 1941). As mentioned before, some of these techniques were used for experiments with surviving epidermis. All are applicable to postmortem studies. Isolation of nuclei by chemical removal of the cytoplasm was described earlier (Part III, Chapter 2, A, 4) in connection with micrugical techniques.

Different components of an organ may be unequally resistant to decomposition after death. The hard substance of a bone is more resistant to decomposition than its soft inner and outer linings, blood vessels, and bone marrow cells. Therefore, in order to obtain a clean, hard bone (as used for museum specimen), the fresh bone is placed in water until all soft parts have decayed and can be scraped off readily. After drying, the remaining fatty components are extracted by fat solvents. Finally the bone is dried and bleached. This whole process is known as maceration. The macerated bone consists of a dried connective tissue framework impregnated with its original calcium salts.

If a fresh bone is placed in a fixative such as formalin, its soft parts will be preserved. Then treatment with acids can be used to dissolve the calcium salts. The decalcified bone maintains its original shape, but is flexible instead of rigid. A decalcified bone tied into a knot is shown in Ham's "Histology" (1957, Fig. 166). Such a bone is to be stored in a moist condition in order to preserve its flexibility. A different type of treatment is incineration. If this is done carefully, the ashes show enough cohesion to maintain the form of the original bone. The incinerated bone is very brittle; it requires gentle handling and dry storage. A special method is used to dissect those bones of an adult skull which are joined by the sutures. After maceration, the cranial cavity is filled with wet rice or peas, all openings are plugged tightly, and the enclosed material is allowed to swell until the sutures of the skull burst apart.

Collagenous and elastic fibers, which frequently occur together, can be separated either mechanically or chemically. By teasing under a dissecting microscope one can observe that the elastic fibers behave like rubberbands and the collagenous fibers like cotton threads. Treatment with pepsin in acid solution digests the collagenous fibers, but leaves the elastic fibers intact. Differences between the two types of fibers can also be demonstrated by staining or by electron microscopy, as will be mentioned later.

A combination of mechanical and chemical separation is necessary in special cases. As an illustration, a procedure may be mentioned which is used in human pathology. A fresh brain is placed in a bottle with water and shaken in a machine for several days. Then the brain substance decomposes chemically and disintegrates mechanically while the vascular tree is preserved. Pathological dilations (aneurysms) of small arteries or veins can be detected with this technique.

To complete the description of dissection techniques, I mention processes which may be called *metabolic dissection*. Anyone who has to dissect human beings or animals prefers lean individuals to obese ones. Even in well-nourished, though not obese, rabbits or rats it is very difficult to separate the diffuse pancreas from the adjacent fat tissue. To identify small nerves in a mass of fat is near to impossible. Richter (1950) solved the problem, at least for rats, in the following ingenious way. The animals were fed a dextrose-thiamine diet for 40–60 days. Slow loss of weight took place resulting in extreme leanness without signs of nutritional deficiencies in teeth, skin, hair, or bones. In these rats the abdominal fat had disappeared so that autonomous nerves could be recognized which are too delicate to be dissected out under ordinary conditions. According to Hartroft (1954) the so-called Rappaport units of the liver became very distinct in rats with cirrhosis resulting from choline-deficient diet. As Hartroft puts it, these units were dissected out by the cirrhotic process (my Fig. 43b).

It is not unusual that normal structures are accentuated by pathological processes. Some segments of the renal tubules in man can be distinguished by their swollen cells when certain disturbances of protein metabolism occur (lipoid nephrosis). In human patients who died with an excessive content of fat in their blood (lipemia), I observed an accumulation of visible fat drops in the impulse-conducting bundle of the heart although there were few in the ordinary muscle fibers. Consequently, the conducting bundle was much more conspicuous than under normal conditions.

3. *Microscopic Procedures*

a. *Microscopes, illumination, photomicrography.* My late colleague Alan F. Kirkpatrick introduced a talk before the New York Microscopical Society (1955, unpublished) by projecting a picture of the human brain. He wished to emphasize that one's brain is the most important tool in microscopy. Among frequent errors in judgment I mention the failure to use low power devices, such as hand lenses or dissecting microscopes, a failure which results in frustrating gaps between the macro- and microscopic analyses. The emotional appeal of high magnifications is greater than that of low magnifications. Therefore, in medical schools the student microscopes never lack oil immersion objectives for magnifications close to 1000 times, but they do not always include the equipment needed for magnifications of less than 100 times. One excuse may be that for magnifications from approximately 100 to 1000 times, similar condensors and light sources can be used, while in the range of 5 to 20 times a different type of illumination is required.

Anyone who wants to acquire a solid background in microscopy should study Shillaber's "Photomicrography in Theory and Practice" (1944). Photomicrography not only produces permanent records and makes structures visible which are outside the range of the human eye, but it is also the best training ground for the use of the microscope and its accessories. Because the book appeared in 1944, phase contrast microscopy could not be considered. The effect of infrared radiation is briefly described in the text, and the use of polarized and ultraviolet light is mentioned only in the Glossary. The book does not include color photography or stereoscopic pictures. In other words, the book is restricted to what Shillaber rightly calls basic material. This basic material is handled in a clear, thorough, and practical way and illustrated with unusual skill. The different types of microscopes and lamps and the techniques of correct illumination are discussed with a masterly combination of theoretical and practical considerations. There is a special paragraph on the effect of dirt on seven parts of the optical system, showing where the dirt is especially harmful. Shillaber is careful to state that dirt on the mirror may absorb appreciable amounts of light, but will not cause image deterioration. Very useful detail is found in the discussion on quality and cleaning of slides. Figure 147 of Shillaber shows a dark-field photomicrograph of a microscope slide with innumerable white spots which were probably devitrified areas since all cleaning methods failed to remove them. He points out that such a slide is unfit for dark-field work although it might be used for the most exacting bright-field work. The principles of sectioning, staining, and mounting procedures are described by Shillaber in their application to industrial fibers, metal particles, biological material, etc. I wish to emphasize that the understanding of microscopic techniques in biology is greatly enhanced by some familiarity with procedures used for mineralogical and industrial material.

Historically, Mann's 1902 "Physiological Histology" is remarkable, both for its assets and for its liabilities. It carries the subtitle, "Methods and Theory." The author tried to explore the chemical basis of fixation and staining. His avowed aim was to substitute rational thinking and experimentation for unsystematic recommendations of new procedures. In a way, Mann postulated the development of what is now known as cyto- and histochemistry. The value of Mann's book is limited considerably by his expressed refusal to examine fresh living and dead material. This means that he abandoned the crucial test for each fixing and staining technique. Belling's book, "The Use of the Microscope" (1930), is written from the point of view of a botanist. It contains an unusual amount of detailed practical advice. I heartily endorse Belling's recommendation for more extensive use of water immersion objectives; this recommendation is also made by Cowdry (1952, p. 205). Comparable to Belling's book is the more modern "Précis de Microscopie" by Langeron (1949). It impresses me as being a book in which practical problems of biological microscopy are handled with remarkable skill; it is one of the few books in which field finders are described. A brief but clear introduction to electron microscopy is included. Both Shillaber's and Langeron's books offer tabular comparisons of the characteristics of objectives manufactured by the leading optical firms. Shillaber's table lists ten American, German, and British firms; Langeron mentions five firms including a French firm. Introductions to special fields of microscopy are found in different articles written by specialists in Glasser's "Medical Physics," Vol. I, 1944, and Vol. II, 1950; microscopy in ultraviolet light, electron microscopy, and many other items are discussed and illustrated adequately. Cowdry's "Laboratory Technique" (1952) gives good descriptions but does not have any illustrations. In McClung's Handbook (1950), some parts contain illustrations, e.g., Bennett's comprehensive article on microscopy in polarized light.

I will limit myself to a brief survey of fundamental features of microscope models and illumination devices. Microscopes can be divided into monocular and binocular types. Many stands are equipped for a change from one to the other. The term monocular microscope refers to the presence of one tube. At a given moment one objective and one eyepiece are used with the tube. The term binocular microscope refers to models with one or two objectives. In the first type the picture is produced by one objective and then split by prisms into two pathways which lead to two eyepieces. This type produces an inverted image of the object. In the other system, two objectives are mounted on two separate converging tubes, with an eyepiece on each. Here the prisms are arranged in such a way that the observer sees the object upright. This type which is known as a dissecting microscope is used for magnifications not more than approximately 30 times. The purpose of binocular microscopes is to provide stereoscopic vision. This will be discussed in connection with three-dimensional morphology (Part III, Chapter 6, B, 1). Monocular microscopes are necessary for the projection of pictures on a screen or photographic emulsion. This includes photomicrography and spectromicrography. Studies in polarized light also require monocular microscopes.

The light which enters the objective of a microscope is either reflected from the surface of the object or transmitted through the object. Obviously transmission of light is possible only through transparent objects, whereas reflection can be obtained from the surface of both opaque and transparent objects. Light of any color or wavelength can be procured by means of special lamps, solid and liquid filters, or monochromators. Polarized light was mentioned previously. If ultraviolet light of medium or short wavelengths is to be used, glass must be replaced by quartz throughout the path of the light. Structures which are to be analyzed in ultraviolet light can be made visible indirectly, either by fluorescence or photography. Dark-field illumination and phase contrast microscopy have extended the use of the light microscope considerably. (For relative merits of dark-field and phase contrast microscopy see Bessis, 1956, p. 36.) Finally, the electron microscope has added new orders of magnitude to the kingdom of structural analysis.

Microscopic visibility depends on resolution and contrast. Resolution refers to the ability to recognize two structures as two distinct entities; for instance, two dots as two dots. It is the amazing resolving capacity of the electron microscope which distinguishes it from light microscopes. Contrast refers to any optical differences between adjacent structures, such as differences in color or illumination. In a histological section which is enclosed between a slide and a coverglass, only those structures are visible whose refractive index is sufficiently different from the mounting media, such as air, water, or resins. If an unstained section has been dehydrated and is mounted in Clarite between slide and coverglass, the section may be practically invisible since the refractive index of the dehydrated section is close to that of Clarite. If a similar section has been mounted in water between slide and coverglass, many structures may be clearly recognizable in the section. However, when high resolving power is required, Clarite is superior to water as a mounting medium.

The preparation of biological material for microscopic study consists of sampling, fixation, embedding, sectioning, staining, and mounting. The elements of these techniques will be described briefly in the present chapter. Detailed discussions will be devoted to certain items which are taken for granted in most manuals and handbooks on microscopic techniques. One of these items is sampling for microscopic study.

Some microscopic procedures which are used for the study of dead material have been mentioned previously when the importance of continuity between macro- and microscopic procedures was emphasized, and when conflicting requirements of a postmortem study were discussed. The influence of the size of an organ on the technique of preservation was described in detail. Some of these subjects will be rediscussed here.

b. *Sampling.* If the autopsy of an animal is carried out in a warm room, the time between beginning and end of the autopsy may be important, since the last organs placed in fixatives may suffer from autolysis (decomposition after death). Obviously the time used for an autopsy is less critical in a cool room.

Mistakes in the selection and handling of samples for microscopic study can rarely be corrected. The danger of such mistakes increases with the size of animals. As mentioned before, organs of small animals such as mice are frequently preserved as a whole, and microscopic sections may represent the whole organ at least in one plane. This is not the case with the large organs of a dog. In the study of human corpses or those of fairly large animals, the safest procedure is to remove samples for microscopy at autopsy and to preserve the remaining organ as a whole. If, after study of the original samples, it becomes desirable to examine additional samples, the preserved organ is available. This can be of great value although the fixation of the tissues may not be as good in the stored organ as in the first samples.

Sometimes it is desirable to incorporate an organ into a museum. If this is anticipated, the removal of samples for microscopy should not interfere more than necessary with the important features of the exhibit. There was a period in human pathology when the beauty and intact appearance of museum specimens took precedence over verification by microscopic sections. Therefore the labels on older jars in museums are not always reliable. The modern pathologist does not mind a visible sampling defect on a museum specimen, since such a defect indicates proper microscopic confirmation of the diagnosis. Studies of changes in frequency of pulmonary cancer or influenza pneumonia should not be based on old museum specimens unless the diagnosis is supported microscopically. If sections or records of histological diagnoses are not available, the museum jars should be opened and samples taken for microscopic study, which is perfectly feasible with many old specimens.

The labeling of samples should follow the rules which are observed in chemistry, bacteriology, and other branches of science. Whenever possible, a container (not its cover!) should be labeled before receiving the specimen. The morphologist may be tempted to postpone the labeling of a specimen if its appearance seems to be characteristic enough for identification. This is fallacious, since a sample which looked characteristic when it was placed in a jar, may be followed by a similar sample some hours later. In surgical pathology the postponement of labeling can be catastrophic. Suppose two patients have undergone resection of their rectums because of cancer, and the specimens are sent to the pathologist in order to ascertain microscopically whether the tumors are well inside the cut edges of the specimens. If the accompanying notes of the surgeons are not clearly attached to the respective rectums, confusion of the two is possible and the microscopic diagnosis may be useless.

Sometimes it is necessary to tag a surgical sample so that its orientation in the body is not lost after removal. If a piece of the large intestine has been resected because of cancer, the isolated segment which is sent to the pathologist has to be tagged in such a way that the end toward the cecum can be differentiated from the end toward the anus. In case the pathologist finds that the cancer microscopically extends into one of the two ends of the resected piece, this observation does not help in planning further treatment unless one can tell which end is involved. At autopsies the samples taken from the central nervous system need tags with respect to orientation. As long as the spinal cord is intact, left and right can be distinguished. However, when a small cylinder is removed, one can no longer recognize which cut surface of the cylinder pointed toward the brain. Consequently, it will not be possible to tell left from right. Suppose during the life of the patient or animal, certain unilateral disturbances had been observed. Then, expected correlations with structural abnormalities can be verified only if the orientation is preserved. In other words, the left and right sides of such a cylinder are to be tagged at the moment of removal from the spinal cord.

Distortion of topographic relations between different organs or tissues can

---

FIG. 24. Sampling of abdominal wall of a rabbit with preservation of topographic relations between skin (*1*), subcutaneous fat tissue (2), and muscular layer (3). Purpose of study: determination of visible effects of subcutaneous drug injection. Sample taken 8 days after injection. (Mayer, 1944, unpublished data.)

(a) Dotted oval: place of subcutaneous injection. *S*, sutures made through the whole thickness of abdominal wall to prevent dislocation of layers in injected area, *P*, places for pins by which the preparation is mounted on a corkboard for fixation in formalin. Note unequal retraction of (*1*), (2), and (3).

(b) Cut surface of central portion of (a) after fixation. Pins (black) and sutures (*S*) still in their places. *V*, blood vessel filled with clotted blood.

(c) Microscopic section parallel with cut surface of (b). *V*, vein with clotted blood, *G*, areas of granulation tissue which indicates healing phase following local damage (necrosis). Note: inflammatory processes, indicated by dots, are found between the bundles of muscles in (3). Diagram made from a photomicrograph magnified $\times 42$.

easily occur when parts are removed from their natural connection. An example may illustrate how problems of this type can be resolved. The task was to study the effect of high concentrations of a sulfa drug when injected under the abdominal skin of a rabbit (Mayer, unpublished data, 1944). The abdominal wall consists of skin, subcutaneous tissue, and muscle. What type of injury would take place in the subcutaneous tissue? Would the injury involve deeper

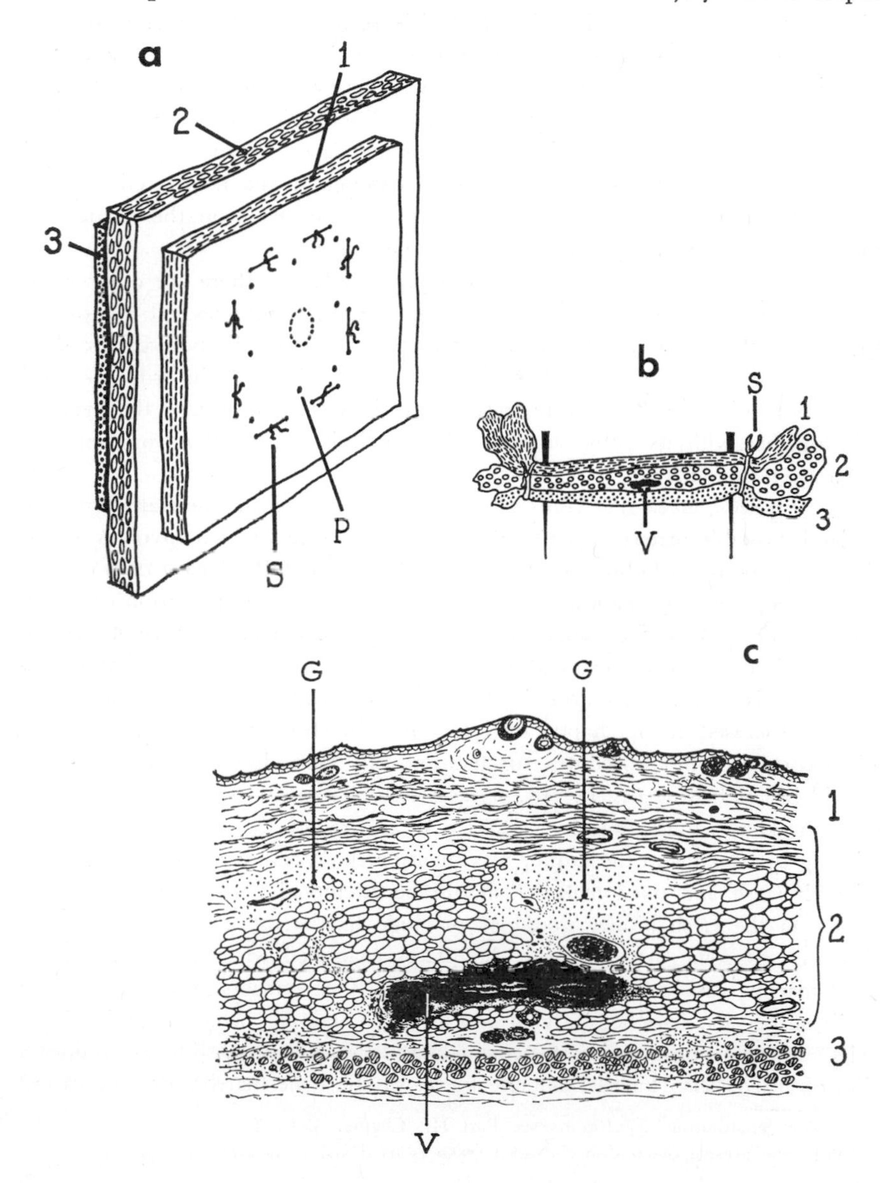

layers such as the muscles? To remove that part of the abdominal wall which has been injected would lead to gross distortions of the relative positions of skin, subcutaneous tissue, and muscle, since these three layers retract differently when cut. In order to avoid this, sutures were made around the injected area so that the threads went through the whole thickness of the abdominal wall. Outside the sutures incisions for removal were made (Fig. 24a). The resulting piece was pinned on a corkboard and fixed in formalin. After fixation, sections vertical to the surface showed the three layers in their proper relationship, and the damaged area in its proper place and with its proper limits (Figs. 24b and c). Maximow (1906) described a slightly different procedure for sampling the loose connective tissue between the muscular layers of the abdominal wall without distorting topographic relations. Through a slit he placed a small frame of cork under the abdominal wall, pinned all layers on the frame, and then excised that piece which was held by the frame.

Besides the task of maintaining topographic relations there are other problems of sampling which are related to the fixation techniques. It is possible to remove, at autopsy, the fresh hypophysis of a rat from the base of the skull and to place the isolated hypophysis in the fixative. However, in order to avoid mechanical injury to the delicate fresh organ, it is safer to remove the hypophysis together with its bony support, place both in the fixative, and separate them after fixation.

c. *Fixation; smears, sections.* In biology, fixation means the transformation of fresh unstable material into stable material, a process which involves alteration of proteins, including coagulation.[10] Regaud and Policard (1913) made a distinction between fixation of morphological structures and fixation of chemically defined substances. Since these two purposes of fixation have shown increasing interrelations, they will not be separated in the present chapter. If the emphasis is definitely on the determination of chemical properties, the subject will be discussed in the chapter on cyto- and histochemical procedures.

Before selecting procedures for fixation, one must decide whether fixation should be applied at all. In certain cases examination of fresh material[11] supplies the key to the problem on hand. As an example of this type, the study of drug precipitates in the kidney was mentioned. Whenever possible the study of both fresh and fixed material should be attempted. Study of drug precipitates in the fresh kidney does not preclude fixation of parts of the kidney that were at some distance from the places examined when fresh. Similarly, bone marrow can be divided in various ways. One portion may be used as a moist suspension for the study of motility and staining behavior in surviving cells. Another portion of a suspension may be spread on a slide to produce a film of dried cells. Finally, suitable pieces of the bone marrow may be placed

[10] For denaturation of proteins see Part III, Chapter 3, C, 2.

[11] In the present discussion the term fresh is used synonomously with unfixed.

in fixatives for embedding and sectioning. The study of fresh material can cover the range from low magnifications under the dissecting microscope to high magnifications under the phase contrast microscope. Since the use of the latter has been described sufficiently in many recent publications, I will give an example of the use of low magnifications. In order to examine the degree of calcification in the so-called epiphyseal line of a young rat much information is obtained by splitting a fresh femur or tibia lengthwise and examining the cut surface under the dissecting microscope in reflected light. Calcification can be tested by scratching with a needle. It is difficult to estimate calcification in fixed and stained sections, since at least partial decalcification is necessary to allow the use of a microtome knife.

As mentioned earlier, small objects can be fixed either by immersion in the fixative or by injection. Large organs need injection to obtain satisfactory fixation.

In most cases fixation is one of the steps which precede sectioning. However, suspensions of isolated cells such as bacteria, protozoa, blood cells, or cells exfoliated from the linings of body cavities can be fixed and examined without sectioning, by spreading a drop of the suspension on a slide or coverglass. For the study of protozoa (Chatton and Lwoff, 1936) the organisms are fixed and suspended in agar or in gelatin; then thinly spread drops or films of the suspensions are placed on slides for subsequent consolidation and examination. For most other purposes, cells are suspended in body fluid (blood, lymph) or saline when spread on a slide or coverglass. With this technique, fixation is accomplished by evaporation and drying. The evaporation is usually accelerated by gentle heating. If a fresh drop is spread into a thin film by means of a second slide or coverglass, the preparation is called a smear. Instead of fixation of a moist smear by drying in air, one can use alcohol and ether for dehydration and fixation. There are marked differences between the appearance of stained blood cells in a smear and in a section (Figs. 53 and 83).

Chemical fixatives as well as heating are liable to alter and dislodge components of living cells and tissues. To avoid this, two types of procedures were designed in which fresh samples are frozen as rapidly as possible. The first type comprises freeze-drying and freeze-substitution. In freeze-drying (Gersh, 1932) the frozen samples are dehydrated in a vacuum at temperatures below $-30°$ C., whereas in freeze-substitution chilled alcohols or glycols are used to replace the water (Blank *et al.,* 1951). Finally, the blocks are embedded for sectioning. Although interesting cyto- and histochemical results were obtained by these techniques, they proved to be not only very cumbersome, but also burdened with the uncertainties connected with dehydration and embedding. Procedures of the second type, known as cold knife and cold microtome (cryostatic) methods, allow transfer of sections, without thawing, from the cold knife to a cold slide. In the cold knife technique (Schultz-Brauns, 1929) an ordinary freezing microtome is used at room temperature, but the blade of the knife is sprayed with an additional CO_2 jet. With the cryostatic method, microtome and knife are enclosed in a refrigerated box. This method was developed by Linderstrøm-Lang and his associates (1934) for the study of enzymes. Details will be given in the discussion of quantitative histochemistry (Part III, Chapter 7, D, 4). The cryostatic method is relatively simple. Depending on the purpose, eventually the sections obtained may or may not be dehydrated.

The freezing-drying and cryostatic methods were developed on a rational basis in the modern era of histology and histochemistry. In contrast, the enor-

FIG. 25. Effect of two different fixatives on a dog's thyroid gland, which is in the storage phase. Diagrams made from photomicrographs (E. Mayer, 1947). Approximate magnifications: A and C, ×150; B and D, ×300; nuclei not drawn to scale. (A) and (B), Formalin; (C) and (D), Bouin's fluid (picric acid, formalin, and acetic acid). Shading represents colloid which forms a coherent ball in (A), whereas it is fragmented in (C). This indicates high concentration of colloid. Note: Lining cells of alveoli are extremely shallow in (A) and (B), thus accentuating the storage phase.

mous number of traditional chemical fixatives has a more or less empirical background. Fifty of the most common fixatives are listed in Cowdry's "Laboratory Technique" (1952). The list includes formalin, ethyl alcohol, various acids, solutions of metal salts, and a number of mixtures of such fluids known under the names of the inventors. To give some examples: Bouin's fluid consists of picric acid, formalin, and acetic acid; Orth's fluid of potassium bichromate and formalin; and Zenker's fluid of potassium bichromate, mercuric chloride, and glacial acetic acid.

The choice of the fixative depends on the plans for subsequent embedding and staining. If staining of fat is intended, dehydrating embedding procedures must be avoided, because dehydrating fluids are fat solvents. Similarly, fat-dissolving fixatives cannot be applied. To some extent the fixative determines the appearance of nuclei, cell bodies, and proteinic tissue fluids, such as blood and lymph plasma. As an illustration I use the microscopic study of two thyroid glands, one in storage phase and the other morphologically activated.[12] Of each thyroid one slice has been fixed in formalin and an adjacent slice in Bouin's fluid (picric acid, formalin, and acetic acid). After fixation the slices

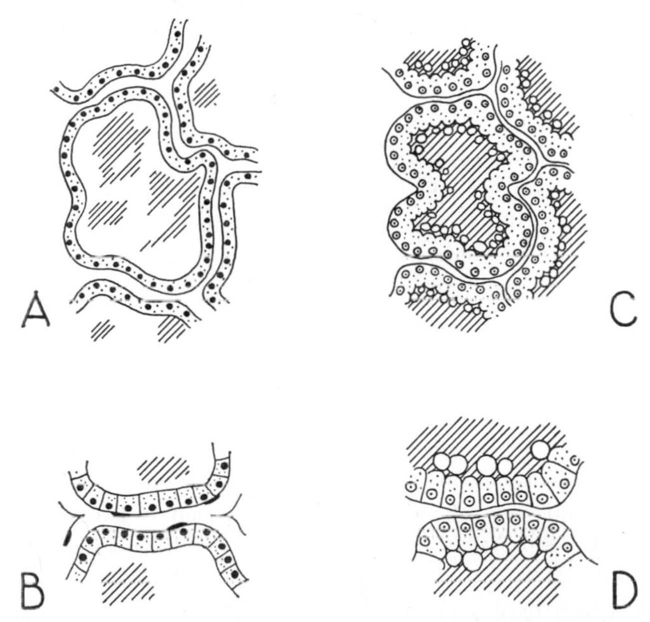

FIG. 26. Effect of two different fixatives on a dog's thyroid gland, which is morphologically activated (functionally activated or inactivated). Diagrams made from photomicrographs (E. Mayer, 1947). Approximate magnifications: (A) and (C) ×150; (B) and (D), ×300; nuclei not drawn to scale. (A) and (B), Formalin; (C) and (D), Bouin's fluid (picric acid, formalin, and acetic acid). Shading represents colloid which forms shreds in (A) and (B), and shows peripheral vacuolization (scalloping) in (C) and (D). This indicates that the colloid is less concentrated than in Fig. 25. Note: Lining cells of alveoli are extremely high in (C) and (D), thus accentuating the morphological activation.

were dehydrated in alcohol, cleared in chloroform, embedded in paraffin, sectioned with the microtome, and stained. Figures 25 and 26 show the results in a diagrammatic way (diagrams based on photomicrographs in Mayer's 1947 paper). Formalin fixation makes the cell bodies appear shallow as compared with their counterparts fixed in Bouin's fluid. This relation is observed both in the storage phase and activated phase. In the formalin-fixed section of the

[12] The picture of morphological activation can be associated with increased or decreased functional activity as described in Part II, Chapter 2, G.

activated gland the cells appear higher than in the Bouin fixed section of the storage gland. The so-called thyroid colloid, a proteinic fluid which fills the cavity of the follicles, coagulates under the influence of fixatives. It is more concentrated in the storage phase than in the activated phase. In the storage phase the coagulated colloid fills the cavity of the follicle snugly after formalin fixation (Fig. 25A and B) whereas some retraction is produced by Bouin fixation. The retraction is indicated by peripheral vacuolization or scalloping (Fig. 25C and D). With formalin fixation the ball of coagulated colloid appears solid, whereas it breaks up into numerous fragments with Bouin fixation. In the activated Bouin-fixed gland, fragmentation of the colloid ball is rare and peripheral vacuolization is most pronounced (Fig. 26C and D). Evidently this is the way in which Bouin's fluid causes a dilute protein solution to coagulate if the solution is contained in a closed cavity. In the formalin-fixed section (Fig. 26A and B) no scalloping is visible since the dilute colloid retracts to irregular shreds with this fixative. The maximal height of cells and the greatest distinctness of cell boundaries is produced by Bouin's fluid in the activated gland (Fig. 26D). The formalin-fixed section of the activated gland (Fig. 26B) shows cells lower than in Fig. 26D, but higher than in Fig. 25D and much higher than in Fig. 25B. In my example the properties of the two fixatives complement each other. It is peculiar that the two fixatives produce changes of the dead thyroid structures which could also be the expression of different conditions during life. When fixed in formalin, the structures change in the direction of the storage phase, whereas fixation in Bouin's fluid modifies the structures in the direction of morphological activation.[13] Partial dislocation of thyroid colloid by diffusion currents of the fixative will be discussed later (Part III, Chapter 3, C, 2).

Fixatives which have a pronounced color have the disadvantage that they must be washed out to avoid interference with some stains. However, they have the advantage that their penetration into the specimen can be easily verified. I mentioned before the problem encountered in fixing the testis of an adult rat because of the impermeability of its capsule (Part III, Chapter 2, B, 1). Figure 27 illustrates what happens if the best procedure is applied, namely, removal of one polar cap. Because of its yellow color the penetration of Bouin's fixative can be readily determined. That portion of the testis which is most distant from the cut surface remained white, a result which meant that it was not fixed.

[13] In his study of guinea pig thyroids Wilcke (1935) described the fact that formalin fixation produced a picture similar to the storage phase while fixation with sublimate plus trichloroacetic acid produced a picture similar to the activation phase. We did not know Wilcke's paper when we observed that, in a certain experiment, one thyroid lobe of each beagle seemed to be in the storage phase, while the other lobe seemed to show activation. It turned out that this astonishing result was caused by our routine of fixing one lobe in formalin and the other in Bouin's fluid (Mayer, 1946, unpublished data).

The middle zone was fixed properly and not damaged by the cutting of the cap. The zone adjacent to the cut surface was properly permeated by the fixative, but it was riddled with holes as the result of the extrusion of seminiferous tubules. Only the middle zone is suitable for histological study.

In some cases the fixation also produces the desired stain. Potassium dichromate and silver nitrate belong in this category. The medullary cells of the adrenal are stained yellowish brown if the fresh material has been fixed in a fluid which contained potassium dichromate, e.g., in Orth's or Zenker's fluid. If the material has been fixed in other fluids such as formalin or alcohol, subsequent treatment with potassium dichromate will also produce the stain. Fixa-

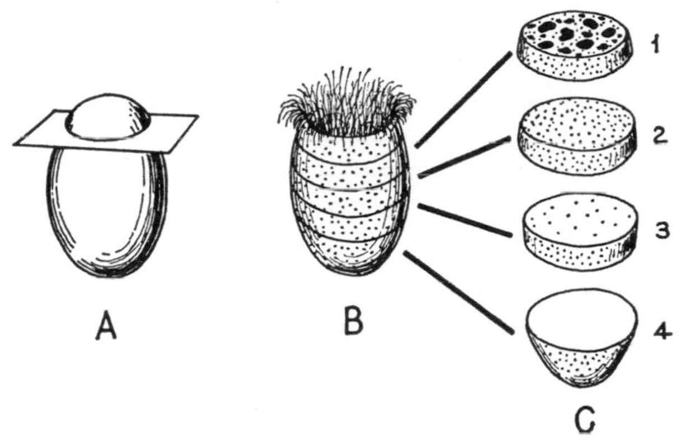

FIG. 27. Compromise technique for fixation of testes in adult rats. The capsule is not permeable to fixatives, and the contents are too soft to be handled in the fresh condition. Use of a colored fixative (e.g., one containing picric acid) allows determination of parts reached by the fluid; they are indicated by dotting. Slightly larger than natural size. (A) One pole of fresh testis is cut off. (B) Seminiferous tubules protrude from the cut surface. (C) Slices at different levels: (1) properly permeated by fixative but riddled with holes where tubules extruded; (2) mechanically intact, portion near slice (1) is properly fixed; (3) and (4) mechanically intact but not adequately fixed. Result: only slice (2) is useful.

tion and staining are also joined in one procedure if fresh tissues are placed in $AgNO_3$. If the mesentery of a small animal is treated in this way, the outlines of the cells which coat the two surfaces are impregnated with $AgNO_3$ and stained black after reduction (Fig. 51a). In a formalin-fixed mesentery, $AgNO_3$ stains the connective tissue fibers while the outlines of the coating cells remain unstained (Fig. 51b).

In studying nervous tissue one selects adequate fixatives for optimal presentation of different structures. Some fixatives are best for Nissl bodies in the nerve cells, others for neurofibrils, some for intact myelin sheaths, and others for myelin sheaths in early stages of disintegration.

The preservation of special, chemically defined substances may require special fixatives. It was mentioned before that staining of fatty substances is possible only if fat solvents have been avoided during fixation and embedding. Osmium tetroxide is the only liquid fixative which stabilizes fatty substances in such a way that they are no longer dissolved by fat solvents. This permits the study of fatty substances in paraffin-embedded material. Osmium tetroxide also fixes protein structures as other liquid fixatives do. It is to this dual capacity that Zeiger rightly ascribes the superior ability of osmium tetroxide to preserve delicate structures irrespective of subsequent dehydrating and embedding procedures (Zeiger, 1938, p. 23). Disadvantages of osmium tetroxide are its poor penetration and its interference with most staining techniques. Glycogen or calcium deposits, of course, can be stained only if they have been preserved throughout all preceding steps. Some of these points will be mentioned again in the chapter dealing with histo- and cytochemistry. The pH of the fixative has a definite effect on the result of some staining procedures (Petrunkevitch, 1937). Possible dislodging of structures by the diffusion of fixatives will be discussed in the chapter on artifacts. Diffusion currents of fixatives are avoided when the sample is exposed to vapors instead of being submerged in or perfused by fluids. Vapors of osmium tetroxide or formalin can be used for the fixation of smears of isolated cells and also for the fixation of very thin slices.

A peculiar competition of procedures is to be faced in the study of bone marrow in the long bones of small animals. Fixatives cannot reach the marrow unless its bony case is decalcified, but the process of decalcification is relatively slow, so that in the meantime the marrow is subject to postmortem decomposition. A procedure which allows sectioning of whole femoral marrows of rats and guinea pigs without decalcification has been described by Mayer and Ruzicka (1945).

d. *Embedding and the sectioning of embedded material.* Fixation of biological samples is usually followed by so-called embedding procedures. These procedures serve two purposes. The first purpose is to endow the sample with a consistency which is suitable for sectioning. The sample should be resistant enough to be mounted on a wooden block and clamped in such fashion that it will not be displaced when touched by the microtome knife.[14] Although fixation improves the consistency of fresh samples, additional freezing or embedding is necessary for cutting thin sections. The second purpose of embedding is to maintain the topographic relations of all parts in thin sections, and particularly to prevent the loss of such structures which, in the plane of sectioning, are not attached to adjacent parts. Dislocation and loss of parts can be caused by the action of the microtome knife, or by shrinking and expanding

[14] In the early days of microscopic techniques, sections of fresh or fixed material were made with a straight razor. To facilitate the holding of the unembedded sample it was wedged into a piece of elder pith or of amyloid liver (a pathological condition frequent at that time but almost extinct now). These techniques were known as encasing.

effects of staining and dehydrating fluids used after sectioning. The most common embedding procedure is paraffin embedding. The fixed sample is dehydrated by increasing concentrations of alcohol. From absolute alcohol the sample is transferred into chloroform, xylol, or other solvents which mix readily with paraffin. These solvents are known as clearing media because they make the sample more or less transparent, which is helpful for orienting it in the paraffin (for other uses of clearing agents see article "Clearing" in Cowdry's "Laboratory Technique," 1952). From a mixture of a clearing agent and paraffin the sample is transferred into melted paraffin and kept in a 56° C. incubator until permeation with paraffin is complete. Finally the sample is placed in an open chamber containing melted paraffin and is oriented for sectioning. The chamber is submerged in cold water. In this way a solid block of paraffin is obtained which encloses the sample. Such blocks can be stored indefinitely at room temperature. Identification tags must accompany the sample from the fixative to the paraffin block. Prior to sectioning, the paraffin block is trimmed into a rectangular shape and mounted on a carrier of wood or other hard material by means of melted paraffin. Then the carrier can be held by steel clamps on the table of the microtome.

The table of the microtome advances automatically toward the plane of the knife after each section so that sections of identical thickness are obtained. Either each single section is picked up from the knife, or a number of sections are allowed to stick together by their paraffin edges, thus forming a ribbon of paraffin. In this ribbon the tissue sections are located at regular intervals and in their proper sequence. Serial sections of this type are necessary to follow a microscopic structure through three dimensions.

The paraffin sections, single or as a ribbon, are placed on the surface of a warm water bath to remove folds. From the bath they are transferred to glass slides which, as a rule, are coated with a film of glycerin and egg white. Sections affixed in this way to the slides can be deparaffinized safely by xylol and subsequently transferred to the different solutions needed for staining and final mounting under a coverglass.

For certain purposes it is preferable not to affix the section to a slide prior to staining. By means of a glass rod single sections can be transferred from one fluid to another. This is done routinely with frozen sections, and for special purposes with paraffin sections. Sections handled in this way are called floating sections. They are subject to mechanical injury and loss of components. Their advantage is that, if shrinkage occurs, the original ratio of spaces to structures will be maintained, whereas this ratio is frequently distorted in affixed sections.

In order to illustrate the possible loss of structures from sections, the diagrams of thyroid sections, Figs. 25 and 26 will be used again. The homogeneous balls of colloid seen in Fig. 25A adhere to the follicular walls and therefore might not be lost in unembedded frozen sections or in floating paraffin sections that are deparaffinized and stained prior to mounting. In Fig. 25C the colloid ball had been fragmented; obviously most of the fragments would have been lost if the material had not been embedded and the paraffin section had not been affixed to a slide for staining and mounting. Similarly, the shreds of colloid seen in Fig. 26A would certainly have been lost without embedding. The colloid with marginal vacuolization (scalloping) seen in Fig. 26C might stick to the follicular walls in thick sections but would probably be lost in thin sections. This point is also clearly illustrated in Fig. 40d, which shows a diagram of a renal corpuscle. The ring-shaped cross section through a protruded tubule (*P*) would probably have been lost if the section had not been embedded and affixed.

The completeness of permeation with paraffin depends on the size and texture of samples. The distribution of paraffin crystals around the different histological structures can be determined in polarized light as shown by Richards (1944, p. 755, Fig. 8).

Paraffin embedding can be handled in various ways ranging from mass production to elaborate individual care for each single sample. For mass production, efficient apparatus are available which consist of a timer, a series of containers, and cranelike devices which transfer the samples from one container to the next. Thus dehydration, clearing, and embedding are executed in the proper sequence and at intervals set by the timer. Mechanical procedures of this type are useful in hospital laboratories where many biopsies have to be processed within a short time. As a counterpart I mention the careful manipulation of individual samples in studies of Heuser and Streeter (1929; techniques given in detail by McClung, 1950, p. 168).

In order to investigate early stages in the development of pig embryos, Heuser and Streeter fixed and dehydrated the material in small glass dishes with frequent and very gradual change of solutions. Instruments heated just above the melting point of the paraffin were used to orient the specimens before the paraffin was allowed to solidify. Tools and specimens were observed under a dissecting microscope.

The general principles of embedding techniques have been covered by my description of the paraffin method. The reader of the present book will now be in a position to understand other embedding techniques as well. Gelatin embedding is a procedure without dehydration. It is used for frozen sections. Favorable material can be cut into gelatin sections as thin as 5 μ. In many cases cellodin embedding competes with the paraffin technique. The paraffin technique has the advantage that ribbons can be cut for serial sections and that sections as thin as 1 μ can be obtained. When electron microscopy made it necessary to cut sections as thin as 0.1 μ, special embedding media had to be designed. I mention *n*-butyl methacrylate, a thermoplastic material introduced for this purpose by Newman *et al.* (1949). The sample is placed in the liquid monomer. By means of a catalyst and incubation, the monomer is polymerized so that the sample is embedded in a solid matrix.

The problem of making ultrathin sections for electron microscopy led to interesting developments besides the introduction of more suitable embedding media. Microtome knives of steel had proved satisfactory in making sections of 1 or 2 μ thickness; therefore, the improvement needed to make thinner sections seemed to be higher speed of cutting. Microtomes were constructed in which the knife was carried by a rotating arm. The knife-bearing wheel made 57,000 revolutions per minute, and the cutting speed of the knife was 1100 feet per second (Gessler and Fullam, 1946). However, the cutting problem was solved in an entirely different way. Glass fragments with a sharp edge proved superior to any steel knife (Latta and Hartmann, 1950). Instead of the enormous speed of the steel knife, the glass fragment moves at a slow rate. The condition of the glass edge and its movement through the sample are observed under a microscope. This technique was improved by using diamond instead of glass fragments (Fernández-Morán, 1953). The advance of the block which carries the specimen needed a finer control when sections of a uniform thickness of 0.1 μ were to be obtained. It has been possible to reduce the advance of the block to approximately one-tenth the advance on traditional microtomes, either mechanically or by thermal expansion after each section. My description is based on the 1953 paper by Porter and Blum. Sections for electron microscopy are mounted not on glass slides, but on fine wire-mesh screens.

Whereas ultrathin sections of a very small area are needed in electron microscopy, there are certain morphological problems which require particularly thick sections (Fig. 43a, b) and large areas. Golgi's and Cajal's classic methods of silver impregnation of nerve cells are intended for sections 30–100 μ thickness. The regional histology of the brain was developed in such sections (see Fig. 49a and b). Sections through a whole human brain were made of celloidin-embedded material with the use of giant microtomes. For various histotopographic purposes, Christeller (1927) made frozen sections, without embedding, through whole human organs. For the study in reflected light of whole human lungs, Gough and Wentworth (1948) designed a technique in which gelatin-embedded frozen sections are cut at a thickness of 400 μ; such sections are mounted and stored on sheets of paper (for description of technique see also Carleton, 1957). These items will be mentioned again in a chapter which deals with topographic morphology. Hard material, such as teeth or bones, can be ground to thin slices suitable for microscopic study.

e. *Staining procedures and other techniques for increasing contrast.* In biological microscopy staining serves two purposes: to increase the optical contrast between different structures, and to provide information on the chemical or physical nature of structures. In most cases the two purposes are interwoven.

Simple inorganic reactions which produce a characteristic color may be applicable to histological sections: iron in pigment granules can be demonstrated by the Prussian blue or Turnbull blue reaction. Whenever staining procedures were invented to increase the contrast between adjacent structures, sooner or later attempts at chemical interpretations of the staining processes followed. Carmine and hematoxylin were introduced to make nuclei more conspicuous than they are in unstained preparations. To accentuate the contrast, the cytoplasm (cell body) was counterstained with picric acid or eosin, respectively. These empirical discoveries suggested that different biological structures may have different affinities to the various dyes. Ehrlich (1879) introduced the classification of dyes as basic and acid dyes, and of substances as basophilic, acidophilic, and neutrophilic substrates. Since most dyes are neutral salts, the terms basic and acid refer to the radical (anion or cation) with which the staining power resides. There is a tendency of some substances to attract the acid radicals of acid stains, and of other substances to attract the basic radicals of basic stains. Thus, a classification of biological substrates according to their staining behavior is justified if far-reaching chemical implications are avoided. Various items of cyto- and histochemistry will be mentioned in a later chapter.

Conn's handbook (1946) is a comprehensive survey of biological stains, including their history and theory. Among the numerous reviews on this subject, that of Singer (1952) is devoted to one of the most important and difficult areas. It deals with the factors that control the effect of acid and basic dyes on protein-like material in tissue sections. Much insight into these factors had been gained by the procedure of Singer and Morrison (1948). The substrates which they used for their staining experiment were not tissue sections containing a mixture of unknown protein-like substances, but a simplified model, consisting of a film of well-defined protein, pure fibrin. Figure 28 shows that within a pH range above the isoelectric point of fibrin (near pH 6), increasing alkalinity of the staining solution produces more binding of the basic dye, methylene blue. In a pH range below the isoelectric point of fibrin, increasing acidity of the staining solution causes more binding of the acid dye, orange G.

Stains for fatty substances work in a special way. Dyes such as Sudan III are more soluble in fat than in alcohol. If sections which contain fatty structures are placed in a solution of Sudan III in 70% alcohol, the dye moves from the alcohol into the fat; the fat is not washed out by limited exposure to dilute alcohol. This is a purely physical staining process. Physicochemical principles of staining are presented in Michaelis' 1920 analysis of the state of general theories of histological stains. He distinguished three mechanisms of staining: ionic reaction between two solutions resulting in a colored salt; adsorption of a dissolved dye on more or less solid material, such as carbon, without any exchange of ions; and finally, adsorption of a dye solution on solid material of the kaolin type, with ion exchange and fixation of the salt on the solid partner. Other valuable discussions of physicochemical problems of biological stains are found in Zeiger's 1938 monograph.

The use of biological stains profited in two ways from the development of industrial dyes for textiles, leather, and paper: a great variety of synthetic organic dyes became available and progress was made in the theory of staining. Readers not familiar with industrial dyes are referred to publications by Royer *et al.* (1945, 1947); they contain instructive illustrations and discussions of surface staining and penetrating staining of natural and industrial fibers, and touch upon other important aspects of staining.

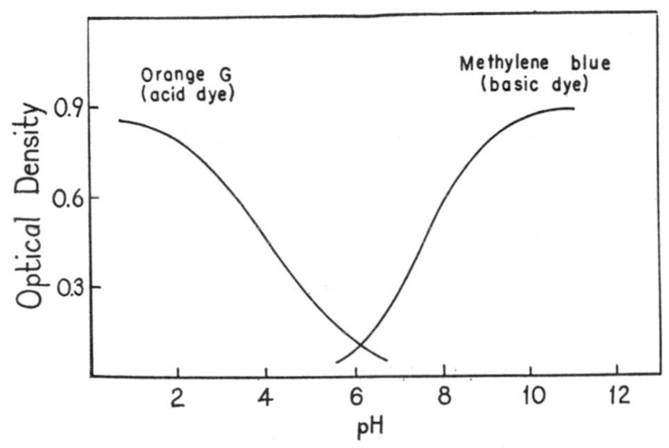

FIG. 28. Staining of a film of pure fibrin with an acid dye (orange G) and a basic dye (methylene blue), as controlled by the pH of the staining solution. The formalin-fixed fibrin film serves as a simplified model of ill-defined protein mixtures in histological sections. The isoelectric point of fibrin is close to pH 6. The optical density is directly proportional to the amount of dye bound (moles of dye per milligram of fibrin $\times 10^6$). Graph supplied by Dr. M. Singer, 1959.

The present chapter will deal with some elementary principles of traditional staining as used in dead biological structures. In a later chapter, staining procedures will be discussed again in connection with chemical identification and microscopic localization of material in tissues and cells (Part III, Chapter 7).

A suitable example of traditional procedures is staining with hematoxylin, a bluish violet, water soluble, natural dye, which was introduced by Waldeyer in 1863. If a section is placed in hematoxylin, the nuclei stain first, but gradually the stain extends to the cytoplasm (cell body), muscle fibers, and other structures. Finally the section appears more or less uniformly stained. If, after staining, dilute acid (e.g. 2% HCl in 70% alcohol) is applied to the section, the different structures lose their stain in a sequence inverse to that of taking it: those which stained last are destained first. Consequently, the nuclei retain the stain when most other structures have lost it. This process is known as differentiation. Contrast between nuclei and other structures can be obtained in two ways. Either the period of staining is increased until the nuclei have become sufficiently distinct while the other structures are stained only lightly if at all (progressive staining); or all structures are stained maximally and differentiated subsequently, as described above (regressive staining). To observe the process of differentiation at high magnifications, water immersion objectives are indispensable.

How do we know that the structures which stain readily with hematoxylin are nuclei? The answer is that this staining property is shared by all variations and phases of structures which are reasonably considered to be nuclei. Hematoxylin stains the vesicular nuclei of egg cells, nerve cells, epithelial mosaics, and embryonic tissues as well as the solid nuclei of mature spermatozoa, polymorphonuclear leucocytes, and normoblasts (a stage in the development of red blood corpuscles). Some of these nuclear types are represented in Fig. 61a–d. When nuclei undergo mitotic division, they reorganize themselves into fibrils and little rods (Fig. 61, M1). These structures, known as chromosomes, have a particularly pronounced affinity to hematoxylin, which incidentally accounts for their name meaning "stainable bodies." If nuclei break down into irregular pieces, the fragments also retain their tendency to stain with hematoxylin.

Nuclei can be stained with many other dyes besides hematoxylin: carmine is one of the oldest nuclear stains. Since those carmine solutions that are innocuous fail to stain the nuclei in living cells, and all hematoxylin solutions kill and fix living cells, how can one be sure that fixation and staining have not produced some distorted picture which has little to do with living nuclei?

At the present time several methods are available by which nuclear structures can be made visible in living cells. Phase contrast microscopy shows that the nuclear structures in living cells are by and large comparable to those seen in fixed cells stained with hematoxylin. During the 19th century such detailed observations in living cells had not been available. However, *outlines* of nuclei were identified with sufficient certainty to support the nuclear interpretation of hematoxylin stain. For general relations between living and dead structures and a more detailed discussion of so-called artifacts see Part III, Chapter 3, C.

The next question is: How specific is hematoxylin stain? Bacteria and calcium salts may stain with hematoxylin under the same conditions and with the same intensity as nuclear material does. No confusion is possible if the nuclei show their characteristic membrane and inner network, or if they are seen in mitosis with distinct chromosomes. Unusually dense nuclei, or fragments of nuclei, may resemble spherical bacteria (cocci) or particles of calcium salts. Great regularity in shape, and a uniform size of a few microns in diameter, would be criteria in favor of bacteria. Solubility of the doubtful structures in acid would suggest calcium salts (Figs. 55 and 56) or some unknown substances rather than fragmented or clumped nuclei.

In many cases the application of more than one stain is useful. If the objects of primary interest such as nuclei, or bacilli, have been stained by one dye, the second stain is referred to as counterstain. The most common stain after hematoxylin is eosin. This dye imparts a pink color to the cytoplasm (cell body), including red blood corpuscles, to muscle fibers, collagenous fibers, fibrin, and structureless ground substances. In other words, in sections stained with hematoxylin and eosin the nuclei appear blue and most other structures pink. The hematoxylin and eosin stain, abbreviated H and E, is widely used in normal and pathological histology. There is no doubt that eosin is necessary to show granules of eosinophile leucocytes and some other special structures. In many cases the counterstain with eosin does not add much to the structural contrasts obtained by hematoxylin alone, and in some instances the contrast is *decreased* by eosin. If the medullary cells of the adrenals have been stained yellow by the use of potassium dichromate in the fixative, hematoxylin stain gives these cells an olive green color, a good contrast to the blue nuclei. This contrast is spoiled by additional eosin stain since it makes the bodies of the medullary cells appear muddy brownish red and all other cells of the adrenal pink. Similarly, brown pigment granules are obscured by eosin stain. I am aware that the authority of "H and E" will not be shaken: it seems to be supported by an international conspiracy.

If used in their proper place, dual and triple stains are extremely useful. In the study of blood vessels the combination of a nuclear stain with a stain for elastic fibers is indispensable. Triple stains which show nuclei, cytoplasm, and collagenous fibers in different colors have helped in solving many problems. However, during certain periods in the development of normal and pathological histology the attractive appearance of colorful sections interfered with analytical thinking. Not only students, but also professors, were sometimes unable to recognize histological structures in the unstained condition. It may be well to remember that a remarkable number of histological structures have been discovered without staining.[15] The histologists of the 19th century who introduced carmine and hematoxylin as nuclear stains had been accustomed to identify nuclei in unstained preparations. When the concept of contrast was discussed in the introduction to this chapter, I mentioned the fact that structures in an unstained section may be visible if the section is mounted in aqueous media (water, levulose, glycerin), but not if it is mounted in oily or resinous media (cedar oil, Canada balsam, Clarite). The early microscopists studied their objects in aqueous media.

Besides the staining of the object, there are numerous procedures by which microscopic contrast can be increased. Colored filters are interposed between the source of light and the objective of the microscope mainly for photomicrography, but they can also be used to increase the contrast in direct observation.

[15] My teacher, D. von Hansemann, told me that as an assistant of Rudolf Virchow he was ridiculed by the master for playing with stains.

Photography in ultraviolet light makes contrasts visible which are not seen in ordinary light. Narrowing of the diaphragm, oblique illumination, dark-field illumination and finally phase contrast microscopy are procedures that improve contrast. For an interesting combination of staining and filters as used in the study of industrial fibers, see Royer *et al.* (1954, p. 25, "Optical Staining").

In electron microscopy, the equivalent of staining is known as shadow casting or shadowing. If a stream of evaporated metal hits an obliquely placed smooth surface, the metal will condense on the surface, but obstructions on the surface will be coated on one side only. Suppose a bundle of connective tissue fibers is exposed to the metal particles. Then, the uncoated or shadowed side of each fiber will be more permeable to electrons than the metal-coated side. The result is an increased contrast between two adjacent fibers and, in addition, a gain in estimating depth, i.e., the third dimension (Dempster and Williams, 1952, p. 312). Shadow casting with metals has been combined with the use of removable Formvar casts to produce resistant replicas (see Zworykin and Hillier, 1950, p. 527). Robinow (1956) observed an astonishing amount of detail in electron micrographs of flagellates which had not been shadowed or impregnated with any electron-dense material. It turned out that the houses of these flagellates contained so much iron that their natural condition was extremely favorable for electron microscopy.

I have described staining procedures after embedding and sectioning, since this is the most common sequence. However, staining may precede embedding and sectioning. This sequence is known as *en bloc* staining. Small embryos are stained with nuclear dyes, embedded as a whole, and cut in serial sections. In larger embryos either cartilage or bone was stained selectively, whereupon these components showed distinctly in total mounts after clearing. This method will be mentioned again in connection with three-dimensional studies. Occasionally, *en bloc* staining is also applied outside embryology. Certain techniques used for silver impregnations of neurons (Golgi, Cajal) are based on the treatment of fixed pieces of brain which are several millimeters thick. The block is embedded and sectioned after the silver impregnation.

The same dye may produce different results when applied to living or preserved material. Examples were given previously (Part III, Chapter 2, A, 5). I mention a practical point which is frequently forgotten. A staining procedure which is very satisfactory for direct microscopic observation may not be the most suitable one for photomicrography. Sometimes it pays to prepare special sections for photomicrography.

f. *Mounting.* Most microscopic preparations are placed on a slide for handling, examination, and storage. If such a preparation is covered by a thin layer of transparent fluid and a coverglass, one says that it has been mounted. Mounting media are either water-miscible or resinous liquids. Most resinous media change gradually from a liquid to a solid state, usually as a result of evaporation of the solvent. Typical representatives of this class are solutions of gun dammar, Canada balsam, or Clarite in xylol. Gelatin dissolved in glycerin (glychrogel) is a water-miscible mounting medium which is widely used. It is liquid at approximately 37° C. and is applied for mounting while still warm. When the preparation is kept at room temperature, the glycerin-gelatin becomes semisolid (jellied) after a day. However, unlike resinous mounts it will melt when stored in a warm place or exposed to the heat of a projectoscope.

A permanently liquid mounting medium such as glycerin requires sealing of the coverglass. This is done by painting a frame of paraffin or suitable varnish on the edge of the coverglass and the adjacent surface of the slide so that the space between coverglass and slide is closed.

Mounting serves two purposes: to facilitate the examination of the specimen and to preserve its structure and stains. For examination of unknown material the mounting should be reversible. Water or saline are suitable for study of biological material for short periods. For long-range preservation resinous mounting media are safer, by and large, than the water-miscible ones. Histological stains such as eosin are apt to diffuse gradually from the section into water-miscible media; this happens even with glycerin-gelatin (glychrogel), which gives the impression of a solid state some days after mounting. Some aniline dyes fade in sections mounted in acid Canada balsam, but not neutral balsam or gum dammar. Methylene blue changes into the colorless leucobase by *reduction*. Therefore a section stained with methylene blue is likely to fade, though the stain may persist in parts close enough to the margin of the coverglass to be oxygenated from the air. The Sudan stain of fat drops is apt to crystallize out in the mounting media (glychrogel), which entails destaining of the fat drops; it is difficult to control this disturbance.

The role of mounting media in microscopic observations including photomicrography cannot be overrated. In order to obtain satisfactory results one has to consider the design of the objective, the mounting medium, and the medium between coverglass and objective. If air objectives are used, aqueous or resinous mounting media remove disturbing reflection and diffraction. To utilize the high resolving power of oil immersion objectives, media of the resinous class are required. The mounting in a resinous medium must be preceded by dehydrating and clearing processes. This means that the aqueous staining and differentiating fluids are to be replaced by alcohol of increasing concentration and finally by xylol, toluene, or benzol if the mounting medium is to be Canada balsam, Clarite, or gum dammar in xylol. As mentioned previously, unstained transparent structures may hardly be visible in mounting media of the Canada balsam type; there will be little contrast between adjacent structures if all of them have refractive indexes similar to that of the mounting medium. Lists of mounting media with their refractive indexes are given in various books. I mention Lillie's "Histopathologic Technique" (1954) with one table of resinous media and one of water-miscible media; solvents and other properties are mentioned besides refractive indexes. Shillaber's 1944 book on photomicrography lists some mounting media which are important for metallographic and industrial microscopy.

The coverglass has various functions. If the specimen is to be examined at moderate magnification in a drop of water with a few millimeters' working distance between the front lens of the objective and the surface of the drop, disturbing condensation of water may form on the lens. A coverglass will prevent this. With high magnifications and working distances of two millimeters or less, water immersion objectives can be used which dip into the drop of water covering the specimen. Water immersion objectives can also be used with coverglasses. Then a drop of water connects front lens and coverglass. Another function of the coverglass is to protect the specimen and mounting medium against dirt and mechanical injury. The thickness of the coverglass is an important point in microscopy. The objectives of most microscope companies are corrected for coverglasses from 0.16 to 0.18 mm. thickness. For demanding work with oil immersion objectives, the refractive index of the coverglass should not differ too much from that of the oil. To avoid the problems of coverglasses, stained blood smears are examined mostly without mounting medium and coverglass. This is particularly convenient in routine examination with oil immersion. Covering of blood smears by a resinous mounting medium and coverglass promotes fading of methylene blue-containing stains. However, the stain can be regenerated by oxidation, by mere removal of the coverglass.

Coverglasses facilitate the marking of particular areas or items in microscopic specimens. The marking can be done by scratching with a diamond or by applying India ink with a fine nib. Two methods are available for relocating an item on a slide irrespective of the presence or absence of a protective coverglass. Calibrated mechanical stages permit characterizing any area through the numbers of the abscissa and ordinate of the stage, but they depend on the use of the same microscope for relocation of a particular field. I am partial to the field finder which is a slide of standard size with a network and numbers on it.[16] By substituting a field finder for the slide with the microscopic preparation, preserving the position on the stage, a field can be characterized by the numbers of the field finder. By means of the recorded numbers and the field finder the interesting area can be identified under any microscope.

---

[16] A picture and description of Kazeeff's 1939 field finder is found in Langeron's "Précis" (1949, pp. 287-288). In the United States the Gurley-Lovins field finder is supplied by W. and L. E. Gurley, Troy, New York.

Chapter 3

Interpretation of Morphological Observations

The present chapter will deal first with the fate of living structures, or the chronological interpretation of morphological data. Then the appearance of structures in living and dead material will be compared. This leads to an analysis of the various ways in which procedures can interfere with the purpose of a study. Attempts will be made to define natural and artificial conditions. Chapter 4 will be devoted to natural and artificial units.

Let us begin with the *dynamic interpretation of structural data.* The prerequisite for such interpretations is the transformation of static observations into a chronological sequence. The rules for establishing a chronological sequence are probably known to most morphologists but have never been published.

The fate of living structures, such as organisms, organs, and cells, can be observed directly in some cases, whereas in other cases special procedures are necessary. The special procedures may involve the use of tags to ensure the identity of objects under changing conditions. The banding of migratory birds is a characteristic example of tagging. Without banding it is impossible to determine the routes of migration of the flocks or to verify the return of individual birds to the nesting area. It is possible to reconstruct, without tagging, the migrations of salmon and eel since conspicuous morphological changes, such as increase in size, are correlated with the changing localizations. Individual fish cannot be traced without tagging. Before discussing tagging procedures, I will describe other techniques by which the fate of living structures can be followed. The art of tracing without tracers should be appreciated even in the era of radioactive isotopes.

A. Procedures for Following the Fate of Biological Structures without the Use of Tagging

As an introduction to procedures of this type, I use again the development of the frog's egg and early embryo as illustrated in Fig. 3a–l. Changes in shape and surface structures (Fig. 3a–g) can be seen directly in the same living egg. In order to study changes of the interior structures it is necessary to make sections (Fig. 3h–l). For this purpose one has to kill a number of embryos at different stages. The stages can either by identified on the basis of external criteria, or they can be expressed in terms of the intervals which have elapsed since fertilization, provided the fertilization time is known. In the study of embryos which are enclosed in an opaque eggshell the start of incubation is the zero point. The development of the chick embryo in the egg had been known for a long time. Eggs were opened after different periods of incubation, and in this way the story was put together.[1] On this basis one could conclude that a chick embryo on the fifth day of incubation would be in a more advanced stage than an embryo on the fourth day.

The study of *whole embryonic organs* suggests, in most cases, the *probable* interpretations of time sequence. Let us assume that sections represented by pictures 3j, 3k, and 3l were available without information concerning the age of the embryos from which they were obtained. One would be inclined to interpret 3j as the earliest stage of the three, 3l as the most advanced, and 3k as the transition between the two others. It would appear unlikely that the simple neural tube of 3j should develop into the complicated eye cups with lenses seen in 3l, and from there into the less complicated eye vesicles seen in 3k. However, complicated organs may develop into simpler structures. This occurs in parasites in which the larvae are more highly differentiated than the mature animals.

In embryos of certain invertebrate animals it has been possible to identify individual cells through many stages. By studying the embryos microscopically in their chronological order, one could even follow the cells until they have produced the primordia of various organs. For illustration, see the embryology of rotifers in L. H. Hyman's "Invertebrates," Vol. III, Figs. 49 and 50 (1951).

It is unsafe to reconstruct the *development of tissue cells* solely on the basis of transitional pictures without known chronology. My Fig. 29 shows diagrams of the development of blood cells. In group (I) four transformations of white cells are possible on the basis of nuclear shapes. Group (II) shows granulated white cells in which the number of cytoplasmic granules increases with increasing indentation of the nucleus, which again permits four possible transformations. In group (III) cytoplasmic granulation and nuclear indentation vary in opposite directions; therefore no more than two patterns of transformation seem possible. The three phases of red blood corpuscles illustrated in group (IV) allow only one chronological interpretation, namely, that

[1] "Take twenty eggs or more and give them to two or three hens to incubate, then each day from the second onward till the time of hatching, take out an egg, break it, and examine it." (Needham, "Chemical Embryology," Vol. 1, p. 57, 1931, quoting from Hippocrates 400 B.C.). It seems that direct observation of chick embryos through transparent windows in the shell was not used before the 20th century.

the intact nucleus of (D) is fragmented and extruded as seen in (d), resulting in a corpuscle without any nuclear material (δ). It would be unreasonable to interpret the picture (d) as a corpuscle in stage (δ) being invaded by a nucleus, since populations of naked nuclei are not found in the blood plasma.

These examples were given to demonstrate the difficulty in establishing developmental stages of single white cells with the use of dead preparations. In the literature, I found a paper by Boll (1958) in which microcinematography was used to follow the transformation of an individual immature granulocyte (myelocyte from human bone marrow) into a mature leucocyte. Peculiar forward and backward steps in the development of this cell occurred during the four days of observation.

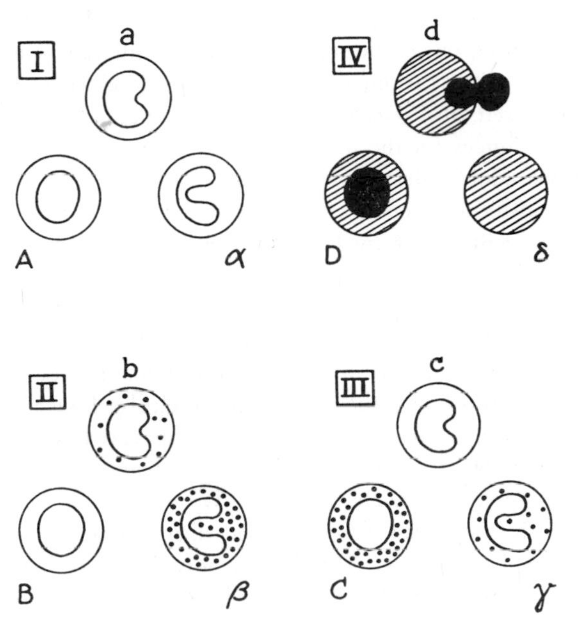

FIG. 29. Developmental interpretations of transitional pictures of blood cells. I, II, III: white cells. IV: red cells.

I. Four transformations possible on the basis of nuclear shapes:

$A \rightarrow a \rightarrow \alpha \qquad A \leftarrow a \leftarrow \alpha$

$A \leftarrow a \rightarrow \alpha \qquad A \rightarrow a \leftarrow \alpha$

II. Four transformations possible on the basis of nuclei and cytoplasmic granules:

$B \rightarrow b \rightarrow \beta \qquad B \leftarrow b \leftarrow \beta$

$B \leftarrow b \rightarrow \beta \qquad B \rightarrow b \leftarrow \beta$

III. Since nuclei and cytoplasmic granules do not vary in parallel, only two transformations are likely:

$C \leftarrow c \rightarrow \gamma \qquad C \rightarrow c \leftarrow \gamma$

IV. Only probable transformation of the pictures shown here:

$D \rightarrow d \rightarrow \delta$

Note: In the maturation of red cells the nuclei may also be dissolved instead of being extruded.

Probable pedigrees of blood cells have been derived from *studies of cell populations* in the blood-forming organs of embryos of different ages. Suppose the cellular composition of the hemopoietic organs changes, along with the development of the embryo, from a pure population type (x) to a mixture of (x) and (y), from here to a mixture of (y) and (z), and finally to a pure cell population (z). Then one may reasonably assume that (x) cells have been transformed into (z) cells through an intermediary stage (y). However, the possibility cannot be excluded that the later types have been produced from new sources unrelated to the earlier blood-forming tissues. New blood cells form all the time in the adult organism, and under pathological conditions great numbers of immature blood cells may form in the most unusual parts of the organism. As long as pedigrees of blood cells depend on the interpretation of transitional pictures, one cannot be surprised that different schools of thought still coexist. One school assumes that all white blood cells develop from a common ancestor cell, the so-called stem cell. Other schools assume two or three different pedigrees for the different mature white cells which can be distinguished in the adult. The relation of the pedigree of the red cells to that of white cells is also controversial (Fig. 85A and A′). There is some hope that more information will be obtained with the use of tagging procedures which will be described later.

Some steps in the development of blood cells are well established. There seems to be no doubt that hemoglobin-containing red blood corpuscles are later stages of cells without hemoglobin, and that the immature cells which contains a nucleus is transformed into the mature corpuscle by loss of the nucleus as shown in Fig. 29, IV. Yet recent papers have appeared with the claim that mature red blood corpuscles can be the product of disintegrating striated muscle fibers without any preceding nucleated stage (Wajda, 1954). The claim is based on pictures which show that fragments of disintegrating muscle fibers can look like red blood corpuscles. This bizarre idea has not been able to shake the generally accepted doctrine that all red blood corpuscles are derived from nucleated forms. I mention the muscle story only to point out where one can be led by an imaginative interpretation of static pictures without being guided by a time table.

Migration and mitotic division have been observed directly and recorded cinematographically in tissue cell colonies *in vitro*. In the superficial layer of blastulas and early gastrulas of amphibian embryos the movements of cell aggregates have been followed after the cells were stained with red and blue vital dyes. This will be described in connection with tagging procedures. At this point let us resume the discussion of studies which, without the use of tagging, revealed the fate of cell populations in the interior of embryos.

Figure 30 shows diagrammatic sections through the spinal cord of chick embryos of different ages (after Levi-Montalcini, 1950). After 4 days of incubation (A), a uniform population of round cells is found in the superficial ventral layer of the cylindrical cord throughout its length. After 5 days (B), the peripheral population has undergone the following changes. Cells have multiplied in two regions (cervical and lumbar), but in one of the two regions

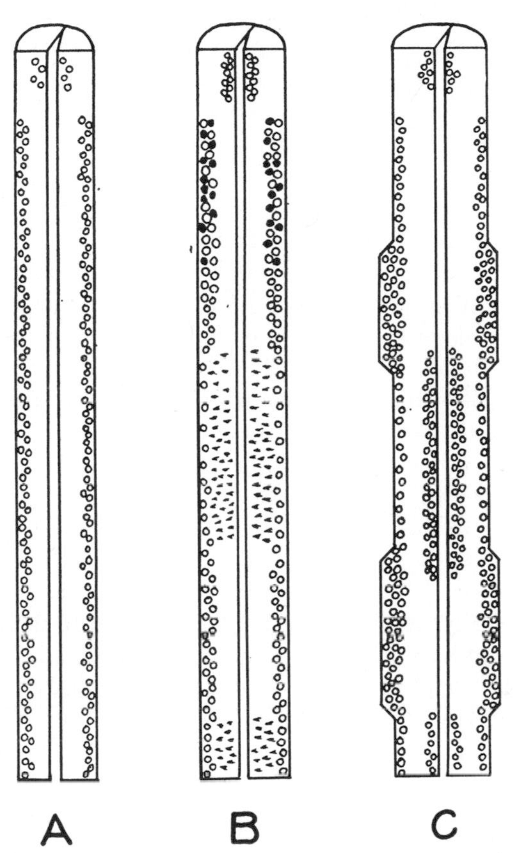

FIG. 30. Fate of populations of embryonic cells followed without tagging procedures. Diagrammatic frontal sections through spinal cord of chick embryo. Parallel lines in center indicate central canal. (A) Four days of incubation: cell population aggregated in periphery. (B) Five days of incubation: peripheral population increased by multiplication but decreased by death of cells and by emigration toward the center. Dead cells are indicated by black dots, migrating cells by stipples. (C) Eight days of incubation: establishment of new population near some regions of the central canal; peripheral populations developed in various ways, depending on regions. The emergence of this structural pattern is connected with the development of different functions. (Modified after Levi-Montalcini, 1950, Fig. 1.)

many cells have died (indicated by black dots in the cervical region). In the region between the two others (the thoracic region) the peripheral population has decreased and the space between the periphery and the center is now filled with a new cell population. This is reasonably interpreted as a migration from the periphery to the center which is completed on the 8th day of incubation (C). Final spatial relations of cells and nerve fibers (not shown in the diagrams) support the interpretation of a migration of cells from the periphery to the central portion of the thoracic region.

The chronological order of the stages (A), (B), and (C) is known in terms of days of incubation. In addition, the volume of the whole cord increases markedly. If the chronological order were not known, the changes in populations, considered by themselves, could also be interpreted in the reverse direction. In the cervical area the small population seen in (C) could be the original one. (B) could again represent multiplication of these cells accompanied by a high death rate, and (A) could be the final stage representing a population somewhat larger than that in (C). Similarly, the migration in the thoracic region could have started from the central population seen in (C) and ended with the peripheral population seen in (A). Here, too, the picture (B) would serve as the intermediary stage. The morphology of the migratory cells in (B) does not indicate the direction of their movements.

In the case described, the developmental migration of cell populations occurred within one organ, the spinal cord. There are instances where populations of embryonic cells originate in one area of the embryo, travel through various regions of the embryo, and settle in a distant place for their final development. A striking example is found in the history of the gonads in the chick embryo. The precursors of the male or female reproductive cells are produced very early in the region from which the head of the embryo originates. When the blood circulation is established, these cells are carried by the blood stream to the primordia of the testes or ovaries which are located in the middle of the body. Here the travelers settle, and thereby complete the construction of the gonads. The unmistakable appearance of the large primitive reproductive cells made it possible to trace their route of traveling (see Willier, 1939, Figs. 6, 7, and 14). Another example of cellular migration across the embryo is represented by the melanoblasts of birds and amphibians. My brief description will be based mainly on the review by Rawles (1955). Melanoblasts are unpigmented precursors of mature pigmented cells which are called melanophores. The melanoblasts originate in a transient embryonic structure near the spinal cord, the neural crest. The fact that the unpigmented cells are precursors of pigmented forms was demonstrated by cultivation *in vitro* of neural crest fragments as well as by implantation of fragments into distant places of an embryo, such as limb buds. In both cases the production of pigmented cells was observed. Ingenious grafting experiments with two different species proved that the melanophores which are eventually found in the skin, and particularly in the feathers of birds, are, without exception, derived from melanoblasts which had migrated all the way from the neural crest. However, the exact pathways of these migrating populations remain obscure since the individual migrating cells lack pigment granules and other distinguishing characteristics (Willier, 1948). These are the limitations of studies of the fate of cell populations with no tags available.

The pigmentation of feathers in fowl can be utilized for various fundamental problems. Weiss and Andres (1952) injected suspensions of embryonic cells of pigmented races into the vascular system of embryos of unpigmented races. The authors reported that the dissociated cells of the donor did not settle at random in the host, but showed regional selectivity related to the places of their natural development in the donor.

The renewal of cell populations in young and adult organisms has been reviewed comprehensively by Leblond and Walker (1956). I mention the systematic studies by Leblond and his associates on the intestinal mucous membrane of rats and mice. The mucous membrane forms villi with narrow tubular spaces between their bases known as crypts (see my diagrams Fig. 33A and B). Leblond and Stevens (1948) observed that the lining cells (epithelium) of the crypts showed more mitotic divisions than the maintenance of the local cell population seemed to require. While counts of mitoses per thousand cells were high in the crypts, they were low on the surface of the villi. This suggested that the lining cells move from the depth of the crypts to the top of the villi, and that the excess is sloughed off at the top. Inspection of the top parts of the villi showed loosened and isolated surface cells. On the basis of experiments with colchicine, which arrests all mitoses in metaphase, the average duration of mitoses in the lining of duodenal crypts was estimated as 1.13 hours. Using this figure, and the fact that without colchicine application 3.07% of cells (on the average) were seen in mitosis, the life span of these cells was calculated to be 1.57 days. The conclusion was that approximately this time was needed by a cell to move from the depth of a crypt to the top of a villus.

These data are given in detail here to show that, under certain conditions, fixed and stained sections can be used not only to establish a chronological sequence, but also to determine the average life span of individual cells and the turnover rate of the cell population. Subsequently, Leblond and his collaborators used radioactive tracers to test their theory of migration of cell populations from the crypts to the top of the villi. This will be described later.

The reconstruction of chronological sequences is of practical importance in the study of parasitic protozoa and helminths. Although the life cycles of many parasitic worms are very involved, they have been clarified with remarkable success. A classic example is the work of Looss on the history of the hookworm *Ancylostoma duodenale* (see e.g., Looss, 1905). The mature stage of the larva, which resembles a threadlike worm, bores its way into the skin of the host, enters the capillaries and veins, travels through the right heart into the lungs, moves from the pulmonary vessels into the air spaces (alveoli) of the lung, travels through bronchi and trachea to the larynx, and crawls over the edge of the epiglottis into the esophagus. Here it is swallowed and thus transported through the stomach until it reaches the first segment of the small intestine, where it attaches itself to the mucous membrane. The worm feeds on the components of the mucous membrane and develops into the final sexual stage. The females produce eggs which are discharged with the feces of the host. In moist soil the eggs develop into larvae which are able to penetrate into the skin of new hosts.

How was it possible to unravel this life cycle, and particularly the fantastic migration from the skin to the intestines of the host? First of all, Looss had accidentally infected himself through the skin of his hands. Then he found that he could use for study an *Ancylostoma* which is a parasite of dogs, but very similar to that of man; dogs could be killed at different intervals after experimental infection. Finally, this work was enhanced by two factors which facilitate studies of parasitic helminths in general: a worm can be recognized readily in any part of the host, and the worms become larger and more mature during their migration. The migrating hookworms acquire oral grinding structures upon reaching the small intestine. Therefore, the morphological changes of the worms help in establishing their route.[2] Readers interested in this type of procedure are referred to a recent well-documented paper by Sprent (1956) dealing with the life history and development of roundworms in the cat.

Helped by a number of favorable conditions, investigators were able to follow the fate of minute parasitic protozoa, such as malaria plasmodia. Blood sampled from the host at different times contained morphologically different populations of malaria parasites. The correct time sequence of these differences could be derived from two circumstances: (1) the great precision of the inner clockwork expressed in the life cycle of the plasmodia, which allowed reproducible sampling from the blood at fixed intervals, and (2) the availability of an external clockwork, consisting of periods of increased body temperature of the host, which coincided with particular phases in the life cycle of the parasites.

In contrast to helminths and malaria parasites, the fate of bacterial populations in the host is elusive; the absence of visible changes in individual bacteria frustrates the establishment of any time sequence. Some types of bacilli produce spores when the population is past its maximal rate of multiplication. The ability to retain stains after treatment with acids seems to increase with the age of tubercle bacilli. However, both the spore formation and the acid fastness are modified by environmental factors, and therefore hardly permit conclusions with respect to the fate of the population in the host. As a consequence, it is also difficult to determine the time sequence of responses of the host to the presence of bacteria. If bacteria are injected into experimental animals, at least the portal of entry is known, although it remains difficult to follow the fate of the injected population of bacteria. In infections of human beings the portal of entry is frequently unknown or hypothetical. If hemolytic streptococci are found on

[2] As mentioned before, morphological changes parallel with age made it possible to reconstruct the migrations of eel and salmon.

the surface of the tonsils, they may be on their way from the oral cavity to the interior of the tonsil and subsequently into the blood circulation, or they may be travelers who left the blood circulation and appeared on the surface of the tonsils. Blood samples of a patient may show certain bacteria on one day, may be sterile the next day, show the bacteria again on another day, and so on. It is very difficult to determine hiding places of bacteria in the host, to find the conditions under which they leave the hiding places, to ascertain why they appear in the blood circulation or in special organs, and why they disappear. One of the first puzzles along these lines was presented when the removal of the spleen from apparently healthy rats produced the presence of numerous *Bartonella muris* bacilli in their blood, whereas such bacilli seemed to be absent from the blood or organs of rats with spleens. Subsequently, at rare occasions, small numbers of these bacilli were found in the blood of rats with spleens. One may conclude that the presence of the spleen keeps the number of bacilli in the blood too low to be detected (Wigand, 1958, pp. 31 and 32). There is still the alternative of unknown hiding places of the bacilli in some organs. It was observed that the organs of rats and mice which had been treated with Roentgen rays or cortisone were flooded with various bacteria. The species of the bacteria and their places of origin were frequently unknown; it was a problem whether these bacteria caused the death of the animals, or were only indicators of their agonal condition.

Evidently various procedures are required to follow the fate of populations of pathogenic bacteria in the host. Samples from fluids or organs can be used for cultivation, for microscopy of stained smears, or for the study of histological sections which are stained for bacteria. Each of these techniques may be applied in a qualitative or quantitative way. In other words, the presence or absence of a bacterial population may be ascertained at intervals, or the changes in population size may be determined. Quantitative determinations in cultures are made by means of plate counts of colonies. These counts are based on the assumption that, in sufficiently diluted samples, each colony develops from a very small number of viable bacteria, possibly from one bacterium. Many publications have been devoted to the trapping mechanism of spleen and liver and other factors by which a bacterial population is removed from the blood. I mention some interesting recent papers: D. E. Rogers (1958) on the cellular management of bacterial parasites, and Janssen, Fukui, and Surgalla (1958) on the fate of *Pasteurella pestis* after intracardial injection into guinea pigs.

For a long time tubercle bacilli seemed to be unsuitable for plate counts. Finally, an adequate technique was developed by Fenner, Martin, and Pierce (1949). With an improved method McCune and Tompsett (1956) studied the effect of various drugs on the fate of tubercle bacilli in mice. Among their numerous interesting results I report one which is pertinent to our present discussion. Mice were treated with pyrazinamide plus isoniazid for 90 days after infection. At the end of this period microscopic examination and cultivation of lung or spleen samples did not reveal any tubercle bacilli. Thirty treated mice were followed for another period of 90 days. During this period no drug was administered. At the end of the drug-free period 12 mice showed tubercle bacilli in their spleens and one of the 12 also in the lungs (McCune, Tompsett, and McDermott, 1956, Table I). The disappearance and reappearance of populations of tubercle bacilli in treated mice again opens the question of hiding places which was discussed previously in reference to the latent infection of rats with *Bartonella* bacilli. In most studies which deal with the fate of bacterial populations there is a certain *discontinuity of data.*

In chemotherapeutic studies, suspensions of tubercle bacilli were injected into the tail veins of mice. From here the circulation transported the whole bacterial population into the lungs. Several observers who injected *finely dispersed suspensions* of bacilli reported that most bacilli disappeared from the lungs 24 hours after injection. Approximately one week later numerous tubercle bacilli reappeared in the lungs (Raleigh and Youmans, 1948; Stewart, 1950). This seemed to be another example of discontinuous data concerning bacterial populations in the host. Recently, however, this time gap was bridged by investigations of D. F. Gray (1959). A fine suspension of tubercle bacilli lodging in the lungs of mice could be detected in comparable numbers on each of the subsequent days, but these bacilli were less acid fast than the injected ones. They were detected by fluorescence and parallel plate counts. Five or six days after inoculation the bacilli increased in number and at the same time in acid fastness. This brings us face to face with the well-known problem of changes in acid fastness of tubercle bacilli as a function of their life cycle and environmental conditions (see, for instance, Hauduroy, 1950; Knaysi, 1951).

Infection with *coarse suspensions* of bacilli produces a form of embolic tuberculosis in mice which allows reconstructions of the fate of the bacterial population. Studies of this type were published by Grün and Klinner (1952) and by E. Mayer, Jackson, Whiteside, and Alverson (1954). A detailed report of the studies by Mayer *et al.* will be given here because of interesting procedural aspects. In chemotherapeutic experiments, the infected untreated mice die after a certain period. An effective antitubercular drug prolongs the survival of infected mice. The experiments were performed to analyze the fate of the injected bacterial population and the course of the disease in infected untreated mice. Pellicles of tubercle bacilli grown on the surface of a liquid medium were harvested and blended. Suspensions made from this material contained closely packed bacillary clumps of various sizes, aggregates of few bacilli, and single bacilli. Clumps were defined as aggregates $20 \times 10\ \mu$ or more in diameter. When such a suspension was injected into the vein of a mouse, the clumps were arrested in the small arteries of the pulmonary circulation (for a diagram of the mammalian circulation see Fig. 84). The term embolus is applied to particulate material which travels in the circulation until it is caught in vessels too narrow to be passed.

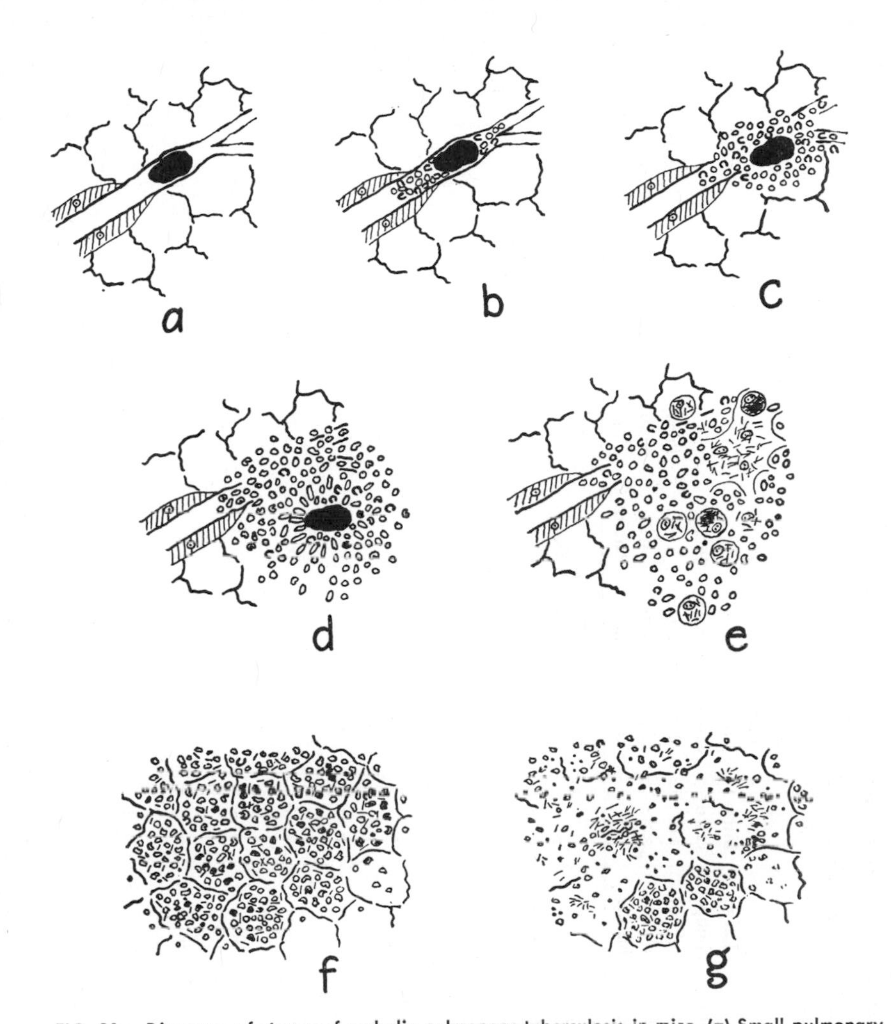

FIG. 31. Diagrams of stages of embolic pulmonary tuberculosis in mice. (a) Small pulmonary artery with typical sudden change from thick wall (shaded) to thin wall; bacterial clump (black) arrested in thin-walled part. (b) Intravascular cellular reaction around bacterial embolus; 15 minutes after injection or later. (c) Bacterial clump plus reacting cells breaking through elastica of vessel; 1 hour after injection or later. (d) Reactive nodule, with bacterial clump in its center, located in the alveolar tissue; usually, 24 hours after injection. (e) Bacterial clump dispersed by mononuclear phagocytes (which are drawn disproportionately large to show bacilli); starts at 24 hours or later, fully developed usually after 4 days. (f) Tuberculous pneumonia; alveolar pattern still preserved; fully developed after 10 days or later. (g) Massive necrosis; alveolar pattern destroyed (except on lower margin of picture); debris of cells and elastica, teeming with bacilli. Fully developed after 12 days or later. *Note:* topographically, stages (f) and (g) are no longer related to embolized vessels, but may develop anywhere in the lungs. (From Mayer, Jackson, Whiteside, and Alverson (1954).)

TABLE 7

Histology of Lungs of 71 Mice, from 5 Minutes to 14 Days after Intravenous Injection of Standard Dose of Tubercle Bacilli[a,b]

Stages:			a	b	c	d	e	f	g	
Time after injection	Series	Number of mice	Large clumps† of tubercle bacilli	Intravascular reaction	Rupture of vascular elastica	Extravascular nodules	Phagocytes with bacilli	Pneumonic plugs	Massive necrosis	Consolidated area, estimated as per cent
5 Minutes	10 L	1	+	–	–	–	–	–	–	0
	11 L	2	+ +	– –		– –	– –	– –	– –	0 0
	16 L	2	+ +	– –		– –	– –	– –	– –	0 0
	*20 L	2	+ +	– –		– –	– –	– –	– –	0 0
15 Minutes	10 L	1	+	±	±	–	–	–	–	0
1 Hour	10 L	1	+	+	+	–	–	–	–	0
	*20 L	2	+ +	+ +		– –	– –	– –	– –	0 0
3 Hours	5 L	2	+ +	+ +	+ +	– –	– –	– –	– –	0 0
	10 L	1	+	+	+	–	–	–	–	0
6 Hours	5 L	2	+ +	+ +	+	– –	– –	– –	– –	0 0
	10 L	1	+	+		–	–	–	–	0
12 Hours	5 L	2	+ +	+ +		+ +	– –	– –	– –	0 0
24 Hours	3 L	2	+ +	+ +	+	+ +	– –	– –	– –	1–25 1–25
	5 L	2	+ +	+ +	+	+ +	– –	– –	– –	1–25 1–25
	10 L	1	+	+		+	±	–	–	1–25
	11 L	2	+ +	+ +		+ +	+ –	– –	– –	1–25 1–25
	16 L	2	+ +	+ +	+	+ +	± –	– –	– –	1–25 1–25
	*20 L	2	± +	+ +		+ +	+ +	– –	– –	1–25 1–25
2 Days	10 L	1	+	–		+	+	–	–	1–25
	16 L	2	± ±	– –		± ±	– –	– –	– –	1–25 1–25
	*20 L	2	– –	– –		+ +	+ +	– –	– –	1–25 1–25
3 Days	16 L	2	± ±	– –		+ +	+ +	– –	– –	1–25 1–25

TABLE 7 (Continued)

Stages:			a	b	c	d	e	f	g		
Time after injection	Series	Number of mice	Large clumps† of tubercle bacilli	Intravascular reaction	Rupture of vascular elastica	Extravascular nodules	Phagocytes with bacilli	Pneumonic plugs	Massive necrosis	Consolidated area, estimated as per cent	
4 Days	3 L	2	+ +	+ −		+ +	+ +	− −	− −	25–50	25–50
	10 L	1	+	−		+	+	−	−	25–50	
	16 L	2	− −	− −		+ +	+ +	− −	− −	1–25	1–25
	*20 L	2	− −	− −		+ +	+ +	− −	− −	1–25	1–25
6 Days	3 L	2	− ±	− −		+ +	+ +	− −	− −	1–25	25–50
	10 L	1	−	−		+	+	−	−	25–50	
8 Days	3 L	2	− −			+ +	+ +	+ −	− −	25–50	25–50
	10 L	1	±‡			+	+	+	+	>50	
	*20 L	2	− −	− −		+ +	+ +	− −	− −	25–50	25–50
9 Days	3 L	2	− −			+ +	+ +	+ +	− −	1–25	25–50
11 Days	3 L	2	− −			+ +	+ +	+ +	− −	25–50	1–25
	10 L	1	−			−	+	+	+	>50	
	11 L	2	− −	+ +		− −	± ±	+ +	± ±	25–50	25–50
	*20 L	2	− −	− −		+ +	+ +	+ +	− −	25–50	25–50
12 Days	8 L	2	− −			− −	+ +	+ +	+ +	>50	>50
	8 L	2	− −			− −	+ +	+ +	+ +	>50	>50
	10 L	1	+§	+§		−	+	+	+	>50	
13 Days	3 L	2	− −			− +	+ +	+ +	+ +	>50	>50
14 Days	8 L	1	−			−	+	+	+	>50	
	*20 L	2	− −	− −		− −	+ +	+ +	+ +	>50	>50

[a] From Mayer, Jackson, Whiteside, and Alverson (1954).

[b] Strain H37 was used in Series 20 (marked with asterisk), and strain D4 was used in the 6 other series. Stages a to g are illustrated in Fig. 31. Explanation of symbols: + present; ± doubtful; − absent; blank, not examined. If 2 symbols are entered, each refers to one mouse.

† Clumps of at least 20 x 10μ in diameter.

‡ One clump.

§ One large clump in remnant of vessel.

By the technique described, embolism of the pulmonary arteries was produced. A total of 71 mice were killed at the following intervals after injection: 5 and 15 minutes; 1, 3, 6, 12, and 24 hours; 2, 3, 4, 6, 8, 9, 11, 12, 13, and 14 days. In survival studies in which the mice died of the infection, 50% survived 10 to 13 days, and none more than 18 days. A variety of phenomena were observed in the lungs of the 71 mice killed at different intervals. It seemed possible to arrange these phenomena in a chronological sequence which is shown in the diagrams of Fig. 31 and explained in the legend.

Has the sequence of events presented in Fig. 31 been proved or is it merely probable? The chronology offered here is based on three arguments. (1) If the seven pictures of Fig. 31 represent a chronological order at all, the only conceivable sequence is from (a) to (g), since the following interpretation of the process is obvious. The embolic clump, together with the reacting intravascular blood cells, break through the thin part of the arterial wall and thereby move into the lung tissue. The bacterial clumps are dispersed by phagocytes in which the bacilli multiply (as demonstrated in explants *in vitro*). Thus the infection spreads until so much lung tissue is consolidated or destroyed that the mouse dies from lack of respiratory surface. (2) The stages shown in the seven pictures of Fig. 31 were used as the headings of seven columns of Table 7 with (a) on the left and (g) on the right. Presence of a picture was entered as a plus sign, absence as a minus sign. The individual mice were listed in the order of increasing intervals after injection, so that the mice killed after 5 minutes were at the top of the table, and those killed after 14 days at the bottom. The result was a clear trend of plus signs from the left upper corner to the right lower corner of the table. In other words, as the intervals after injection increased, the plus signs moved from column (a) to column(g). Two minor exceptions could be readily analyzed. (3) *All morphological studies were paralleled by survival studies* in the following way. Each mouse of a batch received injections with the same dose of the same suspension of tubercle bacilli. Usually, half of the batch was used for morphological study, and the other half for observation of survival. The survival data obtained are shown in Table 8, in which the 130 mice observed for survival are marked "Control". Many of the mice planned for morphological study were found dead before the day of their scheduled sacrifice. The 71 mice which became available for morphological study are those listed in Table 7. When experiments of this type were repeated over a period of two years, the survival times as well as the morphological observations remained similar. Therefore, the conclusion was justified that the mice sacrificed for morphological study were representative of those studied for survival. The high degree of reproducibility of observations was the strongest argument in favor of a uniform sequence of events in the lungs of the infected mice. Organs other than the lungs hardly had time to develop tuberculous processes, since they received a very small dose of scattered bacilli in contrast to the massive dose deposited in the lungs. Three independent arguments support the chronological sequence presented in Fig. 31.

Why was such a lengthy discussion devoted to the possibilities and limitations of interpreting static pictures in terms of time sequences? First of all, I emphasized the fact that in some areas of biology the possibilities of chronological study are greater than in others. I compared the favorable position of studies on helminth and malaria infections with the difficulties in investigating bacterial infections. In areas such as the embryology of animals, where the facilities for the determination of time sequences are good, the rules for chronological interpretations are very strict, whereas in unfavorable areas, such as human pathology,

TABLE 8

Survival Time of 130 Mice Infected with Standard Dose of Tubercle Bacilli (Strain D4 or H37), in Parallel with Mice Infected and Sacrificed for Morphological Study[a,b]

				Survival time in days			
					Calculated values[d]		
Date of infection	Strain of bacilli	Series number	Number of mice[c]	Actual range	16% dead	50% dead	84% dead
February 2, 1950	D 4	3 L Control	20	4–18	11	13	15
December 21, 1950	D 4	5 L Control	20	5–15	8	10	12
January 11, 1951	D 4	8 L Control	10	9–18	10	12	15
February 22, 1951	D 4	10 L Control	20	4–15	9	11	14
September 12, 1951	D 4	13 L Control	20	8–18	10	12	15
October 18, 1951	D 4	11 L Control	10	9–17	10	13	17
February 11, 1952	D 4	16 L Control	10	9–17	11	13	15
September 8, 1952	H 37	20 L Control	20	7–18	10	13	16

[a] From Mayer, Jackson, Whiteside, and Alverson (1954).

[b] The series numbers are identical with those in Table 7, in which the morphological observations are recorded. Series 13 L was a survival study only and is not represented in Table 7.

[c] At termination of experiments, all mice had died.

[d] Calculated according to Litchfield's method (1949).

the rules are traditionally lax. In experimental pathology chronological sequences can be established in most cases by killing animals at different stages of the experiment. This was illustrated in the study of embolic mouse tuberculosis, Table 7. Another example was given in an earlier part of the book when the different aspects of sunburn and tanning were discussed. Biopsies from the skin of pigs exposed to ultraviolet radiation gave an insight into the sequence of microscopic developments (Table 1).

In human pathology chronological sequences of morphological processes can be obtained by taking biopsies or Roentgen-ray pictures at intervals. However, the literature is full of hypothetical interpretations without the support of timetables. In pathological morphology cytogenesis, demonstrated or hypothetical, is an important criterion in the classification of abnormal cells, particularly of cancer cells (Part V, Chapter 4).

B. Tagging Procedures for Following the Fate of Biological Structures

I have described a number of studies in which it was possible, without the help of tagging procedures, to follow the fate of cell populations in the embryonic and postembryonic organism and of parasitic helminths, protozoa, and bacteria in their respective hosts. Now I shall discuss the use of tagging as a means of following the fate of living structures. As stated previously, tags ensure the identity of objects under changing conditions. Rules for the use of tagging procedures have been developed systematically in the field of tracer isotopes. The physical foundations, the possible uses and limitations, precautions for workers, costs of isotopes, and many other items can be found in excellent books and reviews, e.g., the article by J. Sacks, "Tracer Techniques: Stable and Radioactive Isotopes" (1956). Three fundamental principles apply not only to the use of tracer isotopes, but also to the use of any other artificial tags such as vital staining of cells or attaching of metal clips to animals. (1) The tag should be sufficiently permanent to be clearly recognizable throughout the experiment. (2) Tags should as much as possible remain on the objects for which they are intended. If subsequent transfer of tags to other objects occurs, the new recipients must be identified. (3) The tag should not interfere with the purpose of the experiment. This means that the tag should not alter the reactivity of the object to stimuli which are important in the experiment; the living object should neither suffer nor benefit from the tagging procedure in any way relevant to the particular experiment. Principle (1) is less flexible than principles (2) and (3). As long as the nature and degree of exchange or interference of a tagging procedure is known, it may still be useful.

I will give examples of the use of visible artificial tags and then review some characteristic studies with radioactive tracers. Finally, natural tags will be discussed briefly.

Tagging of domestic animals is an old custom. Cattle have been branded and sheep have been dyed to mark them as the property of this or that owner. For biological research the tagging of fish with metal clips and the banding of birds are used in order to follow their migrations and to determine their life span. Remarkable discoveries were made concerning the life of worker bees when color tags in conjunction with observation frames were applied. An observation frame is a beehive modified in such a way that each cell can be observed through a pane of glass. Numerous individual bees can be identified by an ingenious combination of color spots. When individual workers were followed from hatching throughout their 30 days of life, it was found that each of them devoted the first phase to cell cleaning and feeding of larvae, the second phase to comb building and nectar storage, and the third phase to foraging for nectar, pollen, and water (Rösch, 1925, 1927). Morphological changes of organs are correlated with these functional changes.

Visible tags for the tracing of *cell populations* in embryos and adult organisms require a detailed discussion. The application of vital dyes by Vogt (1925) in studies of amphibian embryos revolutionized the ideas on mechanisms of differentiation. Prior to Vogt's work it was assumed that *local differences in cell multiplication* take the lead in embryonic differentiation. This idea had found its expression in the "principle of differential growth" suggested by W. His (1874) as the mechanism which produces folds in the sheets of cells of which

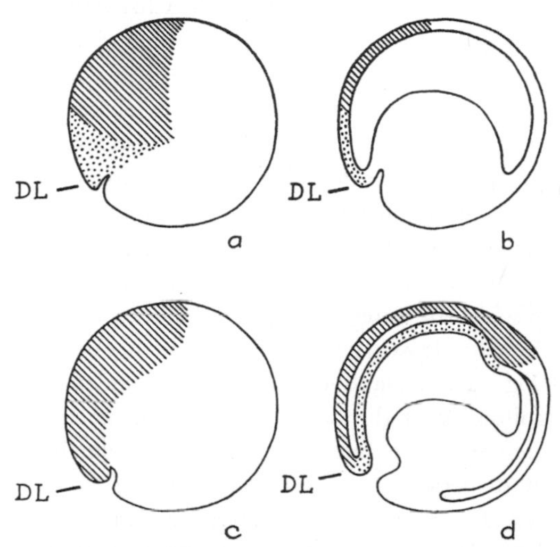

FIG. 32. Migration of cell populations (morphogenetic movements) during gastrulation in amphibians (cf. Fig. 3f and f′). (a) and (b) earlier stage, (c) and (d) later stage; (a) and (c) surface views (of spheres); (b) section through (a); (d) section through (c); both sections in the plane of the paper crossing *DL*, the dorsal lip of blastopore. During the development from the earlier to the later stage the shaded area has changed its shape and has extended to *DL* while the dotted area has disappeared from the surface. As (d) shows, a folding process has taken place at *DL* in the course of which the dotted material formed a second layer inside of the shaded area. Potentialities: the shaded area is presumptive neural plate and the dotted area is presumptive chorda-mesoderm. Note: The amount of yolk decreases from (b) to (d) by absorption into the migrating and multiplying cells. Yolk-rich parts are indicated by large white areas in the lower halves of the diagrams (b) and (d) (area Y in Fig. 3f′).

early embryos are composed. Vogt's observations led to the conclusion that in blastulas and early gastrulas of amphibia the *migration of cell aggregates* is the leading factor in differentiation. The main points of Vogt's procedure follow. Small cubes of agar were soaked with neutral red or Nile blue sulfate and gently pressed against selected areas of the embryos. The dyes diffused into the cells which composed these areas and, without causing injury, remained in them long enough for subsequent observations (local vital staining). It turned out that the red- or blue-tagged cell aggregates moved as a whole to various areas of the embryo (my diagram Fig. 32). Stained cells which had moved into the interior of the gastrula could be located in sections. These migrations of cell groups proved to be related to the distribution of potential organs and, therefore, were called morphogenetic movements. The dorsal lip of the blastopore, the region of Spemann's organizer, plays an important role in controlling these movements. At the stage shown in Figs. 32a and b, the moving cell groups converge toward the dorsal lip; for an illustration of the convergent streams see Rugh's manual, 1948, frontispiece and p. 222.

In the adult organism migratory cells have been tagged by vital stains for various purposes. I mention a pioneer study by Christeller and Eisner (1929) concerning the distribution of leucocytes (white blood corpuscles) in the circulatory system. By injection of turpentine under the skin of a dog, a massive accumulation of leucocytes was produced (sterile abscess). These leucocytes were stained with trypan blue. A suspension of blue leucocytes was injected into the femoral artery of another dog. When the receptor dog died a few minutes after injection, it turned out that some of the foreign blue leucocytes had been arrested in the capillaries of the leg, but many had passed through these capillaries, had traveled to the right heart, and from there to the lungs. All blue leucocytes which reached the lungs seemed to have been arrested in the pulmonary capillaries, since no blue leucocytes were found in organs other than the lungs and the injected leg. The technique of Christeller and Eisner was not suitable for an extended follow-up of the fate of the stained leucocytes. It seems to me that this study suggests some interesting questions which could be answered after relatively short periods of observation. There is no difference known between the caliber of capillaries of an extremity and that of pulmonary capillaries. However, the pulmonary capillaries are more tortuous than those in any other part of the body. If this factor was decisive, the blue leucocytes should be retarded only temporarily in the pulmonary circulation: they should be washed out from the pulmonary capillaries within 24 hours. The blood contains more CO_2 and less O_2 per unit of volume in pulmonary arteries and the first part of the capillaries than in other parts of the body. Does this regional peculiarity enhance the adhesion of the blue leucocytes to the wall of pulmonary capillaries? This could perhaps be decided by examining the adhesiveness of stained and unstained leucocytes to capillary walls under different CO_2 and O_2 tensions. Such considerations seem to be particularly relevant in the light of studies by Lanman *et al.* (1950), who found that transfused leukemic leucocytes of man are eliminated from the general circulation of the recipient by being arrested in the pulmonary capillaries. The authors discuss possible variations in stickiness of either leucocytes or the endothelial wall. Ambrus *et al.* (1954) studied the fate of untagged leucocytes in perfused heart-lung preparations by counts at short intervals. They found that the lungs rapidly removed white cells from the circulating blood, thus confirming the results of Christeller and Eisner. Stored leucocytes were released from the lung when blood, which contained few leucocytes, was introduced into the perfusion system. Incidentally, the experiments of Ambrus and co-workers did not shed any light on the mechanisms by which the lungs arrest white blood cells.

Farr (1951) made an attempt to determine the fate of rabbit lymphocytes by tagging them with a vital nuclear stain, 3,6-diamino-10-methylacridinium chloridine, which produces fluorescence in ultraviolet light (DeBruyn, Robertson, and Farr, 1950). Farr obtained a suspension of lymphocytes from a subcutaneous lymph node of a rabbit, stained them, and injected approximately forty million of these lymphocytes into a vein of the same rabbit. When animals were sacrificed after 12 hours, labeled lymphocytes were found in the bone marrow, lymphatic tissue, intestinal mucosa, and other connective tissues. Attempts were made to account for the forty million lymphocytes injected by estimating the total weight and volume of bone marrow and lymphatic tissue in a rabbit and by calculating how many tagged lymphocytes should be found in each section of a volume of 5 $\mu \times 2$ mm. $\times$ 2 mm. Unfortunately, complete counts of tagged cells per section were not possible owing to a difficulty inherent in this technique: under the influence of ultraviolet light the fluorescence of the dye faded so fast that identification was feasible for too short a time. In spite of this limitation Farr was able to detect nuclear fluorescence in myelocytes in the bone marrow 12 hours after injection. This fluorescence was not present at 2 hours after injection. The identification of myelocytes was based on phase contrast microscopy of the same living cells whose nuclei showed fluorescence. Farr concluded that some of the injected tagged lymphocytes which had reached the bone marrow had developed into myelocytes. The author claimed that transfer of the dye from lymphocyte nuclei to other nuclei was excluded. It seems to me that this study illustrates a point which I emphasized previously, namely, the problem of demonstrating transformations of white blood cells from one type into another.

An ingenious tagging procedure was designed by Rous and Beard (1934) to prove or disprove the alleged transformation of sessile Kupffer cells in the liver into motile cells of the circulating blood, so-called monocytes. It had been proved experimentally that particulate material in the blood is phagocytosed by Kupffer cells, and it was assumed that a very large intake of such material causes the cells to loosen their connnections with neighbor cells, to change from a polyphedral to a spherical shape, and to float away in the blood stream as monocytes. Rous and Beard proceeded as follows. A suspension of fine particles of ferric oxide was injected into subcutaneous veins of rabbits or dogs. It was known from previous observations that after 2 or 3 days the Kupffer cells would be loaded with particles. Therefore, after this interval, the animals were anesthetized and the liver was prepared for perfusion. After the blood was rinsed out, perfusion with a mixture of gelatin and Tyrode's solution was started. Changing perfusion pressures and gentle massage were applied to enhance the mobilization of Kupffer cells. The washings were collected and immediately exposed to a powerful unipolar eye magnet so that the cells containing iron particles were separated from the others. The selected cells were examined microscopically and also studied in tissue cultures. The result was that the mobilized Kupffer cells did not resemble the monocytes of the blood and that there was no support for the assumed relationship between the two cell forms (Beard and Rous, 1934).

In recent investigations the term tagging has usually referred to the application of radioactive isotopes as tracers in chemical and biochemical processes. Studies of metabolism have profited tremendously from tracer techniques. Compounds containing a radioactive isotope are administered to animals orally or by subcutaneous or intravenous injection. If, for instance, radioactive iodine (I^{131}) is administered, one expects its selective accumulation in the thyroid. In the living animal a Geiger counter can be placed over the area of the thyroid. Counts at different intervals after administration of the tracer indicate the rate of uptake, the period of maintenance, and the rate of disappearance of radioactive iodine from the thyroid. Thus both the total turnover and the rate of turnover of iodine can be determined. If one wants to know the distribution of the radioactive iodine within the thyroid, animals are killed at various intervals after administration, the thyroid is removed and fixed, and microtome sections of the gland are prepared. By placing a section close to a photographic film, different areas of the emulsion will be affected in proportion to intensity of the radiation. The picture obtained is called a radioautograph or autoradiograph. Semidiagrammatic presentations of radioautographs of jejunal villi of the mouse are shown in Fig. 33. There are many good descriptions of radioautograph techniques, their advantages and pitfalls. I refer to an article by Leblond and Bogoroch (1952) and a presentation by J. H. Taylor (1956). The metabolic activity of cells is indicated by their rate of handling the tracer. Metabolic tracers will be mentioned in several chapters of the present book, mainly in Part III, Chapter 7, which deals with histophysics and histochemistry. As a ramification of metabolic studies, radioactive isotopes have proved to be useful in the analysis of the fate of cell populations. Studies of this type have been discussed in the 1956 review by Leblond and Walker mentioned previously.

In population studies with radioactive tracers both Geiger counters and radioautographs can be used. The former procedure was applied by Andreasen and Ottesen (1945) in studies of the lymphatic tissue of rats at different ages. This work was based on the well-supported premise that radioactive phosphorus (P^{32}), if administered to an animal, is utilized by actively metabolizing cells for the synthesis of deoxyribonucleic acid in the same way as ordinary phosphorus. Three hours after subcutaneous injection of sodium phosphate containing P^{32} the animals were killed. The lymphoid organs were removed and ground, their deoxyribonucleic acid was extracted, and its radioactivity was measured with a Geiger counter. The greatest radioactivity was found in the thymus of young rats, exceeding that of lymph nodes. This was interpreted as a greater turnover of cells in the thymus as compared to lymph nodes. In subsequent investigations by Andreasen and Christensen (1949) the ratio of dividing nuclei to resting nuclei was determined morphologically. By treatment of minced thymus and lymph nodes with 5% citric acid, most of the cytoplasm was removed from the nuclei (Dounce, 1943). After grinding and centrifugation, a suspension of nuclei was obtained which permitted counting of mitotic and intermitotic (resting) nuclei in a counting chamber.[3] Roughly, the morphological counting results paralleled the data obtained with labeled deoxyribonucleic acid. Thus it was shown by two independent methods that the turnover of lymphocytic populations in young rats is greater in the thymus than in the lymph nodes.

To illustrate the use of radioautographs in population studies, let us return to the investigations of Leblond and collaborators on the turnover of lining cells (epithelium) in the intestine. As described previously, the excess of mitoses in the crypts and the scarcity of mitoses in the top of the villi led to the theory that the cell population migrates from the bottom to the top (Leblond and Stevens, 1948). In a preliminary study Leblond, Stevens, and Bogoroch (1948) injected P^{32} into rats and obtained radioautographs of the duodenum which indicated some migration of the cell population. A final test of the theory became possible with the advent of radioactive tritium (H^3) as a label. If tritium-labeled thymidine is injected into animals, it is taken up exclusively by deoxyribonucleic acid (DNA). The labeled thymidine enters the places where new DNA is synthesized, particularly chromosomes just before mitosis. The radioautographic reactions are limited to the area of nuclei. Leblond and Messier (1958) injected tritium-labeled thymidine into mice and killed the animals 8, 24, or 72 hours later. The results are shown in diagrams Fig. 33A and B, which I made from the photomicrographs of the authors. At 8 hours after injection radioautographic reactions were concentrated over the lining cells of the crypts of the jejunum (Fig. 33A). Most mitotic nuclei and many non-dividing nuclei were labeled. At 72 hours the upper third of the villi showed intense radioactivity over many nuclei of the lining cells (Fig. 33B). Some of the sloughed off cells on the top were still labeled. There was a marked decrease in radioactivity from the upper to the lower parts of the villus. Intermediary stages were observed at 24 hours after injection. These studies left no doubt that the lining cells migrate from the crypts to the top of the villi, as it had been inferred on the basis of indirect procedures and calculations.

Radioactive tracers have been used extensively to determine the turnover time of mature red blood corpuscles in the circulating blood and also to esti-

[3] Evidently the chromosomes of the different mitotic phases were held together by some central cytoplasm which was not dissolved by the citric acid (personal communication from Dr. S. Christensen, 1960).

mate the whole life span of red cells from their immature stages in the bone marrow, through the mature phases in the circulating blood, to the disappearance of old cells from the blood. The renewal of tissue cell populations will be discussed again in another chapter, which deals with the variations in volume, weight, and composition of organs (Part V, Chapter 3).

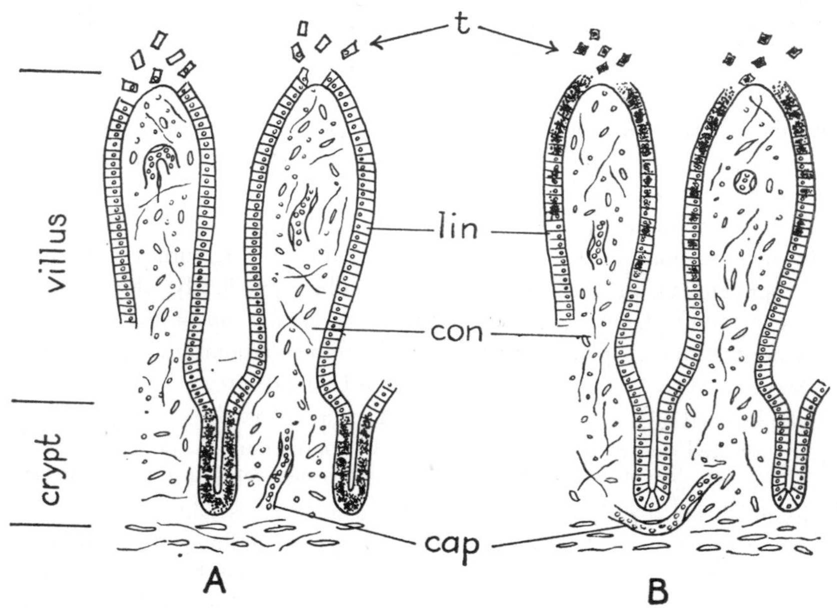

FIG. 33. Migration of cell populations which line the jejunum of mice demonstrated by radioautographs. Two villi (fingerlike projections of mucous membrane) are shown in each diagram. The levels of villi and crypts are indicated on left margin; *lin*, layer of lining cells; *con*, connective tissue; *cap*, capillaries; *t*, tops of villi with sloughed off cells. (A) Eight hours after injection of radioactive tritium (H^3)—labeled thymidine. Black dots indicate uptake of radioactive material in the lining of the crypts where mitotic activity is maximal. (B) Seventy-two hours after injection. Radioactive material has disappeared from the level of the crypts and its maximal concentration is now in the tops of the villi. Interpretation: since exchange of radioactive material across cells of the lining layer is excluded, the change in location of the radioactive material is ascribed to migration of those cell aggregates which had taken up H^3 during the first hours after injection. (Diagrams made from radioautographs published by Leblond and Messier, 1958.)

It was mentioned earlier that without the use of tagging it is usually impossible to follow the fate of bacterial populations in a host. The fate of embolic clumps of tubercle bacilli in mice was described as an exception. Radioactive tracers have been used extensively for metabolic studies in bacteria, but rarely for following the fate of populations in the host. There are a few studies in which tubercle bacilli tagged with radioactive phosphorus were identified in the infected animals. However, Sternberg and Podoski (1953) expressed the opinion that much *in vitro* work is needed on tagged bacteria, especially on transfer of tracers to the media, before identification in the host can be claimed with reasonable certainty.

Finally, I would like to recount interesting examples of natural tagging. The first one concerns the experimental infection of canary birds with malaria parasites (*Plasmodium cathemerium*). Hegner and Hewitt (1937) reported that young stages of the parasites, the merozoites, invade a certain immature stage of red blood cells. At this stage the cytoplasm of the red blood cell does not yet contain hemoglobin, and the nucleus shows a netlike structure. While the parasite within the red cell reaches the phase of mature schizonts, the red cell becomes mature, too. The mature red cell shows hemoglobin in its cytoplasm and has a dense structureless nucleus. Diagrams of these changes are seen in

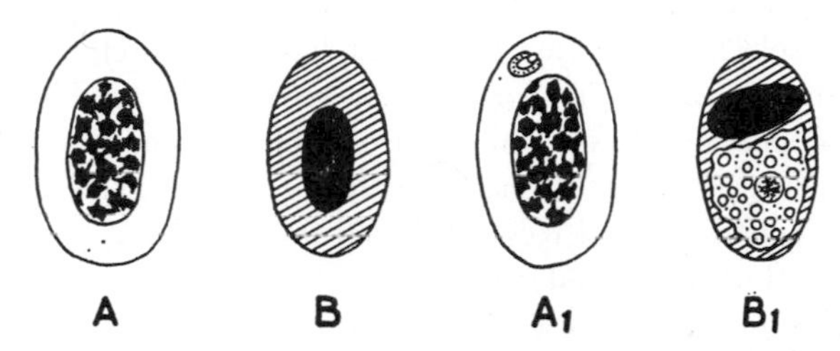

FIG. 34. Maturation time of red blood corpuscles of canary birds determined by infection with malaria parasites. (A) Early stage of red blood corpuscle, with network structure of nucleus and no hemoglobin in cytoplasm (unshaded). (B) Later stage of red blood corpuscle, with dense, structureless nucleus and hemoglobin in cytoplasm (indicated by shading). (A_1) Same stage as (A), but infected with ring-shaped parasite (merozoite). (B_1) Same stage as (B): parasite has reached the phase of mature schizont which occupies the larger part of the red blood corpuscle, crowding the nucleus to one side. Particles in the circle are malaria pigment. Interpretation: Since the development of schizonts from merozoites is known to occur in 24 hours, the change of red blood corpuscles from stage (A) to (B) also takes 24 hours (confirmed *in vitro* in uninfected red blood corpuscles). (Hegner and Hewitt, 1937; pictures drawn from colored plate in Hewitt, 1940).

Fig. 34. By the methods described previously, it was determined that the merozoites need 24 hours to develop into mature schizonts. Evidently, a similar period is necessary for the red blood cells to change from the immature stage A of Fig. 34 to the mature stage B. Thus a visible clock had been placed in the red blood cells which indicated that 24 hours elapsed between their two stages. Subsequently the authors verified that the rate of maturation of infected red blood cells in the circulating blood does not differ from that of uninfected red cells observed *in vitro.*

The second example refers to the use of a tag based on the interaction between red blood corpuscles and sera which belong to different human blood groups. Ashby (1919) discovered that in mixtures of corpuscles of different groups it was possible to separate the corpuscles by adding a serum that agglutinates the corpuscles of one kind, leaving the others unagglutinated. After a recipient has been transfused with blood of a group other than his own, samples of his blood, treated with a serum that will agglutinate his own corpuscles but not the transfused corpuscles, show unagglutinated corpuscles in considerable numbers. Counts of unagglutinated corpuscles at different intervals after the transfusion indicated that the donor's corpuscles survived in the recipient's blood for longer than a month.

Our last example represents a combination of one artificial and two natural tags. It is taken from a study by Crosby (1957) on human siderocytes. Some individuals with disturbances of their blood-forming system have a varying percentage of red blood corpuscles which contain one or several abnormal granules. In fixed blood smears the granules give a positive iron reaction, therefore these corpuscles are called siderocytes. Blood from donors with siderocytes was transfused to recipients whose spleens had been removed, and to recipients with spleens. The transfused red blood corpuscles were identified by tagging with radioactive Cr^{51}, or by their behavior in the blood group test of Ashby, described previously. In addition, the transfused siderocytes were recognizable by the presence of their peculiar granules. The results were as follows. In the splenectomized recipients both the transfused normal corpuscles and the transfused abnormal siderocytes persisted for some time. In the recipients with spleens the transfused red blood corpuscles remained, while the siderin granules disappeared. The author concluded that probably "the spleen was capable of removing the inclusion body without destroying the red cell that contained it." I felt that Crosby's paper should be mentioned here because the radioactive chromium served as an artificial tag, the blood type of the transfused cells as a permanent natural tag, and the iron-positive inclusion bodies as a natural tag which could be removed from red blood corpuscles without visible injury to the corpuscles.

C. Comparison of Living and Dead Material. Natural and Artificial Conditions

The concepts of natural and artificial conditions are relative. Domesticated animals live under conditions which are considered to be artificial as compared to the environment of the corresponding wild types. Wildlife can again be graded: depending on the criteria used, conditions in unprotected woods may or may not appear more natural than those in a reservation. With respect to domestic animals, conditions on the farm are considered natural if compared to those in the laboratory. This point is of practical importance in studies of infectious diseases such as coccidiosis in fowl and histomoniasis in turkeys. When a certain treatment proves successful under laboratory conditions, the next step has to be a field trial. This means that the new treatment is tried out on farms or under conditions simulating those of a farm.

In physiological experiments compensation for the deviations from natural conditions is possible, to some extent, by a study of the same problem with several procedures which have different advantages and disadvantages. This principle applies not only to the study of living material, but also to the study of dead material. Therefore, in our discussions on sampling and fixation the usefulness of complementary procedures was emphasized.

In traditional morphology certain complications have arisen from an idea which we have already rejected, namely, that the study of live material as such is always superior to that of dead material. Everybody will agree that dead is opposite to living and artificial opposite to natural. Yet, some biologists seem to equate living with natural and dead with artificial. This confusion is expressed in the frequent statement that such and such morphological structures probably were not present in the living organs, but were produced by death *or* the technique of investigation. This definition of histological artifacts was still in the second edition of "Webster's New International Dictionary" (1959, p. 157), but the third edition (1961, p. 124) gives an improved definition, with emphasis on inconstant occurrence of appearances.

In "McClung's Handbook of Microscopical Technique" (3rd ed., 1950) the material which is classified as fresh corresponds with what we call living material. Most dead material discussed in McClung's book is fixed material, and dead unfixed material is rarely mentioned. The present book covers the study of dead material both in preserved and nonpreserved conditions, and, as a consequence, includes natural changes which occur after local or general death. Identical structural changes may be caused during life by one agent, and after death by another agent. For instance, the proximal convoluted segments of renal tubules in man are destroyed by certain poisons such as mercury bichloride. After death, the same segments are particularly susceptible to digestion by intracellular enzymes: this refers to kidneys which were not damaged during life. The pathological condition which developed during life is called necrosis, and the condition which developed after death is called postmortem autolysis: the two morphological pictures are indistinguishable.

Unfortunately many investigators use the terms natural and artificial in an *absolute* sense. They seem to feel that, eventually, with the progress of biology, the true, real, natural condition of the objects will be revealed. I prefer to approach this problem operationally, by defining as natural the condition of an animal before the start of an experiment, and as artificial the conditions after the start of the experiment. In this operational definition, natural and artificial are relative terms, since the zero point is fixed arbitrarily. Suppose a rabbit is tied on a board and given anesthesia. Which is the zero point, the tying or the anesthesia? Before being placed on the board the rabbit was moved from its usual quarters to the operation room. This change may have modified the rabbit's reactivity and therefore should perhaps be considered the starting point of the experiment. There is no end of such apprehensions. The only solution is to describe what has been done to the rabbit, but the amount of information needed depends on the nature of the experiment and the results. When results are evaluated, it may turn out that important detail about the history of the rabbit is lacking, but some points which are on record may prove irrelevant. This is why the starting point of an experiment or observation needs a preliminary definition and a final definition; the two may or may not coincide.

Let us assume that a structure in the living animal shows a certain appearance at the start of the study. This appearance may change under the influence of experimental factors (artificial change) or without experimental interference (natural change, e.g., by aging). Death also produces structural changes. Different changes result from different types of dying, such as slow or fast death, death by nutritional, circulatory, respiratory, or other disturbances. Finally, a number of structural changes take place after death. Again, changes during the process of dying as well as changes after death may occur naturally or under experimental influences. Absence or presence of interferences with natural conditions is not, as such, good or bad. Within the frame of each investigation the procedures applied are to be judged by the degree of their usefulness for the purpose of the particular investigation. With this logical background, our present discussion will be organized as follows.

First a number of examples will be given illustrating changes caused by death. Then we will describe interferences with morphological structures which result from the procedures applied by the investigator; this covers the subject matter generally known as artifacts and contaminations. Finally, relations between natural and artificial units will be analyzed.

1. *Changes Caused by Death*

The borderline between living and dead conditions or organisms is not always distinct. The stopping of the heart beat and of respiration are the conventional criteria of death. However, neither the ability of muscles to contract upon stimulation nor tissue respiration nor mitotic cell division stop abruptly when the general circulation and pulmonary respiration have ceased. If parts are removed from a living animal they will survive for different periods, depending on their inherent properties and on their treatment after removal. In the earlier description of isolation preparations for the study of living material many examples were given of parts surviving outside the organism for short or long periods.

There are conditions in living animals which are similar to death, e.g., hibernation and encystment. Microscopic organisms such as rotifers and bacteria may not show any signs of life when desiccated, but may be reactivated by moisture. These borderline conditions are sometimes referred to as dormant life. They play a particularly important role in microorganisms and plants (see Thornton's article "Dormancy" in Growth and Differentiation in Plants, edited by Loomis, 1953). Difficulties in determining whether biological material was alive, dormant, or dead at the time of sampling are obviously more frequent on the microscopic than on the macroscopic level. These difficulties will be discussed in connection with the problems of technical interferences (Part III, Chapter 3, C, 2). Our present discussion will be devoted to *structural changes caused by death* when there is no doubt which is the living and which is the dead condition.

Most observations on macroscopic changes by death were made during autopsies of patients in hospitals. The color of the human skin depends on its vascularization and on the amount and distribution of pigment granules (see Part II, Chapter 2, A). The color which is due to pigment granules does not disappear after death, whereas the color due to vascularization may change or disappear completely. An individual who died at the height of measles or scarlet fever loses the red color of his skin when his circulation ceases. A boil which was conspicuous as a localized swollen and red lump flattens and bleaches after death so that it may be overlooked at autopsy. These examples refer to so-called active acute hyperemia. Passive chronic hyperemia leading to a bluish discoloration of the skin may continue after death. Edema of the subcutaneous tissue may disappear after death if it was of short duration, but may remain after death if it had been of long standing.

In hair-covered animals the inspection of the skin is more difficult than in man. On the other hand, the condition of the fur does not change at death and, therefore, may supplement observations made during life. Sick animals fail to clean themselves. A dirty condition of the fur after death indicates that the animals have been sick for some time.

Among the superficial structures which can be inspected directly both during life and after death, the eye plays a prominent role. When after death the movements of the lids and the secretion of tear fluid discontinue, the cornea becomes dry and wrinkled and loses its transparency. Certain diseases also cause the cornea to become opaque. Therefore, it is important to be familiar with these points.

The state of contraction of muscles is subject to change by death. Muscles which were contracted during life may appear relaxed after death whereas rigor mortis, as the name indicates, produces rigidity of many muscles after death. The heart muscle contracts and relaxes rhythmically during life. The contracted phase is called systole and the relaxed phase diastole. In the dead body the heart may be found in either condition. During and after death a redistribution of blood takes place. The contracted heart is, of course, found empty while the relaxed heart contains blood either in its original liquid condition or clotted after death. In a dead body the redistribution of blood results from the last contractions of the heart and of the arteries. The blood content found in each dead organ also depends on the pressure of surrounding structures and on gravity.

Skeletal and heart muscles belong to the striated type. Smooth muscles are found in the wall of most viscera such as the bronchi, the stomach and intestines, the urinary and genital tracts, and the spleen of some species. The dog's spleen is particularly rich in smooth muscles. Therefore great variations in volume are observed in living dogs, as discussed previously. In the dead dog the spleen may be found at its maximum, minimum, or intermediate volume. In autopsying dogs I found it useful to determine the volume of each spleen under two opposite conditions. Immediately after death the abdominal cavity is opened and the vessels of the hilum are closed by clamps. With the clamps on the hilum the spleen is placed in a graduated cylinder with water. The volume read is the maximum after death. Then the clamps are removed, and the spleen is placed in cold water so that the blood is expelled and the spleen can contract. The volume of the contracted spleen represents the minimum after death.

The capacity to vary their length is much more pronounced in smooth muscles than in striated muscles. This capacity leads to astonishing differences between the length of intestines in living and dead organisms. The length of the intestine can be determined in living human beings by inserting into the mouth the pear-shaped thickening of a radiopaque tube. This thickening is swallowed, passes through the stomach and is transported by peristaltic movements through the intestines until it appears in the anus. The position of the tube inside the intestines can be checked in front of the Roentgen screen or on a Roentgen photograph. In this way, the length of the intestines can be determined during life. In a corpse one measures the length of the intestine by removing it from the mesentery to which it is fixed, and by displaying the isolated intestine on the table. The length measured in the dead intestine on the table is at least twice as long as the length seen in the living by a Roentgen determination (Van der Reis and Schembra, 1924; Miller and Abbott, 1934; Underhill, 1955). The reason for this discrepancy is the following. In the living the degrees of contraction vary from segment to segment, but there is never complete relaxation. The intestine which is removed from the dead body is completely relaxed.

The contents of the digestive tube are subject to changes after death in the stomach and small intestine, but much less so in the large intestine. In a healthy animal the large intestine contains semisolid or fairly solid fecal masses. In rodents these masses form pellets. Whenever pellets are found in a dead rodent one can be certain that this animal did not suffer from diarrhea during the last few days before death. This point can be of considerable importance. In many experiments several animals are kept in one cage. Under such conditions it is difficult to tell during life whether some of these animals had diarrhea and if so which of them. Therefore, desirable information can be obtained from the postmortem examination of the large intestine.

The lungs alternate during life between the maximum volume at inspiration and a lesser volume at expiration. In the intact pleural cavity the lungs do not collapse because of mechanical factors which will be discussed with reference to the pleural cavity as a potential space (Part IV, Chapter 4). In order to ascertain whether an animal died with its lungs in an inspiratory, expiratory, or intermediate position, one prevents collapse of the lungs at autopsy by ligating the trachea tightly prior to the opening of the thoracic cavity. In publications on experimental mouse tuberculosis one frequently reads the statement that the lungs of the infected mice had a much larger volume than those of the uninfected controls. It is, of course, impossible that tuberculous lungs should occupy a volume larger than that of normal lungs in inspiration since both cannot do more than fill the thoracic cavity completely. This is illustrated by photographs in a paper by E. Mayer *et al.* (1954). The misunderstanding came from the fact that tuberculous lungs are stuffed with cells and fluid and therefore cannot collapse when the chest cavity is opened, whereas the air-containing normal lungs do collapse. Moreover very few normal lungs are likely to be in maximum inspiration at death, and, as mentioned before, this phase can be ascertained only if the trachea has been ligated prior to opening of the chest.

After death the walls of blood capillaries develop an increased permeability. Dyes which remain within the bed of the blood circulation when injected into a living organ may leak into the pericapillary tissue in the dead organ (for brain capillaries see Part IV, Chapter 3, C, 2, c).

The consistency of most organs depends, to a great extent, on the local blood pressure. Since this pressure drops to zero after death the consistency of the organs decreases instantaneously. This change is more or less proportionate with the vascularization of the different organs. In addition to this sudden decrease in consistency a gradual softening takes place as the result of postmortem autolysis. Intracellular enzymes are able, after death, to attack the cells in which they were produced. By and large the organs in the abdominal cavity are subject to faster autolysis than organs in other regions. If two organs in the same location show a marked difference in the rate of their autolysis we conclude that their enzyme content was different. An example of this kind is the fast autolysis of the thyroid tissue as compared to the very slow autolysis of the parathyroid tissue. This difference is particularly impressive in those animals in which the parathyroid is enclosed by the thyroid.

Pathological conditions during life may have an influence on the rate of autolysis after death. In a human disease known as acute yellow atrophy of the liver a large number of liver cells disintegrate during life. If, after death, such liver is kept unfixed at room temperature and samples are taken at different intervals, the proportion of decomposed cells increases much faster than in other livers. As a rule, cancer cells seem to be less subject to postmortem autolysis than ordinary tissue cells. This is conspicuous if groups of cancer cells are scattered in the liver or kidney.

In the study of pathological morphology a comprehensive knowledge of changes caused by death is necessary. As mentioned before, certain segments of the renal tubules are equally susceptible to chemical injury during life and to autolysis after death. Since the resulting pictures of renal cells are the same, additional criteria are needed to determine when the destruction occurred. Similarly, circumstantial evidence is needed to separate hemolysis after death from hemolysis during life.

Special conditions may develop if an organism dies a slow death. The blood may coagulate in such a way that so-called agonal thrombi form; they are intermediary between the usual products of clotting during life (thrombosis) and the different types of postmortem clots. In patients dying slowly of a brain tumor, edema of the lungs is observed. This is the result of increasing paralysis of the respiratory center; it is part of the mechanism of dying. Human patients who die of various diseases sometimes develop small thrombi on the valves of their hearts during the last weeks of their lives. These thrombi are not caused by bacteria, but are otherwise very similar to those which form in infectious endocarditis. The formation of small noninfectious thrombi on the valves of the heart is not a disease in the ordinary sense, but rather a symptom of approaching death. The diagnosis of terminal endocarditis is very appropriate; for a discussion of this condition see Gould's "Pathology of the Heart" (1953, pp. 755-756). In experimental animals which received large doses of Roentgen rays or cortisone the organs are frequently found to be flooded with bacteria. Most of these bacteria do not belong to any disease-producing type. The whole mysterious phenomenon was mentioned earlier in Part III, Chapter 3, A.

In this connection I wish to dispel certain erroneous ideas concerning postmortem changes of the bacterial flora. Contrary to a widespread belief, the tissues of a dead body are not readily invaded by bacteria which were present in the intestines or other places during life. Von Gutfeld and Mayer (1932) showed that, at different intervals after the death of human patients, no intestinal bacteria appear on the outside of the intestinal walls (serosa). These authors also demonstrated that no mechanisms are available by which bacteria could possibly be transported or migrate over large distances in a corpse. The conclusion is that bacteriological investigations of the dead body are suited to shed light on the bacterial flora which was present in the living organism; the precautions needed in such studies were described in Part III, Chapter 1, B.

In experimental animals different methods of killing may lead to different conditions at autopsy. The lungs will be found congested when inhalation of chloroform or ether was the method used. Decapitation mutilates the organs of the neck and produces a violent redistribution of blood. Convulsions during agony are apt to cause mechanical injuries which may interfere with the morphological analysis.[4] A minimum of structural disturbances seems to occur if pentobarbital sodium is used as follows. First a small quantity is injected intraperitoneally to produce anesthesia. Then a lethal quantity is injected intravenously.

Environmental stimuli to which an animal was exposed during the last days or hours of its life may produce morphological responses which can still be seen after death. In animals which were subjected to forced exercise before death, the muscles show a much larger number of patent capillaries than the number seen in the muscles of animals killed without preceding exercise (Krogh, 1922). Rabbits which were killed after having been scared by ferreting showed conspicuous morphological alterations of their thyroid glands as compared to rabbits which were killed without being aware of danger (Eickhoff, 1949; Kracht and Spaethe, 1953). When Nissl stated in 1910 that the basophilic structures which he had discovered in fixed and stained nerve cells were equivalent to pre-existing though invisible structures in the living cells, he emphasized that not only histological procedures but also the method of killing and the condition of the animal prior to death had to be standardized. The fascinating story of the subsequent demonstration of Nissl bodies in living cells will be reported in our later discussion of so-called artifacts.

It seems that both egg cells and somatic cells show differences in ultraviolet absorption before and after death of the cells. This has been attributed to redistribution of nucleoproteins (Vlès and Gex, 1928).

2. *Interferences by Procedures Used*

a. *The concept of artifacts in the light of reproducibility, precision, and accuracy. Nissl's equivalent pictures. Valuable observations with uncertain reproducibility.* The procedures for the study of living and dead material which were described in Chapter 2 entailed, almost without exception, some interferences with the natural conditions of the biological object. Some of the interferences were of a negligible degree such as the inspection of the living eye, whereas others produced highly artificial conditions such as the perfusion of isolated organs *in vitro,* or the preparation of fixed and stained microscopic sections. The fact that a procedure creates highly artificial conditions does not necessarily detract from its usefulness. As Zeiger put it very aptly in the preface of his 1938 book: "In the present era of microtechnical experimentation we deliberately produce new structures and alter pre-existing structures, thus using artifacts as tools in biological research."

A certain technique may produce a useful artifact, whereas a seemingly slight

---

[4] When using illuminating gas for the killing of experimental animals one should be aware of the difference between the effect of carbon monoxide in manufactured gas and the effect of the methane in natural gas.

modification of the same technique may produce a disturbing or useless artifact. If a proper amount of colchicine is administered to an animal, mitotic divisions in all cells which are reached by the colchicine are arrested in the metaphase. Therefore, histological samples taken some time after colchicine administration can be used for *cumulative counts of mitoses.* In their study of mitotic activity in the small intestine of rats, Stevens and Leblond (1953) observed that 4 hours after colchicine administration the arrested metaphases could be identified. However, 6 hours after administration of colchicine clumped and disintegrated nuclei made the counting of metaphases impossible. For a survey of colchicine techniques see Levine (1951).

If the artificial conditions which an experiment creates can be reproduced every time the experiment is repeated, then we are faced with *potentially useful artifacts.* If the conditions of an experiment cannot be reproduced, the data obtained are usually valueless. Then, the decision whether the uncontrolled variations were natural or artificial is important only for selecting better animals or designing better procedures. An example of a useful, reproducible artifact is the staining of nuclei with hematoxylin in sections of fixed, killed material. An illustration of a useless artifact is given in Fig. 35. It shows an unstained section of the formalin-fixed spleen of a dog. The section contains a neat pattern of black dots, which are precipitates of formalin with hemoglobin. The dotted outlines correspond exactly to the walls of the sinusoids. In other words, structures which are known to exist during life, have been made visible in an attractive picture. Yet, this technique is useless, since formalin precipitates cannot be reproduced at will.

The concepts, *reproducibility, precision,* and *accuracy,* are not always clearly understood. A procedure is reproducible if it yields similar results upon repeated application. The term accuracy refers to results obtained with two or more different procedures. The degree of accuracy increases with the number of independent reproducible procedures which yield the same results. There is no absolute accuracy.

This is the operational definition of accuracy. Frequently one encounters an unoperational definition referring to absolute accuracy which can never be achieved and, therefore, is called "the true value." While the concept of true value is misleading in most areas, it is useful or innocuous when referring to specified standards. The standard may either be available, such as a standard solution made of known components, or it may be fictive. An example of a fictive standard is the true mean used in statistics, which is a theoretical value referring to the mean of the universe, i.e., of all samples in the world.

The term precision refers either to the reproducibility or to the sensitivity of a procedure. Reproducibility was defined above. Sensitivity concerns the units of measurement: the smaller the units which allow reproducible measurements the greater is the sensitivity of the method applied.

In biological work reproducibility, precision, and accuracy play the same role as in physics or chemistry. However, special difficulty has been produced in biology by a vague use of the artifact concept. As pointed out earlier it is unfortunate that the terms natural and artificial are used in an absolute sense. With respect to morphological procedures the idea of *the* natural (true, real)

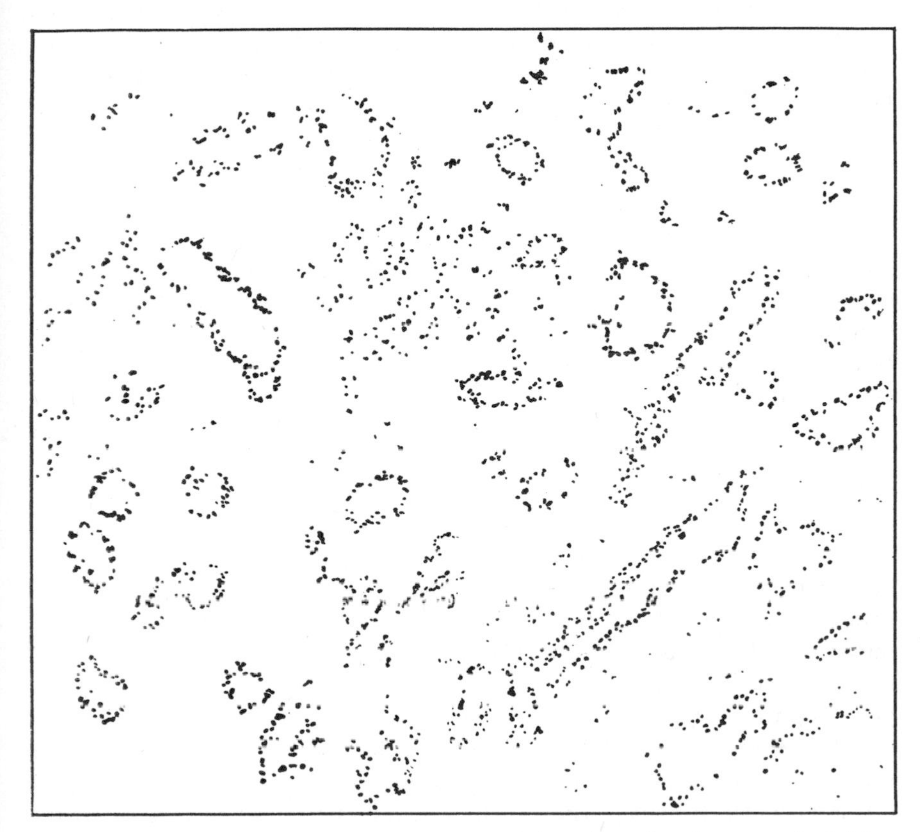

FIG. 35. Formalin pigment outlining sinusoids of the spleen of a dog, a result of interaction between formalin (the fixative) and hemoglobin. Although structures known to exist during life are outlined by the black dots, this is not a useful technique since it is not reproducible. Drawing, approximately ×200, traced from a photomicrograph (Mayer, unpublished data, 1944).

structure is as unoperational as the use of accuracy in an absolute sense. In a recent discussion of electron microscopic pictures of the cytoplasm a distinguished cytochemist expressed regret that "the unpleasant problem of the possible artifacts" had to be raised. I do not find it unpleasant to determine which procedure causes the minimal departure from the initial state of a biological structure. It simply means that the same structures should be examined with two or more independent procedures so that accuracy can be determined operationally. Comparisons between the appearance of certain structures during life and after killing by fixation will be illustrated later by examples from light and electron microscopy. Let us clarify the operational concept of accuracy by a relatively crude example. In Bassett's (1943) study of the vascular pattern of the rat ovary it was important to determine the stages of the estrous cycle. Therefore he used three independent criteria: the vaginal smear, the copulatory response, and the histological appearance of the sectioned ovaries. Dependence was placed only on those animals in which the three criteria were in reasonable agreement. From our point of view a satisfactory degree of accuracy was achieved with respect to timing of critical stages.

The confusion caused by a vague use of the artifact concept is the result of two misconceptions: (1) absolute naturalness, which we have just discussed, and (2) the idea that findings in dead biological material cannot be as valuable as findings in living material. In order to dispel the second misconception an example from experimental embryology may be helpful; it is taken from Holtfreter and Hamburger's chapter in "Analysis of Development" (1955). The experiments to be discussed deal with the so-called inductors, which were described in Part II, Chapter 3, C (see also Table 2). Certain areas with known inductive power were killed by different chemical and physical procedures, and subsequently tested by grafting them into the blastocoel of early *Triton* gastrulas (see my Fig. 4). If neural structures and eyes with lenses formed, the presence of inductive power was proved. Inductive power was retained by some methods of heat killing, but destroyed by others. Most surprising were the results of a supplementary experiment: areas of the ectoderm and entoderm which are noninductors under living, natural conditions acquired inductive capacity by some of the killing procedures. This phenomenon proves that inductive power does not depend on the living or dead condition as such but on physicochemical properties which may be present in either the living or dead condition.

Another example may illustrate how useful the discovery of structures in dead material can be, although, for a long time, no identification of these structures may be possible in living material. It was in such a situation that F. Nissl's systematic approach proved most constructive (see Nissl, 1910). In 1892 he had discovered special structures in the cytoplasm of nerve cells, which disappeared upon experimental destruction of the axon. Since these structures, subsequently named Nissl bodies, could be seen only in killed and fixed cells, after special staining procedures, many investigators would have stopped at that point. Nissl, however, demonstrated the reproducibility of his findings. Variability in size and number of Nissl bodies proved to be related (a) to the different types of nerve cells, (b) to the intact or damaged condition of axons which belong to the nerve cells, and (c) to the condition of the animal immediately before death (e.g., disappearance of Nissl bodies in animals exhausted by exercise). Nissl concluded that the fixed and stained bodies were *equivalent* to some structures present in the living tissue, although invisible with the methods available. As Zeiger (1938) pointed out, Nissl's concept of *equivalent pictures* should be applied broadly: all pictures resulting from fixation and staining are useful if they can be considered as reproducible equivalents to some living structures. At this stage the question remains open how much *structural resemblance* may exist between the fixed-stained preparation and the living structure.

Subsequent investigations concentrated on the chemical composition of the Nissl bodies. In nerve cells which had been fixed in fluids, such as alcohol, or by Gersh's freezing-drying method without the use of chemical agents, the absorption of ultraviolet was studied with the microspectrophotometric technique of Caspersson. It turned out that the Nissl bodies consisted mainly of ribonucleic acid, a result which was supported by enzymatic procedures. We mention some of the publications which represent these developments: Landström, Caspersson, and Wohlfart, 1941; Gersh and Bodian, 1943; Hydén, 1943; for a historical summary with illustrations see Caspersson's 1950 book. At this point the chemical nature of the Nissl bodies was determined, but not their presence as discrete morphological structures in living cells. There was still the possibility that material which was finely dispersed in the living cell formed clumps when the cells were killed. Peterson and Murray (1955) followed the development of Nissl bodies in explanted neurons of spinal ganglia of chick embryos. Explants from a 6-day embryo cultivated 8 days *in vitro* showed, when fixed and stained, the Nissl substance in the form of fine particles. Fixed and stained explants from 9-day-old embryos showed, after 15 days of cultivation *in vitro,* either heavier particles or the classic flake-like Nissl bodies throughout the cytoplasm of the nerve cells. These studies were completed by microscopic observations of living cells. Deitch and Murray (1956) also cultivated neurons of chick embryo spinal ganglions *in vitro.* With phase contrast microscopy the more mature living cells exhibited discrete flakes which corresponded in location to the Nissl bodies. Subsequently Deitch and Moses (1957) examined with ultraviolet photomicrography the same type of cultured cells during life. There was a satisfactory correlation of ultraviolet absorption, ribonuclease extraction, and staining. Deitch and her collaborators rightly concluded that Nissl bodies pre-exist in the living neuron as discrete aggregates, are essentially unchanged after good fixation, and contain high concentrations of nucleoprotein. I am sure that this chain of splendid investigations would have been impossible if the interest in the Nissl bodies had vanished during the long period when they could not be identified in living cells or in fresh dead material.

Since microscopic observations and interpretations depend to a great extent on photomicrographs it is important to realize that the same material may yield different pictures with different photomicrographic techniques. This is illustrated in Shillaber's (1944) book by a nonbiological example. When a fine wire mesh is photographed in transmitted light, the resulting silhouette shows two-dimensional black stripes which include white rectangles. In reflected light the same wire mesh shows a three-dimensional appearance of intersecting rods with reflecting convex surfaces. Shillaber emphasizes that neither of the two pictures (his Figs. 211 and 212) can be said to be exclusively correct. For the purpose of determining mesh size, transmitted light alone is enough; but for recording the amount of wear on the surface, reflected light is necessary. This leads us to a seemingly trivial but important statement, namely, that relevance is as important as reproducibility, precision, and accuracy. In each biological or non-biological study, one may like to use all procedures available in order to obtain the highest degree of accuracy. In most studies, however, it will be necessary to restrict oneself to those procedures which appear *relevant to the particular problem.* It is, for instance, not practical to investigate the general problem of growth. The question is which changes of living material one wants to investigate. One may measure increase in volume, in cell number, in wet weight, in dry weight, in total protein, in nucleic acids, etc. None of these measurements is more correct than the others. Attempts at correlating several of the measurements are, of course, very useful (see Part V, Chapter 3, A and B).

Many of the so-called artifacts are irrelevant. In preparing paraffin sections the fat is dissolved. Therefore, holes are seen in the places where fat drops had been. This is, for most purposes, an innocuous or irrelevant artifact.

In the earlier part of this discussion I stated that observations or experiments are valueless unless there is a known degree of reproducibility. However, there are exceptions to this rule. As an example I mention early experiments on artificial parthenogenesis in vertebrates. Bataillon (1910) pricked unfertilized frog eggs with a needle that had been dipped in blood. As a result a number of eggs underwent cleavage and some of them went through advanced embryonic stages. Very few reached the tadpole stage. F. Levy (1920) used the same technique on approximately 200,000 eggs. A small proportion of them developed into tadpoles, usually with some malformations. Two of the tadpoles completed their metamorphosis. Thus it was proved that mature frogs can develop from unfertilized eggs once cleavage has been initiated. The fact that Levy succeeded in obtaining two mature frogs from 200,000 eggs with artificially induced parthenogenesis does not indicate how frequent the success would be in another series of 200,000 eggs. Incidentally, in most later studies of experimental parthenogenesis the number of embryos that reached final stages remained a small percentage of those that started to develop. Several factors are known to which the disturbance of the parthenogenetic development can be attributed (see Tyler's article in "Analysis of Development," 1955).

Recent experiments with transplanted nuclei of embryos yielded important results in spite of a low degree of reproducibility. Briggs and King (1953) removed the nucleus from the frog's egg after having activated it by the prick of a glass needle, as described above. With micropipettes it was possible to transfer nuclei from cells of later stages into the enucleated eggs. This technique permitted the determination of the role of nuclei in differentiation, a problem which heretofore had been considered insoluble. Conclusive data were obtained although transfers were successful with no more than 34% of blastula nuclei and 15% of gastrula nuclei.

b. *Technical interferences in studies of living material.* In keeping with our program, the discussion of interferences with morphological procedures will be preceded by a discussion of interferences with various physiological procedures. In this way I hope to avoid the impression that the question of artifacts is a peculiarity of morphological techniques.

Special precautions against interferences are taken in psychosomatic experimentation. The stomach reflexes of the Pavlov dog can be studied only if the animal is unaware of the presence of an observer and is properly insulated from uncontrolled stimulation of its sense organs (for illustration of isolation arrangement see, e.g., Winton and Bayliss, 1955, p. 467).

Elimination experiments such as the removal of the adrenals involve a number of factors which may obscure the results of the mere removal of the organs. Therefore sham operations in control animals are necessary. Such animals receive the same restraint and the same anesthesia as the experimental animals, their skin and abdominal wall are cut open, and the adrenals are exposed to the mechanical injury connected with the preparation of their removal. In other words, the sham-operated animals are subjected to the same abnormal stimuli as the experimental animals, except for the final removal of the adrenals. Sham operations were suggested by Claude Bernard in his "Introduction à l'étude de la médecine expérimentale" (1865; p. 181 of the 1952 edition).

Physiologists have been aware of the various interferences produced by isolating surviving organs. In 1920 Belt, Smith, and Whipple took the pessimistic stand that isolated, perfused organs deteriorate and die so fast that no reasonable experiment is possible with such techniques. It was an important contribution of the Carrel-Lindbergh technique (1938), which was described previously, to show that after several weeks of perfusion an isolated cat thyroid contained enough living tissue to produce colonies of multiplying cells *in vitro.*

In some cases an isolated organ or piece of an organ is most useful immediately after isolation, whereas in other cases the living material needs some time to adjust to the new condition. In the method of O. Warburg (1923) slices of a living organ maintain a constant metabolism during the first half hour after isolation; after half an hour *in vitro* the results lose their reproducibility. Luft and Hechter (1957) reported that excised adrenals of cows lost their reactivity when they were tested by perfusion immediately upon arrival in the laboratory. The time which had elapsed since the death of the cow was one and one half hours (dissection and transportation from the slaughter house). However, after 30 to 60 minutes of perfusion in the laboratory, the glands started to respond to adrenocorticotropic hormone (ACTH) by synthesis of corticosteroids. Attempts were made to correlate electron microscopic pictures with the different stages of biochemical reactivity; this was done by taking biopsies from the glands which were used for perfusion. Markowitz *et al.* (3rd ed., 1954, p. 276) describe that the isolation of bowels by the method of Puestow in dogs causes a severe inflammatory reaction of the isolated segments. The bowels recover satisfactorily within a period of 10 to 16 days, and thus become suitable for studies of reflexes and secretion.

Under special conditions the removal of two important organs interferes less with the health of the whole animal than the removal of one organ. A famous example of this paradoxical occurrence is the so-called Houssay dog. If the pancreas is removed, the dog develops high blood sugar concentrations and dies from severe diabetes. If, in addition to the pancreas, the hypophysis is also removed, the animal can be kept alive for a considerable period. This is a result of the antagonistic functions of the two organs. The hypophysis increases the blood sugar concentration, while the islets of the pancreas decrease it. In the absence of both organs the blood sugar remains within values that are compatible with the life of the dog. Such a dog is weaker than a control dog and not able to meet unusual physiological demands (for references see Houssay *et al.*, "Human Physiology," 1955, p. 420).

Interferences on the macroscopic level are important in experimental physiology and pharmacology. Elimination of the thyroid function can be achieved by surgical removal of the gland. This is relatively simple in some species, but more difficult and uncertain in other species. Loss of some parathyroids can hardly be avoided when surgical methods are used. The thyroid function can be suppressed by administration of the so-called antithyroid drugs such as propylthiouracil. Surgically inaccessible thyroid tissue is reached by this procedure, but a complete elimination of thyroid function is not possible in all species with the use of chemical suppression. The third method is the application of radioactive iodine which is stored in thyroid tissue. Sufficiently large doses destroy all thyroid tissue and therefore the whole thyroid function, but little is known concerning side effects of this type of irradiation.

In order to illustrate the idea of complementary procedures with their respective advantages and disadvantages, I will now give a detailed discussion of the determination of blood pressure in rats. During each contraction of the heart, or systole, the blood pressure reaches a maximum, called the systolic pressure. When the heart muscle changes to its relaxed phase, or diastole, the blood pressure drops to a minimal value, the diastolic pressure. These pressures can be measured by inserting a cannula into an artery and by connecting the cannula to a manometer through a tube filled with saline. If the manometer is attached to a moving chart, a graph is obtained showing (1) the pulse rate, i.e., the alternation of systolic and diastolic pressure as a function of time, and

(2) the values of the pressures in manometric units. In small animals such as rats the pulse rate is so high that special manometers with very small inertia and special equipment for optical reading (Hamilton, Brewer, and Brotman, 1934) are necessary in order to follow the rapid alternation of systolic and diastolic pressure. Manometric measurements are *direct determinations* of blood pressure which, from the standpoint of hydrodynamics, represent the most reliable type of procedures. From the biological point of view these procedures entail a number of factors which are liable to interfere with the natural circulation of the rat: (1) the animal lies on its back for the duration of the observation; (2) it is restrained in its movements; (3) general or local anesthesia is applied to expose the artery surgically; and (4) the artery is manipulated by ligatures, insertion of the cannula, and removal of the nerve-containing external layer.

Indirect determinations of blood pressure are possible which interfere much less with the natural conditions of the rat. One of the indirect procedures is based on changes in volume of a rat's foot which are produced by changes in local blood pressure (Kersten, Brosene, Ablondi, and SubbaRow, 1947). The rat is held, in a more or less natural position, on the top board of a box, one hind foot across a hole in the board. Under the hole is a photocell, and on top of the foot is the source of light. The light which is transmitted through the foot is measured by a so-called densitometer (galvanometer attached to the photocell). Around the lower part of the leg a hollow rubber cuff is wrapped which can be inflated and which is connected with a manometer. By moderate inflation of the cuff the veins are compressed, and stronger inflation also compresses the arteries. As a result of vascular compression the foot swells. More light is absorbed by the swollen foot than by the foot in its original condition. This change is read in the densitometer. When the same rat is used both for indirect densitometer measurements and for direct manometric measurements, a satisfactory correlation between the two methods is found. Thus the densitometer values can be transformed into manometric units. The advantages of the indirect method are the following: (1) the posture of the animal during observation is fairly natural, (2) no anesthesia is needed, (3) no surgical disturbances are produced in the blood vessels, and (4) blood pressure can be determined repeatedly at various intervals; thus changes in blood pressure can be studied which are produced by aging or by the administration of drugs or special diets over long periods of time. When the direct manometric determination is used routinely, each animal is killed after a few hours. It may be possible to maintain a rat for several weeks while the cannula remains in the artery, but this procedure would require special studies of interferences. In larger animals, such as the dog, it is easy to cannulate the same artery repeatedly. For this reason, direct manometric measurements of blood pressure can be used without difficulty in the same dog over long periods of time.

It is remarkable in how many ways the choice of procedures is governed by the size of animals. This was discussed previously in connection with the sampling of organs, or pieces of organs, for microscopic study. The sampling of blood from living animals is another case. Blood samples in the quantity needed for most studies can be taken from a dog at fairly frequent intervals without causing anemia, since the samples form a negligible proportion of the total blood volume of the dog. In the rat the minimum amount of blood needed for a morphological or biochemical examination represents a tangible loss to the animal. In young rats frequent sampling of blood leads to anemia in spite of seeming smallness of samples. This is particularly important in the study of new drugs. While blood samples may be taken from the same dog twice a week, the same rat can be bled only every two months. Therefore different rats which have been treated in the same way must be available for blood samples after one week, and after two, three, or more. weeks.

In Part III, Chapter 3, B, tagging of whole animals was mentioned. Possible disturbances caused by tagging deserve a place in our present discussion. Different species of birds respond differently to trapping and banding. Blue jays avoid the trap after having been caught once; they learn how to obtain food from the trap without being caught. When black-capped chickadees have been caught and banded, they allow themselves to be trapped again; it seems that food is their only concern and that they are not bothered by the handling (data from de Kiriline, 1946). In most species of migratory birds, banding has evidently not prevented birds from reaching their destination. It is difficult to tell what interferences, if any, are produced by the presence of a band in birds. Tagging of fish has been done in various ways. Bass have been tagged by fixing numbered clips both on the upper and lower jaws. This may restrict the opening of the mouth to a slight extent. Special studies have shown that this tagging of bass does not interfere with their longevity, although it does interfere with their increase in size and body weight (Schumacher and Eschmeyer, 1942). When worker bees are studied in an observation frame, individuals can be tagged by different color markings, as described previously. However, the marking procedure makes the bee unacceptable to the other bees so that it will be prevented from entering the hive. This interference can be compensated by placing small drops of honey on the fresh color marks. With this precaution the marked bee is not objectionable (after von Frisch, 1931).

Interferences on the microscopic level have many aspects. When microscopic structures are studied in fixed and stained preparations, the first question is: "Were these structures present in the living material?" The use of living structures as standards needs two qualifications. First of all, most living microscopic structures are not static, but change in the organism as a function of time and environment. Secondly, observations of living microscopic structures are hardly possible without the production of some alterations. Bělař (1930) mentioned changes which occur at great speed in living cells as a result of mechanical injury such as teasing apart a group of cells or puncturing a cell by microdissection. He emphasized that the mere spreading of living cells in a drop of fluid on a slide causes delicate structural changes. Some microscopic structures are more sensitive to changes in environmental temperature than others. Heated stages and incubator-like boxes for microscopes have been constructed to control these factors.

Photomicrography with visible light does not produce, as a rule, much injury if heating is avoided. Photomicrography in ultraviolet (UV) light is apt to damage living cells. The first picture taken in UV may show the natural conditions if the exposure was completed before visible responses of the cells took place. Subsequent photographs, however, may show alterations caused by the first exposure (illustrated in Fig. 20 of Caspersson's 1950 book). For this reason Caspersson declared living cells to be unsuitable for microspectrophotometry. However, P. M. B. Walker (1956) differentiated between two degrees of damage: (1) damage which makes it impossible to follow the fate of an individual cell over a considerable period of time, and (2) damage which prevents precise measurement of ultraviolet-absorbing material at a given moment. Since degree (2) rarely occurs, Walker concludes that, as a rule, it is possible to make valid microphotometric measurements on living cells in the ultraviolet, although often it may be necessary to discard a cell after measurement. As I mentioned earlier, UV is necessary for observations with fluorescent dyes, but some of these dyes fade rapidly when exposed to UV. This is the case with the diaminoacridines which were introduced by DeBruyn, Robertson, and Farr (1950).

Vital staining of nuclei with diaminoacridines was described previously in connection with Farr's work on the fate of lymphocytes (1951). In our general discussion of tagging procedures on the microscopic level by vital stains or radioactive isotopes (Part III, Chapter 3, B), it was stated that these procedures should cause a minimum of interferences with the structures and functions of cells. It is not always easy to verify the presence or absence of interferences on the microscopic level. In examining the cytological action of methylene blue, Ludford (1935) proceeded as follows. Fibroblasts of chicks and rats were cultivated in serum (hanging drop method). When a substantial zone of migrating and multiplying cells had developed, methylene blue was added to the medium and allowed to remain in it for five minutes. The action of the methylene blue took place while the culture was in the incubator. Then, the staining of mitochondria was verified microscopically. After this, the dye was washed off with saline, fresh drops of serum were added, and the cultures were returned to the incubator and examined at various intervals. Twenty hours after staining, the experimental colonies showed an inhibition of cell migration and multiplication, although the individual cells appeared morphologically similar to those of the control colonies. In other words, the mitochondria had been stained while the cells were healthy, but somehow the methylene blue damaged the cells during the five minutes of vital staining.

Radioactive isotopes can be used either for marking or for destroying microscopic structures. If radioactive iodine (I^{131}) is administered to an animal or human being, the thyroid cells store it selectively and radioautographs show the distribution of the I^{131}. If large amounts have been deposited, the thyroid tissue is destroyed by radiation. This destruction is utilized in certain experiments and in the treatment of patients with excessive thyroid function.

Several examples of natural tagging were mentioned earlier (Part III, Chapter 3, B). One of them will be rediscussed here with respect to the question of natural versus artificial conditions, namely, the experimental use of malaria parasites as tags in red blood corpuscles of canary birds (Fig. 34). In this case, a timetable was supplied by the fact that immature parasites infect immature red cells, and that the rate of maturation of the parasites is known. Hegner and Hewitt (1937), who arrived at a maturation period of 24 hours, questioned whether the maturation of the red cells is perhaps accelerated or retarded by the presence of the parasite. They studied uninfected red blood cells *in vitro* with the result that within 24 hours the frequency of young red cells decreased from 6% to 0.4%. In order to obtain a large initial number of immature cells, some birds were made anemic with phenylhydrazine so that a flood of regenerative young cells was obtained. In this experiment the young cells decreased from 33.4% at sampling to 0.2% after 24 hours survival *in vitro*. Which of the conditions of the experiments above are natural and which are artificial? The infection of canary birds with malaria parasites as used here is produced in the laboratory, but the relations between red blood cells and parasites are probably the same as those that occur in natural malarial infections of birds. The uninfected red corpuscles studied *in vitro* came from their natural environment, namely, the circulating blood of healthy canary birds. The *in vitro* condition is, of course, artificial. The pretreatment with phenylhydrazine created an anemia, an abnormal state of the blood, and in addition, the red blood corpuscles were observed under unnatural conditions *in vitro*. Yet, the phenylhydrazine experiment confirmed the two others with respect to the period of 24 hours needed for maturation. Obviously *natural and artificial conditions may be equally useful*. This statement holds only if the conditions of experimentation and observation have been properly specified, as it was in the study of Hegner and Hewitt.

The explantation of fragments obtained from living organs was described previously. The production of a colony of somatic cells *in vitro* (tissue culture) was illustrated in Fig. 21. The cells on the surface of the fresh explant are dead or damaged as a result of mechanical injury which is unavoidable in isolation procedures. It takes a certain amount of time until the cells from the interior of the fragment migrate through the necrotic crust. This is known as the lag time. When the viable cells have reached the culture medium they migrate and multiply freely for a considerable period. If a well-developed colony is transferred to a new medium, it is usually divided into equal halves, as shown in diagram 4 of Fig. 21. By this act, a number of cells are crushed, but the injury is better controlled than the injury connected with the original explantation. Consequently the lag time and other quantitative criteria of the development of each colony are more or less predictable. This is one of the reasons why transferred cultures (passage cultures) are more suitable for quantitative work than fresh explants. The different factors which are involved in the transfer of tissue cell colonies have been analyzed by E. Mayer (1933, 1935). Fresh explants should not be used for any experiments concerning the media, since the substances which are liberated from the dead and dying cells are largely unknown.

In order to obtain reproducible results, uniform cell strains *in vitro* are desirable. Two factors can interfere with the cultivation of uniform strains: (1) the embryo juice which is added to the media for promotion of cell migration and multiplication may contain intact cells which contaminate the experimental strain; (2) since most explanted fragments contain a variety of cell types, more than one cell type may survive, in spite of serial transfers. Contamination by extraneous cells can be reduced by filtration of the embryo juice. Bryant, Earle, and Peppers (1953) have described filters with pores small enough to retain cells but wide enough to allow the passage of large molecules on which the activity of the embryo juice depends; to make the embryo juice filterable its viscosity had to be reduced by brief treatment with hyaluronidase. The problem of obtaining uniform somatic cell strains has been solved by cultivating single cells which were isolated by means of micropipettes, as described in Part III, Chapter 2, A, 4 (Sanford, Earle, and Likely, 1948). However, the cultivation of strains obtained from single cells does not prevent the occurrence of variations, i.e., of cell types which deviate from the majority. This segregation of new types in a seemingly homogeneous population has been known for some time in bacteriology. Another analogy between tissue cell populations and bacterial populations *in vitro* is the lag time after transfer into a new medium. While in the case of tissue cell colonies mechanical injury seems to account for the lag time, unknown biochemical factors have been invoked in the case of bacterial populations. It is customary to state that the lag time represents a period of adjustment to the new environment. One may question whether the term adjustment, as used in this connection, has any operational meaning.

In concluding our discussions of procedural interferences with living structures I refer the reader to McIlwain's book "Biochemistry and the Central Nervous System" (1955). This author emphasizes that the study of isolated cerebral fragments yields systems which are artificial but allow a wide range of useful experimental studies. He also explains that a number of metabolic problems could not be studied in cerebral preparations with intact cellular structures but required ground and centrifuged homogenates. In other words, it may be necessary to destroy cell structures in order to discover activities of systems inside the cells (McIlwain's book, p. 60). It seems to me that this is one of the best recent analyses of useful artifacts.

c. *Technical interferences in studies of biological dead material.* Let us first consider the *macroscopic level.* Changes of macroscopic structures which are produced by the death of the organism were mentioned earlier. The main examples were contractile organs such as the heart, the arteries, the intestines, and the urinary bladder; in some species, especially in dogs, the spleen belongs in that group. Some of the macroscopic procedures described in Part III, Chapter 2, B, 2, are not without pitfalls. The caliber and number of vessels which are filled by an injection of colored masses depends on the state of the animal before death and on the presence or absence of rigor mortis. In experimental animals, it turned out that injection immediately after death, i.e., before rigor mortis set in, stimulated the vessels to contract, particularly when glycerin-containing masses are used. To counteract this effect vasodilators have been injected either before death (during anesthesia), or after death, as an admixture to the injection masses (Gatenby and Beams, 1950). By such treatment one obtains a maximum of injected vessels, but no information on the state of vessels during life without artificial interferences. Not only the injection of hollow structures with colored masses, but also ordinary dissections may yield results which do not correspond to the natural conditions. These problems are of importance in the field of pathological morphology, and particularly in medicolegal pathology. Suppose there is the suspicion that a human patient had a pneumothorax (air in the pleural cavities) resulting in the collapse of one lung or both lungs. How can that be ascertained at autopsy since any opening of the chest will collapse the lungs when the conditions were normal? This difficulty can be resolved by dissecting the skin from the thoracic wall and holding the skin up so as to form a pocket with the thoracic wall. Water is poured into this pocket, and intercostal spaces are punctured by a knife under water. Under normal conditions few air bubbles if any will appear, but with a pre-existing pneumothorax there will be a conspicuous amount of bubbles. It is difficult to determine after death how variable the volume of the lungs was during life. By means of Moolten's (1935) apparatus the lungs can be fixed in an inflated state, but the degree of inflation produced by the investigator may or may not be the natural inspiratory maximum.

Ulcers of the stomach or intestine that were near perforation before death may be ruptured at autopsy by manipulations far from the ulcers. Thrombi and emboli may be extracted inadvertently from blood vessels unless the vessels are dissected in their natural situation. Pus and other abnormal fluids are apt to trickle from the thoracic to the abdominal cavity when the diaphragm is cut. Therefore, the contents of the abdominal cavity must be ascertained while the diaphragm is intact. The proper procedures for obtaining bacteriological samples at autopsies (von Gutfeld and Mayer, 1932) were described in Part III, Chapter 1, B.

Procedural interferences on the *microscopic level* were mentioned repeatedly in connection with the relative advantages and disadvantages of different histo- and cytological methods (Part III, Chapter 2, B). Here some of the points of the earlier discussion will be repeated, but the emphasis will now be on the interferences.

The first step in preparing fixed sections is sampling. The possible distortions of original conditions which may occur at sampling are more or less identical with the distortions produced by macroscopic procedures: disruption of topographic relations, dislodging of material, loss or redistribution of blood, and mechanical injury (see e.g., Fig. 27, rat's testis). Fixation may lead to the swelling of some structures and the shrinking of others, but different fixatives may produce opposite effects. This is illustrated by the diagrams of thyroid sections, Figs. 25 and 26. Important material may be dissolved by the fixative. Some fixatives, especially those that contain sublimate, form precipitates in the sections. Special techniques are available to remove such precipitates. Granules of formalin pigment are shown in Fig. 35; their appearance and solubility are similar to those of malaria pigment, a circumstance which can lead to confusion if only formalin-fixed material is available. Some fixatives make the application of stains difficult.

One of the main concerns of cytologists has been the change of living structures by different fixatives. When phase contrast microscopy extended the visibility of structures in living cells, a new appraisal became necessary. Buchsbaum (1948) cultivated *Salamandra* macrophages *in vitro,* took phase contrast photomicrographs of individual living cells and replaced the medium by a fixative. Then the fixed cells were stained with hematoxylin and photographed. In this way a number of fixatives were studied.

Embedding procedures cause a filling of natural spaces with the embedding material; if the embedding material cannot be removed from the sections, it may take some of the histological stains and thereby disturb the interpretation. The process of sectioning with a microtome may involve mechanical injury, such as tearing, scratching, or cracking. More treacherous is another effect of sectioning, the distortion of shape without any of the obvious injuries mentioned. Olszewski (1952) described how the sections of the brains of cats or monkeys were flattened in the direction perpendicular to the edge of the knife. Olszewski's Figs. 4a and 4b illustrate the marked difference in shape of two adjacent sections from the same block. The only technical difference was that after the cutting of the first section, the block was rotated 90 degrees relative to the edge of the knife.

Staining procedures are rarely specific in the sense that one procedure stains one type of structure only. As a rule, a variety of structures take the same stain. The technique of differentiation, which was described previously, produces the preferential stain of some structures over others. If variable structures are stained, differentiation may produce any result which is desired by the investigator. The only protection against arbitrary handling of differentiation is the use of relatively constant structures as standards. If, for instance, the highly variable colloid of the thyroid is stained with the red dye azocarmine and the blue dye aniline blue, many combinations can be obtained at will. However, the nuclei in the same section should always stain red and the collagenous fibers should always stain blue, while the cytoplasm should be practically colorless. If these three structures show their standard staining behavior, the staining of the colloid in the same section is no longer arbitrary: the degree of differentiation is defined. In the staining of bacteria, the comparison with known standards has been traditional. In order to determine the effect of Gram stain (see, e.g., Dubos, 1945, p. 72ff.) on unknown bacteria, one stains known Gram-positive and known Gram-negative bacteria together with the unknown sample. This routine of bacteriological staining is sometimes neglected in histological staining.

The last steps in the preparation of stained sections are mounting and sealing. Different mounting media may increase or decrease the transparency of the sections in an undesirable way. The mounting media may gradually cause the fading of stains; or fading may result from oxidation if the sections are not sealed airtight.

If there is any reason to assume that certain structures seen in a fixed and stained section were produced by the method applied, usually more than one of the technical steps can be accused. It may be the combination of a particular fixative, a particular embedding technique, and a particular stain which is responsible for the final structural result. This will be illustrated by a number of examples. In these examples two questions will be posed: (1) To what extent does the procedure applied interfere with the purpose of the investigation? (2) Does the procedure create unknown gaps between the living structure and the fixed and stained preparation? In some cases it will be possible to find satisfactory answers, but in other cases the questions will remain open.

The well-known dye, hematoxylin, was used to discuss general techniques of staining and differentiation (Part III, Chapter 2, B, 3). What are the relations between pre-existing structures of the sample and hematoxylin-stained structures seen in the fixed preparation? In the early era of microscopy nuclei had been discovered in unfixed and unstained preparations, but not more than their outlines and some vague inner structures had been seen. In a fixed, hematoxylin-stained section more nuclei are visible than in an unstained section. Yet, no one considers the excess of stained over unstained nuclei to be misleading artifacts, since the reason of the increase is known. Because of the improved contrast by the stain most nuclei are visible even when located in several layers, whereas in an unstained section many nuclei are obscured by structures above and below. During mitosis the nuclear material is rearranged in the shape of chromosomes (Fig. 61). They were discovered in fixed and stained preparations, and it took a long time until improved methods allowed their recognition in fresh and living cells.

Suppose irregularly distributed blue dots, a few microns in diameter, are seen in a hematoxylin-stained section, but no similar structures are seen in the unstained section. Let us also assume that dilute HCl destains the blue dots as well as the nuclei in the section. To find out what these dots are, we stain a similar section with a hematoxylin solution which has been filtered immediately before use. If staining with the filtered solution produces none of the blue dots in the section, the dots evidently were particles of the dye. If the blue dots also appear when the freshly filtered solution is used, we conclude that the blue dots are *equivalent* to pre-existing structures which cannot be demonstrated in the unstained preparation. The concept of equivalence as introduced by Nissl was defined previously. There are various dotlike structures which stain with hematoxylin and can also be detected in unstained sections: fragments of nuclei, spherical bacteria (cocci), and particles of calcium or of calcium-containing material such as bone. The possible identification of such dotlike structures was discussed earlier in connection with the question: how specific is hematoxylin stain? The chemical determination of nuclear substances and calcium will be mentioned again in the chapter on cyto- and histochemistry.

The history of cytological and histological discoveries started with an era in which an amazing number of microscopic structures were discovered without the use of staining procedures, or with a casual use of dyes. The next era was characterized by improvements of microscopes and microscopic techniques. This included a prodigious development of staining prescriptions. However, some of the fine structures that were discovered by fixation and staining were suspected to be products of the techniques applied and were usually called artifacts. In the third, and present, era, systematic attempts are made to demonstrate the same structures with different techniques, and, if possible, to identify them in the living material. Contrary to expectation, most of the structures which had been discovered in fixed and stained preparations proved to be pre-existing in living material when appropriate methods were applied.

One of the most impressive vindications of old-fashioned fixing and staining procedures is the story of the Nissl bodies which we have reported in detail. Two other cytoplasmic structures, mitochondria and the Golgi apparatus, had been discovered in fixed and stained cells, but were suspected by many investigators to be artifacts. Subsequently the mitochondria proved to be stainable with Janus green in unfixed cells and to be visible, without staining, in dark-field microcinematographic pictures of living chick fibroblasts cultivated *in vitro*. For some time the Golgi apparatus could not be demonstrated in fixed and stained preparations of tissue cultures (Walker and Allen, 1927; Macdougald and Gatenby, 1935). The absence of the Golgi apparatus in cultivated cells allowed two interpretations. Either the conditions *in vitro* caused some damage to the cells, expressed by the loss of intracellular structures such as the Golgi apparatus, or the fixation of structures in cultivated cells was superior to that in cells of organ samples from whole animals, since no dissection had to precede the fixation of cells *in vitro* and penetration by the fixative is particularly efficient in the flat, loosely arranged cells of tissue cultures. According to the second interpretation, the Golgi apparatus was a fixation artifact which was avoided when cultivated cells were used instead of organ samples. In later studies optimal cultivation techniques were applied to well-established strains, with the result that a distinct Golgi apparatus was seen in most cells (Macdougald, 1937). Since there is a death rate even in the healthiest cell population, some cells showed a disintegrating Golgi apparatus. Whereas the role of mitochondria in cell metabolism is clearly established, the functional role of the Golgi apparatus, or Golgi complex, is still "one of the dark points of cytology," as Brachet puts it in his 1957 book (p. 62). Neither structure is an artifact.

It was described previously that sheets of flat cell aggregates, when killed and fixed in $AgNO_3$, show a pattern of black lines after reduction of the $AgNO_3$ to metallic silver (Fig. 51a). Since these lines run at some distance from the nuclei and enclose similar areas of cytoplasm they were interpreted as cell outlines (von Recklinghausen, 1863; see also Ranvier, 1889). There was, however, the alternative that the silver precipitated along special chemical zones rather than along pre-existing morphological structures. The question was finally decided by a study of Robinow (1936). He used explants of rabbit kidneys, which produce flat sheets of cells when cultivated on coverglasses (hanging drop cultures). A group of well-identified living cells was microphotographed with dark-field illumination. Within five seconds after exposure[5] the coverglass with the cell colony was placed in an $AgNO_3$ solution; the reduction to silver was carried out as usual. By comparing photomicrographs of a group of living cells with the silver picture of the same group of cells, the following result was obtained. The dark-field photomicrograph of the living cells showed white lines which corresponded exactly with the black lines of the silver-impregnated cells. There was no doubt that the silver had precipitated along pre-existing structures of the living cells. A three-dimensional analysis of these structures will be given later.

Robinow observed perforations of the silver-positive lines resulting in pro-

---

[5] Personal communication from Dr. Robinow (1957).

trusions of cytoplasmic masses. One had to arrive at the paradoxical conclusion that some cytoplasm was located outside the cell boundaries. Similar observations were made by Bartelmez (1940) in the inner lining of the uterus. This material was obtained surgically from macaque monkeys. Three to five minutes after removal, the fragments of fresh material were ready for microscopic inspection. Comparisons were made with material fixed immediately after removal. Tongues of cytoplasm protruding beyond the terminal bars (cell boundaries) and clear vesicles forming a dome on each cell were seen both in the fresh and in the fixed preparations. These bizarre structures would have been considered to be fixation artifacts if they had not been present in the fresh preparations as well.

The preparation of ultrathin sections for electron microscopy demands special embedding and cutting procedures. The sections are not mounted on the smooth surface of a glass slide, but on the perforated surface of wire screens (see Part III, Chapter 2, B, 3). It is not known what temperatures are produced in thin tissue sections which are in the path of the electron beam, but they may not be far from 100° C. In view of this brutal treatment of delicate sections, the electron microscopist is permanently faced with the question familiar to the light microscopist: To what extent do the dead structures seen in the electron microscope correspond with living structures? An early comparison, before the availability of phase contrast microscopy, was made by Porter, Claude, and Fullam (1945). As an example of a recent analysis, I mention the paper by Fawcett and Ito (1958). In their electron microscopic study of testicular cells of guinea pigs these investigators found a large number of hitherto unknown structures consisting of undulating double lines. Then they examined the same type of cells with phase contrast microscopy, while the cells were still alive. The living cells, suspended in fluid, were at least 10 μ thick. Since the resolving power of a light microscope is incomparably smaller than that of an electron microscope, one could not expect to see the double-line structures with any distinctness. However, a number of single-lined structures were seen with the phase contrast microscope which were arranged and spaced similarly to the double-lined structures and also matched them well in relative size as compared to the nuclei. The conclusion was justified that these structures were identical and that therefore the electron microscopic pictures corresponded to pre-existing structures in living cells. In addition, the study under the light microscope (phase contrast) allowed an analysis in the third dimension:[6] the structures in question proved to be cytoplasmic membranes of which the cross sections appeared as double lines in the electron microscope (see my Fig. 57G).

The original discovery of paired membranes was made by Sjöstrand and Hanzon in 1954 in an electron microscopic study of exocrine cells of the pancreas. By examination of living pancreas cells in polarized light these investigators

[6] The problem of vertical resolution (focal depth) in electron microscopy will be discussed in Part III, Chapter 6, B, 1, c.

confirmed the view that the electron microscopic structures corresponded with structures present in the living cytoplasm.

Certain structures in fixed and stained sections are known to be results of the histological technique. Ham's textbook "Histology" (3rd ed., 1957, pp. 24-26) gives instructive photomicrographs of scratches produced by nicks in the microtome knife and of folds produced by the mounting process. Other cases are more involved. Structures which obviously did not pre-exist in the living material may require a special study: it may be necessary to demonstrate *how* the histological technique produced these particular structures. Figure 25 illustrates the fact that thyroid sections fixed in one fluid show cracks of the colloid, whereas such cracks were absent in sections of the same thyroid fixed in another

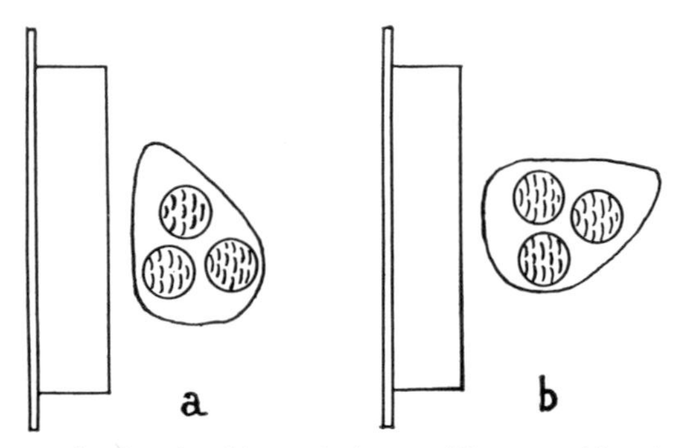

FIG. 36. Diagram showing thyroid sample in two different positions (a and b) relative to microtome knife (the paraffin block which encloses the sample is omitted). In both positions the ripples which indicate cracking of colloid are parallel with edge of knife. Evidently the cracks were produced by a shoving action of the microtome knife. Knife, thyroid cross section, and follicles are not drawn to scale.

fluid. The only step in the procedures which could possibly account for the cracking was the sectioning with the microtome knife. Some investigators ascribed the cracks to nicks of the knife. This interpretation was unlikely in view of the regular pattern of the cracks and was finally excluded by a special experiment illustrated in Fig. 36. When the paraffin block was cut in different positions, it turned out that the cracks always ran parallel with the edge of the knife. This means that the knife exerted a shoving action on the colloid so that it cracked if brittle enough. A shoving effect of the microtome knife as seen in brain sections was mentioned before (Olszewski, 1952). It seems that the brittleness of the colloid is an expression (1) of a relatively high concentration of proteins, and (2) of a rapid coagulating effect of some fixatives (Mayer, 1949). Incidentally, the knife produces what one may call *latent* cracks in the colloid, since the cracks are not visible in the paraffin sections before the sections are spread on a slide. Under certain conditions, nicks of a knife can interfere with staining properties. Yegian and Porter (1944) observed in sections that tubercle bacilli lost their acid-fastness (resistance to destaining by acid) where they had been hit by a nick of the microtome knife.

The diagrams Fig. 25 and Fig. 26 illustrate the fact that the histological picture of the thyroid, including the colloid, is determined both by the functional state of the gland and by the technique applied. The origin of cracks of the colloid has just been analyzed. There is still another variable property of the colloid which will be discussed now, namely, its multiple staining. If a section is stained with the red dye azocarmine and the blue dye aniline blue, the colloid may appear red in some follicles and blue in other follicles. An intermediary violet stain of colloid also occurs. Finally, the colloid may be stained part red and part blue in the same follicle, usually with a fairly distinct demarcation between the two areas. This last condition is seen in the photomicrograph Fig. 37. The fact that the blue part is nearer to the surface (capsule) and the red part nearer the interior of the thyroid is not accidental, but is the rule, as expressed in the diagram Fig. 39. This regular pattern suggested that the diffusion of the fixative from the surface of the organ into the interior was responsible for the dual staining.

In order to verify this possibility an experiment was designed in which the fixative entered the organ from one side only (Fig. 39; after Mayer, 1949). One of the almond-shaped lobes of the thyroid of a dog was cut into halves by a cross section. In the surface of a large paraffin block a hole was made which had the shape of half an almond, but was slightly wider than the thyroid. The thyroid was placed in the hole, and the gap filled with Vaseline petroleum jelly so that only the cut surface of the thyroid remained free. Then the block with the thyroid was submerged in Bouin's fixative, which consists of picric acid and formalin. Under these conditions the fixative could enter through the cut surface only. After the usual fixation time the gland was removed from the paraffin block and split longitudinally. As one would expect, the fixative produced a graded yellow staining, with a maximal intensity near the cut surface and practically no staining of the tip area opposite the cut surface. Paraffin sections were prepared parallel to the plane of the longitudinal splitting. Staining with azocarmine and aniline blue gave the result shown in the diagram Fig. 39. Traveling in the direction of the diffusion current of the fixative, one encounters first a zone of follicles with dual staining in which the blue part of each colloid ball is located toward the cut surface. The next zone shows predominantly violet colloid, and the last zone unstained colloid. This experiment proved that the multiple stain was a direct effect of the diffusion currents of the fixative. Earlier investigators had interpreted the dual staining as an indicator of different iodine concentrations in the colloid, or of different chemical types of colloid. In view of the experiment described, the multiple stain is to be ascribed to differences in protein concentration. Mayer and Fürstenheim (1930) had arrived at the same conclusion when multiple stains, similar to those of thyroid colloid, were seen in accumulations of clotted protein in cysts of kidneys and ovaries. I also found that clotted plasma in blood or lymph vessels, or proteinic casts in the tubules of the kidney may show multiple staining indistinguishable from that of thyroid colloid. Figure 38 illustrates such casts in the kidney of a rat. The marginal vacuolization (scalloping) of the casts adds to the similarity with thyroid colloid (renal "colloid casts," S. Weiss and F. Parker, 1939). Since this particular rat had been treated with a drug which suppresses the synthesis of iodine-containing hormone

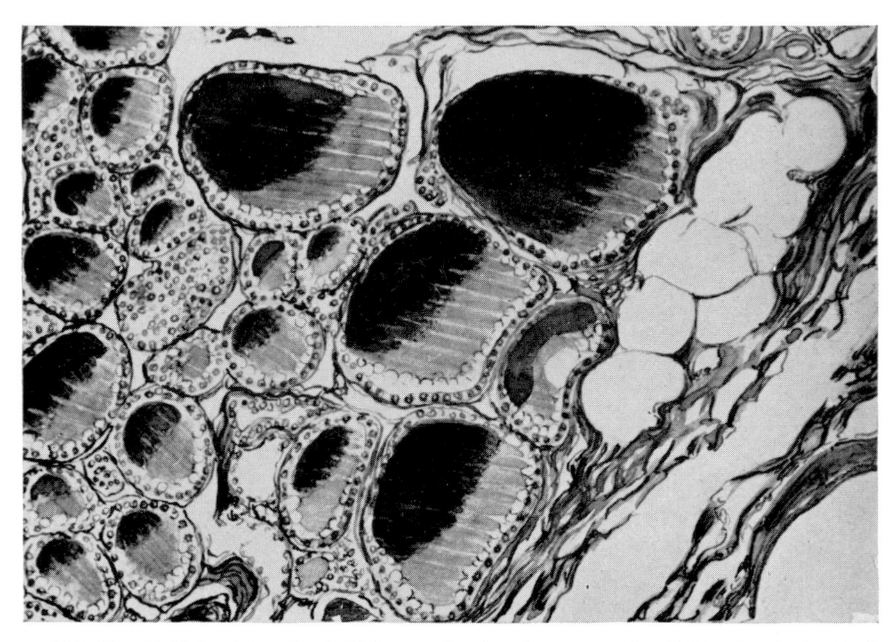

FIG. 37. Multiple stain of colloid in thyroid of a dog. Parts of colloid stained blue (with aniline blue) appear gray in the photograph, whereas parts stained red (with azocarmine) appear black. Fixation with picric acid, acetic acid, and formalin; paraffin embedding, "azan" stain. Photograph made from colored drawing by E. Piotti. Magnification approximately ×200. Interpretation: The fixative, which entered through the capsule (on the right side of the picture), carried the proteins of the colloid toward the center of the gland so that in each follicle the dilute colloid coagulated near the capsule and the concentrated colloid near the interior. At coagulation the colloid retracted from the follicular wall, resulting in scalloping. (Mayer, unpublished, 1949.)

in the thyroid, there seemed to be a possibility that the kidneys had taken over the thyroid function. For this reason, stored samples of these kidneys were analyzed for iodine content in Dr. E. B. Astwood's laboratory (Boston, Massachusetts), but their iodine content proved to be very low. The major part of urinary protein in rats is similar to serum α- and β-globulin (Sellers, Roberts, Rask, Smith, Marmorston, and Goodman, 1952). In the thyroid colloid, globulin forms the bulk of the proteins. It is an open question whether in all protein masses which are found inside natural cavities, the phenomena of marginal vacuolization and multiple staining are tied to the presence of globulins.[7] The conclusion from all these studies is that the staining properties of the thyroid colloid are unspecific indicators of varying protein concentrations and are unrelated to the iodine content of the thyroid colloid (Mayer, 1949). It seems to me that this interpretation is well supported by recent observations in rats by Hooghwinkel, Smits, and Kroon (1954), who found that in each follicle the

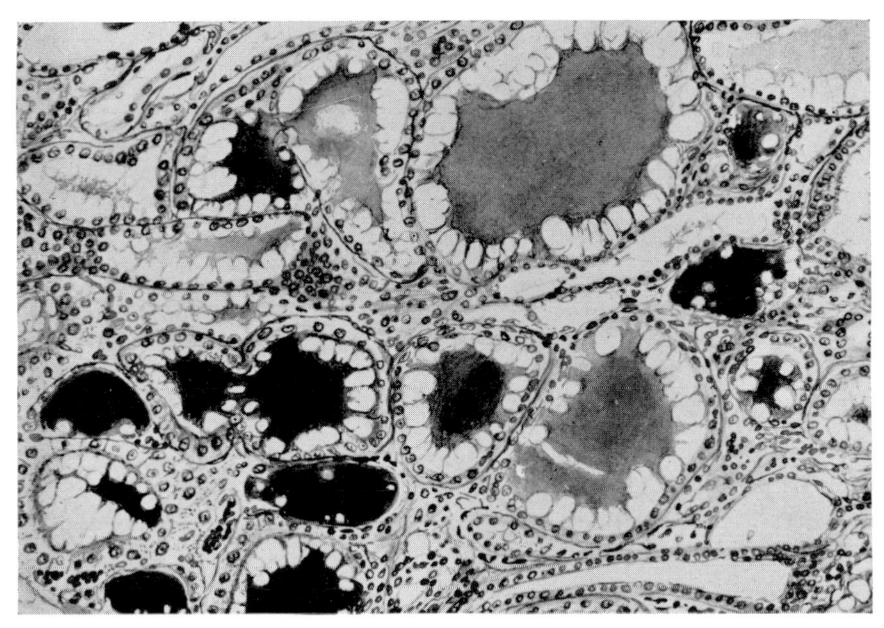

FIG. 38. Multiple stain of protein casts in dilated tubules of a rat's kidney. Technique as in Fig. 37. Blue-stained casts appear gray in the photograph, red-stained casts appear black. Note pronounced scalloping. Photograph made from colored drawing by E. Piotti. Magnification approximately ×200. Interpretation: Multiple stain indicates different concentrations of proteins. Similarity of observations in thyroid colloid and renal casts makes it improbable that the phenomena in the thyroid are related to iodine content. (Mayer, unpublished, 1949.)

radioactive iodine or the PAS (periodic acid-Schiff reaction) positive material was concentrated in a crescent-shaped peripheral portion of the colloid. This is illustrated in their Figs. 2 and 3, which are very similar to my Fig. 37. In summarizing, I may state that the role of the histological techniques in producing different histological pictures of thyroid colloid has been demonstrated with satisfactory completeness.

In other instances it has not been possible to resolve the relations between

[7] I do not know whether colloid-like clotted protein found outside the thyroid contains PAS positive material as the thyroid colloid does.

puzzling histological pictures and the procedures applied. Structural variations in the kidneys of dogs may serve as an example. The main components of the mammalian kidney are the glomeruli and tubules. The glomeruli consist of a tuft of blood capillaries from which certain components of the blood leak into the space of Bowman's capsule. The resulting fluid is the so-called glomerular urine. From Bowman's capsule a funnel-like opening leads into the first segment of the tubular system. Well-known variations in the shape of the glomeruli

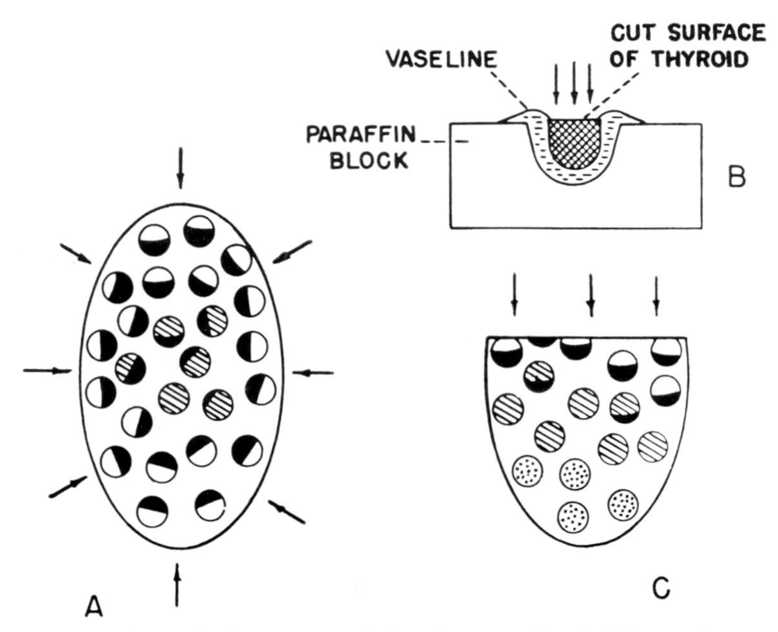

FIG. 39. Experimental evidence that multiple stain of thyroid colloid is a result of diffusion of fixative. Arrows indicate entry of fixative into the sample. (A) Diagram of a complete section through a dog's thyroid. Outline indicates capsule. Figure 37 corresponds to any peripheral area of this diagram. (B) Half of a fresh thyroid (crosshatched) is placed in a slightly larger depression of a paraffin block. The space between thyroid and paraffin is sealed with Vaseline petroleum jelly (marked by stipples). Upon submersion of the whole preparation in fixative, the fluid can enter from the cut surface only. (C) Section vertical on cut surface of (B) after fixation and staining as in Figs. 37 and 38. Symbols of colloid stain in follicles (circles in diagrams): white for blue stain; black for red stain; shaded for violet stain; dotted for gray (unstained). (Mayer, unpublished, 1949.)

and the lining of Bowman's capsule are shown diagrammatically in Fig. 40a–c. An interesting variation which has received relatively little attention is the protrusion of the first tubular segment into the space of Bowman's capsule, as illustrated in Fig. 40d–f. By means of serial sections it was demonstrated that the protrusions result either from telescoping or invagination of the first tubular segment (Mayer and Ottolenghi, 1947). Some investigators who saw protruded tubules in cross sections only (Fig. 40d), without visible connection with the tubular system, were mystified by the ring-shaped structures in the space of Bowman's capsule. The protrusions of tubules were probably mentioned first by Kölliker in 1867; they were forgotten and rediscovered repeatedly. Mayer and Ottolenghi, who investigated various aspects of the problem, found protrusions not only in experimental dogs, but also in healthy control dogs of different ages. A variety of procedures was used: different ways of killing the animals, different fixations, frozen and paraffin sections, etc. None of the techniques could reasonably be accused of causing the protrusions of tubules. On the other hand, it remained unknown whether tubular protrusions are present

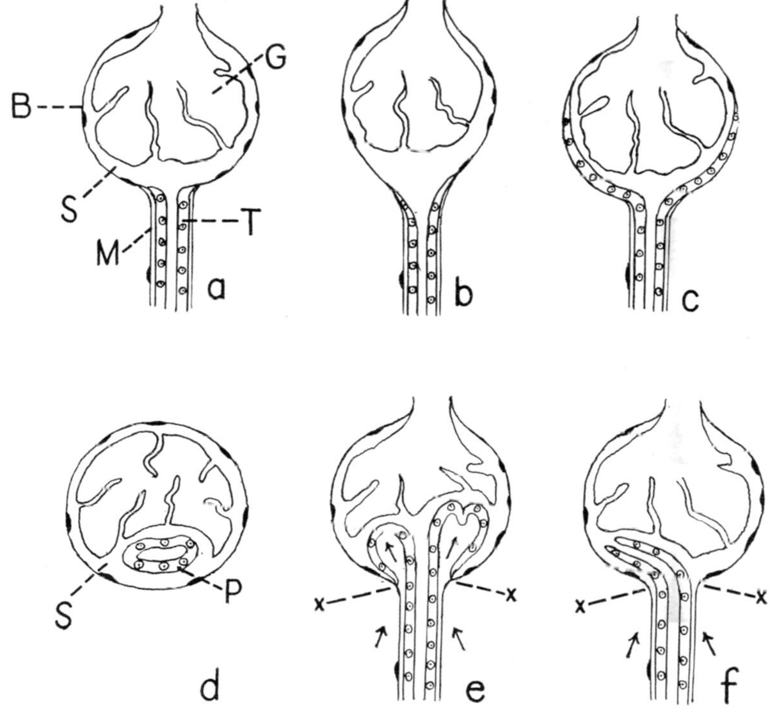

FIG. 40. Diagrams of variations of renal structures in dogs and cats: a–c, variations occurring during life; d–f, variations of unknown origin, either occurring during life or produced by morphological technique.
(a) Statistically normal glomerulus: *B*, lining of Bowman's capsule; *S*, space of Bowman's capsule; *G*, glomerular tuft of capillaries; *T*, wall of the tubular segment which is next to the glomerulus; *M*, basement membrane forming a sheath around the tubule. (b) Pear-shaped variation. (c) Increased height of cells which line Bowman's capsule. (d), (e), and (f) Protrusion of tubular segment (*T*) into the space of Bowman's capsule. (d) Cross section through a piece of tubule (*P*) in space of Bowman's capsule (*S*). (e) Invagination of tubule, the end of which is still attached to Bowman's capsule (places of attachment indicated by two crosses). (f) Telescoping of tubule; attachment to Bowman's capsule lost. In (e) and (f), arrows indicate hypothetical forces which pushed the tubule into the space of Bowman's capsule. (Modified from Mayer and Ottolenghi, 1947.)

in living dogs. Nor was it possible to decide whether agonal redistribution of blood in the kidneys may play a role in the formation of tubular protrusions.

A detailed report on the protrusions of renal tubules was given because this story helps to clarify the use and misuse of the term artifact. Evidently, it is not justifiable to dispose of bizarre or mysterious histological structures as artifacts as long as the question is open whether these structures were or were not produced by technical interferences.

d. *Concluding remarks on artifacts and contaminations in different branches of science.* The concepts of artifacts and contaminations are used to indicate limitations in the validity of observations or experiments. If the limitations can be specified, the terms *artifacts* and *contaminations* are acceptable. Let us begin with so-called contaminations. If monochromatic light of a certain wavelength is desired, the admixture of a small amount of light of other wavelengths is sometimes called a contamination which, in this case, is a graphic and unambiguous statement. In chemistry and biochemistry, the term *impurities* is used rather than contaminations. There are well-developed procedures for the purification of compounds. The relative amount of impurities can be determined or estimated. In some cases it is sufficient to remove the impurities, but in other cases it is necessary also to identify them. In bacteriology, the concept contamination is less useful. As a rule, it is qualitative and means that it has not been possible to obtain the desired pure cultures. Then it remains an open question whether the original sample contained the undesirable admixtures, or whether poor technique allowed other bacteria to enter the culturing media. In their bacteriological study of human corpses, von Gutfeld and Mayer (1932) had to overcome preconceived ideas concerning contaminations.

The concept of artifacts in biological investigations was the main subject of the preceding discussion. It was emphasized that physiological procedures, including morphological techniques, are liable to change the conditions of the object from those which existed at the beginning of the experiment or observation. *In physics and chemistry the concept of artifact is hardly needed.* So-called absolute standards are clearly recognized as agreements between investigators. No one will claim that the absolute zero point of temperature is more natural than zero based on the melting point of ice. The condition of a gas at a given time is characterized by temperature and pressure, but no temperature (or pressure) is considered more natural than another. Yet, physicists and chemists are concerned with procedural interferences as much as biologists. Let us look at the way in which organic chemists deal with various interferences in the determination of a melting point, which is an important characteristic of a crystalline substance. I cite two examples from Morton's "Laboratory Technique in Organic Chemistry", 1938. (1) The melting point of a crystalline mannite as observed under the microscope was 150° C. in unwashed tubes, whereas in clean tubes the melting point was 165° C. (2) The rate of heating may exert an influence on the melting point of the sample: the melting point of glycocholic acid could be changed from 132° to 178° C. by varying the rate of heating. Other technical problems concerning melting point determinations are mentioned by Morton. Concerning substances which decompose upon heating, the determination of a melting point may or may not lose its usefulness as a characteristic of a compound. For a brief presentation of this particular problem see the article "Micro- and Semimicro Methods" by Cheronis (1954, pp. 180 and 181).

An interesting problem of interference developed in connection with the advent of isotopes as tracers. According to the review by Sacks (1956), reaction kinetics are modified by the greater mass of a heavy isotope. For instance, a bond $C^{12}—C^{12}$ will rupture more readily than a $C^{12}—C^{13}$ or a $C^{12}—C^{14}$ bond. Sacks points out that this isotopic mass effect is not likely to interfere with studies of metabolic mechanisms, since such studies are usually qualitative or roughly quantitative, the reaction rate being irrelevant.

Protein chemistry is the one area of chemistry in which the concept of artifact has had a major role. Since proteins are products of living organisms, the term *denaturation* was introduced to denote changes caused by procedures of investigation. The natural state of proteins appeared inaccessible, and the deviations from this state remained hypothetical. In its modern usage, the term denaturation is entirely operational and has lost the implication of non-reproducible artifacts. The current *theoretical definition* of denaturation is based on Roentgen diffraction analysis. Such analysis indicates unfolding and uncoiling of protein molecules which results from definite chemical and physical factors. Uncoiling of protein molecules may very well occur within living cells (Lawrence, Miall, Needham, and Shen, 1944, p. 261). In current *laboratory language,* denaturation refers to a set of phenomena which are produced by handling protein samples from natural sources, such as egg white, blood plasma, or large living cells. Some effects proved to be reversible whereas others did not. It was found that a protein retains its solubility if it is first dried then heated to 100° C. but that it coagulates irreversibly if heat is applied in the presence of water (Anson, 1938, p. 407). In other words, seemingly rough procedures may not cause irreversible changes under all conditions. As a rule, rough procedures cause greater deviations from the original state of the sample than those caused by delicate procedures. No single criterion is a sufficient means to characterize a protein as denatured (Neurath, Greenstein, Putnam, and Erickson, 1944, p. 162). The characteristics of denatured proteins are listed in the textbooks of biochemistry (see, for instance, West and Todd, 1955, pp. 351–354).

Studies devoted to the determination of the natural conditions of hormones have particular aspects which deserve a brief discussion here. Hormones are substances produced by living organisms and have characteristic biological effects (see Part II, Chapter 3, G). Some of the hormones are proteins, other are not. Suppose a raw sample has been obtained by extraction from an endocrine gland which is known to secrete a certain hormone. Then the different steps of chemical treatment are accompanied by biological tests. The products of each new step are tested as to whether the known activity of the hormone is still present, has decreased, or increased. If there is an increase in activity per weight unit, one assumes that the raw material has been purified by the chemical procedures applied. However, the biological assay does not always help in deciding which of the new products is the more natural one. Two different products may be equally active. This is illustrated in a paper by Shepherd *et al.* (1956) on isolation, purification, and properties of β-corticotropin. By relatively gentle treatment a form with more amide groups was obtained, whereas somewhat rougher treatment produced a form with fewer amide groups. Both forms showed similar biological activity. Since it was unlikely that amide groups were added to the molecule under the isolation conditions used (alkali treatment), the most probable interpretation was that the loss of amide groups represented a step away from the natural form of the hormone.

In summarizing, one may say that procedural interferences, or artifacts, are not a peculiarity of biological techniques but are unavoidable in all branches of science. In biological as well as in nonbiological studies, *artifacts are reasonably classified* as follows: (1) reproducible or not reproducible; (2) reversible or irreversible; and (3) useful, irrelevant, or harmful with respect to the purpose of the particular investigation.

A useful artifact may, by a relatively small change of technique, turn into a harmful artifact. This may be illustrated by a microscopic study of nonbiological material described in Shillaber's book (1944, pp. 507-511). The problem is how to photograph a fine precipitate of "calcium carbonate" (probably calcite) mounted in Canada balsam. The particles are hardy visible in Canada balsam since their average refractive index is close to that of the mounting medium. If, prior to mounting, the particles are coated by a thin layer of selenium metal, they become visible with great distinctness. If, however, there is a flaw in the coating technique, bare areas around each particle appear and unevenness of coating makes the microscopic study useless. Three pictures by Shillaber show the poor visibility without artificial coating, the effect of perfect coating, and the effect of deficient coating. This example is comparable to the effect of colchicine, which arrests mitoses in metaphase: without colchicine the mitotic count may be too low for significant determination; after a proper interval between colchicine administration and killing of the animal the cumulative count becomes useful; after an excessive interval counting is spoiled by nuclear destructions (see Part III, Chapter 3, C, 2, a).

Chapter 4

Natural and Artificial Units

A. Units Smaller Than One Organism

In an earlier chapter various units of reactivity were discussed (Part II, Chapter 2). A motor unit was defined as a group of muscle fibers which is under the control of one motor neuron, and it was reported that the number of muscle fibers per motor unit varies considerably according to the function of the different muscles. The thread-shaped actomyosin molecule was mentioned as the contractile unit of the muscle on the molecular level. Unusual relations between nerve cells, nerve fibers, and muscle were discovered in squids. A large number of nerve cells (100 to 1500) of ordinary size fuse their processes to form a single giant axon, which may be as much as 800 μ in diameter. Each nerve consists of one giant fiber and a number of small fibers (1 μ in diameter). Hence one giant fiber with a large area of circular muscle forms the motor unit. In other animals, invertebrates as well as vertebrates, a single giant nerve cell gives rise to one giant axon. I have followed Prosser's (1950) presentation of this subject.

One of the items of my Table 2—genetics—contained a particularly interesting unit of reactivity, the gene. I reported the history of the transformation of the gene from a formal abstraction into a chemically defined substance. The gene started as a theoretical unit of hereditary reactivity and is now characterized by deoxyribonucleic acid (for four competing definitions of the gene see discussion of Table 2). I compared this development with the transformation of the atom concept after the establishment of the periodic table of elements. The general philosophy of units in natural science is fascinating, but it is beyond the scope of this book. However, the *choice of units* is an important *procedural step* in each study and is therefore one of our subjects. The order of magnitude of the item to be studied requires a proper order of magnitude of the units of measurement. It is not always relevant whether the units which are used for a particular purpose are natural or artificial. Taxonomic units will be discussed in Part V.

Artificial units smaller than one organism are obtained by isolation procedures. The isolated systems, described previously (Part III, Chapter 2, A, 3) covered a wide range: perfused groups of organs, perfused single organs, partially isolated slices of cerebral cortex *in situ,* completely isolated tissue slices *in vitro,* and finally explants cultivated *in vitro.* Some of the earlier discussions will be supplemented here by descriptions of artificial and natural units which are larger than one organism.

B. Artificial Units Larger Than One Organism: Cross Circulation, Parabiosis

Perfusion procedures, which are related to those described before, are used in so-called *cross-circulation* techniques. Let us start with an experiment which was designed to determine the nervous control of blood pressure. One dog was used to perfuse with its blood the head of a second dog. The second dog's head was severed from all connections with the trunk except spinal cord and vagus nerve. In addition, the suprarenal vein of the second dog was anastomosed with the jugular vein of a third dog. The blood pressure was measured in the femoral arteries of each of the three dogs. We refer the reader to a brief presentation of these experiments in Winton and Bayliss' book (1955, especially their Fig. II,25). It seems that, on rare occasions, two dogs with cross circulation were kept alive for 6 days (Firor, 1931). Temporary cross circulation between two cats was used by Lawrence, Ervin, and Wetrich (1945) in order to study the life cycle of white blood cells. In each experiment the carotid artery of a normal cat was joined to that of a leucopenic cat, i.e., a cat in which the production of white blood corpuscles had been suppressed by Roentgen irradiation of the bone marrow or by a particular disease (feline infectious agranulocytosis). When joined to a normal cat the leucopenic cat showed a more or less normal white blood count. The count dropped very soon when the exchange of blood with the normal partner was discontinued. Counts and calculations showed that the white cell population in the blood is replaced approximately one and one-half times in 24 hours, and that a white blood cell is present in the blood circulation for 7.2 hours. Studies by Sarnoff *et al.* (1958), in which cross circulation was combined with perfusion of an isolated heart, were mentioned earlier (Part III, Chapter 2, A).

Experiments based on the broad union of two or more animals are known as *parabiosis.* In rats, two different techniques of parabiosis have been developed. In one of them, skin and muscles of the abdominal wall are sewn together in such a way that a communication between the abdominal cavities of the two partners is produced (coelio-anastomosis); the intestines can be transposed at this occasion. In the other technique, much larger areas of skin and muscle of the two partners are joined together, but the body cavities remain separated.

Early investigators demonstrated direct union of blood vessels in parabiotic rabbits, with rapid passage of soluble substances, and also of large bacteria, from one partner to the other (Sauerbruch and Heyde, 1908). Subsequently, it was found in rats that some hormones of the hypophysis influenced the partner, while others did not, and that testicular and ovarian hormones, produced by one partner, had very little effect on the other. What appeared to be differences in the rate of transmission of various substances could not be correlated with any known properties of the respective substances such as molecular size. This seeming selectivity was ascribed to a mysterious "parabiotic barrier." The barrier idea became untenable when the highly efficient exchange of red blood corpuscles between parabiotic rats was documented numerically by Van Dyke, Huff, and Evans (1948). These investigators used five parabiotic pairs, three of them with and two without coelio-anastomosis. Rat erythrocytes, which were tagged with Fe^{56}, were injected into a tail vein of one partner, and samples from both were taken at intervals for measuring radioactivity. The average time for complete mixing of the added cells in the five pairs was 3 hours and 43 minutes, with a range of 1 hour and 40 minutes; maximum and minimum times were observed in two of the pairs with coelio-anastomosis. It was calculated that, during each minute, 0.64% of the total number of red blood corpuscles flow from one partner to the other. In order to determine the speed of capillary union, 8 other pairs were used. Prior to parabiotic junction, one of the prospective partners was pretreated for 2 weeks so that its red blood corpuscles were tagged with Fe^{56}. Then, each radioactive partner was joined to an untreated littermate. The results were as follows. On the first and second postoperative day very little exchange occurred; gradually, the exchange increased so that the bloods were completely mixed between the third and fourth day. The conclusion was that the capillaries began to unite between the second and third day.

In his 1952 review of parabiosis, Finerty described how it became eventually possible to account for seemingly unequal transmission of substances, such as hormones, from one partner to the other. Some substances when injected into rat *A* were destroyed rapidly in the organs of rat *B,* or in the circulation even before reaching the organs of rat *B.* Other substances were found to be more resistant, so that they could be identified in rat *B.*

The studies which I have described originated from the parabiotic techniques. I will now report the application of parabiosis to a special problem of dynamic morphology which could not be tackled with other techniques. It was known that in the liver of young adult rats mitoses are extremely rare. At a given moment no more than 0.005 to 0.01% of the liver cord cells are found in mitosis. Removal of a large part of the liver in rats was followed by marked increase of mitoses in the remaining portion. (Because of the stalklike bases of liver lobes in rats, several lobes can be removed without injury to the remainder). In order to determine whether local or systemic factors control the mitotic activity in the liver, Bucher, Scott, and Aub (1951) made a number of parabiosis experiments, with and without anastomosis of peritoneal cavities. A special point of their technique was the fusion of spleens in some of their experiments. Of two parabiotic rats, one partner was subjected to removal of the major part of the liver. When the parabiotic twins were sacrificed 48 or 72 hours later, the intact liver showed a sixfold increase of mitotic frequency. The next step was to prepare parabiotic triplets. The two external partners were deprived of 80% of their liver each. Twenty-four hours later the triplets were killed. The intact liver of the central partner showed a fiftyfold increase in mitoses. All these experiments proved to be sufficiently reproducible. Parabiosis as such did not change the mitotic rate in the liver as compared with that in single rats, and neither did sham operations on the livers have any effect on mitoses. Besides the mitotic counts, the livers were studied for possible changes in their chemical composition. The liver material obtained from one partner at hepatectomy was compared with the intact liver of the other partner at sacrifice. No significant changes were found in the concentration of water, total nitrogen, alkaline phosphatase, deoxyribonucleic acid, or ribonucleic acid. The authors concluded that as indicators of starting regenerating activity, the chemical criteria listed were less sensitive than the mitotic count. The most important conclusion, however, was that the mitoses in the liver were evidently initiated by chemical alterations in the blood rather than by any local factors. Islami, Pack, and Hubbard (1959) could not confirm the increase in mitoses in the liver of the intact partner of a partially hepatectomized parabiotic twin. These investigators did not use parabiotic triplets, nor did they produce a union of spleens. Evidently, parabiosis techniques are far from standardized. This is also evidenced by the high mortality rate reported by Finerty and Panos (1951); their data will be given below in connection with so-called parabiotic intoxication.

Numerous parabiosis studies deal with the action of sex hormones in postembryonic life; they were reviewed by Finerty (1952). Parabiosis has also been used successfully in the study of sex development in *embryos* (see Parabiosis in Hamburger's 1942 manual and in Burns' 1949 review). By fusing two amphibian embryos, important information was obtained concerning the interaction between the genetic determination of sex, i.e., the establishment of gonads under chromosomal control, and the hormonal determination of sex, i.e., endocrine effects of the established gonads. In any species with a natural 1:1 sex ratio, there is a 50% probability that parabiotic twins will be combinations of a genetic male and a genetic female. In their subsequent development, any significant deviation from this ratio can be ascribed to hormonal factors. If, for instance, 75% female-female combinations develop, a third of them can be ascribed to the transformation of genetic males by female hormones.

Remarkable hormone studies have been made in insects by the artificial union of two or more individuals. Certain bloodsucking insects *(Rhodnius prolixus)* have a long rod-shaped head which lends itself to partial or complete amputation by microsurgery. The decapitated insect can survive 6–10 months. The anterior part of the head produces a hormone that promotes molting, while a gland in the posterior part inhibits molting. Since molting is associated with metamorphosis (gradual maturation), the hormone of the inhibiting gland is also called the juvenile hormone. At the critical period the balance changes in favor of the molt-producing hormone. These mechanisms were demonstrated by decapitation experiments and by joining the head regions of two individuals

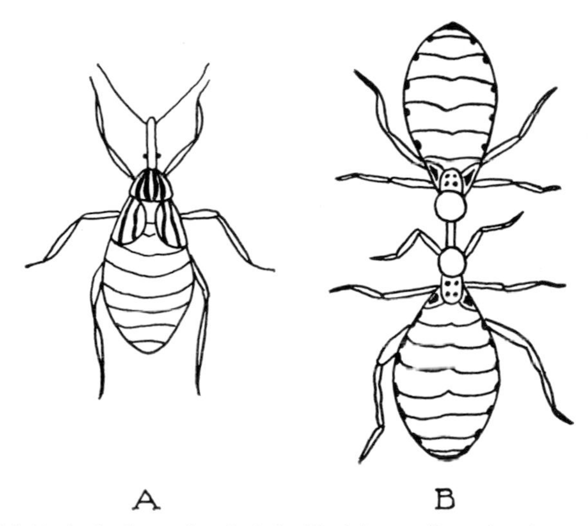

FIG. 41. Telobiosis (end-to-end union) in *Rhodnius prolixus,* for the study of developmental hormones. (A) Nymph (17 mm. long from tip of head to tip of abdomen) drawn 2 days before molt which produced a female. (From Brumpt, 1949, Volume II, Fig. 851). (B) One young individual decapitated before critical period: left by itself it would not have molted. Another young individual decapitated after the critical period: this one would have molted since its tissues contained a sufficient quantity of molt-promoting hormone. Joined together, both of them molted. (From Wigglesworth, quoted after F. A. Brown's article in Prosser's "Comparative Animal Physiology," 1950, Fig. 280 A).

in different stages. The joining of heads, or of anterior ends of decapitated individuals, is known as telobiosis. In Fig. 41 one of the young individuals was decapitated before the critical period: left by itself, it would not have molted. The other individual was decapitated after the critical period: this one would have molted, since its tissues contained molt-promoting hormone in sufficient strength. Joined together, both of them molted (Wigglesworth, quoted after Brown, 1950).

In mammals, experiments with cross circulation of parabiosis involve problems of compatibility between partners. Finerty and Panos (1951) produced 78 pairs of *non-littermate,* 31-day-old rats, using parabiosis without coelio-anastomosis. Death of one partner occurred in 50 pairs within 3 weeks after operation. Mortality was similar in groups treated with antihistaminics or cortisone and in untreated groups. In 22 pairs, signs of injury were evident for some time, but recovery occurred. Six pairs showed no signs of injury. In another series 10 *littermate* pairs were joined; of these five suffered death of a partner. Different hypotheses on parabiotic intoxication (injury by parabiosis) are discussed in Finerty's 1952 review. This subject is related to the problem of compatibility in transplantation of organs from one animal to another, which will be mentioned later (Part V, Chapter 2, C, 2). The observations of Finerty and Panos show that parabiotic pairs represent units of considerable variability.

C. Choice of Units Irrespective of Naturalness or Artificiality

In his analysis of the embryology (ontogeny) and evolution (phylogeny) of the vertebrate skull, de Beer (1937) was faced with the problem of identifying the morphological units of the skull. He found it impossible to offer a satisfactory embryological definition of "a cranial bone" if separate centers of ossification were used as criteria (de Beer, pp. 503–504). I mentioned the importance of the choice of units at various occasions. Electrical stimulation of an isolated muscle fiber was compared to stimulation of a whole muscle to illustrate the fact that the choice of different units of reactivity can result in qualitatively different responses to similar stimuli (Part II, Chapter 2, E).

For our purposes the term *natural unit* is applied to biological material which has been subject to a *minimum of interference* by observation or experimentation, and the term *artificial unit* refers to a unit obtained by *marked procedural alterations* of the original condition. As in earlier parts of this chapter, we define natural and artificial as relative terms. For instance, the wet weight of the whole liver of a rat is a natural unit, whereas a slice of liver weighing 2 mg. as used in the Warburg technique (Minami, 1923) is an artificial unit. In either case, it is necessary to describe the procedures by which the units were obtained. This is obvious in the case of the slice, but it is equally important with respect to the wet weight of the whole liver. One should state how the animal was killed, how the liver was dissected, and whether or not the blood was removed by rinsing.

It was mentioned before that two seemingly similar interferences may prove to be of unequal importance. In the heart-lung-kidney preparation, the heart of the dog can be replaced by an artificial pump (Winton and Bayliss, 1955, p. 243), but a dog's lung had to be included in the blood circuit to maintain blood flow and urine production in the isolated kidney (Bainbridge and Evans, 1914; Starling and Verney, 1925). Winton and his collaborators analyzed the various disturbances in the kidney caused by the absence of a lung; no circulatory disturbances occurred in the hindlimb when it was perfused without lungs (Wittaker and Winton, 1933). On the basis of all these experiences it finally became possible to maintain isolated kidneys in a functioning condition for several hours, without any lungs in the perfusion system. Couch, Cassie, and Murray (1958) perfused the kidney in a siliconized glass and Tygon system at a temperature of 25° C, using heparinized, oxygenated blood of the dog from which the kidney had been isolated. I give these details in order to illustrate the fact that, in ordinary isolation preparations, the kidneys and the lungs form a biochemical unit. This kidney-lung relation is not visible under natural conditions in the intact organism. Obviously, the question of natural or artificial units is less important than the recognition of the units which are of interest in a particular study.

The *individual organism* and the *cell* are probably the two biological units with the widest range of applicability. Some comments on these two units will be appropriate before we start our discussion of such units which are intermediary between organism and cell. The choice of the most useful unit is a problem which occurs mostly in studies of tissues, organs, or systems.

The *individual organism* is easily defined in vertebrates, but can mean many things in some classes of invertebrates. A comprehensive survey of different criteria of individuality is found in L. Loeb's 1945 book "The Biological Basis of Individuality." Questions of compatibility in cross circulation and parabiosis which were covered in our previous discussion are part of the problem of individuality. A few other aspects of this problem will be briefly mentioned here. The difference between the individuality of a vertebrate and the individuality of some invertebrates is well illustrated by the different principles of their respective motor systems. As von Uexküll (1921) puts it, the sea urchin is a republic of reflexes: when the sea urchin walks, the legs move the animal; when the dog walks, the animal moves the legs. The concept of individuality is hardly applicable to colonial animals. Therefore, a member of an anatomically continuous colony is called a zooid, which means something resembling an animal. Compound forms or colonies occur both in noncellular organisms, such as *Zoothamnium* and *Dendrosoma,* and in multicellular organisms such as coelenterates. A living stalk or bridge connects the different members of the colony. In the case of the sessile corals the living stalks and bridges fill a complicated system of channels inside rocklike masses. In the Siphonophora, which are motile colonial coelenterates, the zooids show an amazing polymorphism, i.e., an elaborate separation of functions with a corresponding variety of structures. The different specializations are floating, propulsion, palpation, mechanical protection, attack of prey and enemies (nettle organs), intake and digestion of food, and reproduction. All these zooids are in open communication through the cavity of a common hollow stem, the coenosarc (after Parker and Haswell, 1940).

A temporary loss of individuality occurs in protozoa which form feeding associations. This phenomenon was observed mainly in Heliozoa (*Actinospherium* and *Actinophrys*). In order to cope with a prey, such as a daphnia or alga, which is larger than an individual heliozoan, several heliozoans fuse their cytoplasm around the victim until it is enclosed in a common vacuole. When the digestion of the prey is completed, the fused mass separates into individual heliozoans (Biedermann, 1911, Vol. 2, p. 287).

In plants the question of individuality is frequently unanswerable. If two tree trunks of the same species develop close to each other and share their root systems, it is equally justified to talk about one or two trees. Moreover, the regenerative power of plants is very high. In some species, such as poplars, isolated twigs (cuttings) are able to regenerate a whole tree. Poorly defined individuality and almost unlimited regenerative ability are characteristics of so-called open systems.

Colonies of cells cultivated from explanted fragments of animal organs are considered as artificial organisms by some investigators. This matter will be discussed in the next chapter which includes tissue cultures as living models.

The *concept of cell* is equally applicable in vertebrates, invertebrates and plants. Difficulty is encountered in microscopic organisms which contain organs with different functions but do not show any subdivision in cells. These organisms are generally known as Protista or unicellular organisms. Some biologists, including me, prefer the term noncellular organisms which was suggested by Dobell (1911), since it expresses the fact that these creatures *build their organs without the use of cells as building stones.*[1] It is remarkable that some multicellular and noncellular organisms exist which are similar in size and show a similar endowment with digestive, excretory, locomotor, and other organs. As a representative of a large, richly organized, noncellular organism I mention *Metadinium medium*, a ciliate, which may reach a length of more than 250 μ. Comparable in body size and number of different organs is the multicellular rotifer *Taphrocampa annulosa*, with a length of 200 μ (data on *Metadinium* from Kudo's "Protozoology," 4th ed., 1954, p. 820; data on *Taphrocampa* from Pratt's manual, 1948, p. 273).

One might expect nature to use cells as building stones for all organisms above microscopic size. However, large noncellular organisms are known, such as *Acetabularia*, an umbrella-shaped alga, which is 2 to 3 cm. tall and contains only one nucleus (Hämmerling, 1934). Some multicellular organisms consist of, or contain, cells of gigantic size, which as a rule possess more than one nucleus. The freshwater alga *Nitella flexilis* consists of a chain of cells. The individual cells may attain a length of 12 cm. (Osterhout, 1927-1928). The cytoplasm

[1] This argument is based on the definition of an organ as the structural carrier of one or more definite functions. Those biologists who define an organ as a structure which is composed of cells, of course, cannot accept Dobell's suggestion; they use the term organelles to characterize functional units which are not composed of cells.

of each cell is enclosed in a definite wall; numerous nuclei are distributed in the cytoplasm.[2] Very large cells occur also in animals. Muscle cells of large nematodes may be 10 mm. long (L. H. Hyman, 1951). Various structures with numerous nuclei, but without visible partitions in the cytoplasm between the nuclei, occur in plants, invertebrates, and vertebrates. The question of functional significance of visible cell boundaries and other cytoplasmic partitions will be discussed in later parts of the book. In mammals the lining of the placental villi is a well-known example of multinucleated unpartitioned cytoplasm (Fig. 54a).

As a rule, multicellular organisms can regenerate a substantial number of their cells after loss or damage. However, not all types of cells can be regenerated in metazoans, and there are some metazoans which cannot regenerate any cell at all. In the first group are the nerve cells of adult vertebrates which seem to have lost their ability of dividing. The second group is represented by the rotifers which are known as having "constant cell numbers." It would be more correct to refer to the numerical constancy of their *nuclei,* since each organ of a rotifer may contain areas with and without cytoplasmic partitions. Not only the total numbers of nuclei are constant in these animals, but also the numbers of nuclei per organ. Those species of rotifers which have been studied contain a total number of 900 to 1000 nuclei. In *Epiphanes senta* (older name *Hydatina senta*), which is approximately 600 μ long, the total number of nuclei is 959; the peripheral nervous system contains 63 nuclei. These figures are attained during embryonic development and do not change later. If one of the nuclei is lost, it cannot be regenerated. Data are taken from Martini's classic publication on *Hydatina* (1912) and from L. H. Hyman's book "The Invertebrates" (1951). Martini compared (on p. 631) the exceedingly rigid plan in the embryonic development of rotifers to the manufacturing of machines according to a fixed blueprint. This comment of Martini seems to anticipate certain points of our discussion on organism and man-made machine (Part II, Chapter 3, H). In some organisms, special parts of the body have a fixed number of nuclei, while the rest of the body is less regimented. Leeches, for instance have a constant nuclear number in parts of their nervous system (Van Cleave, 1932). This adds to the difficulty of comparing rigid and flexible blueprints of nature.

How rigid are the numerical rules which were claimed for nematodes and rotifers? Shull (1918) and Van Cleave (1922) supplemented Martini's study of *Hydatina* by determinations of variability. Of 770 vitellaria (organs associated with ovaries) Van Cleave found 767 with the statistically normal number of 8 nuclei; two had 10 nuclei, and one had 12 nuclei. In each of 435 gastric

[2] If one places more emphasis on the numerous nuclei than on the common wall, one will interpret each unit of *Nitella* as equivalent to several cells. However, such difference in definition would be of little consequence.

glands Van Cleave counted the same number of nuclei, namely six. Shull (quoted after Van Cleave) found a slightly greater variability. I agree with Van Cleave that nuclear counts in total mounts of the rotifers, as used by Martini and Van Cleave, are more reliable than the counts in serial sections made by Shull. In any case the data obtained by various investigators form a satisfactory basis for the concept of cell-constant organs.[3] I did not find statistics on the total number of nuclei of rotifers, but the total counts must be fairly constant since they are the sums of organ counts.

The biological significance of cells varies not only from class to class, but also within each multicellular organism. The reproductive cells, i.e., spermatozoa and ova, differ from all other cells of the body, which are known as somatic cells. The somatic cells can be classified according to different criteria such as their morphological pattern of association, their origin and potential future, or their function. This will be the subject of Part V, Chapter 2.

As stated above, the choice of units is a problem with which we are faced mainly on *levels between organism and cell.* Our first example is the *muscular tissue.* Small and large units of *skeletal* muscle and their relations to units of motor nerves were discussed in Part II. The *problem of units in smooth muscle* was analyzed by Bozler (1941). The main points of his paper can be summarized as follows. There are two classes of smooth muscles. Those in class 1 are "multiunit." As in skeletal muscles, their contractions are under definite control of motor nerves. In this class are the smooth muscles of the blood vessels and of the nictitating membrane of the eye.[4] Smooth muscles in class 2 do not show any indications of organization in units. In class 2 are, e.g., the smooth muscles of the intestines: they contract rhythmically by virtue of their own automaticity, thus resembling the heart muscle (which is striated). Particularly interesting is Bozler's observation that the guinea pig uterus alternates between class 1 and class 2 behavior. In the anestrous phase, the spontaneous activity consists of weak, incoordinated contractions in all parts of the uterus; electric reactivity is low, and responses elicited by electric stimuli are not propagated. In this phase the uterus is a multiunit organ. After injection of estrogens the uterus becomes a giant unit, as shown by the following criteria. Its spontaneous activity now consists of rhythmic, coordinated contractions of the whole organ, and responses to electric stimulation are conducted to areas distant from the place of stimulation.

---

[3] One may suggest the term "organs with very low variability of nuclear numbers" to satisfy the perfectionist.

[4] The *precision* of muscle-nerve relations as expressed in the motor units of skeletal muscles (Part II, Chapter 3, E) is not attained by smooth muscles in any organ. Since in the smooth muscles of the nictitating membrane the units vary with the frequency of stimulation, Rosenblueth concluded that it is not possible to define units in these smooth muscles (Rosenblueth, 1941).

Traditionally the *blood* was considered as one organ (or tissue) and the bone marrow as another organ. However, much is to be said in favor of Boycott's (1929; quoted after Wintrobe, 1951, p. 350) suggestion that one summarize, under the name of erythron, the sum total of all circulating red blood corpuscles and of the organs from which they arise; the latter comprise the bone marrow in its different locations.

The establishment of new macroscopic units of the *human lung* proved to be advantageous in the treatment of patients. The human left lung consists of two lobes, and the right lung of three lobes. It was described in Part III, Chapter 2, B, 2 (Fig. 23) that modern injection techniques led to the recognition of bronchovascular segments as units which are smaller than a lobe. This information made it possible to limit surgical removal to those segments of a lobe which were affected by a pathological condition. Previously, a whole lobe had been the minimum of resection, no matter how small the pathological structure was (after Lindskog and Liebow, 1953).

A special discussion was devoted to the fact that many structures have multiple functions (Part II, Chapter 2, G). Muscle and bones were given as illustrations of structures which serve both locomotion and metabolism. Some of the organs with different functions can be considered as members of various systems. The liver is a good example. First of all, the liver occupies a key position in the metabolism of carbohydrates, proteins, and fatty substances. By the excretion of bile into the intestine, the liver takes part in the digestion of fat. Since the liver is a great reservoir of blood, it plays a substantial role in the mechanics of the circulatory system. Finally, the liver is part of the hemopoietic system which produces and destroys blood corpuscles. Some of these functions of the liver are clearly associated with peculiarities of its microscopic structure whereas others are not. At this point let us deal with *the three different units of the liver* (diagram, Fig. 42), which have been propounded from different points of view. All of them proved useful in studies of the functional architecture of the liver.

Before describing the different units, I wish to emphasize the fact that the three-dimensional pattern of the liver architecture is not entirely clear, although the relations between the different microscopic components are well established. One may consider the liver as a system of two interlaced sponges, one consisting of the liver cell cords, and the other of the sinusoids, which are thin-walled vessels comparable to capillaries but more variable in caliber. The traditional term liver cell cords refers to the picture seen in sections. Three-dimensional reconstructions by H. Elias (1949) showed that the liver cord cells represent the building stones of anastomosing plates which are one or two cells thick and form partitions between the sinusoids. The liver is characterized by a dual blood supply. In the diagram Fig. 42 both the arterial supply and the supply through the portal vein are indicated. The arterial blood has the general function of oxygenation and nutrition as in other organs. By far the greater volume of blood is that which is carried through the portal vein from the intestinal tube and the spleen. Through the blood, which comes from the intestines, the absorbed food reaches the liver as raw material. It is transformed

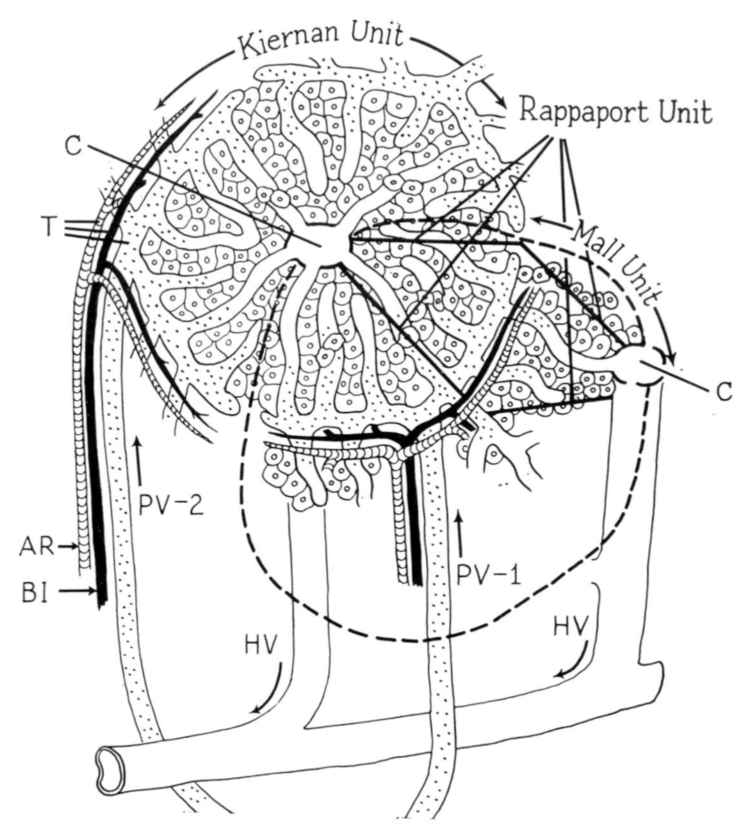

FIG. 42. Diagram of three useful histological units of the mammalian liver. Blood from the digestive tube and spleen reaches the interior of the liver through branches of the portal vein (dotted), *PV-1* and *PV-2*. As the portal blood passes between the rows of liver cord cells (trabeculae), it changes its composition; this is indicated by the transition from dotted to blank spaces. The altered blood is collected in the central veins (C) which lead to the hepatic veins (blank), *HV*. *AR,* Hepatic artery (cross hatched); *BI,* bile duct (black); *T,* triad of branches of portal vein, artery, and bile duct. *Kiernan unit:* the "classic" lobule, a circular or roughly hexagonal area with central vein (C) in the center, and the triad (*T*) at the periphery. *Mall unit:* the "alternate" or portal lobule, a circular or polygonal area (marked by interrupted line), with the triad in the center, and several central veins at the periphery. *Rappaport unit* or "*liver acinus*": smaller than the two other units. Four index lines point to the outline of a "liver acinus"; in the present case it resembles a double cone (heavy straight lines), with a central vein at each apex, supplied by the same subbranch of a portal vein (e.g., branch of PV-1). Frequently the elementary acinus is pear shaped. A Mall unit represents a cross section through the "complex acinus" of Rappaport. (Diagram modified after Rappaport, Borowy, Lougheed, and Lotto, 1954.)

by the liver cord cells in such a way that it can be assimilated by the other cells of the body. The blood from the spleen carries products of the breakdown of the red blood corpuscles; in the liver some of these products are transformed into bile (see Fig. 87C). The bile is secreted through the bile ducts into the duodenum. In the diagram Fig. 42 the change from dotted to blank spaces indicates how the portal blood changes its composition as it passes between the rows of liver cord cells. It is collected in the central veins which are the sources of the hepatic veins. The main branch of the hepatic vein leaves the liver to join the inferior vena cava which carries the blood to the right atrium of the heart (see general diagram of circulation Fig. 84). In Fig. 42 the unit which Kiernan described in 1833, or the "classic" lobule, appears as a circular or roughly hexagonal area. The distinctness of this area varies from species to species. The name central vein refers to this unit since the blood from the sinusoids collects in the center of the unit. Branches of the portal vein, the hepatic artery, and the bile duct system form the so-called triad which is located at the periphery of the Kiernan unit. The areas of the triads are known as the portal spaces. In 1906 Mall published results of injections of the portal vein and the hepatic vein, pictures of corrosion preparations of the portal and hepatic trees, and a transparent model of the dog's liver. On the basis of his studies, Mall suggested replacement of the classic lobule by a unit that was characterized by a central location of the triad and a peripheral position of the traditional central vein. The size of the Mall unit, known as "alternate" or portal lobule, is similar to that of the Kiernan unit. Much smaller functional units were proposed by Rappaport, Borowy, Lougheed, and Lotto in 1954. They injected one branch of a portal vein with a red fluid and another branch with a black fluid. Great care was taken to have both injections made with the same pressure. The result was that the areas filled with the red or black fluids did not correspond to either Kiernan or Mall units. This is clearly illustrated in the colored Plate I, Fig. 3 of Rappaport *et al.* The units which these authors reconstructed were small, berrylike masses grouped around a common stalk. Each stalk consists of branches of portal vein, hepatic artery and bile duct. Evidently the Rappaport unit resembles the Mall unit insofar as the triad is in the center and central veins are at the periphery. However, the Rappaport unit is not only much smaller than the Kiernan or Mall units, but has also an entirely different form. Although its shapes vary considerably, one may compare the Rappaport unit to a double cone with one central vein at each apex. On this basis, six halves of Rappaport units occupy the space of one Kiernan or Mall unit.

It would be difficult to compare, point by point, the advantages and disadvantages of the three units. An excellent presentation of this matter is found in Ham's textbook (3rd ed., 1957). There is one aspect that I wish to emphasize here. Different pathological conditions that affect the liver are likely to accentuate one unit more than another. Moreover, one should not lose sight of species differences. In human beings and dogs, increased resistance in the inferior vena cava produces congestion of the hepatic veins which results in dilation of the central veins. This makes the centers of Kiernan units more conspicuous than they are normally. In the human liver a certain type of visible fat accumulation is located mainly in the periphery of each Kiernan unit. In rats, Hartroft (1954) increased the distinctness of Kiernan units experimentally by administration of a small dose of carbon tetrachloride 24 hours before sacrifice. One of the assets of the Mall unit is that its architectural center, the triad, is also a center of biological activities. The number of lymphocytes in the area of the triad (portal space) shows considerable variation. In human beings this area may contain large accumulations of lymphocytes unrelated to any known disease. If the human liver is involved in diseases of the lymphohemopoietic system, such as lymphatic leukemia or Hodgkin's disease, the characteristic morphological variations develop in the connective tissue of the triad (see also Part III, Chapter 6, on topographic relations). Liver cirrhosis consists mainly of an increase in connective tissue at the expense of liver cord cells. In the transition from fatty liver to cirrhosis, strands or septa of connective tissue develop which extend from one portal space to another (Popper and Schaffner, 1957, Figs. 112B and C). One may say that these strands, which originate from the centers of the Mall unit, finally accentuate the outlines of the Kiernan unit. The Rappaport unit received remarkable support from Hartroft's (1954) study of dietary cirrhosis in rats. As Hartroft puts it, the cirrhotic process, at a certain stage, has dissected Rappaport's structural unit. The result is illustrated by my Fig. 43b, and the corresponding picture of a normal liver for comparison shown in Fig. 43a (both from Hartroft); I mentioned this observation previously as an example of metabolic dissection (Part III, Chapter 2, B, 2). I trust that I have proved my point: none of the three units of the liver is more natural or more artificial than the other, since each of them emphasizes different aspects of the dynamic morphology of the liver.

The *human hand* may serve as our last illustration of alternative units. One hand is a unit, two hands are another unit, and various parts of one hand form different units. As Goldstein (1939) pointed out, a right-handed person who is suddenly prevented from using his right hand by injury, paralysis, or loss can immediately use his left hand for writing. The conclusion is that the left hand had learned to write in some fashion, while seemingly the right hand only had been trained. In other words, both hands plus their controlling areas in the cerebral cortex form a unit. This is readily demonstrated by the ease with which everybody can use both hands in the simultaneous drawing of mirror images of simple structures. Let us now consider the finger units. Different fingers cooperate in various ways depending on the task. For delicate picking, thumb and index are used mostly; the middle finger may assist the index, or substitute for it. In clenching a fist, or similar movements, four fingers cooperate in opposing

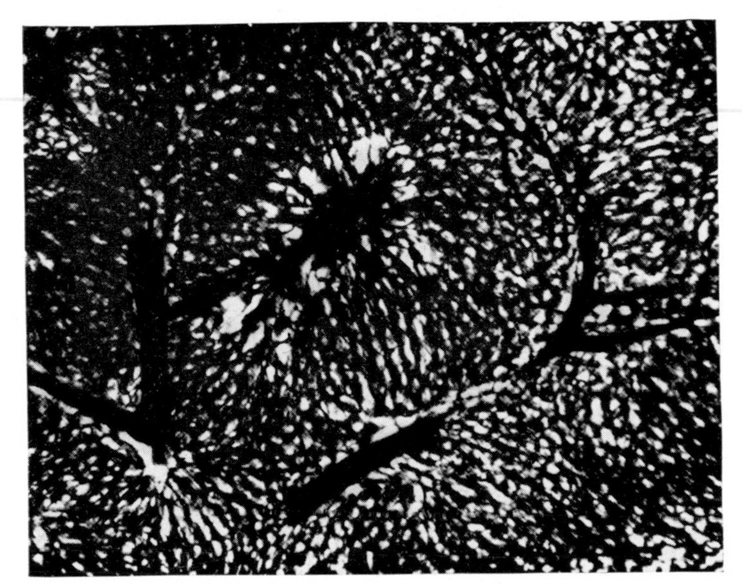

FIG. 43a. Thick section of a rat's liver, which was injected with India ink. The circular structure, which occupies a large part of the picture, is a *Kiernan unit;* its solid black center is a central vein. Section 100 μ thick; approximate magnification $\times$200. Pattern is accentuated by a minimal dose of carbon tetrachloride administered 24 hours before sacrifice of the rat. (From Hartroft, 1954, Plate 1, Fig. 1.)

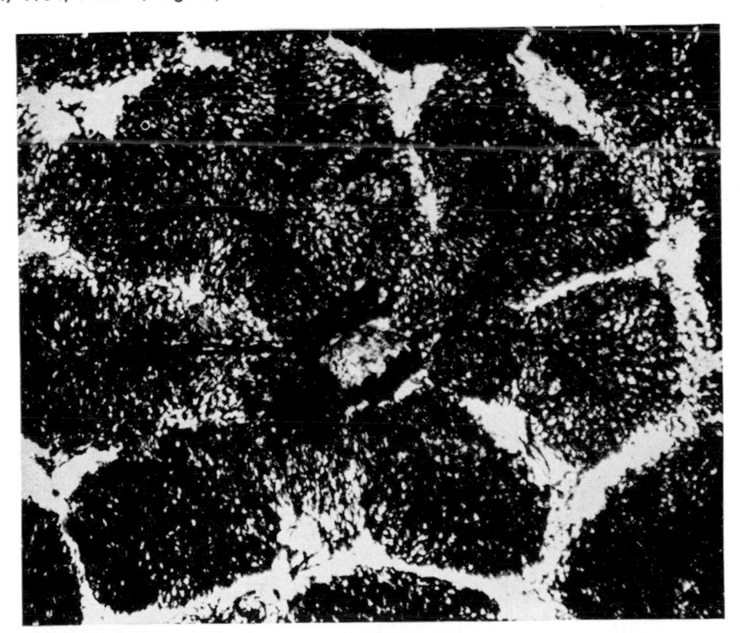

FIG. 43b. Thick section of a rat's liver in an early stage of dietary cirrhosis produced by choline deficiency. Injection with India ink as in Fig. 43a. The cirrhotic process has dissected a group of *Rappaport's structural units.* Approximate magnification $\times$100. (From Hartroft, 1954, Plate 2, Fig. 3.)

the thumb. Anatomically, the sheaths of the tendons of the flexors communicate with each other in such a way that a suppurative process frequently spreads from the thumb directly to the little finger and vice versa. Contrary to expectation, studies of the patterns of papillary ridges (fingerprint patterns) did not reveal any simple correlation between fingers of one hand or the corresponding fingers of both hands. The pattern of living capillaries in the nail wall varies between fingers but proved to be constant in each finger for at least a year (see Part III, Chapter 2, A, 1). At this occasion I mention "The Principles of Anatomy As Seen in the Hand," by F. W. Jones (1920). In his book, the author discusses well-known problems of dynamic morphology, but his approach is fascinating since he always uses the hand as the starting point.

The alternate use of both hands, particularly in ambidextrous individuals, brings us to the subject of *rotating units.* Winton and Bayliss examined the question how muscular forces can be sustained for an appreciable time as required in most skeletal movements. The answer was that most movements involve only submaximal contractions and that this is produced by tetanizing different parts of the muscle in rotation. Consequently, each group of fibers is regularly rested so that small tensions can be maintained for some time without fatigue of the muscle as a whole (Winton and Bayliss, 1955, pp. 332-333).

It is possible that in some species the units of the kidney, called nephrons, function in rotation. Observations of glomeruli in the living frog seemed to indicate such behavior. Nothing is known about rotation of function between the units of the liver. The recognition of rotating units is important with respect to the determination of the reserve power of an organ. If unusual demands are made on an organ, more of its units may work simultaneously than under ordinary conditions. There are, of course, many other factors by which the capacity of an organ can be increased in emergencies, for instance, increased blood supply.

D. Associations of Several Organisms, without Anatomical Continuity; Ecological Units

Anatomical combinations of two or more organisms were described before. Colonial coelenterates were given as an example of natural associations, and artificial unions were represented by cross circulation, parabiosis, and telobiosis. Associations of organisms without anatomical continuity are known as packs of wolves, herds of deer, flocks of birds, schools of fish. For more stationary phases the term colony is used, for instance with respect to beavers, penguins, and nesting herons. The states of insects, such as bees, ants, and termites, are colonies composed of various classes. Each class has its structural and functional characteristics. This polymorphism differs from that of the siphonophore colonies by the absence of anatomical continuity. Finally, groups of organisms and their natural environment are considered as *ecological units,* e.g., a lake with its content of plants, fish, insects, protozoa, bacteria, etc. Among handy textbooks of ecology I mention those by R. Hesse (1924) and G. L. Clarke (1954). A masterpiece of popular and, at the same time, well-documented presentation is the book "Basic Ecology" by R. and M. Buchsbaum (1957). Associations between individuals of different species are called symbiosis and commensalism if none of the partners is injured, whereas it is called parasitism if one of the partners is the victim. Although parasitology has developed as a special field, it overlaps, of course, with ecology. Symbiosis between two different species can lead to such close morphological association that the impression of one organism is produced. Thus lichens were considered to be a special class of plants until it was discovered that they were symbiotic composites of fungi and algae. Certain species of protozoa, sponges, coelenterates, and flatworms regularly contain green algae (see Buchner, 1921). Artificial symbiosis between plant cells and animal cells was produced by Buchsbaum (1937). Colonies of embryonic chick cells were cultivated by the hanging drop method. When a satisfactory degree of uniformity had been obtained by repeated transfers after explantation, a colony was divided in halves (see my diagram Fig. 21). The control half was placed in the usual media consisting of plasma and embryo juice. The experimental half received the same media with the addition of algae. The effect of this procedure was measured by the length of the period for which the chick cells or algae remained in a good condition without being transferred to fresh media. It turned out that both partners remained longer in a good condition when combined than when cultivated separately. This benefit for both partners was observed only when light was available to the cultures during incubation. In the dark, chick cells seemed to be damaged by the presence of algae. Buchsbaum concluded that, in the light, a "photosynthetic mechanism" played a role.

On the borderline of natural and artificial associations are the herds of laboratory mice which were used for epidemiological studies. One may say that laboratory mice always live under artificial conditions. However, in these epidemiological studies more individual hosts were kept in direct contact with each other than in the usual type of bacteriological experiments. As an example, one of the papers by Topley, Greenwood, and Wilson (1931) may be reported here. When a standard dose of a particular strain of *Salmonella typhimurium* (*Bacterium aertrycke*) was injected intraperitoneally into individual mice, most mice died within one or two weeks. However, when a population of 100 uninfected mice, kept in a large cage, was exposed to infection by the addition of 25 infected mice, most of the original 100 mice survived for the duration of the observation, i.e., 60 days. After this period the surviving mice were sacrificed. Approximately 40% of them showed *S. typhimurium* in their spleens. The epidemic had spread widely, but its killing power was relatively slight, although bacilli of the same type displayed a strong killing power in individually infected mice.

E. Orders of Magnitude; Relation of Cell Sizes and Animal Sizes; Symmetry and Asymmetry in Units and Composites

Orders of magnitude are a subject which is inseparable from the problem of units. Comments on different orders of magnitude were made by Galileo, in his "Discorsi" of 1638, with respect to static properties of animals. He stated, e.g., that a small dog can carry on its back two or three dogs of similar size, whereas a horse can hardly carry another horse (quoted from von Oettingen's German translation, 1917, p. 109). The second chapter of D'Arcy Thompson's "Growth and Form" (2nd ed., 1942) is entitled "On Magnitude." It gives a fascinating survey of the *mechanical* potentialities and limitations connected with magnitude. I mention here one well-known example of the role of magnitude in the *metabolism* of animals: the energy per unit of weight that is needed to maintain the body temperature is larger in small animals than in larger animals, since volumes decrease in third powers, but surfaces decrease in second powers.

A comparison of cell sizes and animal sizes, based on studies of various investigators, led Edmund B. Wilson (1934) to the conclusion that corresponding cells of a large and a small animal are of a similar order of magnitude and that, therefore, the great differences of body size are wholly due to variations in the number of cells. The following calculation is found in D'Arcy Thompson's chapter "On Magnitude." Thompson compared the dimensions of a whole elephant and mouse with the dimensions of motor nerve cells from the spinal cord of the two animals. Let the mouse values be 1. Then the following approximate ratios are obtained: linear dimensions of nerve cells 1:2; volumes of nerve cells 1:8; linear dimensions of whole animals 1:50; volumes of whole animals 1:125,000. Consequently, in corresponding parts of the nervous system, the elephant has 15,000 times more cells than the mouse.

Orders of magnitude in relation to the respective techniques have been illustrated in many useful diagrams. An excellent chart by Bear (1952) shows the structural elements of collagenous fibers on the levels of bright-light microscopy, dark-field microscopy, electron microscopy, and Roentgen-ray diffraction (my Fig. 78).

In the present book, orders of magnitude were discussed repeatedly in connection with procedures. I have described how the techniques of observation, dissection, and sampling for microscopic study depend on the size of the object

(Part III, Chapter 2, B). Procedures for the examination of the heart of a mouse were compared to those which are adequate for the heart of a dog. Figure 27 illustrates a problem of fixation which is peculiar to the testes of adult rats, or animals of similar size, but does not exist in animals with either smaller or larger testes. In later chapters, magnitudes or units will be mentioned again in connection with items such as living and dead models, topographic relations, the classification of cells and organs, and certain problems of regeneration.

At this point I will discuss how *symmetry and asymmetry depend on units as well as on their patterns of combination.* Planes of symmetry may change if two identical units are put together. Figures 44a and b, which are taken from an unpublished study by Billings and Mayer, show the symmetry relations[5] of two types of building blocks. The block in Fig. 44a (I) has three planes of symmetry, perpendicular on each other. If two such blocks are put together in different ways (II and III), one of the compound structures will again show three planes of symmetry, while the other structure has only two. In Fig. 44b, we start with a block which has only one plane of symmetry (I). If two of these are joined in different ways (II and III), one compound structure has only one plane of symmetry, while the other has three planes of symmetry. In other words, initial planes of symmetry, as found in one unit, can increase or decrease if two identical units are fused. The increase or decrease depends on the pattern of the union.

What we have described in Figs. 44a and b is nothing but an explicit presentation of Pasteur's statement (1860) that one can build a spiral staircase with symmetrical or with asymmetrical building blocks.[6] The importance of the whole story is the following. There is a widespread belief that one should attempt to trace asymmetry of whole organisms back to the asymmetry of their molecular building stones. As a distinguished biochemist put it in a conversation with the author: "Why are you surprised at the asymmetry of the human body? Does the body not consist of asymmetrical protein molecules which are all levorotatory?" Gray, Cox, Worthing, Everhart, and MacDonald (1942) injected optically active compounds under the blastoderm of chick embryos with the expectation that the presence of asymmetrical molecules in these compounds might change the direction of rotation of the whole embryo. Rotation refers to the following phenomenon. On the second or third day of incubation, chick embryos change from a symmetrical to an asymmetrical position. After the third day, under ordinary conditions, approximately 95% of them are found lying on their left eye, the others on their right eye. Gray and his associates

[5] For bilateral and radial symmetry of animals see Part III, Chapter 6, A, 1, b.

[6] In their article on X-ray analysis and protein structure, Crick and Kendrew (1957) differentiated between a crystallographer's brick and a chemist's brick. To illustrate the different levels of complexity, the authors used units of one mermaid and of several mermaids (their Fig. 2).

claimed that the frequency of abnormal rotation was significantly increased, e.g., by injection (prior to turning) of the unnatural isomers of malic and tartaric acid, while the usual frequency was not altered by injection of the natural isomers. They examined all embryos after 72 hours of incubation. However, *direct relations* between molecular asymmetry of the compounds injected and the direction of rotation of the embryo became highly improbable when I demonstrated that the turning of a chick embryo is reversible, i.e., it may lie

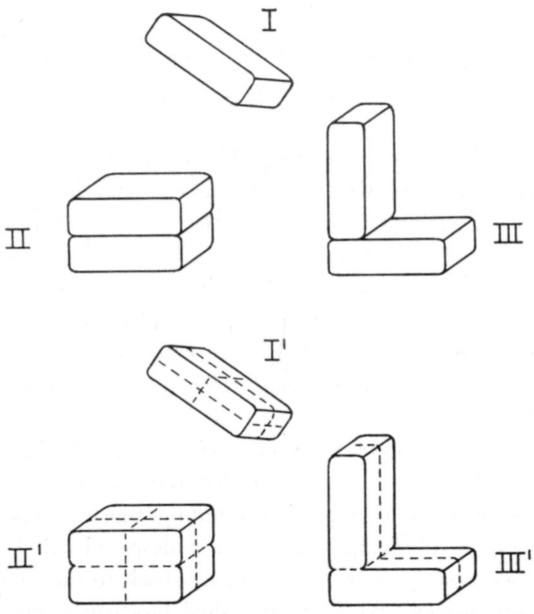

FIG. 44a. Relations of symmetrical building blocks (units) and their composites. I and I′: Unit block with three planes of symmetry perpendicular on each other. II and II′: Composite of two blocks; also three planes of symmetry. III and III′: Another composite of same two-unit blocks; only two planes of symmetry. Each plane of symmetry (indicated by interrupted lines) divides the object into identical halves. (From B. H. Billings and E. Mayer, unpublished data, 1941.)

on the right eye at one time and on the left eye at another time. This was established by observation of the living embryo through transparent windows in the shell (Mayer, 1942).

Wilhelm Ludwig, in his 1932 monograph on right-left problems in animals, gives an excellent analysis of the reasons why it is absurd to connect the asymmetry of multicellular organisms with the asymmetry of their molecules. As an example he uses certain species of snails with sinistral (S) and dextral (D) individuals. If the molecules of which the snails consist were dextrorotatory in the D individuals and levorotatory in the S individuals, this constitution would include their digestive and metabolic enzymes. Since many enzymes can attack only the levorotatory or only the dextrorotatory form of a certain substrate, either the D or the S snails would have grave metabolic difficulties.

I may add that some human beings would be faced with the same difficulties if the symmetry of their body had anything to do with the symmetry of their molecules: approximately one in 10,000 individuals has a situs inversus (mirror-imaged position) of viscera.

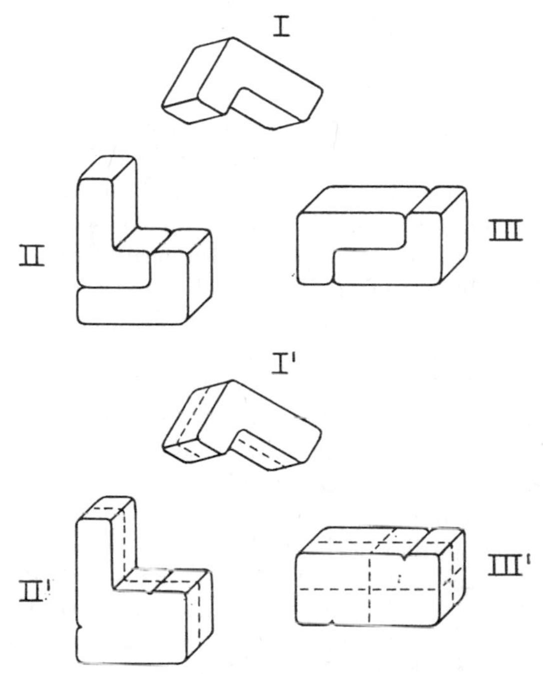

FIG. 44b. Relations of asymmetrical building blocks (units) and their composites. I and I': Unit block with one plane of symmetry. II and II': Composite of two blocks; also one plane of symmetry. III and III': Another composite of the two asymmetrical blocks; three planes of symmetry. Each plane of symmetry (indicated by interrupted lines) divides the object into identical halves. (From B. H. Billings and E. Mayer, unpublished data, 1941.)

There is an interesting study in which the authors have handled the different orders of magnitude in a perfect way. Kemp and Engelbreth-Holm (1931) explanted pieces of a 7-day-old chick embryo which was a double monster. It had two bodies but only one head. In the cultivated cell colonies, approximately 500 mitoses were examined. Five of these mitoses were tripolar, instead of the usual form with two poles. For a comparison, mitoses of colonies cultivated from normal embryos were counted. In 10,000 mitoses of these colonies, only one tripolar mitosis was found. The authors discussed their observation in the light of certain known embryological facts. If ova are fertilized experimentally by two spermatozoa, they show a number of multipolar mitoses during their first cleavage divisions. In subsequent cell divisions, only regular bipolar mitoses are seen. Yet many of the resulting animals have local malformations, or are monsters. The authors rightly concluded that the disturbance of development which expressed itself by the formation of a double monster probably also expressed itself by the formation of tripolar mitoses. In other words, *no direct connection* was claimed between abnormality on the cellular level and abnormality of the whole organism.

Embryonic development offers many illustrations of the fact that symmetry relations may either continue or change from one level of complexity to another. Figure 3 shows diagrams of the frog's egg with a type of cleavage which is radially symmetrical around a vertical axis. The radial symmetry is distinct until the 4-cell stage (Fig. 3c). Beginning with the 8-cell stage the pattern of cleavage shows so many variations that none of them seems to resemble the traditional diagram given in Fig. 3d. For a discussion of this discrepancy, see D'Arcy W. Thompson, 1942, pp. 604-607. The final bilateral symmetry of the embryo is clearly established in the late gastrula stage.[7] Another very common type is the spiral or oblique cleavage, in which the subsequent planes of cell division are not vertical on each other. As a rule, the spiral asymmetrical pattern of cleavage changes in later stages into a bilateral symmetrical pattern (for reviews see Edmund B. Wilson, 1934; Costello, 1955). In other words, most types of animals which go through spiral cleavage end up as bilateral symmetrical forms, e.g., the earthworm and all other Annelida. However, some animals with spiral cleavage show spiral structures in their adult bodies. In snails the direction of spiraling of the shells is always identical with the direction of spiraling in the early cleavage. Additional details will be given in a discussion on the origin of polarized axes in organisms (Part III, Chapter 6, A, 1).

A perpetuation of properties through numerous levels of complexity is perfectly compatible with our building block model. For instance, in Fig. 44a, the step from one block (I) to a composite of two blocks (II) can be repeated in such a way that a composite of thousands of blocks results which still shows three planes of symmetry. The muscle represents such a case: contractility is a characteristic of the whole muscle, of its fibers and fibrils, and also of its smallest known building block, the actomyosin molecule (see Part II, Chapter 2, E).

[7] An element of early bilateral symmetry is introduced by a transient landmark, the "gray crescent," not shown in my diagrams. This structure does not seem to influence the pattern of cleavage.

Chapter 5

Models

A. Different Concepts of Models

The term model has different meanings. In natural science, including engineering, it refers to a *simplified imitation* of an original object or process. A model of a ship represents certain features of a ship. If the original is large, as in the case of an ocean liner, a smaller model will be constructed. If the original is small, e.g., a liver cell, an enlarged model is made.

The role of models in mathematics and science has been discussed in numerous publications. A chapter on models in Beckner's 1959 book covers different concepts of the term model, characteristics of a useful model in biology, and particularly the explanatory value of models. The title of Beckner's chapter is very adequate: "Models in biological theory." By contrast, I might have entitled my present chapter, "Models in biological experimentation." An example may show the difference. Beckner gives much space to Rashevsky's (1948) mathematical model of a "neural net," which was designed to explain the interaction of several stimulus-response relations in the nervous system. On the other hand, Beckner does not mention experimental models of nervous conduction such as Lillie's passivated iron-wire and the large cells of algae used by Osterhout. The reverse emphasis will be found in the present book. Let us differentiate between (a) analogies which are not meant to be tested, (b) mental models, (c) material models, generally known as physical models, and (d) substitutes in which an accessible object is studied in place of an inaccessible object. The inanimate and living models which will be the substance of this chapter belong, of course, in class (c), the material or physical models.

(a) *Analogies.* Analogies can be helpful provided they are not taken too seriously. When Oliver Wendell Holmes stated that nature proceeds like a glass blower from bubbles and vesicles, he probably referred to the basic processes of embryonic development (in "Elsie Venner," p. 31 of the 1889 edition). One could go one step further pointing out that both the blastula stage of the frog's egg and the glass bubble originate from a solid ball. In an earlier part of the book (Part II, Ch. 2, H) the complicated gelatinous house of the tunicate *Oikopleura* was described. According to A. C. Hardy (1956) the house is produced in the following way. From its surface the animal secretes a thin elastic envelope around itself. Then the tail is used to separate the envelope from all around the body. Finally, the tail's undulating movements stir the water inside the envelope until the envelope is inflated like a balloon. It seems to me that this fabrication of the *Oikopleura* house is quite comparable to glass blowing.

In his "Introduction to Muscle Physiology" (1941), Fenn stated that the orderly processes involved in the contraction of muscles offered opportunities for research that were not present in other tissues. He illustrated his point by a graphic analogy: muscle is to any other tissue as a game of chess is to a game of checkers. A detailed discussion of "chess as a model of the battlefield of life" is found in Lotka's "Elements of Mathematical Biology" (1956, pp. 342 ff.). Lotka lists six points of resemblance between life and a game of chess. Five of them are "restrictive elements" such as the laws which define centers of influence in collision and the particular movements permitted for different chessmen. Positive principles which determine actual events are represented, according to Lotka, by tropisms in organisms and by the ultimate aim of checkmate in chess.

It may seem unnecessary to mention false analogies. However, there is a classic example of a false analogy which still appears in current publications. At the beginning of this century, Robertson (1908) and Ostwald (1908) published their discovery that many "curves of growth" were S-shaped and therefore resembled the curve of monomolecular autocatalytic reactions. The conclusion was that such a physicochemical mechanism was somehow the basis of growth. This far-reaching generalization has fascinated many investigators although it suffered from two fatal weaknesses. First of all, there are incomparable types of growth, such as the increase in weight of a rat and of a pumpkin, the numerical increase of a drosophila population in a bottle, and the increase in area of a cultivated cell colony *in vitro*. Secondly, all physical, chemical, and biological processes which start slowly, continue at a steady rate for some time, subsequently slow down and finally stop, produce an S-shaped curve if the changes are plotted cumulatively. Other physical, chemical, and biological phenomena can be plotted as a straight line, and still others as a parabola. Obviously, these groups of graphs represent merely formal resemblances. Examples of S-shaped curves of two different biological phenomena were shown in Figs. 5 and 7a, one of them representing muscular tension as response to electric stimuli and the other per cent survival of infected mice treated with an antibiotic.

(b) *Mental models.* Mental models are man-made designs used to shed light on *general principles or procedures* of biological studies. Claude Bernard discussed how the elimination of one muscle in an animal may fail to produce any visible disturbance when other muscles compensate for the loss. From this one might conclude that the eliminated muscle had no function. To illustrate the fallacy, Bernard used a mental model. Suppose a column has a base which consists of a number of bricks. If, in a series of experiments, a different brick is removed each time, the stability of the column will always appear unchanged. The false conclusion would be that the bricks are not needed to support the column ("Introduction à l'étude de la médecine expérimentale," 1865; p. 181 of the 1952 edition).

In Ham's "Histology," a diagrammatic model shows how a variety of stimuli can stimulate afferent (sensory) neurons (Fig. 278 of the 3rd edition, 1957). This instructive model consists of an electrical circuit which can be completed by touch, pressure, heat, cold, light, sound, or a chemical agent. The unspecific starting mechanisms remind one of a model which has actually been constructed, namely Lillie's (1918) wire model in which an electrochemical wave is started by a variety of environmental changes. Lillie's model will be described later.

In Part II of this book a model (Fig. 1) illustrated the usefulness and limitations of optical procedures, and also the change from random distribution of particles to a distant pattern. The model was a hollow rubber sphere filled with water in which paraffin-coated magnetic and nonmagnetic rods floated; the two types of rods could be distinguished by tags, and their distribution could be altered by the action of a magnet.

The value of these mental models cannot be tested, of course, by any specific biological observation or experiment. Other mental models may be subject to testing. As an example I mention the comparison with a ratchet or catch mechanism used by several investigators in discussing the properties of muscle fibers. A bivalve mollusk, such as *Pecten,* closes its shells when it is removed from the water. There is enough time for the experimenter to push a piece of wood between the shells before they close. If, after some time, one twists the piece of wood out forcibly, the shells remain motionless like the jaws of a vice from which an object has been removed. It is impossible to widen the gap between the shells, but with gentle pressure one can easily narrow the gap. In the new position it will again be impossible to increase the distance between the shells. It is this behavior which suggested a catch mechanism ("Sperrung," von Uexküll, 1912). Figure 45 illustrates a catch or ratchet mechanism in which the upper rod can slide in the direction of the arrow. By this movement the total length of the model is shortened. The upper rod cannot move backward. A reversal of movement is possible only if the upper rod is lifted by the full depth of a tooth so that the teeth of the two rods no longer catch.

The idea of a catch mechanism is based on the assumption that no measurable energy is consumed during the prolonged contraction of the closing muscle of a bivalve mollusk. This muscle seems to maintain large tensions for long periods without fatigue. The data on *Pecten,* the quotation of von Uexküll, and my Fig. 45 are taken from Bayliss, "Principles of General Physiology" (1915, pp. 534-538). In a more elaborate model of a tonus muscle, Bayliss shows a combination of a catch mechanism with two electrical circuits in which the two keys correspond to nerve centers (his Fig. 170). He finds the catch theory applicable not only to the smooth muscles of many species of invertebrates, but also to some smooth muscles of mammals. Winton and Bayliss (1955, pp. 356-357) have made an important objection to the catch theory: intermittent electrical changes can be detected in the muscles of bivalve mollusks all the time they are maintaining tension. This observation is incompatible with the claim that the tension is maintained without energy consumption. More circumstantial evidence may be collected in support of or against the catch theory of smooth muscle. The rigid ratchet cannot be expected to be more than a mental model, but it has played a useful role in the formulation of questions concerning the action of smooth muscle.

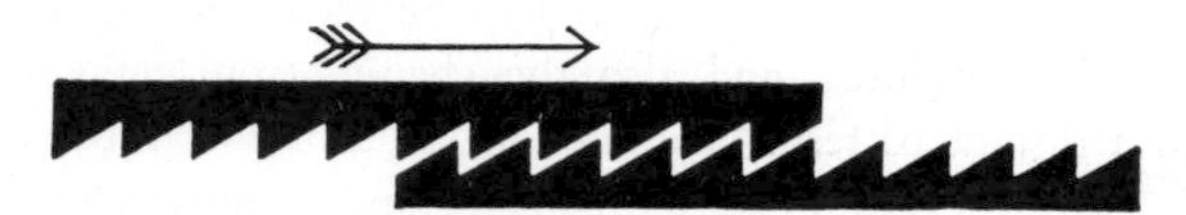

FIG. 45. Diagram of a catch or ratchet mechanism as a model of smooth muscle action. The upper rod can slide in the direction of the arrow. A reversal of this movement is possible only if the upper rod is lifted the full depth of a tooth. (From W. M. Bayliss, "Principles of General Physiology," 1st ed., 1915, Fig. 169.)

Recent studies of *striated muscle* led H. E. Huxley (1957) to the comparison with a "sliding-filament" model. In one of his papers (1958) the idea was expressed that there may be a ratchet device in the linkage between detailed molecular changes and the contraction of the muscle. The description and diagram (Huxley, 1958, p. 80) seem to suggest that the teeth of this ratchet device are somewhat flexible. If studies of this type should eventually succeed in demonstrating the properties of the ratchet teeth, the mental models may lead to the construction of material models.

One of the most famous conceptual models is the lock and key comparison which Emil Fischer (1894) used for enzyme-substrate relations, and which Paul Ehrlich (1900) applied to the relations between antigens and antibodies (cf. Part II, Ch. 2, C). In this comparison, the partners did or did not fit into each other, but they were not supposed to mold each other. Recently, the concept of template has been used extensively to picture both the fitting of finished products and the mutual adjustment of partners by molding.

P. Weiss published a number of mental models in connection with his studies of "molecular ecology," a term which refers to the interrelations of molecular populations in cells and their interaction with the environment. As an introduction to this field, I recommend again his 1947 paper, "The Problem of Specificity in Growth and Development." Weiss suggests the application of basic concepts of modern immunochemistry to every selective mechanism, be it gene action, embryonic differentiation, reactivity to sensory stimuli, or reactivity to drugs. A modernized key-lock diagram (Fig. 1 in Weiss' paper) illustrates configurations in adjacent surfaces of two cells, or a cell and its surrounding medium. In a later paper, Weiss (1950, Fig. 3) used a particular form of a key-lock diagram as a model of possible molecular mechanisms in embryonic induction. This diagram shows molecular rearrangement resulting from contact with different organic surfaces. Weiss and Kavanau (1957) published a paper in which template and antitemplate models were presented in mathematical terms. Unfortunately, this paper is beyond my mathematical capacity. Finally, I mention a diagram of Weiss (1955, Fig. 144) which represents the intricate web of interrelations in the embryonic development of nervous structures. It seems to me that this diagram is very helpful in formulating specific questions and in taking an inventory (1) of well-established data and interpretations, (2) of partly solved problems, and (3) of matters which are in an entirely hypothetical stage.

Rashevsky's work is concerned with mathematical models of various biological phenomena. The reader is referred to his 1940 and 1948 books. I mention his calculations of the physicochemical conditions under which drops divide (published 1928 and 1949). The calculations were supposed to shed light on the factors that govern cell division. Swann and Mitchison (1958, pp. 122-123) have expressed considerable doubt that these mathematical models can be tested by any experimental work. The only experimental attempt in this direction was made by Buchsbaum and Williamson (1943). Their observations in dividing sea urchin eggs yielded data which were in remarkable agreement with those values which were postulated in Rashevsky's "diffusion drag" theory of cell division. Yet, Swann and Mitchison did not accept these results as supporting Rashevsky's theory since numerous other theories could account for the observations of Buchsbaum and Williamson.

(c) *Material models.* I use the term material models in contrast to mental models.[1] Material models can be inanimate or alive. Much has been learned on the transmission of stimuli in nerves both through Lillie's inanimate model consisting of wires and acids, and through Osterhout's live model consisting of long cells of algae. Finally, Young's discovery of the giant axon of the squid made it possible to test directly on nerves those conclusions that had been reached indirectly by models. Detailed descriptions and references will be given in discussions on inanimate and living models. Here some general principles on the proper use of models may be mentioned.

[1] In the literature the term physical model sometimes refers to what I call material models, and sometimes to the use of physical procedures in contrast to chemical or biological procedures.

A model is most useful if it can be tested against the original object of study. A model is supposed to present existing problems in a simpler form. In their study on heart-lung-kidney preparations, Starling and Verney (1925, pp. 321-322) pointed out that such isolation preparations are grossly artificial, but make problems accessible which cannot be handled under the complicated conditions of the whole dog. Starling and Verney call the study in the whole dog the analytic method, and the study of isolation preparations the synthetic method of experimentation. I touched upon these aspects repeatedly in the discussions of natural and artificial conditions and of interferences by procedures used (Part III, Chapter 3, C).

One should be prepared for the possibility that the model may introduce complications which were not seen in the original. The heart-lung-kidney preparation illustrates such a case. As described before, it seemed for some time impossible to simplify this arrangement by using a mechanical oxygenator instead of living lungs, until, finally, the biochemical relations between lung and kidney were recognized (see chapter on units). This shows how a systematic analysis of complications in a model can lead to the discovery of mechanisms that occur in the natural organism.

One of the most important procedural aspects of models is *scaling*. It seems that the theory of predicting the performance of full-scale ships from the results of model testing was conceived about two centuries ago, but was not developed into a practical device until 1872 when William Froude constructed the first model basin in England (from an article on the David W. Taylor Model Basin, *Science,* Vol. 107, p. 614; 1948). The scaling of a model in a biological study may be illustrated by Mitchison's and Swann's paper on the cell elastimeter (1954). The mechanical properties of the surface of the sea urchin's egg were determined as follows. A glass micropipette with an opening smaller than the diameter of the egg was put in contact with the surface of the egg. If hydrostatic suction was exerted through the pipette, the egg's surface bulged into the pipette (Mitchison's and Swann's text figure 6, which shows the bulge of the egg, resembles my Fig. 1c). Since the sea urchin's egg does not collapse after puncturing, Mitchison and Swann concluded that it may have elastic properties like a tennis ball, not like a thin-walled rubber balloon or a drop of fluid. Analysis of the problem in terms of a thick membrane proved mathematically intractable. Therefore, the only way of testing the hypothesis was by the use of rubber balls as models. Each ball had a stalk to inflate it with air or water as desired. Open-ended glass tubes were connected to a pump and manometer on one end. The other end was pressed against the rubber ball, and the place of contact was greased for sealing. The rubber balls were 100 mm. in diameter, with walls from 0.8 to 18 mm. thick. The quantitative relations between hydrostatic suction on the one hand, and size and configuration of bulges on the other hand, proved to be relatively simple: over a considerable range of wall thickness, the pressure-deformation curves were linear. The remaining question was whether the observations in rubber balls of 100-mm. diameter were applicable to the sea urchin eggs measuring 0.1 mm. in diameter. Dimensional analysis showed that it was permissible to scale down 1000-fold, from the model (1000) to the egg cell (1). Thus the hypothesis was confirmed that the mechanical properties of the egg's surface were those of a thick-walled rubber ball.

(d) *Substitutes.* In the discussion of the factors which determine stainability of histological sections, I mentioned the work of Singer and Morrison (1948). These authors felt that the studies of the complicated mixtures of proteins in sections should be preceded by experiments with a single protein of known structure and satisfactory purity, but in a state comparable to that of proteins in sections. Since gelatin is a degraded protein and wool is too highly specialized, Singer and Morrison decided that a solid fibrin film was the best substitute for a histological section. Some of their results were illustrated in my Fig. 28. Other investigators have used gelatin instead of the proteins which are present in histological sections. I reported earlier that the staining properties of colloid in thyroid follicles were recognized to be unspecific and independent of the presence of iodine, since similar staining characteristics were observed in clotted proteinic masses which were enclosed in small cavities of organs other than the thyroid (see Part III, Ch. 3, C, 2, c). This interpretation was supported by the use of gelatin drops of different concentrations: all staining phenomena in question could be reproduced in the gelatin models (Mayer, 1949).

Two examples may illustrate the use of substituting models in studies with living organisms. The first example refers to cytogenetic control of differentiation. This problem requires studies of nucleic acids and enzymes in living systems of the order of magnitude of a cell. The biochemical mechanisms mentioned can only be studied, as yet, in bacteria, whereas the most striking feature of the process of differentiation are observed in metazoa (Ephrussi, 1956). Following Lederberg's 1952 review, one may say that bacteria and protozoa can serve as models substituting for direct approaches to the problems of differentiation in somatic cells of metazoa, whether they are normal cells or tumor cells. Owing to the present state of techniques, cell genetics is best studied in organisms whose germ plasm and somatoplasm are not irreversibly separated. However, the actual processes cannot be learned from the models, but "can only be told by looking at the embryo or tumor itself" (pp. 403 and 412 of Lederberg's article).

My second illustration of substituting models is taken from cancer research. Early experimental cancer studies in mice were handicapped by the rarity of spontaneous tumors in mice obtained from dealers. This difficulty was overcome by the introduction of tumor transplantation (Morau, 1894; Jensen, 1903). From the tumor of one mouse a large number of tumors was produced in other mice.

Transplanted tumors served as substitutes for spontaneous tumors. An additional supply of tumor mice became available through the isolation of carcinogenic polycyclic hydrocarbons from coal tar. Subsequently, systematic studies with spontaneous tumors were made possible by inbreeding strains of mice with constant (high or low) incidence of spontaneous tumors.

One may extend the concept of substituting models to all experiments which are carried out in animals in order to obtain information on human beings. For ethical reasons new drugs for the treatment of human diseases are investigated first in experimental animals such as mice, rats, dogs, and monkeys. For reasons of economy drugs to be used in large domestic animals are frequently explored in small, expendable laboratory animals. Again the greatest problem in such studies is the transfer from the model to the original target. Will responses observed in one species occur in a similar way in another species? Administration of 5-(4′-methylphenylimino)-9-dipropylaminobenzo[a]phenoxazine, briefly benzophenoxazine, can cause hemolytic anemia. Using comparable doses it was possible to rank the sensitivity of four species as follows. The greatest susceptibility to hemolytic anemia was observed in man; dogs were second, rats third, and monkeys fourth, signs of anemia being hardly perceptible in the monkeys (J. T. Litchfield, E. Mayer, and L. H. Schmidt, unpublished data, 1953). Species as taxonomic units within one genus will be discussed later in connection with problems of classification (Part V).

B. Inanimate Models of Biological Phenomena

Three-dimensional reconstructions made from flat sections are sometimes referred to as models. They have been indispensable in descriptive embryology. Various shapes of closely packed tissue cells will be described in Part IV. Here it may suffice to mention models which were used to imitate the shapes of stacked cells. On mathematical grounds D'Arcy W. Thompson (1917, p. 339) had postulated that solids with 12 faces should be the prevailing shape of any material which is closely packed and is molded under pressure. Contrary to this expectation, F. T. Lewis (1943, and other papers) found that stacked cells of plants and animals have an average of 14 faces, with a narrow range of variation. (Wax reconstructions by Lewis are reproduced in my Fig. 67a and b.) When Marvin (1939) compressed buckshot in a brass cylinder, most of the solids obtained had 14 faces. A froth of soap bubbles was studied by Matzke. The bubbles exhibited, on an average, 14 faces, but the pattern was not the regular (orthic) pattern of plant cells. None of 600 central bubbles examined had the form which Thompson postulated (see Matzke's 1950 summary).

The colloidal properties of cells have stimulated many investigators to construct models. As one of the earliest examples I mention Pfeffer's cell model of 1877 (reprinted 1897, Fig. 9).

Lehmann (1910) considered the spontaneous divisions of his "liquid crystals" as models of cell divisions. Various mixtures of proteins and lipoids, with or without the addition of inorganic salts, have been prepared in order to imitate all kinds of cellular phenomena including cell division (Herrera, 1932, with references to his earlier work; Crile *et al.,* 1932). At this occasion I wish to mention the instructive diagram of a mayonnaise dressing in Sharp's "Introduction to Cytology" (2nd ed., 1926, Fig. 1), which illustrates the interfaces between water-soluble and water-insoluble phases in cells. The mitotic division of nuclei has been imitated in the well-known experiments of Leduc (1911). A layer of salt solution was spread on a plate of glass. When a drop of India ink was allowed to fall in the middle of the fluid layer, the ink particles moved in such a way that two poles with connecting threads of ink developed. D'Arcy Thompson (1942, Figs. 88 and 102) shows a mitosis (metaphase) from the trout's egg, and a very similar artificial mitosis after Leduc. Although Thompson shuns the term model, his book is full of mathematical and mechanical models. For instance, photographs of splashes of milk drops (taken by H. E. Edgerton) are reproduced by Thompson because of their resemblance to the fluted cuplike structures of *Campanularia* (protozoa with a stalk). A comprehensive review of older model techniques was published by Rhumbler in 1923.[2]

Which of the numerous inanimate models found in the literature have led to crucial biological experiments, or have, at least, lent important support to the studies with living material?

As an example of successful use of models I refer to Holtfreter's (1948) comparisons between artificial films of lipids or lipid mixtures and living cells, particularly with respect to permeability of membranes to ions and the projection of pseudopodia from the cytoplasm. Holtfreter concluded that the most important kinetic elements of the cell membrane are the lipids rather than the proteins. In connection with the problem of scaling I mentioned the study of Mitchison and Swann (1954) on the mechanical properties of the surface of the sea urchin egg. Rubber ball models were not only useful but played a decisive role in this investigation.

A remarkable model was the artificial *Torpedo* (electrical ray) constructed by Cavendish in 1776. A cover of leather, soaked in salt water, enclosed a wooden board supporting a plate of pewter, which was connected to Leyden jars by glass-insulated wires (Fig. 3 on p. 200 of Maxwell's 1879 edition of Cavendish's papers). The leather surface imparted electric shocks when touched.

Lillie's wire model has made a definite contribution to the study of nerve impulse conduction. According to his 1918 paper, an iron wire was passivated

---

[2] Walker and Allen (1927) were able to make colloidal models containing lecithin and cephalin which resembled the Golgi apparatus. They drew the peculiar conclusion that there was no Golgi apparatus in living cells.

by placing it in strong nitric acid. If stored in a dish with dilute nitric acid and left undisturbed, the wire remained bright and unaltered for an indefinite time. If, then, one end of the wire was touched with a piece of ordinary iron, the bright metallic surface suddenly darkened through formation of oxide. This change was connected with effervescence which could be observed traveling over the entire length of the wire. The velocity of transmission was determined as 100 or more centimeters per second in this experiment. Besides by touching one end of the wire with a piece of metal, the wave of activation could be induced mechanically by bending or tapping, chemically by contact with a reducing substance such as sugar, or electrically.

One may summarize these phenomena as follows (Höber, 1945, p. 353): an electrochemical wave of alternating polarization and depolarization, and alternating oxidation and reduction, travels along a wire of passive iron, coated with an iron oxide film; the wave can be started by any environmental change which is applied to one end of the wire. Lillie's model added an important link to the chain of discoveries (made during the 19th century), which proved that the spike potential of a stimulated nerve is transmitted with finite velocity. For a brief presentation of the historical developments see D. P. C. Lloyd (1949, p. 9). Important present research can be traced directly to experiments of Lillie which were published in 1925. When a passive iron wire is covered with a glass tube broken at regular intervals, the activation does not spread over the whole wire but jumps from one break to another. Lillie considered the possibility of similar jumps in living myelinated nerves from one node of Ranvier to another node. Therefore, Tasaki (1939) credited Lillie with the creation of the theory of electro-saltatory transmission.[3] Tasaki's work will be mentioned in our general discussion of fibers (Part IV, Chapter 3, A).

There are not many inanimate models which proved as useful as the examples described here. I find it difficult to evaluate the collodion tube model which was prepared by Schade *et al.* (1928) in order to study physicochemical properties of living capillaries. Blood and serum were circulated through collodion tubes which had diameters of several millimeters. The tubes were immersed in different fluids so that variations in permeability could be determined. Extracorporeal circulation through dialyzing collodion tubes was introduced by Abel *et al.* (1913) to eliminate toxic substances such as salicylic acid from the blood of dogs. At the present time cellophane tubes are in clinical use for removal of poisons from the circulation and for replacement of some lost glomerular functions (Merrill, 1960).

Some models supply a fresh approach to an old problem. In an article on

[3] Rosenblueth and Wiener (1945) stated that Lillie's iron wire model has not made any contribution to neurophysiology since the phenomena observed in the passivated wire are not entirely understood.

the thermoelastic effect in muscle, Fenn used three models as a basis for discussion: a steel spring, gas compressed by a piston in a cylinder, and a rubber band, each model being stretched by a weight (Fenn, 1945, his Fig. 51). For general relations between rigid and colloidal structures see Part II, Chapter 2, H.

Recently, Jerome Gross (1956) suggested that the behavior of collagen units could serve as a model of the morphogenesis of subcellular structures, such as cell membranes, chromosomes, mitotic spindles, or endoplasmic reticulum. The production of such structures, their dissolution and their reconstruction from high polymer building blocks, may prove to be controlled by physiochemical mechanisms comparable to those which produce the relatively simple collagen system.

Finally, there is a type of inanimate model which one may call a didactic model. Models of this kind are used in laboratory courses of physiology and pharmacology. I mention an example from the manual of Sollmann and Hanzlik (1939, p. 125). Two jars of water and a soft rubber tube filled with water are used to illustrate the transportation of fluid by the peristalsis of the intestines. Each end of the rubber tube dips into one of the jars. By placing a hand around the tube and closing the fingers one by one (milking movements), water is pumped from one jar into the other. My Fig. 1, the water-filled rubber ball with floating magnetic and nonmagnetic particles, represents a didactic model insofar as it illustrates the possibilities and limitations of optical procedures.

C. Living Models of Biological Phenomena

1. *Algae as Models of Physiology of Nerves*

In the development of the electrochemical theory of nerve impulse conduction, an inanimated and a living model have been useful. The inanimated model is Lillie's iron wire which was described above. The living model consists of large cells of algae, which were introduced by Osterhout in 1927 for the study of internal-external potential differences. As described previously (Part III, Chapter 4), the freshwater alga *Nitella flexilis* is a chain of large cells. Each of them is enclosed by a distinct cell wall and may reach a length of 12 cm. On the inside, the cell wall is coated with a thin layer of cytoplasm which contains several nuclei. The center of each unit is occupied by a large space filled with fluid (cell sap). It was possible to isolate intact single cells from the center of the chain. As Osterhout pointed out, one can study such isolated cells in a well-defined environment, in contrast to the ill-defined conditions of cells or nerve fibers in tissues. By placing small drops on different spots of the cell wall and inserting electrodes into the drops, potential differences between several spots could be measured. A marine alga, *Valonia macrophysa,* was used for supplementary studies. The cells of *Valonia* are approximately

5 cm. long, but thicker than those of *Nitella* and therefore more suitable for the insertion of capillary electrometers. The combined studies in the two algae led to the conclusion that, across any one spot of the cell, there is a measurable potential difference based on the presence of five layers: external fluid, three layers of cytoplasm, and the central sap (Osterhout and Harris, 1927-1928). These observations in cells of algae lend substantial support to the membrane theory of transversal migration of ions as the source of action potentials in nerves. Direct experiments with nerves became possible after the discovery of the giant axons of the squid by Young (1936). Into the core of such axons, which measure 500 μ or more in diameter, Hodgkin and Huxley (1939) inserted microelectrodes without causing appreciable injury. With a second electrode placed on the outside, potential differences between interior and exterior could be determined both in the resting and in the electrically stimulated axon. Subsequent developments were reviewed by Hodgkin (1951). This article contains instructive diagrams of techniques for large or small nerve fibers and also comments on the transformation which the "electrochemical molecular theory" of Bernstein (1894) underwent as a result of later experiments.

2. *Isolation Preparations as Models: Explants*

Isolation preparations were described in connection with morphological techniques for the study of live material and in the analysis of natural and artificial units (Part III, Chapter 2, A and Chapter 4). One group of isolation methods deserves an extended discussion here, namely, the explantation techniques. They comprise cultivation of fragments or primordia of organs, with maintenance of organization, and cultivation of cell populations, especially cell colonies, with loss of organization of the original explant.

There is no doubt that explantation techniques have been extremely useful in the studies of embryonic primordia and early stages of embryonic organs. Two examples were given earlier: differentiation of a piece of ectoderm into neural tube (Fig. 19) and continued differentiation of a femur isolated from a 5½-day-old chick embryo (Fig. 20). A third example may be recapitulated here which illustrated procedures for following the fate of embryonic cell populations (Part III, Chapter 3, A). The problem was how to prove that the pigmented cells of feathers (melanophores) are derived from unpigmented cells (melanoblasts) which form in a transient embryonic structure near the spinal cord, the neural crest, and from here migrate to the skin and the root of the feathers. Two experiments proved this theory: (1) transplantation of neural crest fragments into various places of the same embryo, including places remote from the skin, and (2) cultivation of neural crest fragments *in vitro*. Both the transplanted and the explanted unpigmented tissues produced pigmented cells. The explantation experiments were carried out by Dorris (1938) in order to decide a pre-existing clearly defined question. As mentioned before, it was even possible to demonstrate that all melanophores in the feathers of birds are derived from unpigmented immigrants from the neural crest. This was done by grafting experiments between two different species. One may say that in various areas of experimental embryology the combined attack by transplantation and explantation techniques has proved to be particularly successful.

It is the *cultivation of tissue cell populations* which presents a confusing picture to the outsider and therefore will be rediscussed here. The cultivation of cell colonies from explanted fragments of an organ is illustrated in Fig. 21. This procedure was developed in Carrel's laboratory (Carrel, 1912; Ebeling, 1913). Since an unlimited number of transfers (passages) of colonies from one medium-containing vessel to another had become possible, a very unusual living model was established: the individual cells are natural cells, but their association in colonies is artificial (Mayer, 1937).

Like tissue cells in the intact organism, the cells of colonies *in vitro* are able to migrate and to multiply by mitotic division. This permits maintenance of strains in passages with continuous reproduction of colonies after each passage. Cell populations which were cultivated from explanted fragments of some animal tumors have maintained their ability to produce tumors when injected into animals. After various periods of cultivation some normal cell strains were transformed into cancer cells, as revealed when the cells were grafted into animals and produced invasive tumors. This is another striking resemblance with normal tissue cells in the organism which, on various occasions, are transformed into malignant tumor cells.

Certain cytological characteristics persist in cultures, e.g. Nissl bodies and the Golgi apparatus, as described in Part III, Chapter 3, C, 2. Other characteristics disappear, and some are accentuated, *in vitro*. Because of these variations in behavior, identification and classification of cells cultivated *in vitro* may or may not correspond with that of tissue cells in the organism. This matter will be discussed in connection with elementary general structures (Part IV) and the classification of tissue cells *in situ* and *in vitro* (Part V, Chapter 2).

As stated above, the individual cultivated tissue cells are natural, but their population patterns *in vitro* are artificial. However, resemblances can be noted with population patterns of tissue cells in the organism. Metastatic cancer cells settling on the surface of serous membranes tend to form aggregates similar to those of cell colonies cultivated in a plasma clot. Cultivation of tissue cells in liquid media produces dispersed populations comparable to those of circulating blood cells and of ascites tumor cells.

Somatic cells from various tissues of the body adopt new population patterns when cultivated. Depending on their origin and the conditions of cultivation, the cells will be dissociated or connected, loosely arranged or closely packed. On a solid or semisolid support they may form a coherent mosaic or network. Under the conditions of the Carrel technique, connective tissue cells of chicks are arranged as a network of the type illustrated in Fig. 76. Their colony has the shape of a planoconvex lens with a flat extended margin. Among invertebrate metazoa the sponges are known as organisms which, after artificial experimental dissociation of their components (pressing through bolting cloth), form a cell colony if a solid support is available. These colonies have the shape of a planoconvex lens similar to the shape of a fibroblast colony *in vitro*. Eventually, the sponge cell colony reconstitutes a whole sponge with its various structures (H. V. Wilson, 1907; experiments were confirmed by several investigators). The comparison between sponges and tissue cell colonies *in vitro* was made by Harrison in his penetrating paper "On the status and significance of tissue culture" (1928).

Individual circular colonies are obtained by the classic Carrel technique (coverglass or flask cultures), but in some modern techniques the cultivated tissue cells do not form well-defined colonies. One of these techniques is the roller tube method introduced by Gey in 1933. A thin plasma clot is spread along one side of a test tube. The fragments which are placed on or in the plasma, form colonies, but with increasing area the colonies fuse together and a coherent sheet of cells is produced. An entirely different technique was developed in Earle's laboratory (Evans *et al.,* 1951). Large flasks with or without cellophane sheets in them are filled with a liquid medium. As a result of shaking and periodical transfer to fresh media, the cell populations in these flasks form dispersed suspensions. A certain proportion of cells settle on the glass walls or cellophane sheets. At each transfer the suspended cells are readily aspired. In order to permit aspiration of the settled cells, they are detached from their support by scraping; gentle treatment with trypsin facilitates their dispersion. The populations of tissue cells cultivated on the walls of roller tubes or in suspended liquid media are living models not less than the discrete colonies obtained with the Carrel technique. By each of these techniques natural tissue cells are made to live in artificial population patterns.

An amazing variety of biological problems have been tackled by tissue culture techniques, as one can learn from a "Bibliography of the Research in Tissue Culture 1884 to 1950" by M. R. Murray and G. Kopech (1953) and from the Decennial Review Conference on Tissue Culture, held at Woodstock, Vermont, in 1956 (see P. R. White, 1957). This conference had the remarkable feature that some papers and discussions were devoted to the probable *future* of tissue culture techniques as applied to problems of nutrition, morphology, cancer, and viruses. Cytology has profited tremendously from tissue culture techniques. This fact was illustrated in earlier chapters by the examples of the Nissl bodies, the mitochondria, the Golgi apparatus, and the silver-impregnated cell boundaries. The advantages of cultivated cells are (1) the possibility to observe living somatic cells, (2) the flatness of cultivated cells which leads to a favorable display of fine structures, and (3) the certainty that each observed cell is complete. Points (2) and (3) are shared by smears, but injury to delicate structures is unavoidable in smears. Morphological characteristics of cultivated cells and their aggregates will be discussed in Part IV. What concerns us here is *the role of tissue cell populations in vitro as dynamic models.* An attempt will be made now to appraise the contributions which these models have made to the study of natural organisms. The appraisal will be based on the following five criteria: (1) pre-existing biological questions answered by cultivation of tissue cell populations, (2) opening of new avenues, (3) observations not related to problems of natural organisms, (4) observations which eventually may shed light on problems of natural organisms, and (5) failures and doubtful uses of cultivated cell populations as living models.

(1) *Pre-existing biological questions answered by cultivation of tissue cell populations.* In this class are two achievements of general biological importance which were mentioned before: the establishment of the so-called immortal strain (Carrel, 1912; Ebeling, 1913) and the cultivation of populations from a single cell (Sanford *et al.*, 1948). Immortality of cell strains does not mean that one individual cell lives forever. It means that a cell divides into two cells which divide again into four cells, and so forth so that the strain continues indefinitely without the occurrence of corpses. This potential immortality of cell strains had been considered to be a prerogative of bacteria, protista, and the reproductive cells of metazoa (Weismann, 1904, p. 212). The latter were considered to be potentially immortal, since theoretically each ovum and spermatozoon can escape death by producing a new individual. Indefinite cell division seemed to be impossible in somatic cells until the contrary was proved by the 30 years' life *in vitro* of the Carrel strain.[4] Similarly, the multiplication of single cells was believed to be a prerogative of bacteria, protista, and reproductive cells of metazoa. Therefore the cultivation of single tissue cells closed another gap between the somatic cells of metazoa and all other types of cells. An additional resemblance between bacteria and somatic cells of metazoa is the dissociation of clones which is observed in cultures of both forms. A clone is a strain of cells derived from one single cell. Dissociation means the appearance of sublines with different properties, an unexpected phenomenon in sublines of several clones of tissue cells which were cultivated under constant conditions (Sanford *et al.*, 1954; Parker *et al.*, 1957).

Increasing evidences that fundamental properties are shared by bacteria, protista, reproductive cells of metazoa, and somatic cells of metazoa encouraged the use of one of these groups as a model for another group. Generalizations across these groups are, however, possible only with respect to special problems. I mentioned earlier the interpretation of biochemical observations in bacteria and protozoa as models of possible differentiation mechanisms in metazoa.

[4] Other strains have been cultivated in various laboratories for periods of several years, maintaining their ability of cell multiplication.

The segregation of sublines in clones has introduced a new complication into the problems of classification of tissue cells. This will be discussed in Part V.

(2) *Opening of new avenues.* Conspicuous contributions to virus research belong in this class. Since viruses cannot multiply without the presence of living host cells, surviving fragments of organs, or mixed pulps of living, dying, and dead cells had to be used as an *in vitro* replacement for host-to-host transfers. Subsequently viruses were also cultivated in well-defined tissue cell colonies. An interesting early comparison of cultivation of *Hühnerpest* virus in pulps and in cell colonies is found in a paper by Hallauer (1931). For some time it was believed that those tissues which were visibly damaged by a certain virus in the host should be used for cultivation of the virus *in vitro.* Poliomyelitis virus, for instance, was considered to have a special affinity for nervous tissue, because the visible responses in man and monkeys were limited to the nervous system. In 1949 Enders, Weller, and Robbins published the first evidence that this virus multiplied in cultures of cells which were probably unrelated to nerve cells. Step by step these investigators arrived at the result that cultivated cell populations which were evidently derived from the tubules of the kidney were particularly suited for the multiplication of poliomyelitis virus (Robbins, Weller, and Enders, 1952). As a consequence, monkey kidney cultures became the means for a successful development of vaccines.

Important microscopic studies have been made on the interaction between virus and cultivated tissue cells (examples of publications on this subject: Syverton and Scherer, 1954; Bloch *et al.,* 1957). There is no doubt that the remarkable advances of modern virus research would not have been possible without the cultivated cell populations which served as living models of the hosts.

In an unexpected way, hospital pathology benefited from tissue culture techniques. Cultivation of biopsy material made it possible to identify certain tumors which in sections of the same biopsy could not be diagnosed (see Murray and Stout, 1954, 1958). Nervous tissue tends to accentuate its characteristics when explanted. It was only in cultures that sympathicoblastomas could be distinguished from the Ewing tumor or from tumors of lymphatic tissues. Sometimes the explants allowed a diagnosis after a cultivation period of 24 hours.

(3) *Observations in cultivated tissue cell colonies not related to problems of natural organisms.* When Carrel (1923) introduced the concept of "inherent" or "residual growth energy," he referred to the development of a cell colony from an explanted or transferred fragment in "protective media" (plasma) without the presence of "growth promoting media" (embryo juice). Depending on their past history in the organisms or *in vitro,* different strains form colonies with different rates of increase in area. Parker (1929) showed that, after a number of passages with high concentrations of embryo juice, a transfer

to media with very little embryo juice produced different rates of colony development: within a certain period cell strains obtained from the bones of a chick embryo produced much larger colonies than strains cultivated from the heart of the same embryo. These observations were interesting since identical stimuli revealed differences in reactivity of somatic cells which were morphologically indistinguishable but had a different embryological history. The remarkable point is that the imprint of the embryological history was not lost after a number of passages *in vitro* through which the original organization of the explant was completely abolished. Unfortunately, it is not easy to translate these observations into phenomena which occur in the natural organism. Another example in this class is the analysis of the factors which are involved in the transfer of tissue cell colonies as needed for the maintenance of strains in passages (Mayer, 1933, 1935). It was useful to determine the role of three of the factors, namely, change of media, mutilation of cells by cutting, and contraction of the elastic cellular network by detaching it from its support. However, no direct relations to problems of natural organisms emerged from this analysis.

(4) *Observations in cultivated cell populations which eventually may shed light on problems of natural organisms.* In this class are four important areas which have been the object of tissue culture work: regeneration, hormone production, nutrition, and the biology of cancer cells. Fischer (1930) and Ephrussi (1933) studied the *regeneration* of cell colonies after mechanical removal of sectors and segments of different size. As mentioned previously in connection with micrurgical techniques (Part III, Chapter 2, A, 4), similar regeneration studies were made with the use of ultraviolet rays for killing definite areas of a colony (Mayer, 1933). All these investigators observed the marked tendency of the fibroblast colonies to regenerate the circular shape which they had before mutilation and which undisturbed colonies assume with great regularity, irrespective of the geometry of environmental factors. Vogt (1934) compared morphogenetic movements in the early amphibian gastrula (see my Fig. 32) and the movements of cells, which caused regeneration of the circular form after mutilation of a colony *in vitro* in ultraviolet experiments of Mayer (1933).

One of these experiments is illustrated diagrammatically in Fig. 46. When a sector of a fibroblast colony has been killed by ultraviolet radiation, cells from adjacent protected areas migrate in two ways. At the two sides of the dead area, the straight chains of cells bend toward the defect, which means that the cells change their ordinary radial migration to a tangential direction. From the center of the colony, cells migrate radially through the dead area. Finally, the two streams of migration intersect, and a network of living cells results, which, by means of cell multiplication, attains the density of the rest of the colony. By these directed movements the defect is covered and the circular shape of the colony is reconstructed. It was the change from the radial to a tangential direction of migration which reminded Vogt of the morphogenetic movements in the embryo. As Vogt put it: "In spite of the great difference between a cultivated colony of tissue cells and the association of embryonic cells in a blastula, similar mechanisms may be responsible for the directed movements of cells." It seems to me that the factors which direct the movements of cells in either case still

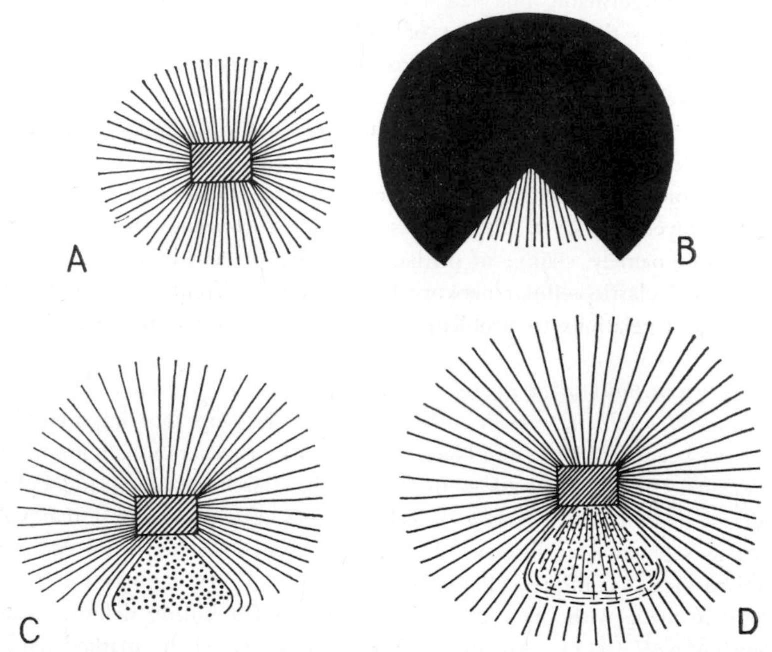

FIG. 46. Cultivated tissue cell colonies as living models of regeneration. Destruction of a sector of a fibroblast colony by ultraviolet radiation, and subsequent restoration of the circular colony form. Shaded rectangle indicates area of the fragment which was transferred to new media. Straight lines indicate usual radial type of cell migration (zone of area increase). (A) Approximately 24 hours after last transfer. (B) Same as (A); whole colony except for one sector protected by opaque varnish. In this condition the colony is exposed to a dose of ultraviolet radiation sufficient to kill the unprotected cells. (C) Approximatey 24 hours after irradiation, varnish removed. Dots indicate area of destroyed cells. At the two sides of this area straight cell chains bend toward the defect: *resemblance with morphogenetic movements in the gastrula of amphibians.* (D) Approximately 48 hours after irradiation. Interrupted lines indicate (1) radial migration of cells from the protected center through the killed area and (2) tangential joining of cell chains which had bent around the edges of the defect. *Circular shape of colony restored.* (Diagrams made after photomicrographs in E. Mayer, 1933, Figs. 21 and 22.)

deserve to be investigated. Studies of directed movements in slime molds (Bonner, 1947) will be reported in connection with transformation of nucleocytoplasmic patterns (Part IV, Chapter 2, E).

The tenacity with which fibroblasts, cultivated in plasma clots, produce and regenerate circular colonies made it unlikely that the circular form could be an expression of randomness. Several investigators concluded that the cultivated cell colonies are not merely aggregates of uniform cells but resemble an organism (Vogt, 1934; Ephrussi, 1935). This point of view is supported by the demonstration of polarized axes in fibroblast colonies which make these colonies comparable to sessile coelenterates or protozoa (Mayer, 1933). Obviously the conditions in the thick and closely packed center of the colony are not the same as in the loose and shallow periphery. This difference leads to visible polarity of individual cells. Fat drops are always most frequent in that part of the cell which points toward the center of the colony. If such cell divides, the fat droplets concentrate in the daughter cell which remains closer to the center of the colony.

The presence of some low-level organization in our living model leads to the question whether specific biochemical activities, such as the *production of hormones,* can occur in cultivated cell colonies. In the organism most hormones are produced by glandlike structures, but hormones can also form in scattered cells which are not arranged in any recognizable pattern (see Part II, Chapter 2, G). However, the environment of such scattered cells is highly organized (vascularization, innervation).

I will limit my discussion to a report on cell colonies which continued hormone production through long periods of cultivation. Waltz *et al.* (1954) explanted fragments from hydatidiform moles. These are pathological transformations of villi of the human placenta known to produce a gonadotropic hormone. Kept in small Carrel flasks on a plasma clot and covered with a cellophane sheet, the explanted fragments showed a minimum of cell multiplication. The resulting colonies were not divided and not transferred at short intervals (different from my diagram, Fig. 21), but were left undisturbed on their support, while the nutrient liquids were renewed three times weekly. Under these conditions, gonadotropin was recovered from cultures after various periods —from three cultures even after more than a year of life *in vitro.* Some cultures showed no increase in gonadotropic activity, if compared to the original explant, whereas others showed tenfold increases, even hundredfold in one culture. In a photomicrograph of one of the hormone-producing cultures taken after four months of cultivation (Fig. 13 of Waltz *et al.,* 1954), I notice a marked variety of cell forms and evidences of morphological organization. It is conceivable that cell colonies which consist of morphologically uniform cells could also produce hormones. I assume, however, that a *minimum of organization* is necessary for specialized biochemical activity. This minimum is probably present in well-defined individual colonies that are kept undisturbed at a low rate of proliferation. Suppose one should attempt to obtain mass production of a specific hormone by cultivating large cell populations in suspensions using the method of Evans *et al.* (1951). I am convinced that such an attempt would fail, since populations of cells which are suspended in a fluid, violently stirred, and transferred to new flasks at regular intervals, would be prevented from attaining the minimum of organization which I consider indispensable for hormone production. Continuous production of hormones has been obtained with the technique known as organ culture. With this technique the explanted fragments maintain a considerable level of organization, as described previously (see Figs. 19 and 20). The ability to maintain hormone production was observed in explants of adrenals (Lewis and Geiling, 1935), of the adenohypophysis (Martinovitch, 1950), and of the parathyroid (Gaillard, 1948). Future work on these lines will probably determine (1) the minimum of organization, and (2) the nutritional factors which are required for continued hormone production *in vitro*.

Studies of *nutrition* in tissue cultures were greatly enhanced by the discovery that cell populations could multiply satisfactorily if supported by cellophane or glass. Consequently, the plasma clot could be abandoned, which had been a complicating factor in biochemical studies. The next goal was the substitution of synthetic, or chemically defined, feeding fluids for the traditional media which contained proteins in the form of horse serum or embryo extract. For some time this subject was handled in the literature in an ambiguous way. Papers were published which made the reader expect that the problem had been solved, while the data showed that cultures could be maintained only if small amounts of horse serum or other proteins were present in the media. Finally, it was proved in two different laboratories that it is possible to keep cell populations in an entirely synthetic medium for periods of several weeks to more than two years. Morgan, Morton, and Parker (1950) used fresh explants which formed colonies on the wall of roller tubes. During the first 3 to 5 days the colonies were fed horse serum plus embryo extract. After this, the colonies were fed the synthetic mixture No. 199 in which they were maintained for an average of 4 to 5 weeks. McQuilkin *et al.* (1957) reported that some strains of cells, which had been cultivated successfully in the usual protein-containing media, required "careful nurturing" for a period of six months before one could be sure that the cells had adjusted definitely to the chemically defined protein-free medium NCTC 109. Cell strains were considered adjusted if they proliferated rapidly and could be readily subcultured. Prior to final adjustment, "critical periods" were observed during which the multiplication rate declined and the cellular morphology changed. Out of 56 strains, 10 strains were able to adjust to synthetic media. The fact that some strains could and others could not adjust is remarkable, since the 56 strains were all derived from one culture which was produced from a single cell.

Evidently there are a number of factors which permit or prevent continuous cultivation in synthetic media. For some time it seemed that cell populations which were suspended in fluid media under permanent agitation could not be supported by chemically defined media without protein, whereas the same strain of cells could adjust to synthetic media if allowed to settle on the wall of the receptacles or on cellophane sheets without being disturbed mechanically (Evans, 1957). Subsequently, these difficulties were overcome and suspended populations could be made to multiply adequately in chemically defined protein-free media (Bryant *et al.*, 1961).

It is difficult to predict how the cultivation of tissue cells in synthetic media may shed light on the nutritional requirements of tissue cells in the organism. Promising attempts have been made to study the amino acid metabolism of cultivated cell populations in synthetic media (Pasieka *et al.*, 1956).

An interesting connection between the cultivation of single cells and metabolic phenomena appears in the ingenious "feeder cell" technique of Puck and his associates (see Puck *et al.*, 1956). It was known that appropriate doses of irradiation inhibit the multiplication, but not the general metabolism, of certain bacteria. Similarly, tissue cells which were exposed to Roentgen rays lose their ability to multiply but retain a lively metabolism. The irradiated cells are used as "feeder cells." Single nonirradiated cells multiply when plated on top of the feeder layer. Otherwise, single cells multiply only when special techniques of cultivation are used, such as the capillary technique of Sanford *et al.*, (1948).

In the field of *cancer research* unexpected phenomena were observed in strains of cells which had been cultivated for long periods of time. I refer to the transformation of normal cells into cancer cells, known as cancerization, and to the transformation of cancer cells into normal cells. In the organism, cancer cells, or malignant tumor cells, are characterized by their ability to invade adjacent normal tissue (invasiveness), and to form subtumors (metastases) in distant parts of the body to which they are transported by the blood or lymph circulation (see Part V, Chapter 4, C, 1). Therefore the presence or absence of cancerous properties in cell cultures *in vitro* is tested by grafting such cultures into appropriate animals.

Up to the present, all attempts have failed to transform cultivated normal cells into cancer cells by the addition to the media of substances which are known to be carcinogenic in animals. At some time, Earle's laboratory reported that normal cells which had been exposed to methylcholanthrene were transformed into malignant cells (Earle and Nettleship, 1943). Subsequently, a number of colonies of the same cell strain, which had been cultivated without the addition of carcinogenic agents, also transformed into malignant cells (Earle, 1944). Finally, those strains which had been treated with methylcholanthrene steadily decreased in malignancy over a period of years. The conclusion was that the *transformation of normal into malignant cells could not be ascribed to the presence of methylcholanthrene* (McQuilkin *et al.*, 1957, p. 885, footnote 5).[5] In the meantime,

[5] Methylcholanthrene produced changes in cellular morphology and rate of cell division in explanted organ fragments, which were cultivated by the watchglass method (Lasnitzki, 1951; Hiebert, 1959). There is no indication that such cultures were inoculated into animals.

other laboratories also reported the transformation of cultivated normal cells into malignant cells under ordinary culturing conditions without the addition of carcinogenic agents. An observation of this kind was published in 1945 by Firor and Gey. These investigators explanted subcutaneous tissue from a rat of a strain with very low tumor incidence. After 4 months of cultivation, which included dividing and transferring of colonies, some batches of colonies showed unusual variability of cells with respect to morphology and rate of division. Grafting of such colonies into young healthy rats produced tumors at the site of the injections. When these tumors were transplanted into a second set of rats, definite signs of malignancy were observed (invasiveness with occasional metastases). Other cell strains which were derived from the subcutaneous tissue of the same rat were transformed into malignant cells two years after explantation. A third group of strains, also of the same origin, were morphologically unaltered at the time of the publication after six years of cultivation. Morphologically unaltered colonies did not produce tumors when grafted into rats.

Such correlation between the appearance of cells and malignant properties is not a general rule. Absence of correlation was, for instance, observed in experiments by Parker (1950, p. 25). Connective tissue from mice of a tumor-free strain was explanted and yielded the usual type of fibroblast colonies when cultivated in standard media. If methylcholanthrene was added to the media, a variety of highly abnormal cells developed (see Fig. 13 in Parker's book). When these abnormal cells were inoculated into mice, no tumors resulted, nor did the untreated control colonies produce tumors. On the other hand, sustained malignancy may be found in cell strains which morphologically appear quite normal. As an example of this kind I mention the mouse carcinoma that was cultivated *in vitro* by Albert Fischer for 14 years without losing its tumor-producing ability when inoculated into mice (Fischer, 1941). After the first 5 years *in vitro* some of these colonies were divided so that one part was injected into a mouse while the other part was fixed and stained. The inoculated parts produced tumors, but the cells were microscopically indistinguishable from normal colonies with the same mosaic pattern known as epithelium (E. Mayer, F. Jacoby, and H. Mayer, 1933, unpublished data).

Let us summarize the observations which resulted from extended cultivation of cell strains on the one hand, and from inoculations into animals as test for malignancy on the other hand.

(A) Inoculation tests
- Normal cells remained normal
- Malignant cells maintained malignancy
- Normal cells became malignant
- Malignant cells lost their malignant potency

(B) Morphology of cells
- Normal cells remained normal
- Abnormal cells remained abnormal
- Normal cells became abnormal
- Abnormal cells became normal

Any of the four possibilities listed under (A) can be associated with any of the four possibilities listed under (B). The morphology of cells *in vitro* does not indicate whether they will produce a tumor when injected into animals. This view was clearly expressed by Westwood *et al.* (1957). These investigators rightly emphasized the necessity of diagnosing malignancy not on the basis of *in vitro* phenomena, but on the basis of interactions between tumor cells and a host, such as invasion of normal tissue by tumor cells and metastasis in distant locations. Unfortunately, some authors use the term malignant, or cancerous, cells when referring to morphology without having tested the malignant properties by inoculation into animals. The problem of correlation between cell morphology and malignancy will be discussed again in Part V, Chapter 4, C, which deals with tumors in the organism. It is difficult to tell whether the appearance, maintenance, and disappearance of malignant properties in cultivated cells eventually may shed light on the mechanisms which, in animals and man, are responsible for the transformation of normal cells into cancer cells.

In experimental chemotherapy of cancer, attempts have been made to screen[6] for active drugs in cultures of cancerous tissue instead of, or in addition to, the usual screening in tumor-bearing animals. In tissue culture experiments, a compound is considered active if it produces visible harmful effects in populations of cancer cells but not of normal cells. As an example of allegedly positive results, I recount a paper by Biesele *et al.* (1951). Fresh explants from several tumors of rats and mice were compared to fresh explants of embryonic tissues of mice and rats. Roller tubes with plasma were used, several fragments being planted in each tube. Twenty-four hours after explantation the cultures were examined microscopically and tubes with satisfactory emigration of cells were selected. Then 2,6-diaminopurine or related substances were added to the media. After 24 hours of incubation in the presence of the compound, the cultures were scored by the following cytological criteria: (1) retraction of cytoplasmic processes, i.e., rounding off of cells, (2) increase in cytoplasmic granularity, and (3) disintegration of cells. The result was that 2,6-diaminopurine damaged the cultures of tumor cells at concentrations which caused little or no damage to embryonic mouse skin or heart cells. However, cultures of embryonic rat skin were more sensitive to this compound than cultures from two of the tumors. It seems to me that it is difficult to draw any clear conclusions from this experiment. Some investigators found it sufficient to observe the effect of chemical agents on cultivated tumor cells only, without testing the effect on normal cell cultures. If there was a positive effect in tumor cell cultures, comparisons with the action of the compound on tumors in animals were made. The different approaches are discussed in Hirschberg's 1958 review "Tissue culture in cancer chemotherapy screening."

[6] To screen means to select for a purpose, but there is no general agreement on the use of the term screening in experimental chemotherapy. Compounds may be selected for testing (a) on the basis of chemical or physical characteristics, (b) on the basis of a known or assumed mode of biological action, or (c) irrespective of (a) and (b).

I am not in a position to state whether experimental chemotherapy of cancer will gain much by *screening* in cultivated cell populations. There is no doubt that the *mode of action* of cancer-inhibiting compounds can be studied effectively in cultivated tumor cells. In human tumors, the study of cultivated explants is almost the only possible approach. I mention a metabolic and therapeutic investigation of human glioblastoma *in vitro* by Murray *et al.* (1954). These authors observed inhibitory effects of 8-azaguanine and the partial reversal of these effects by adenine, adenosine, and guanosine at equal molarity.

(5) *Failures and doubtful uses of cultivated cell populations as living models.* The propagation of viruses in cultivated cell populations has developed on an empirical basis. Therefore, certain failures are not surprising. For more than twenty years attempts have been made to find the proper tissues and media for the trachoma virus. According to the 1958 review by Thygeson, cultures of human corneal tissue show inclusion bodies after the addition of trachoma material, but no one has succeeded in obtaining multiplication of trachoma virus in cultivated tissue cells.

Misleading applications of tissue culture techniques occur mainly in the hands of investigators who underrate the role of vascular, nervous, and endocrine controls in the body. Comparisons between phenomena in cultivated cell colonies and phenomena in a highly organized animal are promising only where neurovascular factors play a minor role. Since the corneal tissue is not supplied by blood vessels, except in its margin, the process of wound healing in the cornea is composed of cell migration and cell multiplication and therefore is comparable to the regeneration of experimental defects in tissue cell colonies. The same is true of superficial wounds of the skin in which nothing but the epidermis is destroyed. The healing of deeper wounds of the skin is a different story. The disruption of blood vessels causes bleeding, and debris and clotted blood fill the defect. These masses are gradually replaced by capillaries which form the so-called granulation tissue. Subsequently, cells from the healthy epidermis around the wound move over the granulation tissue and multiply until the defect is covered. The last phase of this process is the only one which resembles the regeneration of cell colonies *in vitro.* Under these conditions it is somewhat astonishing that substances which promote the migration and multiplication of cells in cultivated colonies were expected to accelerate the healing of fairly deep wounds. Trials with chick embryo extracts were made with negative results (Kiaer, 1927). Even if the results had been positive, a rational bridge to tissue cell colonies *in vitro* would be lacking.

It is difficult to evaluate the application of tissue culture techniques to immunological problems. As an illustration I use sensitization phenomena of the type described earlier in connection with the effects of repeated stimulation (Part II, Chapter 2, F).

If a guinea pig has been sensitized by injection of horse serum, a second injection will produce violent effects on smooth muscles of the bronchi and also on the wall of capillaries. One may ask whether such change in reactivity is a prerogative of highly contractile structures such as smooth muscles and capillaries, or is also shared by cells which are not specialized in this way. To answer this question, the following tissue culture experiments were designed by Meyer and Loewenthal (1928). Guinea pigs were sensitized by an injection of horse serum. Spleen explants from untreated as well as from sensitized guinea pigs were cultivated in media which contained horse serum. The result was that the zones of emigrated cells around the spleen fragments of sensitized animals were not significantly different from those around the explants from untreated animals. Rich and Lewis (1932) made analogous experiments with guinea pigs which were infected with tubercle bacilli and thereby sensitized to tuberculin. Explants from tuberculous and from normal guinea pigs were exposed to tuberculin in the media. In this experiment, a marked difference appeared: the cells which emigrated from the explants of tuberculous animals were badly damaged and much smaller in number than those which emigrated from control explants. This difference in responses in tissue cultures became one of the foundations of Rich's theory (1951, pp. 330 ff., Table XVIII) that there are two types of hypersensitivity, the "anaphylactic type" and the "tuberculin type." However, other investigators failed to confirm such difference. Tuberculin did not kill epithelial cells cultivated from kidneys or livers of sensitized guinea pigs (Jacoby and Marks, 1953). In parallel studies, cells were harvested from exudates induced by tuberculin injection into serious cavities of sensitized guinea pigs. Metabolic determinations and motility tests showed that the macrophages and neutrophile leucocytes from the exudate were not dead: they were hardly damaged (Marks and James, 1953).

Occasionally it has been suggested that the toxicity of new drugs be tested by observing the first emigration of cells from a freshly explanted fragment, or by measuring the increase in size of cultivated cell colonies. Pomerat *et al.* (1946) made the statement that "tissue cultures offer a stringent test of toxic effects." It seems to me that the rules for the possible use of tissue culture techniques in pharmacology and toxicology can be summarized as follows: (1) If it is not known whether a new compound acts on circulatory, nervous, endocrine, or cellular mechanisms, it is necessary to use whole animals. (2) If one wants to know whether a substance has a direct effect on cells which are isolated from circulatory, nervous or endocrine controls, tissue cell populations *in vitro* (tissue cultures) may be suitable.

3. *Ecological Models*

Examples of artificial ecological units were discussed earlier (Part III, Chapter 4). I described the use of populations of mice kept in large cages for epidemiological studies (Topley *et al.*, 1931). These artificial populations served as models of natural populations. Other experimental populations have no natural counterpart; in this group is the artificial symbiosis of embryonic chick cells with algae, which was also described before (Buchsbaum, 1937). An interesting ecological model is represented by the so-called germ-free animals. Under natural conditions every mammal or bird lives in symbiosis with its intestinal bacterial flora. For some time the role of these bacteria in the nutrition of the host could not be studied satisfactorily, since the intestinal flora varies in an uncontrolled way. Finally, it appeared necessary to raise animals without bacteria on their body surfaces or in their respiratory and digestive systems. Mammals had to be delivered by sterile Caesarian section, and birds had to be hatched from uncontaminated eggs. Then the young mammals or birds had to be housed and fed in a sterile way. This technique was inaugurated by Glimstedt in 1932. One difficulty was the supply of the necessary vitamins. This difficulty was resolved, and the method was developed to a high degree of perfection by Reyniers and his associates. An early phase of this work was represented in the 1943 symposium edited by Reyniers, and the present state appears in a 1958 report edited by György.

4. *Models of Pathological Conditions in Individuals*

The study of human diseases in experimental animals belongs in the class of "substituting models." Some models have been designed to make pathological processes reproducible and accessible to quantification. The foreign body granuloma of Meier *et al.* (1950) and the granuloma pouch developed in Selye's laboratory (see, for instance, Selye and Jasmin, 1956) were used to produce local inflammation in the loose subcutaneous tissue on the backs of rats. Meier and his associates placed a ball of cotton under the skin, whereas Selye first injected 25 ml. of air under the skin and then a chemical irritant into the air pocket. The results were well-defined granulomas which could be dissected and weighed. The histology of granulomas will be described in connection with the classification of pathological tissues (Part V, Chapter 4, A). Experiments with embolic mouse tuberculosis made it possible to follow the fate of pathogenic bacteria in the host (Fig. 31, Tables 7 and 8).

D. Models of Nonbiological Natural Systems

In engineering and architecture, models are indispensable. The models are simpler than the planned structures, and, as a rule, on a different scale. As mentioned earlier, great advances in shipbuilding are due to the introduction of the Taylor basins as models of large natural masses of water. It may be useful here to consider also some models which have served as foundations of modern physics and chemistry.

Most readers of this book are familiar with the *models of atoms and molecules* which became important tools in stereochemistry. The tetrahedral model of the carbon atom and the benzene ring are outstanding examples. These models led to well-defined questions which could be resolved experimentally. As a support for the imagination, material models are used in which each carbon atom is represented by a black ball and other atoms by balls with different colors while the bonds are indicated by wires or rods which connect the balls. Wheland (1955, pp. 3 and 4) gives a particularly forceful presentation of the relation between two mental models of the benzene ring, known as the Kekulé structures I and II, and the physical benzene molecule which is a "resonance hybrid" between the two imaginary structures.

Models with a high degree of complexity have been constructed. The Watson and Crick (1953) double-stranded helical model of deoxyribonucleic acid was mentioned earlier in connection with the problems of self-reproducing mechanisms (Part II, Chapter 2, H). With these types of compounds, we cross the borderline between nonbiological and biological systems. Since deoxyribonucleic acid is the most important component of chromosomes, models have been considered which represent both the chemical composition and the visible (electron microscopic) structures of a chromosome. I refer to Kopac's models (1956) which serve as illustrations of topological cytochemistry. In an article by Perutz, which appeared in *Endeavour* (1958), instructive pictures are given of complicated models which form coils, such as the myoglobin molecule, or large packages of "loaves" representing the tobacco mosaic virus.

In nonbiological areas an atom model proposed by Lord Rutherford and developed by Niels Bohr became famous. This theoretical model consisted of a central nucleus surrounded by revolving electrons, thus resembling the solar system. Although subsequently this model underwent many modifications, it has doubtless played a decisive role in the progress of nuclear physics. The application of the theory of scale models to geological problems was treated comprehensively by Hubbert (1937). His article gives the reasons why scale models of geological structures require soft materials, such as half-liquid clay or Vaseline petroleum jelly loaded with iron filings.[7] Bostick (1957) constructed a model in which the evolution of galaxies is simulated. Ionized gas is projected by magnetic forces ("guns") through a vacuum, either across magnetic fields or in the absence of magnetic fields. Several striking resemblances with astronomical spiral galaxies were the result.

---

[7] Dr. C. L. Drake of the Lamont Geological Observatory, Palisades, New York, drew my attention to Hubbert's book.

Chapter 6

Topographic and Three-Dimensional Morphology

In the preceding chapters of the book topographic relations have been discussed in various connections. Most of the embryological examples involved topographic problems, since mapping of potentialities is one of the main tasks of embryology. Many of the morphological procedures, such as surgical exposure of inner organs and micrurgical techniques, had topographic aspects. The importance of preserving topographic relations was stressed in the chapter on the study of dead material (Part III, Chapter 2, B). In the same chapter special techniques for the demonstration of three-dimensional structures were briefly discussed in connection with injection techniques (Fig. 23, bronchial cast). A more detailed and comprehensive treatment of topographic and three-dimensional morphology[1] will be the present task.

Topographic or regional anatomy differs from systematic anatomy as follows. In studying a human arm, one may dissect the blood vessels and nerves with their ramifications, determine the origin and insertion of muscles, and finally isolate the bones to examine their shapes. This is systematic anatomy. In topographic or regional anatomy, one makes a cross section through the whole arm in order to determine the spatial relations between the individual blood vessels, nerves and bones at *the particular level of the cross section.* Corresponding studies on the microscopic level are known as histotopography.

Modern physiology and biochemistry make increasing demands on exact localization. Localization of substances and chemical reactions on the level of cells and tissues is the field of cyto- and histochemistry which will be the subject of the next chapter.

The importance of topography in dynamic morphology has found expression in Woodburne's "Essentials of Human Anatomy" (1957). This book offers a

[1] J. R. Oliver (1954) stated that the newer morphological concepts concerning the nephron have progressed to a fourth dimension. He referred to time as the fourth dimension, in addition to the three dimensions of space, evidently using the language of the special theory of relativity. In this sense, dynamic morphology is four-dimensional.

unique combination of systematic and topographic anatomy, with constant emphasis on functional aspects.

The organization of the present chapter may be compared to geographic procedures. First, the general references of maps to cardinal points, latitudes, and longitudes are stated. Then, the reading of maps and their use in traveling is illustrated by examples. Finally, techniques of surveying and of constructing maps are described.

A. Regional Anatomy and Histotopography

1. *Basic Architecture of the Animal Body*

a. *Topographic planes of the human body and of the dog's body.* Figure 47a shows the planes or orientation in the human body (modified after Woodburne, 1957, Fig. 1). One of the purposes of topographic terminology is to keep the designations independent of changing postures. When a man lies on his back, popular terminology calls the chest "up" and the back "down," but in the standing man one refers to the head as "up" and to the feet as "down." Obviously, such variable terminology would endanger communication in surgery and obstetrics. This uncertainty is avoided by the use of the technical terms of regional anatomy.

Cranial and caudal, ventral and dorsal, medial and lateral are pairs of concepts in which each partner defines the other. The craniocaudal, as well as the dorsoventral, relations refer to unequal poles of an axis. Medial and lateral refer to relative distances from the mid-sagittal plane.

The mid-sagittal or median plane is derived from the sagittal suture of the skull. It is the only plane which divides a vertebrate's external form into two symmetrical halves. Sagittal, transversal, and frontal are three families of planes. The sagittal planes are parallel with the mid-sagittal plane. The transversal planes are perpendicular on the sagittal planes and also on the vertebral column. In an upright man, all transversal planes of trunk, neck, and head are horizontal. In an animal which stands on its four feet (Fig. 47b) the transversal planes of the trunk are vertical; in short-necked animals the transversal planes of neck and head are also vertical, but in animals with a fairly long neck the transversal planes of neck and head may be vertical, oblique, or horizontal depending on posture. The frontal planes, also called coronal planes, are defined as perpendicular on both the sagittal and the transversal planes. Whether frontal planes are vertical, oblique, or horizontal depends on the transversal planes.

The terms medial and lateral refer to relative distances from the median (mid-sagittal) plane. They are applicable to the human foot, as indicated in Fig. 47a, but they are not applicable to the human hand and forearm for the following reason. The arm below the elbow can rotate around a longitudinal axis. If this occurs while the arm hangs vertically, the palm may face any direction, and the place of the thumb may be lateral or medial, ventral or dorsal. Therefore, parts of the forearm and the hand are described by their constant position relative to the ulna, which is the fixed bone, and the radius, which is the rotating bone. Thus, the thumb is on the radial side, and the little finger on the ulnar side. The term dorsal, if applied to hand or foot, does not refer to the planes of the body, but to that surface which is opposite the palm or sole, respectively.

Comparisons between man and four-footed animals, or between different four-footed animals, would be difficult if one depended on terms such as superior and inferior, or anterior and posterior. In man the upright posture is considered standard, therefore the head is up. In four-footed animals the standard posture is standing on four feet. In a short-necked animal, such as a pig, the head is in front, in a giraffe it is up, and in dogs or horses one has the choice. In Fig. 47b, Woodburne's scheme has been applied to a dog, as a representative of four-footed animals.[2]

The term four-footed is used here in a functional sense, and as the concept opposite to upright. Anatomically the forelegs of a dog resemble the arm of a man, and the hindlegs of a dog resemble the legs of a man. The proper terminology of vertebrate comparative anatomy, with respect to extremities, is as follows: *Thoracic limb:* for arms of man and monkeys, wings of bats and birds, and forelegs of all others. *Pelvic limb:* for legs of man, monkeys and birds, and hindlegs of all others.

[2] From the literature I mention two topographic pictures of vertebrates, showing the median (mid-sagittal), the transversal and the frontal planes. One of the pictures represents a mouse (Gray, 1954), and the other a fish (Storer and Usinger, 1957). Neither of the two pictures is intended to supply detailed topographic orientation.

FIG. 47a. The planes of the human body, with terms of direction and orientation. Slightly modified from Woodburne (1957, Fig. 1). Note: (1) It is only in the upright human being that the frontal plane is vertical and is, at the same time, parallel with the vertebral column and the forehead (hence the name frontal). (2) The terms cranial and caudal are preferable with respect to chest and abdomen while superior and inferior are necessary with respect to the extremities. (3) Proximal and distal are terms not indicated in the picture. In the extremities these terms refer to relative distances from the trunk: the knee is proximal to the foot and distal to the hip.

FIG. 47b. The planes of the body of a dog, with terms of direction and orientation. Drawing by A. R. Reynolds, adapting Woodburne's scheme to four-footed animals. Note: (1) As in the human body, the frontal plane is parallel with the vertebral column and perpendicular on the sagittal and transversal planes. However, in the dog the frontal planes change from horizontal to vertical or oblique as the transversal planes change. "Frontal" does not refer to the forehead of the animal. (2) The choice between the terms cranial and caudal, or superior and inferior follows the rule given in Fig. 47a, note (2). (3) Proximal and distal, same as in Fig. 47a, note (3).

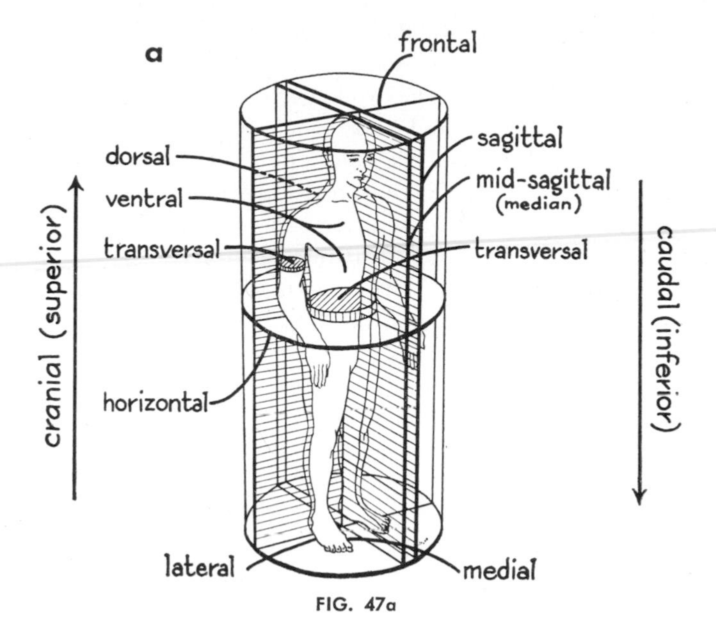

FIG. 47a

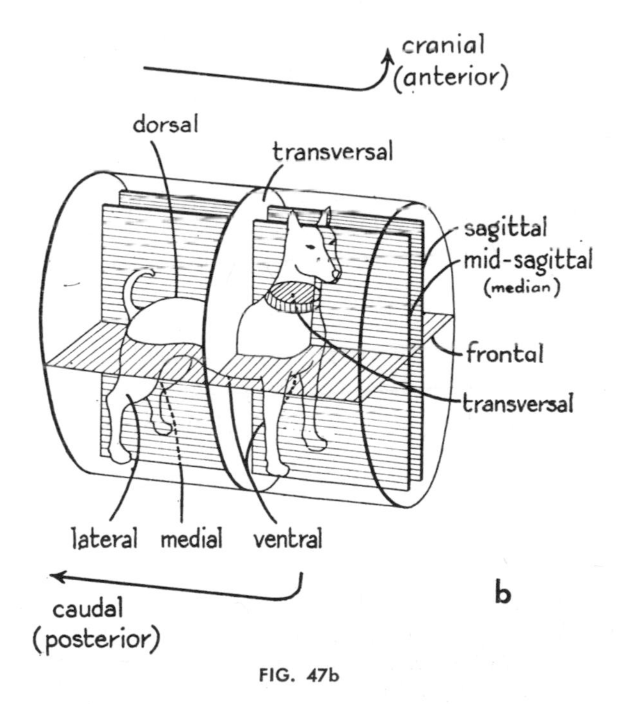

FIG. 47b

I admit that the correct terms "thoracic limb" and "pelvic limb" are not used much. In experimental physiology some terms of human anatomy are carelessly applied to four-footed animals. The two large veins which enter the right atrium are called vena cava superior and vena cava inferior in human anatomy, the former coming from above, and the latter from below (Fig. 84). In a four-footed animal, such as the dog, both veins run horizontally before entering the right atrium, yet they are called superior and inferior by most physiologists. Since hydrodynamics of the blood circulation involve gravity, this is quite misleading. The proper terms are "vena cava cranialis" and "vena cava caudalis" (Bradley, "Topographic Anatomy of the Dog," 4th ed., 1943, p. 56). Another example of misleading topographic terminology is the persistent use of "anterior pituitary" and "posterior pituitary," based on the anterior position of the adenohypophysis and the posterior position of the neurohypophysis in man.[3] In the rat the adenohypophysis is caudal of the neurohypophysis; in dogs and cats the adenohypophysis forms a shell which encloses a large part of the neurohypophysis.

It is almost impossible to achieve complete consistency in topographic descriptions. For instance, the conventional term horizontal has been maintained in Fig. 47a and b, although it is applicable to man and dog only in their standard postures. I have not included the concepts "proximal" and "distal" in Figs. 47a and b. With respect to extremities, proximal means closer to the trunk, and distal farther away. The knee is proximal relative to the foot, but distal relative to the thigh. However, these terms are also used in the description of the circulatory system and the nervous system. Blood vessels which are closer to the heart are termed proximal relative to vessels which are farther away. Similarly, relative distances of nerves from the brain or spinal cord are expressed by the terms proximal and distal. Finally, these terms are used with reference to microscopic tubules in the kidneys: the convoluted segment closest to the glomerulus is called first or proximal, while another convoluted segment is called distal.

b. *Symmetry and asymmetry of animals. Polarized axes and their origin.* A structure possesses *bilateral symmetry* if it can be divided into two equal halves by one plane, and by one plane only. The bodies of vertebrates, as a rule, are bilaterally symmetrical with respect to their external form. In other words, the right and the left sides are mirror-images of each other. Minor deviations from the right-left symmetry of the external form are irrelevant for the purpose of the present discussion.[4]

The similarity of left and right includes organs such as bones and muscles.

---

[3] Pituitary gland and hypophysis are synonymous.

[4] Conspicuous asymmetry of the external form is exceptional in vertebrates; examples are the oblique skulls of whales, and the secondary asymmetry of flatfishes, such as flounders. Different degrees of symmetry and asymmetry are fully discussed in Ludwig's 1932 monograph.

Blood vessels and nerves are fairly similar in the left and right extremities, but not in the chest and abdomen. Inner organs are either pairs, such as the kidneys and ovaries, or single, such as the heart, the stomach and the liver. In paired organs there are different degrees of resemblance between the right and left partners. Kidneys, for instance, are similar in shape, but different in position. The right and left lungs of most mammals have different numbers of lobes. The spinal cord and the brain are organs with a high degree of bilateral symmetry, although functional differences between the right and left cerebral hemispheres are known. The heart shows marked differences between its right and left parts.

A structure which can be divided into equal halves by more than one plane possesses *radial symmetry*. This pattern is found in various invertebrates, such as the cylindrical sea anemones (actinia) and the umbrella-like jellyfishes. Spiral forms (in one plane) and helicospiral forms (three-dimensional spirals) of invertebrates are represented by various types of snails.

Since vertebrates possess bilateral symmetry, their organization is necessarily based on the presence of three axes: a dorsoventral, a craniocaudal (anterior-posterior), and a right-left axis. Evidently, two of these axes have unequal poles: to use popular language, the top is different from the bottom, and the front is different from the rear. What about the right-left axis? As stated before, bilateral symmetry is present in the external form, and in some inner organs but not in others. The statistical norm for man is to have the heart on the left side, the bulk of the liver on the right side, the spleen on the left side, the coecum with appendix on the right side, and so on. Only one in 10,000 individuals has a mirror-image pattern, known as *situs inversus*. In other words, the right-left axis is also polarized, with statistical predominance of one orientation.

The *origin* of the three polarized axes in the embryo is one of the most intriguing problems of biology. Many observations are available which show that three axes are already present in the unfertilized ovum. Particularly in insect eggs anterior-posterior, dorsoventral, and right-left axes can be recognized before fertilization (see Huettner, 1949, p. 9 and his Fig. 22); the position of the eggs in the ovary prepares the axial orientation, and the early axes correspond with those of the future larvae. More evidences for the presence of polarized axes in the cytoplasm of the ovum before fertilization will be mentioned later. At this point, I wish to demonstrate that the *presence* of three polarized axes in the unfertilized ovum is a *postulate in all bilateral metazoa which show a statistical predominance of one type of left-right asymmetry over the other type.* The word "presence" refers here to conditions at a given moment, and the term "postulate" means that this discussion is independent of experimentation.

Suppose a hypothetical spherical egg, which is isodiametric in all respects, is dropped into water. The first point of the egg's surface which touches the water may become the ventral pole of a dorsoventral axis. Let us assume that, after this, some environmental factor initiates a developmental process in any point of the equator, i.e., the largest horizontal circle, and let us label this point the cranial pole.[5] Then, the caudal pole will be on the opposite end of the particular diameter. After the establishment of the dorsoventral and craniocaudal axes, a third axis which is perpendicular on the two others can emerge. This is the right-left axis. Polarization of this axis can possibly be induced from outside, but not a regular predominance of position of the poles. The circularly polarized sunlight which comes through the atmosphere has been invoked as an asymmetrical agent apt to produce asymmetry in random substrates. However, the left and the right side of our free-floating egg have an equal chance of being reached by the circularly polarized light. There is no environmental factor conceivable which could have more effect on one side than on the other. If the right-left axis should develop unequal poles at random, 50% of the animals should have the heart on the left and 50% on the right side. Wherever one of the two positions is the rule, a pre-existing right-left polarity must be postulated, with predominance of one direction.[6] Since there is no right and left without the two other polarized axes, one must assume that the dorsoventral and craniocaudal axes are also pre-existing in the unfertilized ovum. The place of entry of a spermatozoon may determine the position of the axial system inside the egg, but the creation of the axes cannot be ascribed to the entry of the spermatozoon any more than to other external factors.

There are good reasons to assume that the three polarized axes are inherited through the *cytoplasm*. In discussing the acquisition of axes, J. Z. Young (1957) emphasizes the presence of at least one polarized axis in the oocyte (egg cell in the ovary), as indicated by an eccentric aggregation of Golgi bodies. Young also mentions the theory that the animal pole of the oocyte is directed toward the ovarian arterioles, thereby receiving better nourishment and thus becoming the potential head area of the embryo. "Differences in metabolism may produce differences in the activities of the cytoplasm of different parts of the ovum"

[5] Neither in our spherical model egg nor in the actual frog's egg does the early dorsoventral axis represent the dorsoventral axis of the larva and mature animal. Similarly, the "cranial pole" of our model egg does not represent the later head area. However, the assumption that all places of the equator are equivalent is based on observations: on the borderline between yolky and yolkless-pigmented cells, which is more or less parallel with the equator, there is no point at which a blastopore may not form (Needham, 1942, p. 218).

[6] My discussion is limited to the ontogeny (embryology) of asymmetry. The phylogenetic (evolutionary) origin of regular asymmetry in each species is analogous with the question why only one enantiomorph of some asymmetrical compounds occurs in nature. Readers interested in the historical development of the chemical problem are referred to the paper by F. R. Japp, "Stereochemistry and Vitalism," which appeared, with subsequent discussions, in *Nature,* 1898 and 1899, and to P. D. Ritchie's 1933 book "Asymmetric Synthesis and Asymmetric Induction."

(Young, p. 591). According to the interpretation which I present here, environmental asymmetries are apt to determine the position of the polarized axes which pre-exist in the cytoplasm of the oocyte. In order to emphasize the dynamic properties of these axes, many investigators prefer to talk about gradients. Huxley and de Beer (1934, p. 438) described the organization of the egg in its earliest phases as a gradient field, with quantitative differentials extending across the egg in several directions. These authors also stated clearly that the localization of the gradients is not predetermined, but is brought about by agencies external to the egg. Child's general theory of metabolic gradients will be discussed later.

Large ovaries with pronounced regional differences might be promising objects for the study of metabolic gradients in oocytes. According to Laws (1957) the ovaries of mature whales and mares show a conspicuous heterogeneity in the distribution of oocytes: there are many more oocytes in the cranial part than in the caudal part of each ovary. I wondered whether the development of a special vascular pattern could be responsible for this heterogeneity, which is not present in the newborn. However, no marked differences were noticed between the vascular patterns of the cranial and caudal parts of the adult ovary (personal communication by Dr. Laws, 1957).

A number of investigations have been devoted to factors by which the *position* of the polarized axes is determined in the egg prior to cleavage. Among these factors, gravity and the place of insemination were studied most extensively. It was found that in amphibians the position of the dorsoventral axis is stabilized at a time when the position of the right-left axis is still labile (Holtfreter and Hamburger, 1955, pp. 231-232).

Not only the polarized axes of the whole embryo, but also the different axes of its parts are stabilized at different periods. This was demonstrated impressively by Harrison's (1921) transplantation of limb buds of *Amblystoma* (axolotl) embryos. If, at a certain stage, the left forelimb bud of one embryo is transplanted to the right side of another embryo so that the ventral (radial) edge of the limb bud remains ventral, the posterior surface necessarily points in the anterior direction.[7] This wrong position of the anterior-posterior axis of the limb bud is not corrected as development proceeds. If the transplantation is carried out in such a way that the ventral (radial) edge of the transplanted limb bud points in the dorsal direction with respect to the recipient embryo, the posterior surface points in the posterior direction. With advancing development, the ventral (radial) edge of the limb forms ventrally with respect to the body: a normal right forelimb is the result. In other words, at a time

[7] In the forelimb of a swimming urodele, the ulnar side is clearly dorsal and the radial side clearly ventral, since the joints of these animals do not permit functional rotation of the forelimb around a longitudinal axis.

when the anterior-posterior axis of the forelimb is stabilized according to its original location, the dorsoventral axis of the limb is still able to adjust to the topography of a new location (morphogenetic fields).

I stated previously that in vertebrates a right-left asymmetry of inner organs is the rule, and that one type of asymmetry is always found to be present in the overwhelming majority of individuals. In mammals, the arch of the aorta swings to the left side, and in birds to the right side, although the architecture of the heart is identical in mammals and birds (not mirror-imaged). In most families of birds, the right ovary and oviduct are either absent or underdeveloped, but there is no such asymmetry in mammals. W. Ludwig's book on the right-left problem, which appeared in 1932, still presents an unsurpassed collection of data in this field.

In the discussion of orders of magnitude (Part III, Chapter 4, E) I stated the reasons why bilateral asymmetry of organisms cannot be ascribed to asymmetry of molecules.[8] The principles underlying this negative statement were illustrated in Fig. 44a and b, which showed symmetrical and asymmetrical building blocks and the properties of their respective composites. The recognition that morphological asymmetry of organisms cannot be reduced to a problem of stereochemistry need not discourage attempts to trace right-left asymmetry as far back in embryonic development as possible.

Rare instances of situs inversus in man were mentioned before. Experimental production of situs inversus in chick embryos has been reported in the literature, but these claims proved to be without foundation (Mayer, 1942). In amphibians, situs inversus has been produced experimentally by Spemann and Falkenberg (1919). With the use of Spemann's micrurgical techniques mentioned previously (Part III, Chapter 2, A, 4), loops of fine hair were placed around fertilized eggs of triton. The earliest stage used was the two-cell stage, and the latest stage was the blastula (my Fig. 3b–e). The loop constricted the egg in a meridian, i.e., in one of the vertical planes which go through the animal and vegetal poles (*AP* and *VP* in Fig. 3c). Each egg produced two embryos. In the most conclusive series, which consisted of 18 eggs, both partners from each egg were available for analysis. Of the 18 left twins, 17 had the normal situs and 1 a partial inversion of viscera. Of the 18 right twins, 9 showed situs inversus, 8 had a normal situs, and 1 was questionable. The most disappointing result would have been normal situs in all left and right partners. The most gratifying result would have been that all partners of one side showed a normal

[8] In Part II, Chapter 2, H, I mentioned an interesting attempt to find molecular orientation by Roentgen-diffraction analysis of embryonic primordia in which the orientation of the polarized axes was not yet stabilized (Harrison, Astbury, and Rudall, 1940). The results were not conclusive. If any experiment of this kind should yield correlations between molecular and biological orientation, the relation between the two things would still remain obscure.

situs, and all of the other side situs inversus. The fact that the potential left side was evidently more stabilized than the potential right side posed new questions.

A certain insight into the ontogeny of asymmetry was obtained by the study of helicospiral asymmetry[9] in snails. In *fertilized* eggs of snails, the direction of winding can be recognized in the four-cell stage. For instance, in a sinistral (left-handed) species, the four cells are not arranged symmetrically around an axis as they are in the four-cell stage of the frog's egg (Fig. 3c),[10] but are twisted sinistrally. This twist is accentuated by the positions of the mitotic spindles. In Edmund B. Wilson's book (1934, Fig. 465) pictures are given of a dextral snail, *Lymnaea,* with its dextral cleavage pattern and of a sinistral snail, *Physa,* with its sinistral cleavage pattern. The presence of asymmetry in the *unfertilized* ovum was indirectly proved by the breeding experiments of Boycott and his collaborators (1923, 1929). In the species *Lymnaea,* dextral spiraling is the rule, but occasional individuals with sinistral shells occur. Upon cross-breeding of dextral and sinistral individuals, the brood will be uniform, either all dextral or all sinistral. If this brood, the F_1 generation, is interbred, the offspring, F_2, will also be uniform. If the F_2 snails are interbred, Mendelian segregation at the ratio 3:1 will be observed in their offspring, the F_3 generation. Hereditary phenomena which are completely controlled by chromosomes (genes) show Mendelian segregation in the F_2 generation. The delay by one generation as observed in the snails can only be interpreted as cytoplasmic inheritance. This means that the cytoplasm of the ova contains the dextral or sinistral pattern in an irreversible fashion so that it cannot be modified by the fertilizing spermatozoon. However, the genes of this spermatozoon can act on the ova which develop in the ovaries of the F_1 generation.

The fact that snails show the same direction of spiraling in the earliest stages of cleavage, in the late stages of cleavage, and in the shell of the adult animal does not mean that all regions and organs of the snail develop under the control of the spiral pattern. In any snail with pronounced spirality of shell, mantle, and visceral mass ("visceral hump"), the head and the foot, which can protrude from the shell, are bilaterally symmetrical (see H. H. Newman, 1936, p. 290). It seems to be a general rule that, at some stage of development, the spiral pattern of cleavage becomes modified into a bilateral pattern (Costello, 1955, p. 214). I discussed previously the reasons why patterns of symmetry and asymmetry can change readily as the levels of complexity change (Fig. 44a,

[9] A spiral is two-dimensional, and therefore, has no right and left. A helicospiral is three-dimensional. Sinistral and dextral helicospirals are defined in the same way as screws. A screw is dextral (right-handed) if it is forced into the floor by clockwise turning of the screwdriver. Loosely, the term spiral is used for helicospiral.

[10] The radial symmetry of the eight-cell stage of the frog's egg as pictured in Fig. 3d, is a conventional but not realistic diagram.

b). The different species of the class Gastropoda (snails) show numerous transitions between spiral asymmetry and bilateral symmetry: in Fissurellidae, the shell is spiral in young individuals, but not in older individuals.[11] In nemertines (related to flatworms) the four-cell stage of cleavage shows perfect radial symmetry, whereas subsequent stages are spiral (Wilson, 1934, pp. 990 and 1100). The adult animal is bilaterally symmetrical.

Analysis of polarized axes not only is necessary with respect to the architecture of the whole body, but also is important on the level of organs and somatic cells. The polarity of the retina cells was the subject of an earlier discussion (Part II, Chapter 2, G). Their opposite orientation in the human eye and the octopus eye was illustrated in Figs. 11 and 12, and the developmental background of this difference was shown in Figs. 13 and 14. Polarity of cells will occupy us again at various occasions. In glands, for instance, one pole of each cell is in contact with the blood capillary which supplies raw material, while the other pole borders the glandular space into which the finished product is discharged (Fig. 87). Aggregates of cells are polarized in many ways. The epidermis contains keratinized cells in the superficial layers and dividing cells in the basal layers, and the intermediary layers are composed of cells which are on their way to keratinization.

If a mosaic of cells that line the external or inner surfaces of the body consists of several layers, most mitotic activity is usually found in the basal layers. From the position of the mitotic spindle one can tell what the future fate of a cell would have been if the tissue had not been killed for study. When the axis of the spindle is parallel with the surface, both daughter cells will remain in the basal layer. If the axis of the spindle is perpendicular on the surface, one daughter cell will remain in the basal layer, while the other daughter cell moves to the next higher layer. In the case of the epidermis, this next layer is characterized by differentiation into so-called prickle cells. The cells in the prickle cell phase or layer eventually move to the most superficial layers, where they undergo keratinization. If the spindle axis is vertical on the surface, it is also vertical on the supporting connective tissue which carries the blood vessels. Therefore, this orientation of the spindle involves unequal environmental conditions for the two daughter cells. Although it is not known whether the environment of these cells or factors inside the cells determine the course of events, the concept of *differential mitosis* is very proper to describe the phenomenon. The concept was introduced by Koehler (1932) in his studies of scale

[11] In the light of all these variations one can hardly draw far-reaching conclusions from the observations of Alpatov and Nastyukova (1948), who immersed three sinistral and three dextral species of snails in quinacrine (of which Atabrine is the hydrochloride). They found that sinistral snails were killed by concentrations of the *d*-isomer in which dextral snails survived, while the dextral snails were more sensitive to the *l*-isomer (statistically significant differences).

formation and development of pigment pattern in the flour moth *(Ephestia kühniella)*.

Cell colonies *in vitro,* cultivated in a plasma clot on a solid support, may develop polarized axes, as described previously. This is most pronounced in fibroblast colonies which assume the shape of a planoconvex lens. The radial polarity of the colony is reflected to some extent, in the individual cells. As an illustration of cellular polarity, I mentioned the accumulation of visible fat droplets in that portion of each cell which points toward the center of the colony.

Finally, I mention an asymmetrical mechanism in which the direction of asymmetry is reversed by an environmental factor. In *Paramecium aurelia,* a ciliated protozoon, the normal swimming movements are sinistrally helicospiral. Some strains of this species produce the so-called killer factor, kappa. This is a poison fatal to sensitive paramecia of other strains which do not carry the kappa factor. After exposure to individuals of certain killer strains the rotation of the victims *changes from sinistral to dextral spiraling.* This change is most pronounced 4 to 6 hours after exposure. Several days later the animals die (Sonneborn, 1959).

c. *Comparison of basic architecture in vertebrates, arthropods, and worms.* As discussed previously, most technical problems can be solved by nature in various ways (Part II, Chapter 2, G). As an additional illustration I mention here the topographic relations between skeleton and muscles in vertebrates and arthropods. In the extremities of vertebrates the bones are surrounded by muscles (endoskeleton), while in arthropods, e.g., insects and lobsters, the muscles are encased in the chitin skeleton (exoskeleton). This necessarily leads to very different dynamics of locomotion. Yet walking, swimming, and flying are achieved satisfactorily with both schemes.

The general topography of the main visceral systems is different in vertebrates and arthropods. This is seen best in transversal sections through the body. In vertebrates the central nervous system is most dorsal, ventral of it follows the circulatory system (heart and aorta), and most ventral is the digestive system. In arthropods the order is as follows: most dorsal the circulatory system, ventral of it the digestive system, and most ventral the nervous system. Annelids, such as the earthworm, have a general topographic pattern of viscera like that of arthropods, but they possess neither an external nor an internal skeleton. Their body is supported by a continuous mass of muscles which form the body wall and enclose body cavity and viscera.

Besides these differences there is a common feature in the general architecture of vertebrates, arthropods, and annelid worms, namely their *segmental organization.* This type of organization is characterized by a repetition of similar structures along a longitudinal axis (segments or metameres). In the elongated stages of vertebrate embryos, pairs of rectangular masses develop on each side of the notochord, known as somites. The series of vertebrae and ribs, and the serially arranged muscles, develop from the somites. In addition, the nerves which arise from the brain and spinal cord follow the segmental pattern.

In Annelida, of which the earthworm is a representative, the body is divided externally in a number of muscular rings which correspond with a division of the internal parts into a series of segments or metameres. Particularly the excretory system, or nephridia, show this pattern distinctly. Flatworms, such as planarias and flukes, are not segmented. In tapeworms each proglottid repeats, in some respects, a whole organism, but the total chain, or proglottides, is attached to the intestine of the host by one organ, the head (scolex). This gives the tapeworm a peculiar intermediary position between a colony of organisms and a segmented animal. The Arthropoda are a large group of invertebrates with pronounced segmentation. The segmentation is more obvious in the caterpillar than in the butterfly, and more conspicuous in the exterior of a centipede than in that of a lobster. In sessile forms, such as barnacles, the segmental pattern, or metamerism, is greatly obscured in the adult animal.

2. *Various Topographic Relations of Functional Structures*

a. *The topographic dynamic morphology of the central nervous system.* The procedures by which functional maps of the brain and spinal cord were obtained, were described briefly in connection with the determination of functions by elimination of morphological structures (Part III, Chapter 1, A). Additional technical procedures for topographic determinations in the central nervous system will be mentioned later (Part III, Chapter 6, A, 4, b). The present discussion will be devoted to the reading of the maps rather than the making of the maps.

The whole surface of the cortex has been mapped and each area has been characterized by a number. Four of these areas are presented in Fig. 48, two of them motor areas and the two others sensory acoustic areas. In many cases the areas which were determined by functional tests can also be identified by characteristic microscopic patterns (e.g., after silver impregnation of nerve cells). The areas shown in Fig. 48 were selected because of their characteristic cytoarchitectonic differences which are shown in Fig. 49a and b (from Ariëns Kappers *et al.,* 1936). The frame of reference which is used for cytoarchitectonic mapping consists of a system of cortical layers labeled by the Roman numerals *I* to *VI*. The picture of the sensory acoustic areas *41* and *42* shows large nerve cells in zone *III* while no cells of similar size are present in zone *III* of the motor areas *4* and *6*. Additional differences are apparent upon comparison of other zones of the motor and sensory areas.

The legend of Fig. 48 states that destruction of the ventral part of area 6 results in "a form of language disturbance." Language disturbances may consist of inability to form words, inability to understand spoken words, loss of reading or writing ability, shrinkage of vocabulary, and combinations of these disturbances. Since a variety of mechanisms are involved, one cannot be surprised that the different functions which constitute "language" are located in separate areas of the cerebral cortex. In right-handed individuals the left hemisphere contains the important language center of Broca (area 44), which is adjacent (cranial) to the region where line 6 ends.

In Fig. 48 the cortical motor areas are indicated by shading and crosshatching. On the opposite side of the central sulcus *(CS)*, notice an area which contains sensory centers for impulses from the skin, mouth, and skeletal muscles. This

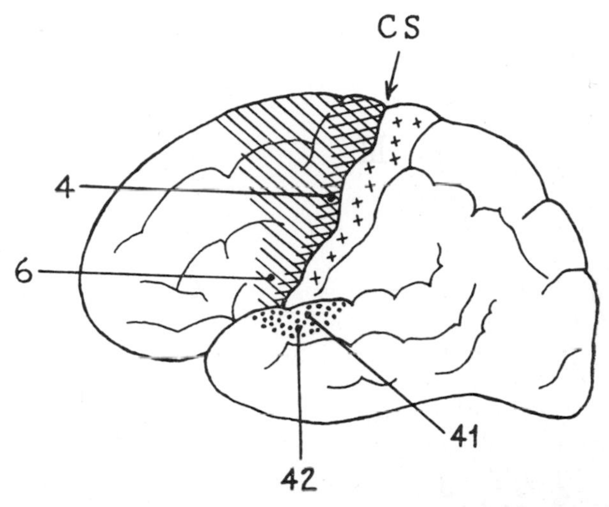

FIG. 48. Partial map of the human brain as an illustration of topographic dynamic morphology. Some cortical areas of the left side of the human brain (simplified from Fig. 68.3 of Best and Taylor, 6th ed., 1955). *4* and *6*: Motor areas, shaded and crosshatched. If that part of the crosshatched area to which the line *4* points is destroyed, the right hand, arm, and shoulder are paralyzed. If that part of the shaded area to which the line 6 points is destroyed, a form of language disturbance results. For cytoarchitecture see Fig. 49a. *41* and *42*: Sensory acoustic areas, dotted. If these parts are destroyed on one side of the brain, hearing is slightly impaired. Bilateral destruction causes deafness. For cytoarchitecture see Fig. 49b. The area marked with small crosses contains sensory centers for impulses from the skin, mouth, and skeletal muscles. It is separated from the crosshatched motor area by the central sulcus (Rolandi) indicated by CS.

area is indicated by small crosses. Since the central sulcus is known as Rolando's sulcus, one also talks of the Rolandic motor area and the Rolandic sensory area (other synonyms: pre- and postcentral gyrus). In the most dorsal part of the motor area is the motor center for the foot, and most dorsal in the sensory area is the sensory center for the foot. In other words, the two centers are just opposite each other across the central sulcus. Going in the ventral direction one encounters a similar paired topography of the centers for the hands. Ventral areas are occupied by the motor and sensory centers for the mouth, which again are opposite each other across the central sulcus. This topographic correspondence of motor and sensory centers seems to be of no functional consequence; evidently, it is merely a result of developmental scaffolding. In the optical apparatus of the brain, there is no vicinity of sensory and motor centers in spite of the delicate balance between sensory and motor functions. The sensory centers are con-

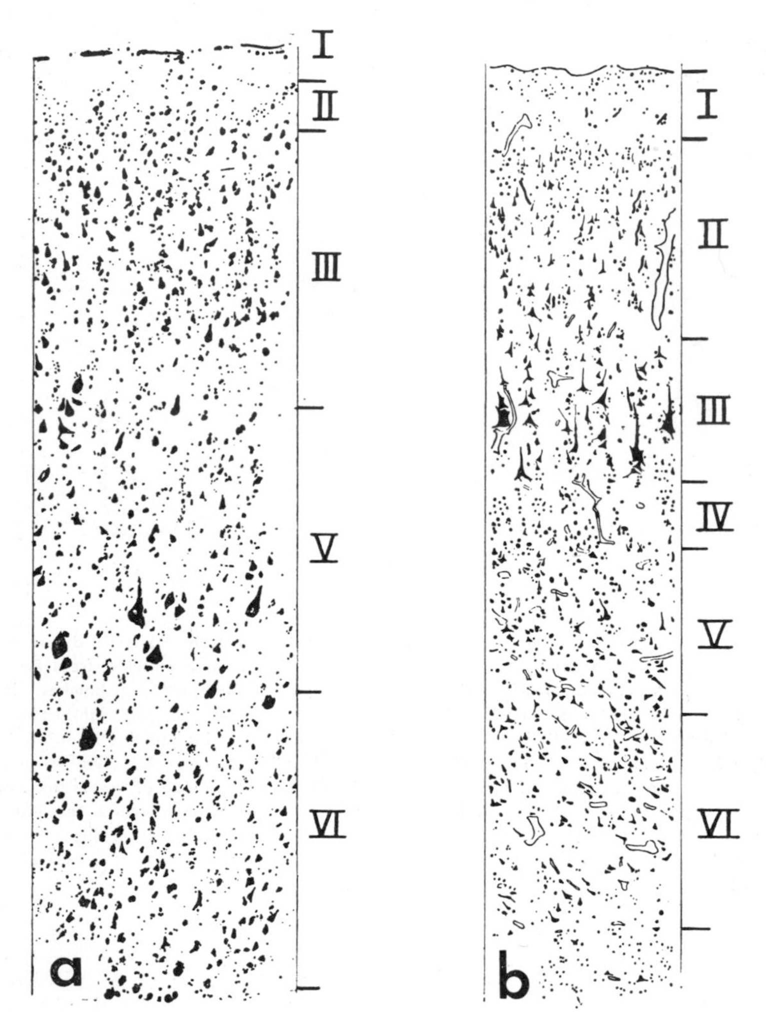

FIG. 49. Cytoarchitecture of cortical areas. Nerve cells appear black after silver impregnation. Roman numerals indicate the layers of the cortex which are distinguished by sizes, shapes, and concentrations of nerve cells. The composition of each layer varies in different areas of the cortex. Magnification approximately ×50. (a) Sample from motor areas of Fig. 48 (after Ariëns Kappers *et al.*, 1936, Fig. 385). (b) Sample from sensory acoustic areas of Fig. 48 (after Ariëns Kappers *et al.*, 1936, Fig. 690A). Note the differences between (a) and (b): presence of a zone *IV* in (b), but not in (a); presence of large nerve cells in zone *III* of (b), but not in zone *III* of (a).

centrated in the occipital cortex, while the centers for the external eye muscles are located in the midbrain and brain stem.

b. *Topographic patterns of various organs and systems.* An organ is defined as a structural carrier of one or more definite functions. *Organs condensed as solid masses* are the liver, the kidney, and the thyroid. The thyroid is part of the endocrine system; the liver, of the digestive system; and the kidney, of the excretory system.

The circulatory system and the nervous system consist of *massive central organs with ramifications* into all regions of the body. The heart is the central pump of the circulatory system. The pulse wave which originates in the heart reaches the most remote peripheral blood vessels. The peripheral nerves conduct impulses to and from the central nervous system which consists of brain and spinal cord. These two organs serve as communication centers on different levels. Sense organs are considered as part of the nervous system.

Examples of *morphologically discontinuous organs* are bone marrow and lymphatic tissue. While, as a rule, the bone marrow is associated with the skeletal system, the lymphatic tissue shows variable associations and localizations. Lymph nodes, which are the best-defined aggregations of lymphatic tissue, are found in special regions of the subcutaneous tissue, in the thoracic cavity, and in the abdominal cavity. Smaller units, known as lymph nodules, occur in the mucous membrane of the digestive tube, in the spleen, and in many other places. Under pathological conditions, bone marrow tissue and lymphatic tissue can occur practically in any place.

Some systems consist of a number of organs with different functions showing various degrees of interdependence. In the endocrine system the structures vary from a solid organ, such as the thyroid, to scattered cells, such as the interstitial cells of the testis (see previous description in Part II, Chapter 2, G). Under these circumstances, the endocrine system as a whole cannot fall into any topographic pattern. Three pairs of endocrine organs present a peculiar topographic puzzle: in each case, two functionally different organs have fused intimately so that the impression of a single organ is given. The adenohypophysis and the neurohypophysis together form a solid structure. The architectural combinations of these two organs vary from species to species. Similarly, the adrenal consists of one organ which shows a mosaic pattern of cells and is known as adrenal cortex and a second organ which consists of a network of nervous cells and is known as the adrenal medulla. In mammals, the nervous (chromaffin) adrenal is enclosed by the glandular adrenal, whereas in fowl the two organs are interlaced; in some amphibians the chromaffin tissue tends to form the external layer, a situation which is reverse of the arrangement in mammals. The parathyroid glands are located near the thyroid gland, and, in some species, even enclosed by the thyroid tissue; the functions of the two organs are unrelated. Finally, the pancreas represents a topographic combination of an endocrine gland with a digestive gland: the islets of Langerhans, which produce insulin, are embedded in the masses of pancreatic acini, which make trypsinogen (Fig. 57C and E). It was this topographic relation which, for a long time, obstructed the discovery of insulin, as recounted previously (Part III, Chapter 1, A).

The *topographic fusion of pairs of heterogeneous organs* appears particularly strange in the light of comparative anatomy. In cyclostomes and teleost fishes, the cortical and chromaffin tissues are separate,[12] the cortical substance being aggregated along abdominal veins and the chromaffin cells along the aorta. In sharks, the chromaffin bodies are arranged in metameric pairs associated with the sympathetic ganglia, and the cortical tissue tends to collect toward the caudal part of the abdominal cavity. (I have followed the description in Atwood's "Concise Comparative Anatomy," 1947.)

3. *Topographic Differences in Nervous, Vascular, Biochemical, and Other Reactivities*

a. *Regional differences of reactivity within one organ; differences between paired organs.* Organs of apparently uniform composition frequently show regional differences in reactivity. The zonation in the rat's thyroid may serve as an introduction to this subject. The function of the thyroid is to adjust the basal metabolism (rate of burning fuel in the resting animal), and the dynamic morphology of the thyroid varies with the environmental temperature. If the rat has been kept at 85° F. or above, all follicles of the thyroid will be in the inactive or storage phase. If the environmental temperature was 65° F. or lower, all follicles are activated. At an intermediary temperature of approximately 75° F., the peripheral follicles are found in the storage phase, while the central follicles show signs of activation. Dempsey and Singer (1946) reported that the binding capacity for methylene blue was different in peripheral and central follicles. They found alkaline glycerophosphatase predominantly in peripheral inactive follicles while acid glycerophosphatase occurred in central active follicles. Certain post-mortem observations also indicate regional differences in enzymatic activity. If a rat which was kept at 75° F. is killed and its thyroid allowed to undergo post-mortem autolysis, this enzymatic process advances much faster in the activated central follicles than in the less active peripheral follicles (Mayer, unpublished data, 1945). Heyl and Laqueur (1935) distinguished three zones in the guinea pig thyroid. They found that only the intermediary zone gave reproducible results in quantitative assays of thyroid-stimulating hormone. The reactivity was too low in the peripheral (inactive) follicles and too variable in the central (active) follicles. It is not known what factors may be responsible for the topographic differences in reactivity in thyroids of rodents.

Whenever topographic differences in reactivity are found within one organ,

[12] Cortical tissue means the histological equivalent of the adrenal cortex of mammals.

it would be reasonable to look first for differences in blood supply. This was done in an interesting study by Barlow *et al.* (1957). The problem was the distribution of isoniazid in the brain of cats. The drug was labeled with radioactive carbon-14 and injected intraperitoneally. Animals were killed at different intervals after injection, and their brains were rinsed to eliminate the blood. Radioactivity was estimated in radioautographs of selected brain slices. In addition, the C^{14} activity was measured in homogenized samples from different areas of the brain obtained by dissection. High concentrations of the labeled drug were found not only in the cerebral and cerebellar cortex, which have a high vascularization, but also in the hippocampus, which has a relatively low vascularization. Low concentrations of the labeled drug were found both in structures with low vascularization, such as the globus pallidus, and in structures with rich blood supply, such as the geniculate bodies and inferior colliculi. Moreover, the concentration of the drug was maintained in the hippocampus for a longer period than in other areas of the brain. The authors concluded that differences in vascularity are not responsible for the unequal distribution of the drug. In the present book topographic differences in the exchange of substances between blood vessels, cerebrospinal fluid, and brain substances will be discussed in connection with intercellular spaces and extracellular fluid (Part IV, Chapter 3, C; Figs. 79 to 81).

In some instances, unequal reactivity on the left and right side of the body has been observed. As an example I mention the effect of anesthetics on the eyes of tadpoles (Politzer, 1931). When tadpoles were placed in water which contained anesthetics, the corneas showed change in mitotic activity, but this change was not necessarily synchronous on the left and right eye. In case statistical analysis should show that the response in one eye is significantly faster than in the other eye, it would still be difficult to discover reasons for this difference. There is one noticeable morphological asymmetry in the tadpole's head, namely the spiracle. This is an opening on the left side of the head which permits water to pass from the gill slits to the outside. Right- and left-handedness in man, the preferential use of some extremities in animals, asymmetry of gait, and similar phenomena offer fascinating but difficult problems. I refer again to Ludwig's comprehensive monograph (1932).

A strange instance of asymmetrical reactivity is observed in human patients suffering from excessive production of thyroid hormone. Although many patients with this condition show a protrusion of both eyes (exophthalmos), a certain number show protrusion of one eye only.

In *medical diagnosis,* the localization of pathological conditions is of great importance. Some diseases of the skin show preference for the flexor side, others for the extensor side, of the extremities. In the conjunctiva of the human eye, swelling of lymph nodules may indicate either a serious or a relatively mild disease. If the swelling is localized in the upper lid, it represents an early stage of trachoma, while in the lower lid it merely means a "vernal catarrh" which appears in spring and disappears in summer.

b. *Factors that control regional reactivity.* Certain *general* topographic conditions control the reactivity of living structures. The spleen can expand considerably because of its location in the abdominal cavity and the possibility of displacing and compressing the intestines. By contrast, the brain which is enclosed in the skull has only a very small space for expansion.

Cell multiplication can be expected only in locations where cells are available which have the ability of dividing. The distribution of cells which are able to divide will be discussed later in connection with the topography of embryonic and regenerative activities.

Inflammatory phenomena are a combination of vascular and cellular responses. Consequently, the topographic distribution of inflammatory processes in the different organs depends on their respective vascular patterns. Since cartilage is not vascularized, typical inflammation cannot occur inside cartilage. The membrane which coats the cartilage, known as perichondrium, is rich in vessels and, therefore, able to produce inflammatory responses. In the central nervous system, various pathological processes are located according to the vascular pattern.

Certain complex pathological processes take a different course in different regions of the human body. Every extraction of a permanent tooth produces a wound with jagged tears in the mucous membrane of the mouth. In spite of the large bacterial population in the mouth, most extractions of teeth are not followed by infection. If a wound of similar shape and size occurs on a finger, there is a high probability that infection will develop. What factors could be responsible for this difference? The only known factor is the vascularization, which is much richer in the mouth than in the finger. However, it would be difficult to demonstrate how vascularity and susceptibility to infection are related. Regional differences in reactivity, which are more readily accessible to analytical studies, will be our next subject.

Changing demands lead to a *redistribution of the blood volume.* Muscular activity directs more blood to the muscles involved. After food intake the digestive organs receive an increased blood supply. High environmental temperature causes dilation of skin vessels which, in turn, is a prerequisite for sweating. Since, under ordinary conditions, the blood volume remains constant, increased flow to one area necessitates decreased flow to another area. I mention Pickering's 1943 papers on circulation in arterial hypertension as an interesting study of flow rates in different organs under normal conditions and in the presence of hypertension (increased blood pressure). General methods of determining the blood flow through the different organs are described in the textbooks of physiology, e.g., Winton and Bayliss (1955, pp. 42 ff). The spleen is one of the large blood reservoirs. Procedures for determining variations in the volume of the spleen were described previously (Part III, Chapter 2, A, 2). In the dog, 20% of the total blood volume can be accommodated in the spleen. The liver is another large reservoir. It seems that in man the skin is a more capacious reservoir than the spleen, while the reverse ratio is found in the cat and dog (Winton and Bayliss, 1955, p. 56).

The concentration of particles in the circulating blood varies from area to area. The concentration of red blood corpuscles, for instance, is much higher in the spleen than in other parts of the circulatory bed. If suspensions of particles are injected into a peripheral vein of an animal, the material will be carried to the right side of the heart and from here to the lungs. Since the capillaries of the lungs are very tortuous, particles are likely to be arrested here. I refer to particles whose diameter is of the order of magnitude of red blood corpuscles so that their passage through capillaries is possible. Depending on various factors, particles may or may not be washed out of the pulmonary capillaries and thus be distributed to other regions of the body. While the lung arrests particles which are transported through the vena cava superior or inferior, the liver retains particles which travel from the digestive tract and the spleen through the portal vein. The fate of clumps of tubercle bacilli after intravenous injection and the distribution of tagged white blood cells were discussed earlier (Part III, Chapter 3, A and B, respectively).

At this point, I mention an important publication by Brickner (1927) on the role of the capillaries in the distribution of colloidal carbon by the blood stream. Various amounts of India ink were injected into the femoral veins of rabbits. Animals were killed at intervals after injection from 1 minute to 120 hours. According to Brickner, the primary distribution of carbon particles in different organs depends on their vascularity, the tortuosity of their capillaries, relative flow rates, and the blood volume which passes through the organ per unit of time. Because of all these factors, the lungs are the most efficient primary catch basin for particles which have been injected intravenously. Some time after injection, organs changed places in their ranking for carbon content. This secondary distribution was largely due to the phagocytosis of particles by the lining cells (endothelium) of the capillaries. Storage by phagocytosis was greatest in spleen, liver, and bone marrow. Brickner points out that it is not necessary to invoke particular stickiness or similar properties of pulmonary capillaries to explain the primary arrest of particles in the lungs. I agree with Brickner's point of view, though regional differences in the reactivity of the capillary walls are known.

Certain agents, which presumably reach all parts of the circulatory system, seem to damage the capillaries in one organ only. Selective bleeding of pulmonary capillaries is seen as a result of kerosene poisoning in rabbits after intravenous, oral, or intraperitoneal administration (Deichmann, Kitzmiller, Witherup, and Johansman, 1944), and after injection of a special component of tubercle bacilli into mice. The component isolated from tubercle bacilli is the "cord factor" (Bloch, 1950) later identified as trehalose-6,6′-dimycolate (Noll, Bloch, Asselineau, and Lederer, 1956).

Regional differences in the reactivity of the skin are a fascinating subject of study. The demarcation of the bearded areas in the face of human males is controlled by genetic and endocrine mechanisms. The difference between the pubic hair pattern in men and women poses innumerable questions of which few have been tackled (see Flesch, 1954). The fact that emotional blushing is limited to several areas of the skin was emphasized by Sir Thomas Lewis (1927). He suggested that the restriction of blushing to the head, the neck, and the upper part of the chest indicated changed reactivity of arterioles and capillaries as a result of chronic stimulation by exposure to air and light.

As stated repeatedly, the recognition of topographic differences should and can be the first step to an analysis of mechanisms. Naegeli *et al.* (1930) observed a patient in whom various areas of the skin were highly sensitive to antipyrine administration. Formation of blisters was part of the response. From one of these areas, which was located on the right side of the abdomen, a piece of epidermis was transferred to the left side to an area which had never shown the abnormal response. The corresponding piece of epidermis from the left side was used to fill the defect on the right side. After healing of the transplants, antipyrine was administered. Again, the abnormal response occurred in that piece of epidermis which had responded abnormally before transplantation. The other piece did not show any visible response. In other words, the change in location had not altered the reactivity of the two pieces of epidermis. Isolated fragments of the hypersensitive skin were also exposed to antipyrine *in vitro:* they responded with formation of blisters. The conclusion from both experiments was that the site of abnormal reactivity was in the cells of the epidermis, and that no neurovascular factors were involved.

Structural differences between different areas of the skin are well known. The skin of the sole of the human foot is thicker than the skin of other places. The attachment of the skin to the underlying structures varies regionally. One can easily lift a fold from the upper eyelid, but it is difficult to do this with the skin of the palm; the dorsum of the hand is intermediary. The concentration of sensory nerve endings for touch is very high on the tongue and palm, and very low on the leg of man. Similarly, the receptors for cold, for warm, and for pain show different concentrations in different regions of the skin. The concentrations of sensory receptors is determined by the so-called two-point threshold, which is defined as the minimum distance at which two stimuli are recognized as separate stimuli. A useful diagram of regional variation in two-point threshold for touch is reprinted by Ruch (1949, Fig. 171). An interesting comparison of density and thresholds of pain spots in different areas of the skin is given by Rothman (1954, Table 1).

The supply of the skin with sensory nerves follows two different patterns. One pattern is based on the course and ramifications of each peripheral cutaneous nerve. Since each peripheral nerve obtains fibers from dorsal roots of more than one spinal segment, and one spinal segment contributes fibers to several peripheral nerves, the segmental pattern is different from the so-called peripheral pattern. Most areas of the skin are supplied by two or three dorsal roots. An instructive diagram of the two patterns of supply is reproduced in the textbook by Winton and Bayliss (1955, p. 431). In a given area of the skin, certain reactivities are controlled by a peripheral cutaneous nerve, whereas other reactivities in the same area are controlled by segments of the spinal cord. Hardy, Wolff, and Goodell (1950) found that increased reactivity to pain after intensive (harmful) stimulation is related to the segmental rather than to the peripheral pattern of supply. In other words, the site of increased reactivity to pain in this case, is not in the pain receptors of the skin but in the spinal cord. A diagrammatic presentation of this relationship is given in Fig. 12 of Hardy *et al.*

As stated earlier, many pathological conditions of the skin have their places of predilection. The characteristic spots (roseoles) which appear in typhoid fever are restricted to the abdominal skin. If psoriasis occurs in extremities, it will be found on the extensor side. The exanthema in the secondary stage of acquired syphilis is located on the breast and flexor side of the extremities, whereas the characteristic skin response in congenital syphilis is located on palms and soles of the newborn. These topographic differences play an important role in medical diagnosis, but I am not aware of many attempts to analyze the factors which are responsible for differences in local reactivities. Such analysis either could proceed along the lines of the hypersensitivity study by Naegeli *et al.* (1930) mentioned above, or a battery of physiological and pharmacological tests could be used to obtain an inventory of neurovascular and cellular reactivities. The way for the latter approach has been prepared by well-established maps of sensitivity to touch, temperature, and pain.

Regional variations of fat deposits are important with respect to the dual function of fat tissue as a storage place for combustible energy and as a mechanical pad (Part II, Chapter 2, G). While it is obvious that the mechanical fat has special locations, such as the sole of the human foot and the socket of the eyeball, it may be somewhat unexpected to find the storage fat subject to topographic rules. Storage fat occurs in the subcutaneous tissue and in the thoracic and abdominal cavities.

In human beings, sex differences express themselves in a different distribution of subcutaneous fat tissue. Accumulation of fat around the hips and ankles is characteristic of females, whereas accumulation in the back of the neck and in the lower part of the abdominal region is characteristic of the male. Age differences also lead to a very different pattern of fat distribution: the general roundness of a well-nourished infant is different from the obesity of an old man or woman. Endocrine disturbances can lead to very peculiar topographic accumulations of fat. Pictures of human patients showing such conditions are found in the textbooks of endocrinology (e.g., Selye's textbook, 2nd ed., 1949).

Experimental studies in animals have demonstrated the control of fat deposits by definite areas of the central nervous system. The stereotaxic apparatus used for such studies will be described later.

As an interesting experimental field for dynamic-topographic studies, I mention the white (yellow) and brown fat in rats. The distribution of brown fat is as follows: large masses in the cranial-ventral part of the thoracic cavity (around the thymus), in the axillary fossae and between the scapulae; small strands along the aorta, sometimes extending from the thoracic to the abdominal cavity (Rasmussen, 1921). All other fat deposits of the subcutaneous tissue and body cavities of the rat are white fat.

Localization of arteriosclerosis is a complex problem. The human aorta is affected by arteriosclerotic changes mainly in its abdominal segment, while the syphilitic changes are found, as a rule, in the thoracic segment. However, the very first part of the thoracic aorta is also subject to arteriosclerosis.

Arteriosclerosis is a pathological condition which involves the inner layer of arteries. Irregular thickening of this layer is produced by deposition of cholesterol and by increase in connective tissue. It is assumed that three factors are involved in causing these changes: one systemic factor, namely increased concentration of cholesterol in the circulating blood, and two local factors which are mechanical injury and the local deposition of fibrin clots. It is the topographic distribution of arteriosclerosis which suggests that mechanical factors are involved in the production of arteriosclerosis. The places where the intercostal arteries leave the aorta belong to these, since one can assume a whirlpool-like movement of the blood when it comes under high pressure from the wide aorta and enters the sudden narrowing of the intercostal arteries.

The impact of the blood column, which at each systole is ejected from the left ventricle, hits a certain area of the aortic arch. Under normal conditions, the impact is dampened by the motility of the aortic arch and the elasticity of the aortic wall. These mechanical advantages are lost in kyphoscoliotic individuals (hunchbacks) in which the arch of the aorta follows the abnormal curvature of the vertebral column. The whole arch is rigidly fixed to the vertebral column, and that part of the aortic wall which receives the impact of the blood column is not motile. It is in this place of the aortic wall that one finds circumscribed arteriosclerosis.

Aside from crudely abnormal conditions, like kyphoscoliosis, there is a great variability in the localization of arteriosclerosis from one human being to the next. Femoral arteries may have the highest degree of arteriosclerosis in some individuals, while others have the highest degree in their coronary arteries, renal arteries, or in one artery of the stomach. Blumenthal, Handler, and Blache (1954) summarized their series of investigations with the statement that the susceptibility to arteriosclerosis and the histological patterns of arteriosclerosis are strikingly different in the following sites: aorta; pulmonary, coronary, renal, splenic, and hepatic arteries; arterial tree of the lower extermities. The 1954 paper by these investigators dealt with particular differences between the arteries at the base of the brain (circle of Willis). I refer to these publications because of their emphasis on variable patterns of arteriosclerosis in relation to regions of the body.

Regional differences in the reactivity of muscles were mentioned repeatedly. In connection with the effect of repeated stimulation by foreign serum, I reviewed the increased reactivity of smooth muscles in different organs of different species (Part II, Chapter 2, F): the smooth muscles which respond with maximal contraction to a second injection of horse serum are located in the bronchi of guinea pigs, in the small pulmonary arteries of rabbits, and in the liver and other abdominal viscera of dogs. Some skeletal muscles respond faster to stimulation than others, and certain skeletal muscles possess layers with different reactivity to electrical stimulation (Part II, Chapter 2, E). Denny-Brown (1929) found that in an extensor muscle with dual composition, the red deep portions show a slow response to electrical stimulation, whereas the pale superficial portions respond fast. In analogous studies with a flexor muscle, Gordon and Phillips (1953) observed that the deeper layers responded faster than the superficial layers.

The *topography of embryonic potentialities* is illustrated in Fig. 3g–l, Fig. 4, and Fig. 32. Different reactivities of adjacent areas in the embryo were beautifully demonstrated by the following experiment by Champy (1922). It was known that the thyroid hormone accelerates the metamorphosis of tadpoles. At a certain stage of the tadpole, the skin at the base of the forelimb is continuous with the skin of the gill pouch. If tadpoles of this stage are placed in water which contains thyroid hormone, the skin of the limb responds with increased mitotic activity and the cells also show morphological indications of metabolic activity. By contrast, the cells of the gill pouch do not show any mitosis but exhibit various stages of disintegration. A sharp line of demarcation separates the two zones with their opposite responses (Champy's Fig. 25). Champy's observation gave a remarkable insight into the finer topography of metamorphotic mechanisms.

Two important theories of embryological differentiation are based on topographic differences in reactivity. One is the theory of metabolic gradients, and the other the theory of morphogenetic fields.

The *theory of metabolic or physiological gradients* was originally conceived by Child with reference to regeneration in planarias and hydras, but subsequently extended to embryonic development. A comprehensive presentation of procedures and interpretations is found in Child's book "Patterns and Problems of Development" (1941). In order to convey the idea, I give a brief description of Child's basic experiments with planarias. The shape of a planaria is comparable to a straight sword, the head of the animal corresponding to the handle of the sword. In the experiments, a number of planarias are cut transversally into five pieces. The heads and tails are discarded. The piece next to the head is labeled *A,* the following piece *B,* and the piece next to the tail *C.* Pieces of each type are tested for their ability to regenerate heads and for oxygen uptake and carbon dioxide production. Regeneration of heads is achieved by almost every *A*-piece, less regularly by *B*-pieces, and rarely by *C*-pieces. Moreover, the heads formed by *A* are all normal, those from *B* predominantly normal, and those from C predominantly abnormal. Oxygen uptake and carbon dioxide production can be determined manometrically by collecting a number of samples of each type. The highest values of O_2 uptake and CO_2 production are observed in the *A*-pieces, lower values in the *B*-pieces, and the lowest values in the *C*-pieces.

The results obtained with fragments were supported by experiments with intact animals. If a whole living animal is immersed in a solution of methylene blue, it stains uniformly blue. When subsequently the planaria is removed from the dye, the blue color fades out first in the head region, then in the middle, and finally in the tail region. Since methylene blue becomes colorless by reduction, the interpretation of the staining experiment is that oxygen was taken from the dye by the living tissue at a rate which decreased from the head to the tail. Whole animals are also tested for sensitivity to respiratory poisons. When a planaria is placed in a solution of potassium cyanide, the head part is the first to die with visible signs of disintegration, and then a "death wave" proceeds along the body to the tail end. The decrease of all these reactivities from the head to the tail led Child to the concept of a physiological or metabolic gradient along the craniocaudal axis of the planaria.

As examples of metabolic gradients in embryonic stages of vertebrates, I mention some observations in gastrulae of amphibians. A decrease of oxygen uptake from the animal pole to the vegetal pole of the early axolotl gastrula was illustrated in my Figs. 8 and 9, after Gregg and Løvtrup (1950). In the frog gastrula, Sze (1953) showed the presence of a double gradient with respect to O_2 consumption. One of the gradients goes from the animal to the vegetal pole, while the other gradient follows a line which corresponds to the later dorsoventral axis of the embryo. It is difficult to decide whether metabolic gradients are the cause or the effect of tissue determination (Holtfreter and Hamburger, 1955, p. 283).

The *theory of morphogenetic fields* refers to the fact that embryonic differentiation depends on topographic interrelations. The interdependent parts constitute the field. As an illustration I recapitulate the experiments of Harrison (1921) in which buds of the anterior limb were transplanted from one side to the other side of axolotl embryos. At a certain stage, the transplanted limb bud reversed its dorsoventral polarity, thus adjusting to the new site. However, a transplanted limb bud of the same stage was not able to reverse its anterior-posterior polarity as the new site would demand. In other words, the imprint of the morphogenetic field in which the limb buds had originated was overruled by the new field with respect to dorsoventral polarity, but not with respect to anterior-posterior polarity.

Julius Schultz (1929) criticized the theory of morphogenetic fields as a false analogy, since, in contrast to physical fields, the morphogenetic fields have no centers, no measurable field strengths, and no lines of force in the proper sense. It is not clear to me what Schultz meant with "lines of force in the proper sense." He certainly underrated the definiteness of morphological fields since the difference between two morphogenetic fields can be expressed in terms of their relative strengths. This was shown in the example of transplanted limb buds. Good reasons for rejecting the criticism of Schultz were also given by Needham (1942, p. 127).

What is the relation between morphogenetic fields and inductors or organizers? One may say that in the field concept the emphasis is on the topographic configurations of any type of factors, while the idea of inductors and organizers implies the assumption of a specific chemical or enzymatic activity of such and such part of the egg, blastula, gastrula, or embryo. The reconstitution of artificially dissociated adult sponges mentioned previously (Part III, Chapter 5, C, 2), and the directed morphogenetic movements, shown in Fig. 32, reveal organizing principles which fall into the category of the morphogenetic fields: everything is dominated by topographic interrelations, but certainly not by one chemically defined substance or one enzyme. Physicochemical factors enter the reconstitution of a sponge only insofar as selective adhesion of cell types is part of the mechanism (Holtfreter and Hamburger, 1955, p. 281).

The search for chemically defined organizers has been replaced by investigations along two operational lines: the study of morphogenetic fields, and studies on the relations between chemical and morphological differentiation. The latter relations were mentioned at various occasions (Part II, Chapter 2, E) and will be discussed again in connection with histochemistry of enzymes.

Morphological differentiation as well as increase in size of primordia and organs involve cell multiplication and cell migration. Both processes are frequently part of the same mechanism, though not necessarily occurring simultaneously. This was illustrated in my Fig. 30, which showed the fate of populations of embryonic cells in the spinal cord of chick embryos (after diagrams of Levi-Montalcini, 1950). I also discussed the publications of Dorris (1938) and Rawles (1955) on the origin of pigmented cells which take part in the development of skin and feathers of fowl. It was shown that the pigmented cells were derived from unpigmented cells which had migrated across the whole embryo.

Many problems of embryonic life are encountered as phenomena of *regeneration in postembryonic life.* Most structures in the adult organism need constant replacement, both by metabolic renewal of subcellular building blocks and by reproduction of entire cells. To replace the aged or lost cells, forerunners of the mature cells must be available which are able to divide. Cells which in the postembryonic organism have kept the capacity of dividing are known in botany as *meristematic cells,* and tissues composed of such cells as *meristems.* For a recent reference, see Thimann's 1960 review on plant growth. It proved useful to apply the same terminology to the cells and tissues of metazoa (Bělař, 1927; Mayer, 1937).[13] The topographic distribution of meristematic cells and tissues is important not only in the study of normal replacement, but also in the study of pathological conditions, such as regeneration after mutilation and the formation of tumors.

In somatic cells, the capacity to divide can be lost by differentiation and aging, or it can be suppressed by mechanical conditions. Bone cells (osteocytes) which are confined to a small space in hard bone might be able to divide into two daughter cells, but there would not be enough room for each of the daughter cells to attain the size of the mother cell. There is some evidence that osteocytes can undergo repeated cell divisions if the surrounding bone is softened by pathological loss of calcium. Nerve cells in the central nervous system of the adult have never been seen in any phase of division: it is taken for granted that they cannot divide. The reason may be mechanical. The physicochemical changes of cytoplasm during cell division involve rounding off of the cell body, which means that ramified cells must withdraw their branches (see Bargmann, 1959, Fig. 86). This is evidently an impossible task for those nerve cells which have very long axons. In the human body, the axon of a motor neuron extends from the spinal cord to the toes. It is not only several feet long, but also fixed and anchored in such a way that a withdrawal is inconceivable. One must assume that cells which are not replaced by cell division are particularly apt to renew their subcellular constituents by metabolic processes.

Continuous replacement of cells occurs on a large scale in the bone marrow, the epidermis, and the lining of the mucous membranes. The bone marrow supplies the circulating blood with replacements for the red and white blood cells which are lost when their life cycle expires. If pathological conditions destroy an unusual amount of red and/or white cells, the bone marrow is able, in many cases, to compensate also for the abnormal losses. Blood-forming bone marrow is present in all bones of a child, but in the healthy adult it is limited

---

[13] His (1901) referred to undifferentiated tissue cells of metazoa as "Keimzellen". This was confusing since "Keimzellen" is the usual designation for reproductive cells.

to the spongy bones. In the long bones of the adult the proximal spongy part usually shows some activity, but not the distal spongy part. If demands for increased activity arise, the inactive fat marrow of the long bones is transformed into active red marrow, according to the following topographic rules. The activation progresses from the proximal end to the shaft and finally to the distal end of each bone. In the shaft, the activation starts at the periphery, i.e., close to the bone, and gradually progresses to the center of the marrow cavity until all fat marrow is replaced by active marrow. When the demand for activity decreases, the process is reversed. The factors which control these topographic patterns are unknown.

The human epidermis consists of a number of cell layers known as stratified epithelium. The deepest layers produce new cells which move to the middle layers. Toward the surface, the cells undergo gradual keratinization, a form of dying. The cells at the very surface are entirely horny, which means maximum protective efficiency. Some inner organs, such as the esophagus, are lined with a cell mosaic of several layers (stratified epithelium) similar to the epidermis, except that there is no keratinization. Again, the meristematic cells are found in the deeper layers. It seems that cells of the middle layers are also able to divide. This ability does not show under ordinary conditions, but appears when there is a sudden demand for more cell production, for instance, in the vicinity of a wound.

The nictitating membrane, or third eyelid, of the rabbit is particularly suitable for such studies, since it is coated with several layers of cells (stratified epithelium) without keratinization of the top layers. Another advantage of the nictitating membrane is that it is easily accessible to observation and, at the same time, protected against undesirable mechanical interferences. Grieco (1939) made wedge-shaped incisions into the free margin of the nictitating membrane of one eye while the membrane of the other eye served as a control. Both membranes were removed 31, 48, or 72 hours after the wound had been made in the experimental membrane. The frequency and distribution of mitoses in the epithelium of normal nictitating membranes and of epithelium not far from the wounds was determined. Sections vertical to the surface showed five nuclear layers. In assigning mitoses to a particular layer, corrections were made for oblique sections.

The results were as follows. In the normal nictitating membrane, mitoses were most frequent in the basal or first layer, less frequent in the second layer, and quite rare in the third layer; no mitoses were observed in the two top layers. In the experimental membranes, the epithelium showed, at some distance from the wound, a more than twofold increase in mitotic frequency in the three basal layers. In addition, a small number of mitoses were seen in the second layer from the top and even in the most superficial layer. Grieco's experiments have demonstrated that in a stratified epithelium of five layers, the cells of all layers have the potentiality of dividing. Under normal environmental conditions, this potentiality is expressed only in the basal layers, whereas abnormal conditions also activate cells of the top layers. Although I am not aware of any similar study in other tissues of adult organisms, I do not doubt that Grieco's observations probably represent a general rule. This generalization is supported, to some extent, by the following experiments in embryos.

Wilson, Hughes, Glücksmann, and Spear (1935) examined the distribution of mitoses in the neural tube of the chick embryo. They found that in the neural tube of the normal embryo all mitoses are located in one layer near the bore. After a certain dose of gamma irradiation mitotic activity is first depressed. A phase of compensatory increase in mitotic frequency follows. In this phase, mitoses are found not only in the usual layer, but also in zones remote from the bore. I conclude that both in adult and embryonic tissues the normal restriction of mitotic activity to a special zone can be broken by an unusual stimulus.

In the liver and kidney of adult human beings and laboratory animals mitoses are hardly seen under normal conditions. However, in the vicinity of areas of destruction regenerative activity is seen: mitoses appear in the liver cord cells or the cells of the renal tubules. Again, it is impossible to tell what proportion of cells is able to divide.

Sheets of one layer of cells, in a mosaic arrangement, line the surface of most mucous membranes. The wear and tear of this layer requires continuous replacement. The question whether all cells, or only a certain percentage of cells, retain the ability of dividing, cannot be answered by mitotic counts.

Exocrine glands consist of the so-called fundus in which the specific secretion is produced, and the duct by which the product is discharged (see Fig. 87A). Schaper and Cohen (1905) found that in such glands the mitotic activity is concentrated in the so-called neck of the gland, which is the zone of transition between the fundus and the duct.

The topographic distribution of meristematic cells cannot be understood without consideration of cell movements. The interrelations between cell multiplication and cell migration in embryonic development were described previously (Part III, Chapter 3, B). Similarly, postembryonic regeneration involves migratory as well as mitotic activities. Among invertebrates, sponges, hydras, planarias, annelids and crustaceans have been studied extensively with respect to regeneration. Regeneration experiments with planarias were mentioned earlier in this chapter in connection with metabolic gradients. In annelids and planarias the meristematic cells are known as stem cells or neoblasts. Experiments of E. Wolff and F. Dubois (1948) showed conclusively that stem cells migrate from an intact area to an area which was mutilated, and that this immigration is a prerequisite of regeneration in the head area.

In vertebrates the cornea of the eye proved to be particularly suitable for studies of regenerative cell division and cell migration, since the cornea is readily accessible to observation and experimentation. In elaborate experiments Friedenwald and Buschke (1944) confirmed the observations of earlier investigators that in the healing of cornea wounds a period of migration of cells precedes the start of mitotic activity.

It is not possible to observe cell multiplication and cell movements in inner organs with techniques as direct as those applied in wound-healing studies of the cornea. However, Leblond and Stevens (1948) used ingenious indirect methods, including calculations, to study the fate of cell populations in the duodenum of rats. I described these studies previously (Part III, Chapter 3, A) and briefly recapitulate their results here: groups of cells which are produced by constant mitotic activity in the deeper parts of the crypts migrate to the surface and thus replace the cells which are lost by wear and tear. My Fig. 33 (after Leblond and Messier, 1958) illustrates how this migration of cell groups was confirmed by radioautographs.

The central nervous system is particularly inaccessible to studies of cell multiplication or migration. As stated before there are good reasons to assume that the nerve cells in the postembryonic central nervous system cannot divide. However, the cells of the supporting tissue, or neuroglia, seem to be able to multiply. Probably some types of the supporting cells migrate under pathological conditions. It is difficult to tell whether in such cells mitotic and migratory activity are associated or not. Because of the limited technical possibilities, any changes in the microscopic structure of the central nervous system are open to various interpretations. Some investigators assume that undifferentiated cells with multiple potentialities (meristematic cells) are still present in the adult central nervous system, while others deny such a possibility. These controversies play a role in the classification of tumors of the glioma group (see, e.g., Bailey and Cushing, 1926).

Cultivated tissue cell colonies which are transferred periodically to new media (passage cultures) can be considered as a continued experiment in regeneration. Such colonies show a close relation between migratory and mitotic activity. In connection with Fig. 21, it was described that the radial extension of a fibroblast colony is due to centrifugal migration, whereas the density of the cellular network is maintained by mitotic cell division. Quantitative studies of the relation between migratory and mitotic activity in different media were made by Jacoby (1937), Willmer and Jacoby (1936), and Jacoby, Trowell, and Willmer (1937); a summary of these studies, including the original graphs, is found in "Tabulae Biologicae," Vol. 19, 1939, "Tissue Cell Colonies *In Vitro*" by Mayer.

4. *Procedures of Regional Anatomy and Histotopography*

a. *Roentgen-ray analysis.* Some principles of Roentgen-ray analysis were briefly stated in Part III, Chapter 2, A, 2, a. The necessity for descriptive statements was illustrated by Fig. 16 showing small radiopacities in the lung which can be produced by a variety of pathological conditions. In diagrams A and B, Fig. 50, an attempt is made to demonstrate the role of topographic relations between several radiopaque and radiolucent structures. Diagram A represents a pair of lungs which have been removed from a human corpse and have been preserved with their air content. Our analysis is limited to the right lung. In A, the Roentgen-ray picture is taken in the ventrodorsal (anterior-posterior)

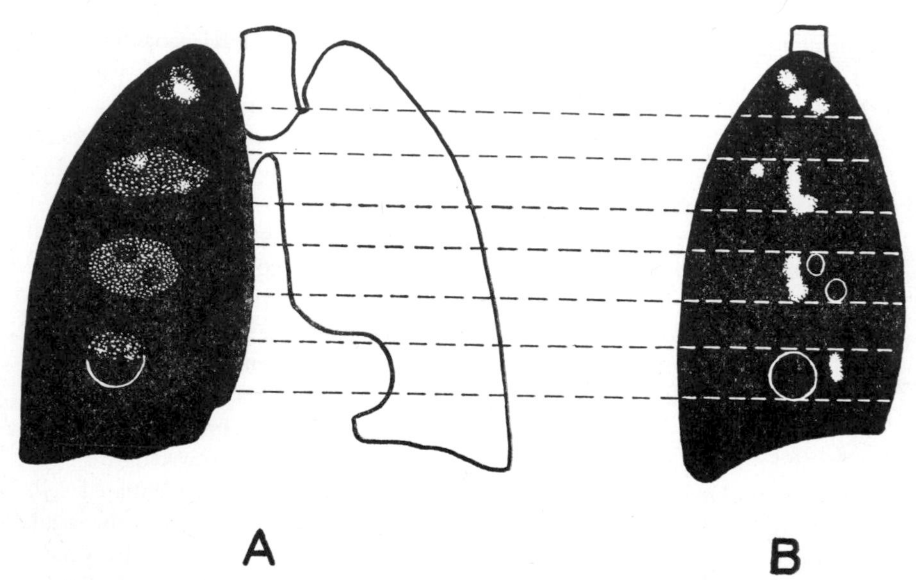

FIG. 50. Diagram illustrating summation and superimposition of radiopacities and radiolucencies in the roentgenogram of an isolated right human lung (modified from Bartone and Grieco, 1955, Fig. 1). The picture A is taken in the ventrodorsal (anterior-posterior) direction: it is a summation of frontal planes. Picture B is taken from the right or left side (lateral projection): it is a summation of sagittal planes. *Radiolucencies.* Air-containing lung tissue is indicated as black areas. Spaces filled with air, without lung tissue, are indicated black with a white outline: they transmit even more radiation than the air-containing lung tissue does. *Radiopacities.* Highest degree of absorption of rays is indicated solid white, while lesser degrees of absorption are represented by white-dotted areas.

direction, whereas in B the picture is taken from the right or left side. In other words, the pictures A and B are converted into each other by rotation of the lung around a vertical axis, through an angle of 90 degrees. Air-containing lung tissue is indicated black in both diagrams. Spaces which contain air but no lung tissue are indicated by black areas with white outlines: these could be dilated bronchi. Air without lung tissue transmits more radiation than air with lung tissue. Solid white areas indicate highest degree of absorption of Roentgen rays, while white dotted areas indicate a lesser degree or absorption. The fact that the white-dotted oval areas in A appear as solid white masses in B leads to the following interpretation. The shape of these structures resembles oval boards whose length exceeds their width and thickness. Therefore, the maximum of absorption occurs when the rays travel lengthwise through the structure as in B where the picture is taken "on edge." If we move from the bottom of B upward, we first encounter a radiolucent spherical structure in the neighborhood of a radiopaque mass. The superimposition of these two gives the picture seen in A, namely a lower semicircle with maximal radiolucency and an upper semicircle which is less translucent than the circle in B but also less radiopaque than the opaque block in B. The next group of structures shows one vertical opaque block and two small structures with maximal translucency. In A, the change of position of the block in conjunction with summation effects leads to the picture of a semiopaque elliptic area (dotted white) with two radiolucent enclosures (black circles).

In the living human being the appearance of A would be quite similar, but the picture B would be complicated by superimposition of the heart, the aorta and the left lung. This is the reason why the original pictures of Bartone and Grieco (1955, their Fig. 1) was changed here to an isolated left lung of a corpse. Historically, the interpretation of Roentgen-ray pictures of the chest was guided by postmortem studies. Large slices through a whole frozen chest of a corpse were compared to the superimposed structures seen in the Roentgen-ray picture.

A later chapter will include comparative studies on the lungs of miners (Rivers *et al.,* 1960) in which chest films obtained during the last years of life were compared with histological sections and chemical analyses of the lungs (Part III, Chapter 7).

In modern Roentgen *diagnosis,* the depth of structures may be determined by stereoradiography: two or more pictures are taken at different angles and can be viewed in such a way that the stereoscopic picture appears three-dimensional. For other methods, such as tomography, see Andrews (1944).

In Roentgen-ray *therapy,* the first concern is to hit the target with a properly aimed Roentgen beam. In addition, it is necessary to determine the concentration of radiation in the target area as well as in the structures through which the radiation has to pass. Three-dimensional mapping is to be combined with studies of absorption of radiation in the intervening layers and the sensitivity of these layers to the dose which they receive. Besides calculations, so-called phantoms are used. These are models which imitate the shape and radiation-absorbing properties of the biological object. Particularly for the treatment of tumors, the "cross-firing" or "multiple portal" method has proven to be very useful. If, for instance, a tumor of the esophagus is irradiated from five different angles, each of the intervening areas through which the radiation passes receives only one-fifth of the dose which the tumor receives. If there is a possibility of overlapping radiation in adjacent "portal areas," this must be taken into account.

A further development in this direction is the "moving field" method. Either the patient rotates in front of the Roentgen tube, or the Roentgen tube revolves around the patient so that the target is always in the center of the beam. Topographic schemes and calculations of doses are described in the literature. I mention Paterson's well-known book "The Treatment of Malignant Disease by Radium and X-Rays" (1948), particularly Chapter 6, "Planning and Prescription of X-Ray Treatment" and Chapter 7, "Beam-Directed Small Field X-Ray Therapy." This subject is also covered in several articles in the first volume of "Treatment of Cancer and Allied Diseases," edited by Pack and Ariel (2nd ed., 1958).

b. *Topographic aspects of clinical and experimental surgery.* As mentioned earlier, regional or topographic anatomy is the basis of clinical and experimental surgery. I refer the reader to Callander's textbook of surgical anatomy (1939) and also to instructive pictures of cross sections through the human arm and leg in Woodburne's book (1957; Figs. 10, 72, 356, and 372). Chest surgery requires familiarity with the topography of thoracic organs. A semidiagrammatic sketch of an intricate operative field in the thorax is given in Fig. 126 of Lindskog and Liebow's 1953 book, which was quoted previously in the discussion on macroscopic units of the lung.

Procedures for localization of functions in the brain were described earlier when combined studies of live and dead material were discussed (Part III, Chapter 1, A).

An example may illustrate the topography of the spinal cord. Different sensory impulses travel on different tracts of the spinal cord. The pathway for pain was discovered accidentally in a patient who lost the sense of pain (and temperature) as a result of a tuberculous destruction of a certain superficial area of the cord. Subsequently, it became possible to relieve patients from intractable pain by surgical cross section through that particular tract. This presentation is taken from Ruch (1949).

Finally, I will recount here a technique for producing well-defined destruction in any desired location of the brain of experimental animals. Since the apparatus allows the adjustment of needles or electrodes in three dimensions, it is known as the stereotaxic machine. I follow the detailed description by Horsley and Clarke (1908), although this was not their first publication on this subject. The principle of the machine is that metal holders and instruments can be moved in such a way that a three-dimensional control of movements is obtained. In a way the machine was a predecessor of the micromanipulators mentioned in Part III, Chapter 2, A, 4. Figs. 6, 7 and 8 of the original paper should be consulted to understand the details. Horizontal movements are controlled by one sagittal guide bar and two transversal guide bars, each of them scaled in millimeters. In order to allow vertical injections, a needle holder is attached by a traveling joint to the sagittal bar. The apparatus is clamped to the skull of the anesthetized animal. Two plugs of the machine are placed into the external auditory meatuses, while special holders rest on the margin of the orbits and on the maxilla. The relations of the various structures of the brain to the principle landmarks of the skull are mapped out in a number of dead animals. With the use of the prepared maps, small holes are drilled in the skull and electrodes are introduced to the desired depth. For electrolytic destructions the necessary values were determined. For instance, a simple anodal destruction 2.7 millimeters in diameter was produced by 3 milliamperes if the duration of the electrolysis was 1.8 minutes. Small groups of cells were destroyed by 2 milliamperes applied for 15 seconds.

In Krieg's (1946) technique designed for the rat, the apparatus is fixed by plugs to the external auditory meatuses as usual, but a special clamp secures it to the incisors. If in rats small hypothalamic areas are destroyed, pronounced obesity develops after a surprisingly short time (Brobeck, 1946). Delgado (1952) used the Horsley-Clarke instrument for permanent implantation of multilead electrodes in the brain of cats. In most experiments, fourteen or twenty-eight electrodes were implanted, but forty electrodes were also tolerated without any visible alterations in the behavior of the animals. The electrodes were used for stimulation, recording, and for local destructions by electrocoagulation. It seems that motor stimulation was achieved in unanesthetized cats without evoking emotional disturbances. An increasing application of the Horsley-Clarke instrument in human patients is noticeable in the literature.

c. *Topographic aspects of anatomical and histological techniques.* Techniques for maintaining the continuity between macroscopic and microscopic analysis were described in Part III, Chapter 2, B, 1 and illustrated in Fig. 24. As mentioned previously, large histotopographic sections were made first by neuroanatomists. With the use of celloidin embedding, whole human brains were sectioned. Sections of this type, stained for myelinated nerves, are illustrated in the textbooks of neuroanatomy. For organs other than the brain, two freezing microtome techniques for large sections were developed, that of Christeller (1927) for transmitted light, and that of Gough and Wentworth (1948) for reflected light (Part III, Chapter 2, B, 3, c). The method of Gough and Wentworth had been used extensively in studies on particulate deposits in lungs of miners. Their method was also used in an interesting study by R. A. Parker (1958) on the topographic distribution of venous congestion in the lungs of patients with mitral stenosis.

B. Three-Dimensional Interpretation of Microscopic Structures

Everybody is used to thinking in terms of three dimensions where macroscopic structures are concerned. All objects of practical life, whether animated or inanimated, are primarily interpreted as three-dimensional. The translation of three-dimensional objects into two-dimensional pictures is standardized by the conventional rules of perspective. Difficulties in three-dimensional interpretation may arise under special circumstances, such as the use of Roentgen rays. This was discussed in connection with Fig. 50.

On the microscopic level the primary approach to the object is in two dimensions. Microscopes were used for a long time before mechanical controls of the focus were developed. Without a micrometer, the determination of thickness of a microscopic object and the analysis of its different strata was restricted to low magnifications. Though improved microscopes facilitated the analysis of the third dimension, this dimension still remained the stepchild of most microscopists. In microscopic sections the usual ratio of thickness to diameter is of the order of magnitude of 1 to 1000. Consequently, the objects which are simultaneously visible in one plane of a section impress the observer more than those which can be perceived by traveling through the different layers of the section. The present discussion is devoted to the three-dimensional interpretation of microscopic structures.

1. *Methods for Visualizing Structures in Three Dimensions*

a. *Wax-plate reconstructions and transparent preparations.* The inner architecture of small objects can be explored by microdissection, by making them transparent, or by cutting them in different planes. If serial sections are mounted in their natural sequence, individual structures can be followed from level to level. The three-dimensional shape of each structure can be deduced mentally from the two-dimensional pictures, or the structure of interest can be reconstructed by the wax-plate method of Born (1883). A brief summary of this method was given in Minot's "Laboratory Textbook of Embryology" (1910, pp. 387 and 388). A modern, more detailed description is found in "Mikroskopische Technik" by Romeis (1948). The basis of the method is to reproduce serial sections in a series of wax plates, maintaining the proportions between magnification and thickness. If each histological section is 20 μ thick and the structures of interest require a magnification of 50 times, each wax plate should be 20 μ times 50 in thickness, which is 1 mm. Upon each wax plate the outlines of the important structures, as they appear in a section, are drawn with a fine steel point. Next, each wax plate is placed on a hard surface, and a sharp knife is used to cut out the outlines of the structures which are to be reconstructed. By piling up the wax pictures in the sequence of the serial sections, one obtains a three-dimensional reconstruction of the object. To ensure correct superimposition one should not depend on the more or less unknown object of study, but should use well-known surrounding structures as landmarks. If such structures are not available, artificial landmarks, e.g., threads of cotton, may be embedded and sectioned together with the biological object. The most precise devices for safe orientation produce on the surface of the rectangular paraffin block vertical grooves and ridges which appear as a characteristic pattern in each section and wax plate.

Pictures of wax-plate reconstruction are to be found in textbooks of embryology. A particularly intricate task of reconstruction was performed by Chi Lan Tsui in her study of the nasal glands of the frog (1935). Wax-plate reconstructions of groups of stacked tissue cells made by F. T. Lewis (1925) are shown in my Fig. 67. Studies in whole transparent (cleared) embryos are less time consuming than wax-plate reconstructions. After injections of opaque material, not only the skeletal system but also the vascular system stands out. Since cartilage stains blue with Victoria blue, and bones stain red with alizarin, cartilage and bone can be distinguished in whole embryos after application of these dyes and subsequent clearing. An appraisal of reconstruction methods is found in de Beer's book "The Development of the Vertebrate Skull" (1937, pp. 14-15).

b. *Total mounts compared to serial sections.* Although serial sections guarantee the maintenance of topographic relations, some structures may be lost or incomplete in serial sections. Mitotic figures are in this class. Since chromosomes vary from rods many microns in length to minute dustlike particles, the counting of chromosome numbers depends on completeness of the preparation. Studies of mammalian chromosomes have been made in total mounts of the amnion since this embryonic membrane is thin enough for microscopic analysis. A very useful method for the study of chromosomes and other components of cells is known as crush or squash technique. If a suspension of cells is crushed between two coverglasses, both coverglasses should be examined to verify the fact that all chromosomes have remained in the preparation intended for study. Even counts of intermitotic nuclei are more reliable in total mounts than in serial sections. Technical differences of this type played a role in the question whether rotifers should be considered to be organisms with a constant number of nuclei in each organ and possibly, in the whole animal (Part III, Chapter 4).

c. *Vertical resolution, focal depth.* It was stated earlier that the study of the third dimension of microscopic objects requires a micrometer as a mechanical focusing device if high magnifications are used. There are, however, certain problems involved in the *interpretation* of the structures which the microscopist encounters as he travels through the different planes of the object.

Figure 51 comprises diagrams of total mounts of a rat's mesentry. As described in the discussion on fixation and staining (Part III, Chapter 2, B, 3) primary application of $AgNO_3$ leads to impregnation of cell boundaries (Fig. 51a), and secondary application of $AgNO_3$ causes impregnation of connective tissue fibers (Fig. 51b). In both preparations the nuclei have been stained with hematoxylin, but only those of the surface layers are shown. In Fig. 52 the same structures are seen in a *section* through the mesentery. Primary impregnation with $AgNO_3$ seems to take place only near the free surface of the preparation. The impregnated surface structures are known as terminal bars. Whether they represent incomplete partitions will be discussed later (Part IV, Chapter 1, C). At this moment let us concern ourselves with the optical phenomena only. Figure

52 illustrates five different positions of an objective. For each position the zones of focal depth, or satisfactory vertical resolution, are marked by a black rectangle,[14] while decreasing distinctness is indicated by tapering black areas.

As magnifying power decreases, depth of focus increases. In other words, more optical planes are visible simultaneously with low power than with high power objectives. However, the low magnification objective has not only less lateral (horizontal) resolving power, but the vertical resolution also decreases in sharpness as more planes are covered.[15]

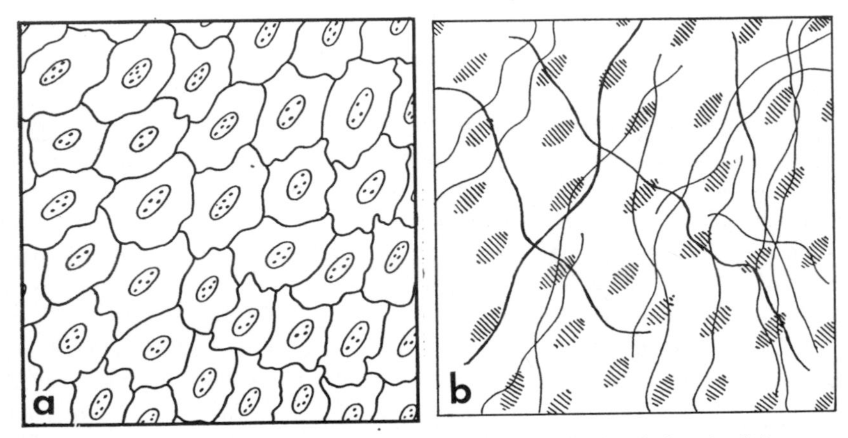

FIG. 51. Diagrams of total mounts of a rat's mesentery. (a) By placing the living mesentery into $AgNO_3$, it was killed and fixed at the same time. The result is silver impregnation of boundaries of lining cells only. (b) By placing the living mesentery in formalin, it was killed and fixed. Subsequent treatment with $AgNO_3$ produced impregnation of connective tissue fibers, but not of cell boundaries. The thickness of the mesentery sample and the optical properties of the objective are selected in such a way that the pictures (a) and (b) are obtained. Nuclei stained with hematoxylin are in focus together with the cell outlines (a). When the connective tissue fibers are in focus, the nuclei are blurred, as indicated by shading in (b). For analysis of focal depth, see section diagram, Fig. 52.

If two objectives with the same magnifying capacity have different numerical apertures, the objective with the lower aperture has a greater depth of focus, but less resolution. These relations are illustrated by Shillaber (1944) in his Figs. 31 and 32 which show photomicrographs of staphylococci, each at the magnification of 1600 times, taken with two objectives with different numerical apertures. One picture has better resolution and the other picture has more depth. The competition between depth of focus and resolving power is a general rule, which also holds if the depth of focus is increased by change of illumination

[14] The zone of satisfactory vertical resolution is generally known as the focal plane. The term focal layer would be more adequate.

[15] Both resolutions are defined as the minimum distances which are necessary to recognize two points as being separate.

such as narrowing of the substage diaphragm: what is gained in depth is lost in resolution. Compromises between requirements of resolution and requirements of focal depth are frequently made in biological photomicrography, depending on the purpose of the investigation.

Dissecting microscopes are characterized by two objectives which are mounted on two separate converging tubes with an eyepiece on each (cf. Part III, Chapter 2, B, 3). The result is stereoscopic vision. As an additional aid to manipulations, such microscopes are equipped with prisms which produce an upright picture. This design enables the biologist to dissect an object under the microscope with

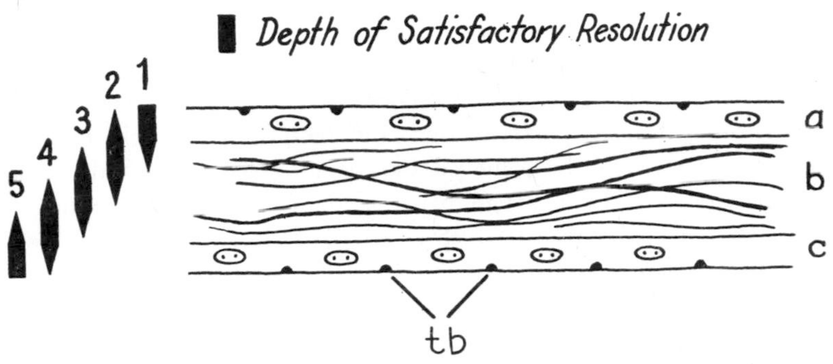

FIG. 52. Diagram of a section through a rat's mesentery to illustrate vertical resolution. The plane of the section is perpendicular to the two layers *a* and *c* which coat each side of the mesentery. The thickness of layers *a*, *b*, and *c* relative to dimensions of nuclei, terminal bars, and fibers has been exaggerated in order to separate focal planes. Layers *a* and *c* show the nuclei seen in Fig. 51a, while the intermediary layer *b* shows the connective tissue fibers of Fig. 51b. The silver-impregnated outlines shown in Fig. 51a appear in the section as black dots indicated by *tb*, terminal bars. It is not known whether or not cell boundaries extend from the terminal bars into the depth of the cytoplasm. Nuclei of the connective tissue layer *b* have been omitted. Symbols on the left side represent five different focusing positions of the objective. The range of satisfactory vertical resolution is indicated by a black rectangle for each position, while decreasing distinctness is marked by tapering black areas.

techniques which are similar to those used for dissecting macroscopic objects. The most important point is the considerable focal depth of dissecting microscopes. The limit of useful depth is reached at a magnification of approximately 30 times (Rosenthal, 1952, p. 205).

Although microscopic measurements are not difficult in planes parallel with the stage, they are quite involved and frequently uncertain in the third dimension. In a transparent object, the vertical distance between two structures is estimated by observing their appearance and disappearance when the focus is moved up and down. The positions are read on the calibrated fine adjustment which lifts and lowers the tube. The difference between the two positions (in microns) multiplied by the refractive index of the object indicates the vertical distance between the two structures.

The variability in thickness in different parts of a paraffin section was determined by Glimstedt and Håkansson (1951). A special instrument (microcator) was used which measured differences of 0.1 μ at 200 mg. pressure or more. In a section intended to be 5 μ thick, mounted on a slide without removal of the paraffin, local thicknesses varied from 2.5 μ to 6.5 μ.

Optical examination of opaque objects requires reflected light. It might be desirable to determine minute height-variations of the relief of a polished surface of steel. For such purpose, a fine wire may be used to cast a microscopic shadow. If the angle of the light is known, the depth of the relief can be calculated in each place (Tolansky, 1954). I do not know whether such techniques have been used to examine biological objects.

In electron microscopy the focal depth is usually 1 μ or more, while the thickness of sections is 0.1 μ or less. Therefore, all structures contained in a section are simultaneously in focus, and, different from the light microscope, vertical resolution cannot be achieved by changing the focus and reading a calibrated micrometer. In his 1953 paper on pulmonary alveolar lining, Low included a diagram (his Fig. 1) which clearly illustrated the difference between focal depth in the light microscope and in the electron microscope. With respect to focal depth, electron microscopy and roentgenography are comparable. Summation and superimposition in roentgenography were discussed previously. Roentgenologic structures which overlap in one plane can be separated by taking pictures at different angles (Fig. 50A and B), which includes taking of stereoradiographic pictures. Similarly, it is possible in electron microscopy to separate structures which overlap in one optical axis by taking micrographs of the object at different angles. Holders have been constructed which permit tilting of the specimen through a known angle around an axis of the microscope (after Zworykin, Hillier, and Vance, 1944, p. 392). Special devices permit not only stereomicrographs but thickness measurements, even in sections (Williams and Kallman, 1955). Measurement of thickness is somewhat easier in structures with a relief, such as blood films (smears). Shadowing techniques for increasing contrast were described earlier (Part III, Chapter 2, B, 3, d). By determining the shadowing angle and the length of the shadow, thickness can be measured.

Bessis' "Cytology of the Blood" (1956) contains instructive diagrams of shadowing techniques (his Figs. 33 and 34) and of the production of replicas from blood cells (his Fig. 45) for electron microscopy. Bessis also illustrates the use of shadowing for light microscopy; he shows side by side the appearance of red blood corpuscles in a conventionally stained film and in a shadowed film (his Fig. 35).

The general importance of *applying electron microscopy at relatively low magnifications* was emphasized early by Burton, Barnes, and Rochow (1942). The deep focus of the electron microscope can be very useful with relatively low magnifications. Zworykin and Hillier (1950, p. 528) showed a replica of the underside of a leaf, containing a lenticel and other structures, taken with the electron microscope at a magnification of 1000 times. Structures in different planes are seen distinctly. Such picture cannot be obtained with the light microscope since objectives which are needed for a 1000 times magnification have a very shallow vertical resolution, as pointed out previously.

When it became possible to take *Roentgen pictures on the microscopic level,* special techniques had to be developed for three-dimensional interpretations and measurements. The basis for stereomicroradiography was similar to that in macroscopic stereoradiography, namely, taking pictures of the object at different angles. Engström (1956, Table II, after Bellman) reports two methods of determining thickness of mica sheets, one by stereomicroradiography with the use of different tilting angles, and the other by light microscopy with conventional micrometric determination of depths. The results of the two methods were in satisfactory agreement.

2. *Application of Three-Dimensional Analysis to Various Structures*

Most textbooks and manuals of histology discuss the three-dimensional analysis of microscopic preparations. Bremer-Weatherford (1944, Fig. 1) shows a vertical rod, a slanting rod, a cube, a cone, a bent rod, and similar simple structures as they appear in three different planes of focusing. Ham (1957, Fig. 8 and Figs. 10–13) illustrates problems of the third dimension not only by simple structures, but also by fairly complex familiar objects using sections through different planes of a hard-boiled egg and of an orange.

Three-dimensional interpretation of microscopic structures may require special histological procedures. In my example of the rat's mesentery a satisfactory three-dimensional analysis was achieved by the combined use of total mounts and sections. In the remainder of the chapter these principles will be applied to a variety of structures on different levels of complexity.

a. *Isolated blood cells in dry and moist preparations.* Figure 53 illustrates differences in appearance of a white blood cell which are produced by different techniques. The cell used in this example is known as a polymorphonuclear leucocyte, which means a cell type with great variability of nuclear shapes. All these shapes have this in common that a number of nuclear segments are connected by bridges which are narrower than the segment. On a solid surface the leucocyte moves around by projecting pseudopodia from the bulk of its cytoplasm. If a leucocyte floats freely in the blood stream without touching the wall of the blood vessel and without being crowded by other blood cells, the shape of the leucocyte will be approximately spherical.[16] On this assumption the left upper diagram was drawn. The appearance of such a leucocyte in sections is shown

[16] Hence, the French designation "globules blancs," which is used interchangeably with "leucocytes."

in the pictures A′, B′, and C′. In contrast to the incomplete pictures in sections, leucocytes in smear preparations are complete, though flattened, as seen in picture D′.

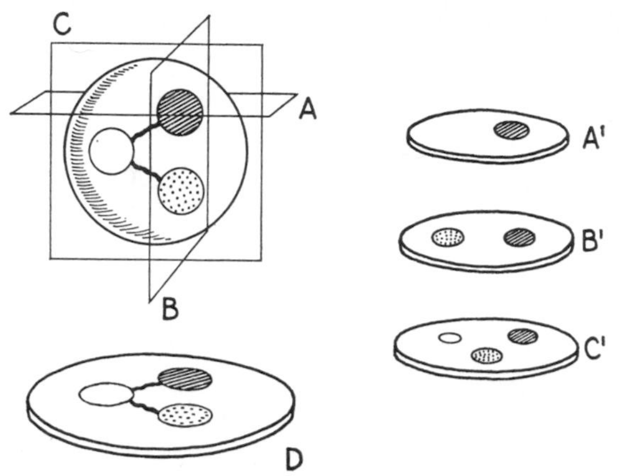

FIG. 53. Diagrams of three-dimensional and two-dimensional appearance of a leucocyte with three nuclear segments. Left upper diagram represents free-floating leucocyte: the various shapes of such a leucocyte are considered as modifications of a sphere. In a histological section there is an equal possibility for the leucocyte to be cut in the planes *A*, *B*, or *C*, or any intermediary plane. The respective sections show one segment in *A′*, two segments in *B′*, and three segments in *C′*. The three segments in *C′* are drawn at unequal sizes since it is unlikely that each of them will be cut through its largest dimension. Bridges between segments are rarely seen in sections. *D* is same leucocyte seen in a smear: the leucocyte is flattened but complete in this preparation; bridges between nuclear fragments are distinct.

b. *Cell aggregates and multinucleated cytoplasm.* A special chapter will be devoted to associations of tissue cells and their various nucleocytoplasmic patterns (Part IV, Chapter 2). Some points of that chapter will be briefly discussed here with respect to three-dimensional analysis.

A mosaic of prismatic cells which are joined together in one layer (Fig. 71 a–c) offers different pictures depending on the direction of sectioning. If the

---

FIG. 54. Three-dimensional analysis of multinucleated unpartitioned cytoplasm.

(a) Diagram of chorionic villus near term of pregnancy sectioned lengthwise (*A*), crosswise (*B*), and tangentially (C). *syn*, syncytial trophoblast (between interrupted horizontal lines); *con*, connective tissue fibers and cells; *cap*, capillary; *tan*, tangential slice of the syncytial trophoblast; the slice (width indicated by vertical interrupted lines) is turned 90 degress to produce the view (C). Note resemblance with C″ in part (b) of this figure.

(b) Diagrams of a giant cell of the Langhans type. In the left upper diagram the masses of nuclei are represented as aggregated in a space of the cytoplasm which can be described as the wall of a hollow hemisphere, or hemi-ellipsoid. *ext cy*, cytoplasm outside nuclear aggregate; *int cy*, cytoplasm inside nuclear aggregate. Sections through the planes A, B, and C will produce the two-dimensional pictures *A′*, *B′*, and *C′*, respectively. The heavy lines which delineate the nuclear masses from the cytoplasm are imaginary. Besides the highly diagrammatic presentation of the tangential section *C′*, a more realistic picture is shown in C″ in order to demonstrate resemblance with C in part (a) of this figure.

section is parallel with the axis of the prism a rectangle is seen, whereas a section perpendicular on the axis produces a polygon. A sponge-like system of cells connected by thin cytoplasmic processes appears as a two-dimensional network in sections (Fig. 73 a, b). Certain structures consist of patches or strands of cytoplasm which contain a number of nuclei without partitions between the individual nuclei; a pattern which offers particular difficulty when seen in two-dimensional sections. Three-dimensional interpretations of such structures are illustrated in the diagrams, Figs. 54a and 54b.

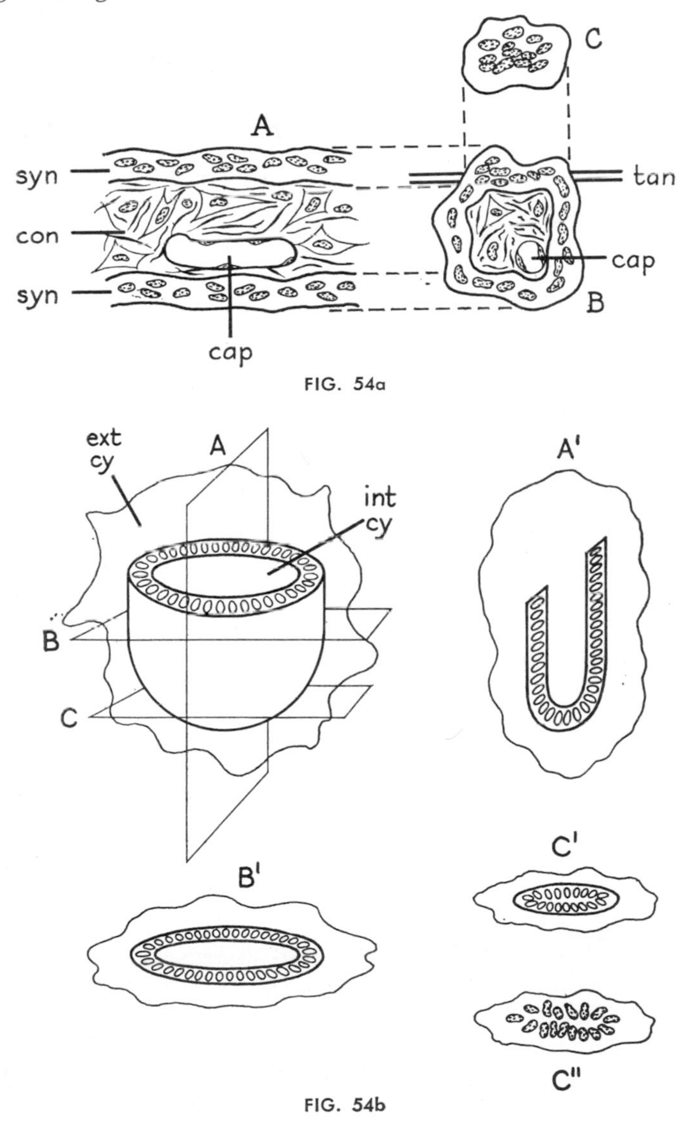

FIG. 54a

FIG. 54b

Three different sections through a chorionic villus (part of the placenta) are seen in Fig. 54a. The longitudinal section *A* and the transversal section *B* show the same structures: a central mass of connective tissue with blood capillaries and a coating by a layer of cytoplasm in which numerous nuclei are distributed. The cytoplasmic layer is characterized by the absence of partitions between the nuclei; it is called the syncytial trophoblast. A tangential section through a villus which cuts through the syncytial trophoblast only is shown in *C*. The structure *C* is a patch of cytoplasm with numerous nuclei and no partitions between the nuclei. If the origin of this structure were not known, it would be called a "multinucleated giant cell."

Figure 54b represents the so-called "giant cell of the Langhans type," which is one of the morphological responses of the host tissue to tubercle bacilli and some other infectious agents. The left upper part of the picture is a three-dimensional diagram of this giant cell. Numerous nuclei are concentrated in a space of the cytoplasm which can be described as the wall of a hollow hemisphere or hemi-elipsoid. The cytoplasm inside and outside this wall is practically free of nuclei. Sections in three perpendicular planes, *A, B,* and *C,* produce the characteristic pictures *A′, B′,* and *C′*. Transitions between these three pictures result from sections in other planes. The tangential section *C′* shows a patch of cytoplasm with a number of nuclei which are not separated by partitions. This is basically the same pattern as that of section *C* in Fig. 54a. In order to make this resemblance more apparent, the extremely diagrammatic features of *C′* in Fig. 54b have been replaced by a more realistic presentation in *C″*.

c. *Fibers.* Fibers will be described in the chapter on Intercellular Substances (Part IV, Chapter 3, A), but some problems of three-dimensional analysis will be discussed briefly here. The appearance of fibers on cross sections and oblique sections is well illustrated in the two books which I mentioned before. Bremer-Weatherford's (1944) Fig. 1 covers simple fibers by diagrams of vertical and slanting rods at different planes of focusing. Ham's (3rd ed., 1957) Fig. 13 shows sections through a cable with insulated wires as a model of nerve fibers which consist of axons with a coating of myelin.

If silver impregnation is applied to thin fibers, they appear black in transmitted light. Thick fibers, however, may appear brown if the impregnation is limited to their surface.

The identification of smooth muscles and nerves in sections requires a certain amount of three-dimensional imagination but depends mainly on criteria such as the distribution and shape of nuclei and the relations to surrounding connective tissue. These points will be discussed in Part IV.

d. *Organs and their components.* In various earlier chapters of this book organs and their components have been subjected to three-dimensional analysis. Vinylite injection of hollow organs with subsequent isolation by corrosion was illustrated in Fig. 23, which shows a cast of a human bronchial tree made by the techniques of Liebow *et al.* (1947). More and Duff (1951) published impressive pictures of neoprene injection with subsequent corrosion as applied to the arterial system of the human kidney. Their preparations range from individual glomeruli to large segments of the arterial tree.

Wax-plate reconstructions of the type used in embryology have also been applied to the reconstruction of adult organs.

e. Pathological variations. Three-dimensional analysis is a basic technique in the study of pathological conditions of the central nervous system. It was stated at various occasions that the dynamic topographic morphology of the brain and spinal cord was established, in many cases, through pathological destruction. Liebow and his associates applied their injection techniques to the study of pathological conditions of the bronchi and of pulmonary circulation. I mention two papers in which pathological conditions were analyzed by means of the methods mentioned: one dealing with the relations of bronchopulmonary venous collateral circulation to emphysema (Liebow, 1953), and the other dealing with bronchial and arterial anomalies connected with drainage of the right lung into the inferior vena cava (Halasz, Halloran, and Liebow, 1956).

f. Cultivated tissue cell colonies. Colonies of tissue cells cultivated *in vitro* shared the fate of histological sections insofar as their third dimension did not attract much attention. However, in the early era of tissue culture a remarkable paper appeared in which fresh explants of chick embryo hearts were subjected to three-dimensional analysis. In 1915 Congdon published a diagrammatic section of such an explant with the plane of sectioning perpendicular on the coverglass. Congdon's Fig. 1 showed the central explant, shaped like a planoconvex lens, and its extension into the zone of emigrated cells. According to Congdon, the central explant has an inactive center and an active superficial zone. He differentiated between a dense membrane of emigrated cells close to the coverglass and a loose network of cells at some distance from the coverglass. This concept of the three-dimensional architecture of a fibroblast colony, cultivated on a solid surface in a plasma clot, was subsequently confirmed in passage cultures (Mayer, 1933). The third dimension of a fibroblast colony proved to be important with respect to the topographic distribution of mitoses and in connection with the effects of ultraviolet irradiation.

F. Jacoby (1937) counted the mitoses in a total mount of a fibroblast colony (strain cultivated in passages). Although the central fragment was not transparent enough to count interkinetic (resting) nuclei, mitoses were identified with certainty. Jacoby counted 684 mitoses in the zone of emigrated cells and 144 mitoses in the central part. In the central part (which has the shape of a planoconvex lens) the mitoses showed the following topographic distribution: near the base 40, near the free surface (convexity) 95, and in the intermediary layers 9.

In studies with ultraviolet radiation the third dimension of the fibroblast colony had to be considered because ultraviolet is readily absorbed by the cell proteins. Consequently it is only in the thin layer of emigrated cells that each cell is reached by a similar amount of radiation. In the thicker central fragment, which consists of layers of twenty or more cells, those layers which are close to the source of radiation are damaged or killed, depending on the dose, while more distant layers are protected (Mayer, 1934). Therefore, a new population of undamaged cells can emigrate from the central area of a colony which has been irradiated with ultraviolet.

Chapter 7

Biophysics and Biochemistry As Related to Dynamic Morphology

A. Biophysics and Biochemistry of Organs in Relation to Histo- and Cytophysics and to Histo- and Cytochemistry

Biophysics and biochemistry are a concern of the present book inasmuch as they are closely related to dynamic morphology. Most techniques discussed in the preceding chapters included some kind of physical, chemical, or physicochemical procedures. I described the different ways of studying a bone: mechanical dissection, extraction of fat, dissolving of calcium salts, or incineration. These are biophysical and biochemical procedures. Microscopic study of dead biological material in sections frequently involves fixation in special fluids. Thin sections can be obtained only when the tissue block is frozen or embedded in special media, such as paraffin or celloidin. The sectioning is done with a microtome, which holds a knife of special shape in the proper position and controls the thickness of the sections mechanically. Since all these steps serve the study of a biological object, they are biophysical and biochemical procedures.

The *application of physical or chemical techniques to tissues or cells* is known as histophysics and histochemistry, or cytophysics and cytochemistry. The same terms are applied to the study of *physical or chemical processes that occur in tissues or cells.* Histochemistry and cytochemistry comprise chemical techniques applied in such a manner that substances and processes can be located on the microscopic level. In bulk biochemistry a dead organ is minced or homogenized, extracted or incinerated, and a living organ is perfused as a whole: no localization in microscopic components of the organ is attempted. With respect to procedures there are two bridges between bulk biochemistry and histochemistry. The first bridge is established by subjecting comparable samples of an organ or tissue both to biochemical analysis by homogenization, extraction, etc., and to histochemical stains or reactions in microscopic sections. Examples of such parallel procedures will be described later (Part III, Chapter 7, D).

The other bridge consists of the use of models. For instance, cigarette paper or pieces of elder pith were soaked with known lipids (Escher, 1919; Kaufmann and Lehmann, 1926), and the different stains for lipids were tested in these well-defined samples. Nucleic acids can be distinguished by the Feulgen method, a color reaction for aldehydes which, after acid hydrolysis, is positive with deoxyribonucleic acid, but not with ribonucleic acid. To produce resemblance with conditions in tissue sections, pure deoxyribonucleic acid was embedded in a mixture of paraffin and agar, so that blocks could be prepared. Sections from the blocks gave a strongly positive Feulgen reaction (Brachet, 1947).

B. The Trend toward the Study of Smaller and Smaller Biological Structures

1. *Old Procedures and New Demands*

Recent developments of biophysics and biochemistry have led to an overlapping of the smallest biological units and large molecules. A rough but very useful classification according to orders of magnitude is given by Engström and Finean in Table II of their 1958 book "Biological Ultrastructure." Their table is reproduced here (Table 9).

TABLE 9
Orders of Magnitude in Chemical Studies of Biological Material[a]

Linear dimension	Weight	Terminology	
1 cm.	1 gram	Conventional biochemistry	
1 mm.	1 mg. 10^{-3} gram	Microchemistry	
100 μ	1 μg. 10^{-6} gram	Histochemistry	Ultramicrochemistry
1 μ	1 μμg. (or 1 picogram or 10^{-12} gram)	Cytochemistry	

[a] From Engström and Finean, 1958.

At the present writing the term "fine structures" is no longer used for light microscopic orders of magnitude but is more or less restricted to electron microscopic levels (= submicroscopic levels). With the increased emphasis on smaller and smaller biological structures it became necessary not only to revise traditional methods, but also to give special attention to aspects that had been considered trivial or unimportant. In many cases rationalization of empirical procedures became desirable. Examples to illustrate relations between the old and the new techniques follow.

a. *Revision of microtome techniques.* Microtome knives had been manufactured in many variations to suit different tasks. For frozen sections and paraffin sections knives were preferred with two plane surfaces, whereas for celloidin sections knives had to be planoconcave. I mentioned the fact that for electron microscopy steel knives were replaced by sharp glass or diamond fragments. The consistency of the embedding media was modified in various ways—for instance, by mixing two types of paraffin having different melting points. New media had to be designed for electron microscopy. Detailed prescriptions for microtome techniques are found in many laboratory manuals, but there are not many articles in which the physical theory of a microtome knife is discussed. A comprehensive presentation of the physical and chemical problems of microtomy is that of Gettner and Ornstein (1956). The theory of knives is summarized in their Fig. 17 and Table III. Gettner and Ornstein also discuss the relation between compression and section thickness (their Fig. 1), and many other items.

b. *Increased importance of water; new aspects.* As a second example of a seemingly trivial item of microscopic procedures let us consider the role of water. In the study of dead biological material, water occupies an important place in dehydration and freezing techniques, but these techniques did not seem to require theoretical considerations. However, when investigation turned to smaller orders of magnitude of biological structures, destructions by ice crystals and distortions by thawing became serious problems. Again empirical histological processes had to be probed by physical analysis. New possibilities seemed to arise with the discovery of vitrification or freezing of biological material without the formation of ice crystals (see Luyet, 1951). The underlying physics were not easily clarified, and the applicability of vitrification in histological work is still very limited (R. J. C. Harris, 1954). The increased interest in the role of water is reflected in the current textbooks of biochemistry. In the chapter on water in Peterson and Strong's "General Biochemistry" (1953), hydrogen bonds between protein and water are discussed. The authors state, among other interesting facts, that a protein molecule may contain several thousand groups which can bind hydrogen (page 10 of their book). A brief theoretical discussion of the aggregation of molecules and their interaction with water is included in Engström and Finean's book "Biological Ultrastructure" (1958). The following items are treated: crystal structure; liquid crystals; transition from water of crystallization to true solution; colloidal solutions; boundary layers; and insoluble monolayers. In living cells, a separate determination of water and dry substance (hydrous and anhydrous mass) was possible with the use of interference microscopy (Mellors, Kupfer and Hollender, 1953).

c. *Adjustment of osmium tetroxide fixation to the purposes of electron microscopy.* The history of microscopes and illuminating apparatus shows a rational development along the principles of physics, although for a long time, the predominant use of such instruments was in biology. There is no break between the 19th century era of light microscopy in biology and the present period of light microscopy. The transition from dark-field to phase contrast microscopy was nothing revolutionary. A new era did start with the invention of electron microscopy which, of course, was an achievement of modern physics. The rationale of the new embedding media which were introduced for electron microscopy has been analyzed by various authors, e.g., in the article by Gettner and Ornstein (1956) quoted above. I discussed previously the problems of determining focal depth in light and in electron microscopy (Part III, Chapter 6, B, 1, c). In electron microscopy most biologists are satisfied with estimating the thickness of the sections by reflection and interference colors. Gettner and Ornstein present these relations in their Table IV, but consider the color readings as too subjective to be reliable. One recognizes here the difference between the highest precision which a tool can yield, and the lower degrees of precision which may be satisfactory in the use of the tool for specific purposes.

It is interesting that one of the oldest fixatives of traditional histology, osmium tetroxide (OsO_4), became the most important fixative in electron microscopy. I mentioned earlier that this is the only fixative which preserves both proteinic and fatty structures (Part III, Chapter 2, B, 3, c). The demands of electron microscopy made a new appraisal of osmium tetroxide necessary. Porter and Kallman (1953) devoted a special study to this subject. The action of vapors of OsO_4 on living cells was observed by placing tissue culture preparations under a light microscope, with or without phase contrast. The cells selected for study were well separated from each other. Neither the outlines of the cells nor the delicate structures in the cytoplasm were altered by the process of killing and fixation. According to Porter and Kallman, one of the great advantages of OsO_4 is that it transforms protein solutions into homogeneous, optically empty gels, while all other fixatives produce irregular coarse coagulation. The physicochemical effects of OsO_4 on fat and proteins remained still obscure, but interesting hypotheses were offered by Porter and Kallman. The disadvantages of OsO_4 as a fixative were stated before: poor penetration of tissue blocks and interference with most of the useful histological stains.

Palade (1952) adapted OsO_4 fixation for the preparation of *sections* for electron microscopy. He used tissue blocks 1 $mm.^3$ or less in volume. He observed that OsO_4 produced acidification of the tissues before the onset of fixation. If acidification was avoided by appropriate buffering, the results were greatly improved. The problem remained to what extent the structures in the sections resembled structures in living cells. Isolated surviving cells in suspensions can be used for comparisons with similar cells in fixed sections (see the study of testicular cells by Fawcett and Ito, 1958). Cells in an organized tissue pattern can be examined in fixed sections, but comparison with the living material is limited to transparent tissues. However, through indirect procedures, Palade arrived at valid reasons why the buffered OsO_4 could be accepted as a satisfactory fixative for sections for electron microscopy. Many important contributions have been made with the use of this technique.

There is the question of reproducibility of OsO_4 fixation. While acknowledging the great advantages of Palade's technique, Low (1953) emphasized "the inherent complexity and irregularity of osmication procedures." Particular difficulties were encountered in applying OsO_4 to pieces of air-containing lung which were the object of Low's study. He found that in some areas of the same block the cell outlines were clearly delineated by osmication, while in other areas the outlines appeared less dark, or not darkened at all. The same variability was observed in the osmication of basal membranes. As Low pointed out, it may be impossible to interpret some structures in a single electron micrograph. He considered suitable for interpretation those structures which occur regularly in a large number of specimens. I wish to add the comment that this type of selection is customary in light microscopy as well as in electron microscopy. For detailed discussion of this aspect see a later chapter which deals with sampling for quantification of morphological structures (Part III, Chapter 8, C, 1).

2. *The Story of the Zymogen Granules of the Pancreas*

A particularly gratifying feature of electron microscopy is that many of its results have been integrated with the results of other techniques. I reported previously that in studies of the synaptic gap the determinations made with microelectrodes in the living animal were in satisfactory agreement with measurements made in electron micrographs of fixed nerve cells (Part III, Chapter 1, A).

The story of the zymogen granules in pancreatic cells illustrates how a discovery that was made in the 19th century was confirmed and expanded by modern cytophysical and cytochemical methods, including electron microscopy. In 1875 Heidenhain observed in experimental animals that the cells of the pancreatic glands contained granules which disappeared shortly after food intake and reappeared a few hours later. Since the disappearance of the granules coincided with the time when proteolytic enzymes appeared in the pancreatic juice, Heidenhain concluded that the granules consisted of precursors of the digestive enzyme. This interpretation was supported by the fact that the granules were restricted to the apical part of each cell, i.e., the part bordering the glandular space into which the secretion is discharged (see my diagrams Fig. 57, F and G). Heidenhain's observations were confirmed by other investigators. However, the interpretation of the granules as precursors of digestive enzymes rested, for a long time, on indirect conclusions.

The first step toward a direct demonstration of the chemical nature of the zymogen granules was taken by Marshall (1954) with the use of immunological and cytochemical techniques, designed by Coons, Creech, Jones, and Berliner (1942). Coons and his associates introduced fluorescent chemical groups into the molecule of pneumococcal antibodies. This did not change the immunological specificity of the antibodies. When the fluorescent antibodies were injected into mice which were dying of pneumococcal infection, the antibodies combined with antigens.[1] In histological sections the places of fluorescence indicated the location of the antigens. Marshall adapted this procedure as follows. Rabbits received injections of bovine pancreatic enzymes which had been obtained in crystallized form. In the serum of the rabbits specific antibodies against the enzymes were produced. The γ_2-globulin fractions of the antisera were labeled by fluorescent conjugates. When thin sections of beef pancreas, prepared by the freeze-drying technique, were placed in a solution of fluorescent antibodies specific for carboxypeptidase, the apex of each cell showed fluorescence. A similar result was obtained with chymotrypsinogen. As controls, other beef pancreas sections were treated with fluorescent-labeled globulin solutions which were obtained from untreated (normal) rabbits and therefore did not contain specific antibodies. In none of the control sections was the fluorescent material concentrated enough to be detectable. In this way it was demonstrated that the precursors of the digestive enzymes were restricted to that area of each cell in which the zymogen granules were located. Since the cytoplasm around the granules showed fluorescence also, this technique did not indicate *exclusive* localization of the enzymes in the zymogen granules.

Hokin (1955) succeeded in isolating the zymogen granules by fractionate centrifugation of homogenates of dog pancreas. The presence of the granules in that particular fraction was verified by light microscopy; the granules were 0.5 to 1.5 μ in diameter. The enzyme activity of the zymogen granule fraction exceeded the activity of the homogenate of the whole pancreas as follows: protease more than fiftyfold, amylase and lipase approximately two- or threefold. More elaborate methods of fractionate centrifugation[2] were applied to the guinea pig pancreas in a study by Siekevitz and Palade (1958). These investigators compared the enzymatic activity of the total homogenate with that of four fractions which contained the following cell components: nuclei, mitochondria and microsomes, identified morphologically, and zymogen, identified by activity. Pellets which contained the different cellular components were studied morphologically with ordinary light microscopy, phase contrast and electron microscopy; for the latter the pellets were fixed in OsO_4. The four fractions were not entirely pure. The zymogen fraction proved to contain mitochondria besides zymogen granules; this was revealed by electron microscopy. The chemical studies which were made of the different fractions included the use of column chromatography for the isolation of chymotrypsinogen from homogenates and the different fractions. The largest amount was present in the zymogen fraction.

I have recounted some of the points of the paper by Siekevitz and Palade to illustrate the ingenious ways in which various modern techniques were combined for a well-defined purpose. To appreciate this fully, one has to read the

[1] For instructive diagrams see Mellor's (1959) article on the fluorescent-antibody method.

[2] For techniques of fractionate (differential) centrifugation, see N. G. Anderson (1956).

paper quoted here and also other publications of the same series of studies.

Palade (1959) summarized the modern investigations on zymogen granules with the statement that Heidenhain's hypothesis has been fully confirmed: the zymogen granules deserved their name, since they contained trypsinogen and chymotrypsinogen, and in addition ribonuclease and other enzymes. These are dangerous products, according to Palade, which could easily wreck the cell in which they are stored temporarily. Are there any protective devices which the cell uses to handle these enzymes without accident? By electron microscopy Palade discovered a barrier, approximately 70 Å. thick, which seems to isolate the enzymes from the rest of the cytoplasm.

3. *Modern Methods of Direct Cytochemical Determination, Especially of Nucleic Acids and Enzymes*

My report on the zymogen granules of the pancreas illustrated how various techniques led to a precise characterization of the granules. When the granules were studied by fractionate centrifugation, they were no longer in their original location, but there was no doubt concerning their identity. In the experiments with fluorescent antibodies, the fluorescence was observed in the apical part of each cell, but fluorescence microscopy could not resolve individual granules. This, however, did not detract from the value of the experiment. What I wish to point out here is the difference between optical localization of substances, and localization by inference.

Physical methods such as microspectrophotometry in ultraviolet light were highly successful in direct cytochemical localization not only within a cell, but even in special areas of the cytoplasm or nucleus. These techniques have been mentioned in earlier parts of the book and will occupy us again in a discussion on relations between nuclear morphology and content of deoxyribonucleic acid (Part IV, Chapter 1, A). Since the majority of publications on cytochemistry of nucleic acids is based on absorption curves obtained by microspectrophotometry, it may be useful to give at least one example of an entirely different procedure.

Goldstein and Plaut (1955) combined various techniques in order to prove *by direct evidence* that the ribonucleic acid which is known to occur regularly in the cytoplasm does not originate there, but is synthesized in the nucleus. Much circumstantial evidence had been collected in support of this possibility (see Caspersson's 1941 theories and Brachet's 1955 review). The interest in this question is associated with the problem of transmission of genetic specificity from deoxyribonucleic acid (located in the nucleus) to cytoplasmic components. Goldstein and Plaut used *Amoeba proteus* in their study. The amebas were fed ciliated protozoa (*Tetrahymena*) which had incorporated P^{32} from their culture media. Nuclei from P^{32}-labeled amebas were transplanted by micromanipulation into amebas which had been deprived of their nucleus, but were otherwise normal. The amebas which contained grafted nuclei were fixed at different intervals after the operation. Fixed and mounted preparations were photographed and subsequently attached to autoradiographic film to which they remained exposed for 2 weeks. Comparisons between the original photographs of the amebas and the radioautographs showed that up to 5 hours after transplantation all significant radioactivity was still within the nuclei, whereas 12 hours or more after transplantation appreciable radioactivity was found in the cytoplasm. Since it was known from experience that under these conditions the radioactive labeling does not occur in deoxyribonucleic acid but only in ribonucleic acid, the authors had proved their point that most of the ribonucleic acid of the cytoplasm is synthesized in the nucleus.

Cytochemical work on enzymes caused the development of a particular technique which calls for the addition of specific substrates to histological sections. This subject was covered in Gomori's classic book of 1952. In addition I mention two different presentations of cyto- and histochemistry of enzymes. The practical aspects are emphasized in the enzyme chapter of Lillie's "Histopathologic Technic and Practical Histochemistry" (2nd ed., 1954). Emphasis on the theoretical side is found in Holt's 1959 article, "Principles and Potentialities of Cytochemical Staining, Methods for the Study of Intracellular Enzyme Distribution."

The *addition of known substrates* to histological sections is necessary for the following reasons. Direct microscopy cannot distinguish between a particular enzyme and other protein, but the activity of an enzyme can be utilized to induce the formation of an insoluble deposit from an appropriate substrate. For some enzymes fresh frozen sections are preferable, while other enzymes tolerate treatment with acetone or alcohol so that fixed material can be used for sectioning. The sections are incubated in buffered solutions of a substrate which is attacked by the enzyme in question. Substances which are set free from the substrate by the action of the enzyme are either directly visible under the microscope or can be made visible by color reactions. In either case one expects to find stained particles in the places where the enzyme was located. Two types of controls are needed: (1) a specificity control, based on the addition of an inhibitor of the enzyme in question, (2) controls for correct localization which indicate whether the technique applied might have caused the enzymes to diffuse away from their original sites.

An example illustrating the application of these principles follows. Suppose a human prostate is to be examined for acid phosphatase. Either unfixed fresh material or acetone-fixed material is used to make frozen sections approximately 10 μ in thickness. The moist sections are placed on slides and allowed to dry so that they will stick to the slides. The slides with the sections are placed in a buffered substrate solution, in this case glycerol phosphate, to which a soluble lead salt, such as lead acetate, is added. If acid phosphatase is present in the sections, the substrate is split into glycerol and free phosphate, and insoluble lead phosphate is formed at sites of action. Finally, the sections are immersed in ammonium sulfide. By this treatment the lead phosphate is converted to sulfide and thus made visible in the shape of black-brown granules. At this point the specificity of the reaction is not yet proved, since axons of nerves or pre-existing calcium deposits may be impregnated with lead and also produce black-brown precipitates. To exclude such unspecific processes, control sections are carried through solutions to which has been added fluorine, which is known to inactivate many enzymes, including acid phosphatase. In these control sections no black granules should appear if the granules which form in the experimental sections were due to liberation of phosphate by acid phosphatase.

There are rare instances in which specificity controls are unnecessary. It seems that the test for cholinesterase by the addition of acetylcholine as a substrate is "absolutely specific" (Gomori, 1957).

Correct localization of enzymes required special attention after it had been observed that diffusion currents may carry enzymes to places different from their natural sites. I mention the study by Martin and Jacoby (1949) on diffusion of alkaline phosphatase. Following a technique devised by Danielli (1946) these authors placed sections of guinea pig liver, which contain very little alkaline phosphatase, underneath sections of organs with large amounts of the enzyme, such as the small intestine. When the combined preparation was incubated with substrate and staining agents, the liver sections became strongly positive. This proved diffusion from the intestinal section. Other ingenious experiments showed transportation from some areas of a section to other areas, and also migration of enzymes from some structures to others. Koelle (1951) published a careful study on the diffusion of cholinesterases in the course of histochemical treatment; he found it possible to modify his technique in such a way that alterations of enzyme localization were reduced to a minimum. The general problem of "diffusion artifacts" was reviewed by Novikoff (1951).

C. Crossroads of Physics, Chemistry, and Morphology on the Light and Electron Microscopic Level

My presentation of the story of the zymogen granules was more or less limited to an old hypothesis which has been verified by new investigations. However, in biophysics and biochemistry, as much as in other branches of natural science, new procedures and discoveries clarify some old problems, but also create many new problems. It seems possible that zymogen granules interact with certain endoplasmic membranes which were discovered by electron microscopy (endoplasmic reticulum, Porter, 1953; α-cytomembranes, Sjöstrand, 1956). The literature on biophysics and biochemistry is increasing at a fantastic rate. In a crude way one may divide the publications into three groups. In the first group, one type of technique is applied to many objects. An example of this group is the book "Biological Applications of Freezing and Drying," edited by R. J. C. Harris (1954), which was quoted earlier. The contributors cover the preservation of viruses, bacteriological media, blood plasma, mother's milk, and foodstuffs; and discuss theoretical aspects, effects of low temperature on living cells and tissues, and application of freeze-drying to electron microscopy and histochemistry. In the second group of publications one problem, or set of problems, is treated by a variety of techniques. A representative of this group is the symposium on '"Subcellular Particles," edited by Nayashi (1959), from which I recounted part of Palade's work. Books of the third group cover both techniques and problems. In this category is "Biological Ultrastructure" by Engström and Finean (1958), from which I reproduced the table of orders of magnitude (Table 9). Engström and Finean summarize and discuss the methods of ultrastructural research in a special chapter. Another chapter deals with the theory or principles of molecular structures. The greater part of the book is devoted to the ultrastructural role of proteins, lipids, carbohydrates, nucleic acids, and mineral salts. I give a few examples to show how, within this chemical framework, microscopic, molecular, electron microscopic, and light microscopic structures are discussed. Viruses, chromosomes, and genes are discussed in the chapter on nucleic acids, according to the recognized key position which nucleic acids occupy in those structures. Retinal cones and rods have been included in the chapter on the role of lipids, since it has been shown that these elements of the retina contain layered lipoprotein structures. The ultrastructure of nerve fibers is also treated in the chapter on lipids. The myelin sheaths are discussed there, as one would expect, but the axons have evidently been placed in that chapter solely because they belong to the nerves. I feel that, according to chemical classification, the axons should have been included in the chapter on proteins. Since flagella are specialized cytoplasm, they are rightly included in the subchapter on fibrous proteins. Under the electron microscope *all* cilia and flagella show an internal pattern of nine peripheral small filaments (or double filaments) plus two central filaments per motile unit. This remarkable fact is mentioned and illustrated by Engström and Finean. Some of their pictures are reproduced in my Fig. 69 (in Part IV). Perhaps the authors should have emphasized that, for the time being, this astonishing numerical law eludes interpretation in terms of protein chemistry. My comments are not meant as criticisms of the book, but are intended to point out the difficulty in organizing the wealth of new cytophysical, cytochemical, and morphological discoveries.

It is interesting to compare the book by Engström and Finean with Brachet's

"Biochemical Cytology" (1957). Both books appeared at approximately the same time. Brachet's book is organized according to the morphological components of the cells. One chapter is entitled, "The Cytoplasm of the Resting Cell," and one "The Nucleus of the Resting Cell." In a chapter on mitosis the nuclear and cytoplasmic mechanisms are discussed together because of their close interactions. A special chapter is devoted to the nucleus and the cytoplasm in embryonic differentiation. In the discussion of all these items Brachet combines the cytochemical with the morphological approach. Occasionally he penetrates to the molecular level including molecular theory. For instance, in the chapter on the nucleus of the resting cell the Watson-Crick (1953) concept of the double-stranded helical structure of deoxyribonucleic acid is included. Only one chapter of Brachet's book has a chemical title: "Nucleic Acids in Heredity and Protein Synthesis." Except for this chapter one may say that Brachet goes from the biological structures to the molecules, whereas Engström and Finean go from the molecules to the biological structures. Brachet shows more electron micrographs of biological structures, but Engström and Finean give more chemical formulas. Because of their difference in emphasis these two books supplement each other in a very useful way.

D. Histochemistry and Quantification

In any new branch of natural science the development of qualitative procedures precedes the establishment of quantitative methods. This applies also to histochemistry, cytochemistry, and ultramicrochemistry. A critical evaluation of quantitative histo- and cytochemical techniques was published by Glick, Engström, and Malmström in 1951. By purely histochemical methods only "a modest degree of quantification" can be achieved, as Gomori emphasized in 1952 ("Microscopic Histochemistry," p. 20).

Under certain conditions, estimates of quantities, or at least relative values, can be obtained by microspectrophotometry, autoradiography, microinterferometry, and other specialized forms of microscopy. The following advantages and limitations of these methods were pointed out by Engström and Finean in their 1958 book. The microscopic methods mentioned permit the study of very small volumes of material, but usually in these small volumes high concentratons of chemical components are necessary to permit analysis. Although very fine qualitative differentiation is possible with these methods, "the accuracy with which individual chemical components can be quantitatively estimated is usually not comparable with that of the standard chemical analysis methods applied to the larger specimens" (Engström and Finean, 1958, p. 6).

Accurate quantitative determinations of tissue components require biochemical techniques which involve homogenizing or dissolving. Such destructive procedures are obviously incompatible with localization of substances on the cellular and subcellular level. The only possible compromise is to correlate quantitative results, which were obtained biochemically, with qualitative or semiquantitative data, which were obtained histochemically. In principle such a dual approach is based on a combination of procedures, as described previously in connection with the identification of the zymogen granules in pancreatic cells. Successful coordination of histochemical estimates and quantitative chemical determinations will be illustrated subsequently by various examples. Reports on parallel use of histochemical and quantitative biochemical methods are rarely found in current periodicals. The trend toward quantification seems to be rather weak in present histo- and cytochemistry for three reasons.

First of all, expensive equipment is needed for ultraviolet microspectrophotometry or Roentgen-ray absorption on the microscopic level. This has been pointed out in manuals of histochemistry. It seems to me that these manuals should at least encourage comparisons of histochemical and quantitative biochemical procedures, since the latter are relatively inexpensive. Secondly, histo- and cytochemists have been concerned with difficult qualitative tasks for a long period. The Feulgen reaction for deoxyribonucleic acid, which was published in 1924 and now is one of the best established and most important cytochemical techniques, was still attacked as chemically meaningless fifteen years ago (Carr, 1945). Elaborate studies were needed to defend the Feulgen reaction against this attack (Dodson, 1946; Brachet, 1947). Thirdly, a large proportion of modern work in histo- and cytochemistry is dedicated to enzymes. It is characteristic that in the manuals by Gomori (1952) and by Pearse (1960), 30-40% of the space is devoted to histochemistry of enzymes. S. C. Shen (1958) pointed out that, at the present time, enzyme-cytochemists are mainly occupied with the following problems: (1) to make more and more enzymes accessible to cytochemical localization, (2) to minimize structural and chemical alterations of cells by the techniques applied, and (3) to increase the resolution of enzyme localization. Quantitative determinations, although desirable for the future, appear less urgent now in the cytochemistry of enzymes.

Under these conditions one can hardly expect many investigators to be interested in combining qualitative or semiquantitative cytochemical studies of enzymes with quantitative chemical determinations. That such combined studies are possible was demonstrated by the investigations on enzymes of the stomach which were carried out by Linderstrøm-Lang, Holter, and Søeborg Ohlsen (1934). These studies will be recounted after a discussion of histochemical and quantitative chemical determinations of glycogen, fatty substances, and inorganic materials. Problems of quantification which arise from spectrophotometric determination of deoxyribonucleic acid in individual nuclei will be treated later in a discussion on nuclear structures and nuclear material (Part IV, Chapter 1, A).

1. *Glycogen*

Of the carbohydrates which participate in energy metabolism, only glycogen can be demonstrated histochemically. Before approaching the problems of its determination on the level of tissues and cells, it may be useful to realize the fact that glycogen metabolism is not necessarily the same in all organs. This will be illustrated by the studies of S. E. Kerr and his associates, in which Pflüger's 1905 procedure was applied to the brain of various animals, and by the investigations of Stetten and Stetten, in which the metabolic fate of glycogen in rats was followed with the use of radioactive tagging and specific enzyme reactions.

Pflüger's procedure begins with the digestion of the organ sample in hot concentrated KOH solution and concludes with the determination of glucose in the hydrolyzate as reducing sugar.[3] Kerr and Ghantus (1936 and 1937) adapted this method to the estimation of glycogen concentration in the brain of rabbits and dogs. Amytal injection was used for anesthesia since this was known to have relatively little effect on the carbohydrate metabolism. Skulls of the anesthetized animals were opened and the brains were frozen by liquid air. The sampling technique described took care of two requirements: (1) maintenance of the circulation in the brain up to the beginning of freezing and (2) fast penetration of the freezing into the interior of the brain. The authors observed that a time lapse of 15 minutes between cessation of cerebral circulation and sampling of brain caused a glycogen loss of 80 to 90%. Lactic acid formed in amounts corresponding to the carbohydrate removed. When all precautions were taken, the glycogen content of the brain was lower than that of liver and muscle, but in contrast to these organs the amount of glycogen in the brain was not changed by fasting, overfeeding, glucose infusion with or without administration of insulin, phlorhizin poisoning followed by epinephrine administration or pancreatectomy. Only overdosage with insulin caused a marked decrease in glycogen. The remarkable stability of brain glycogen was confirmed in perfusion experiments *in situ* by A. Geiger (see his 1958 review). He stated that even Metrazol-induced convulsions, although causing glucose breakdown and increased O_2 consumption, did not produce a significant breakdown of glycogen in the brain.

In a series of studies of an entirely different type, Stetten and Stetten also found that the metabolic fate of glycogen is not the same in all organs, although glycogen samples from different organs yielded only glucose upon total hydrolysis. The authors refer to this phenomenon as "metabolic inhomogeneity"

[3] Seifter, Dayton, Novic, and Muntwyler (1950) introduced an organic reagent, anthrone, for quantitative colorimetric determination of glycogen in KOH digests of organs. I mention this method since it may be applicable also to histochemical estimation of glycogen in sections.

of glycogen. In the glycogen molecule are two different types of linkage (1,4 and 1,6) for which different enzymes are specific. The molecule has a treelike structure consisting of a core which bears the reducing group and carries branches with different tiers (for diagrams and discussion see Larner, Illingworth, Cori, and Cori, 1952). The places of 1,4 linkage are readily attacked by barley β-amylase but the 1,6 linkage which joins the branches to the core is not attacked. To obtain information on the metabolic fate of the tiers and the core of the glycogen molecule, labeled C^{14}-glucose was injected intraperitoneally into rats. By β-amylase degradation it was possible to determine C^{14} separately in the different portions of the glycogen molecule. The penetration of the isotope into the interior of glycogen macromolecules was faster in liver than in muscle. Moreover, the carrier of the tag seemed to favor molecules with a larger molecular weight over those with a smaller weight when the glycogen is located in muscle, whereas the reverse tendency was found in the liver. In addition to these differences within the molecule, the population of whole molecules showed differences in molecular weights. Our brief summary is based on papers in the *Journal of Biological Chemistry* (Stetten and Stetten, 1954; Stetten, Katzen, and Stetten, 1956) and on an abstract in *Science* (Stetten and Stetten, 1956).

Let us now turn to the *histochemical determination of glycogen.* It was known that glycogen hydrolyzed rapidly in the tissues after death. Therefore it was recommended that histochemical tests for glycogen be limited to samples which were obtained either from the living animal or within a few minutes after death of the animal. However, surprising observations were made in human pathology. Some patients who were autopsied many hours after death still showed glycogen in the liver and in the conducting bundle of the heart; in patients who had died of diabetes stored glycogen was seen in certain segments of the renal tubules.

Things became even more puzzling when von Gierke (1929) found enormous accumulations of glycogen in the enlarged liver and kidneys of a child, although there had been a 24-hour interval between death and autopsy. Von Gierke's histochemical diagnosis was confirmed by Schönheimer's chemical determinations in the same material (also published in 1929). When samples of the liver were kept for 6 days at 37° C., their glycogen concentration decreased very little. No indication was found for the presence of a peculiar kind of glycogen. Glycogen isolated from the liver and the kidney of this child was rapidly broken down when mixed with normal liver. Schönheimer concluded that there was a total or partial absence of glycogen-splitting enzymes. For extended studies of this condition, which became known as glycogen storage disease, see Van Creveld's comprehensive review (1939) and the more recent brief review of this subject by Forbus (1952, Vol. 2, p. 583). The excessive glycogen storage does not always involve the same organ; in some cases the heart was the main site of the abnormality.

It had been believed for a long time that alcohol of 80% or higher concentration is the only fixative which allows the histochemical demonstration of glycogen. Vallance-Owen's (1948) experiments showed convincingly that aqueous fixatives do not endanger the preservation of glycogen. It seems to me that *all* fixatives inhibit enzymatic activity and thereby prevent hydrolysis of glycogen. The samples should be chilled rapidly and fixed at low temperature as recommended by Gomori (1952, p. 62). Evidently this precaution retards enzymatic activity until all parts are reached by the fixative. Fixation at low temperature in a fast-penetrating fixative such as Bouin's fluid (picric acid, acetic acid, and formalin) proved particularly efficient in a comparative study by J. Faherty of our laboratories (1960, unpublished data).

Suppose samples from the same rat liver are available both for chemical and histochemical study of glycogen. Is the amount of stainable glycogen seen in the sections correlated with the quantity of glycogen as determined chemically in digests of this liver?

The first systematic study of this question was made by Nielsen, Okkels, and Stochholm-Borresen in 1932 in livers of guinea pigs and dogs. The amount of glycogen seen in sections after staining with Ehrlich's iodine or Best's carmine method was *roughly* indicative of the percent of glycogen per wet weight of liver. On the basis of stained sections it was possible to estimate the following three classes: less than 1%; 1–4%; and more than 4%. This study was well documented by photomicrographs and a table showing each determination.

Evidently unaware of the earlier work of Nielsen and associates, Deane, Nesbett, and Hastings (1946) undertook a study of the same question with refined techniques such as photometric determination of stainable glycogen and physiological standardization of the animals used. These investigators examined the livers of twenty male rats of similar age and body weight. For 5 days before sacrifice the animals were fed for a 2-hour period each day. They were killed in groups of four at 6, 12, 18, 24, and 48 hours after the fifth and last meal. In this way livers with variable but controlled glycogen concentrations were made available, the maximum being in animals sacrificed 6 hours after the last meal and the minimum in those killed 48 hours after the last meal. Samples were taken from the right and left lobes. In the histological sections glycogen was stained by Bensley's modification of the Bauer-Feulgen reagent. The stainable material was measured photometrically and expressed as optical density. Two different methods were used for chemical determination of glycogen in samples of the liver adjacent to the samples used for histology. The values obtained by these determinations are given in a table and a graph. In the graph (their Fig. 5) the percentage of glycogen per wet weight is plotted against optical density of glycogen in sections. The resulting curve shows linear correlation. For a summary of these results, see also our Table 12.

Grafflin, Marble, and Smith (1941) had not found any marked correlation between histochemical stain and chemical determination of glycogen in the livers of guinea pigs. I will report later how their data can be reinterpreted by statistical treatment (my Table 13, with discussion in Part III, Chapter 8, C, 3).

A subsequent comparison of histochemical and chemical estimation of glycogen by Fitzpatrick, Larner, and Landing (1948) showed a satisfactory correlation between the two procedures. However, the threshold of visibility of glycogen in stained sections varied over an astonishing range of chemical concentrations, from 0.04 to 0.6% of the wet weight.

In a later study by Kugler and Wilkinson (1959) glycogen was determined in the ox myocardium with the result that glycogen could be demonstrated histochemically, if its concentration in the heart muscle exceeded 185 μg./100 mg. wet weight. They fixed specimens for histochemical study in ice-cold 80% alcohol. They claimed that 4 to 5 minutes after slaughtering, the samples were submerged in the fixative. In their experience, Gomori's hexamine silver technique was the best procedure, Best's carmine ranked second, and the periodic acid-Schiff reaction was least sensitive.

Parallel use of procedures for chemical quantification and for histochemical localization may shed light on some glycogen problems. Let us return, for a moment, to the quantitative determinations of glycogen in whole brains made by Kerr and his associates. It seems to me that this work should be continued in various ways. Evidently, it would be worth while to apply similar techniques to the study of *distribution* of glycogen in different regions of the brain. The next task would be the *histo- and cytochemical localization* of glycogen. Is the glycogen concentrated in the neurons or in the neuroglia, or is it found in both structures? It seems that in different areas of the human cortex the cell bodies of neurons (perikaryons) occupy not more than 2.85% of the total volume (Bok, 1936, p. 28). Evidently the bulk is composed of neuroglia, and axons and dendrites make up a smaller proportion. Possible differences and similarities of metabolic activity in neurons and neuroglia were discussed by Gerard (1950, p. 203). In view of intimate topographic relations between neuroglia and neurons, the present histochemical methods are hardly sensitive enough to localize glycogen in the two structures. Kerr (1938) pointed out that the glycogen concentration in the whole mammalian brain was found to be less than 0.15% whereas the minimum histochemically detectable glycogen in the liver was, at that time, 0.3% of the wet weight (Nielsen, Okkels, and Stochholm-Borresen, 1932). Subsequent publications give a wide range of minima of histochemically detectable glycogen, varying from 0.02% in some liver samples (Deane *et al.,* 1946) to 0.18% in heart muscle (Kugler and Wilkinson, 1959). The lowest values (0.02%) in the liver may not mean much since other liver samples with 0.08% chemically determined glycogen were negative histochemically (Deane *et al.*).

I do not know of any attempts to study the *brain in situ* for distribution of glycogen in different areas or different histological components. In a recent study by Chain, Larson, and Pocchiari (1960), *isolated slices* of rabbit brain were placed in Warburg flasks containing C^{14}-labeled glucose. After 60 minutes' incubation the glucose disappearing from the medium was accounted for as lactic acid, CO_2, and amino acids. Samples from different regions showed qualitatively similar metabolic patterns but some quantitative differences. For instance, the hypothalamus formed more γ-aminobutyric acid from glucose than the cerebral and cerebellar cortex.

Suppose it remains impossible to increase the sensitivity of histo- and cytochemical methods for glycogen determination. Then, one may consider separate *chemical* determination in neurons and neuroglia after microdissection and separate collection of the two structures in amounts sufficient for analysis. This avenue has been opened by the studies of Lowry, Robert, and Chang (1956). In frozen-dried sections 20 to 25 μ thick, nerve cell bodies were sufficiently visible to be dissected out with fine metal needles. The activity of three enzymes was determined in single cells, and samples from different areas of the brain were compared. To ascertain the reliability of the method, brain homogenates equaling the amounts of a cell body were analyzed. E. Robins (1957) combined these techniques with the quantitative histochemical method of Linderstrøm-Lang and Holter which will be described later. It was possible to measure in single cells several enzymes, including hexokinase, and lipids such as cephalins.

Histochemical techniques were used by Montagna (1949) in his studies of stored glycogen and lipid droplets in the cartilage of the human trachea, epiglottis, external auditory duct, and intervertebral discs. Montagna *et al.* (1951) studied the histochemical distribution of glycogen and lipids in the following components of the human skin (from the mons pubis of males): epidermis, ducts of sebaceous glands, different parts of hair, and associated structures. Continuation of these interesting investigations in parallel with chemical analyses seems desirable and also feasible.

Finally I wish to mention certain relations between carbohydrate metabolism and mitotic activity. In some healthy young rats all liver cord cells were found to be filled with glycogen granules except those cells which were in mitosis (E. Mayer, unpublished data, 1950). J. W. Wilson (1948) observed in the livers of young mice a parallel rise of mitotic activity and of glycogen storage through the night to a maximum between 5 A.M. and 9 A.M. and a subsequent decrease to a minimum between 5 P.M. and 9 P.M. The glycogen was estimated histochemically; Wilson did not indicate whether glycogen was present in dividing cells. Bullough (1949) found that the well-known diurnal variations in mitotic activity in the skin of mice were paralleled by varying glucose concentrations in their blood. In another paper by Bullough (1952) there is an extensive discussion of the relation between carbohydrate metabolism and mitotic activity in the liver. It seems to me that the following combination of procedures would be worthwhile. Glycogen in livers of rats should be determined quantitatively (chemically) under different metabolic conditions, and slices of the same livers should be studied histochemically for distribution of glycogen in dividing and nondividing liver cells. The first phase of such studies should establish the relations between normal livers of adult rats with very low mitotic frequency, and regenerating liver with high mitotic frequency. Since direct surgical reduction of the liver may produce unknown injury, one may prefer the liver regeneration technique of Bucher, Scott, and Aub (1951). As described previously, these investigators removed large parts of the liver in one partner of parabiotic rats, whereupon increased mitotic activity occurred in the intact liver of the other partner. The increase in mitotic rate over the controls was sixfold when parabiotic twins were used, and fiftyfold with parabiotic triplets. For details see Part III, Ch. 4, B.

2. *Histochemistry of Fatty Substances*

In human pathology the amount of visible lipids in the cells of renal tubules has been of considerable interest. Formalin-fixed frozen sections of human kidneys showed very large variability in the amount of intracellular droplets which stained with Sudan III or other fat stains. To what extent is the amount of Sudan-positive droplets in a particular kidney parallel with the total lipids which can be extracted from this kidney? I will briefly report my own studies of this question (E. Mayer, unpublished data, 1922). Kidneys of thirty autopsied patients were examined. Samples of the cortex were fixed for sectioning and staining. The bulk of the fresh cortex was cut into small pieces, and the wet weight of the material was determined. After being minced into small fragments, the material was dried to weight constancy, ground with sand, and extracted with ether in a Soxhlet apparatus. The dry extract was expressed as percentage of wet and dry weight of the cortex. The results can be summarized as follows. The ether-extractable material varied from 10% to 50% of the dry substance. The amount of Sudan-positive droplets in the cells of the cortical tubules varied from a minimum, consisting of very few droplets in a small number of tubules, to a maximum showing every cell stuffed with red drops to such an extent that no fat-free cytoplasm was visible. The correlation between the Sudan-stained material in sections and the ether-extracted material was as follows. All kidneys in which the ether-extracted material was 30 to 50% of the dry substance showed substantial amounts of Sudan-positive drops in the great majority of cells of the cortical tubules. Below 30% ether extract, there was no definite correlation with the amount of Sudan-stained material. However, if the amount of Sudan-positive drops was extremely small, the ether extract rarely exceeded 15% of the dry substance. Most kidneys with 30% ether extract or more were from patients who died from liver insufficiency ("acute yellow atrophy of the liver") or from untreated diabetes mellitus. In both conditions the concentration of fatty substances in the blood is abnormally high (lipemia); and a certain proportion of the circulating fatty material is stored by the kidneys (for documentation of these relations, see Mayer, 1922). Characteristic photomicrographs of human kidneys in which the tubular cells of the cortex are stuffed with Sudan-positive drops are found in a paper by Lucké and Mallory (1946) on epidemic hepatitis (their Figs. 31–33).

In healthy dogs special segments of the renal tubules are filled with Sudan-positive drops, while in cats other segments of the tubules show this condition. Comparisons with chemical analysis are not very meaningful in these cases.

The question has been raised whether the total ether-extractable material on the one hand, and the total Sudan-stainable material on the other hand, have much biological significance, since both materials represent variable mixtures of lipids. In the light of current research it appears more important to correlate quantitative and histochemical determinations of cholesterol rather than of neutral fats. It is interesting to compare the present state of these problems with their state in 1925 when the Deutsche Pathologische Gesellschaft, with competent bio- and histochemists as speakers, devoted a special meeting to cholesterol metabolism. Many of them expressed a critical attitude (a) toward quantification, or even semiquantitative estimation, of cholesterol in histological sections, and (b) toward any attempts to use histochemical stains for identification of those mixtures of lipids which occur in tissues and cells.

In a comprehensive study of these questions Kaufmann and Lehmann (1926) soaked frozen sections of elder pith with single fatty substances or numerous combinations. They found the different staining methods for fatty acids, for phospholipids and cerebrosides, and for cholesterol tolerably useful in pure samples but practically useless in mixtures. Since these discouraging results were obtained in a well defined model, one could not expect much information from application of these stains to complex histological material. The conclusion of Kaufmann and Lehmann is that positive Sudan stain proves the presence of *some* fatty substances, and is therefore useful, whereas qualitative and quantitative identification of fatty substances should not be attempted by histochemical procedures but should be left for chemical analysis. According to Lison (1953, pp. 370-371) some of the results of Kaufmann and Lehmann have been modified by subsequent investigators. It seems to me that the pessimistic views concerning the histochemistry of fatty substances are still justified.

3. *Inorganic Deposits in the Lungs*

Inorganic material may be localized in intact microscopic sections by color reactions, as in the case of iron deposits, or by solubility tests which indicate some calcium salts, or by microincineration of sections with subsequent chemical analysis. In order to determine the absolute quantity of carbon and silica, or their percentage per dry or wet weight of lung tissue, incineration of the lungs and chemical analysis of the incinerated lungs is necessary. As an example of such studies I mention recent publications by Rivers, Wise, King, and Nagelschmidt

(1960) and Rivers and Wise (1960). These investigators compared chest roentgenograms obtained from coal workers within two years of death with the results of chemical analysis of the lungs after death. The substances determined were carbon, quartz, mica, and kaolin. The result was that the mineral and carbon contributed equally to the radiological changes, but weight for weight, mineral contributed about nine times more than carbon. This relation was ascribed to difference in relative absorption of Roentgen rays by carbon and mineral. Large histological sections of the lungs made with the technique described previously (Part III, Chapter 2, B, 3, c) were used to *grade the responses of the lung tissue* to the dust deposits. No attempt was made to rank the amount of dust in these sections. Estimates of dust in sections are difficult because of the mixture of impressive black particles and less conspicuous particles which are not black (personal communication from Dr. D. Rivers, 1960). In their 1945 paper Haythorn and Taylor applied microincineration to sections of silicotic human lungs, but no estimates of quantities in sections were mentioned. Data obtained by chemical analysis of the same lungs were given in detail. In summarizing one may say that any comparison between histochemical and chemical determinations of dust deposits in the lungs of miners would suffer from the uncertainties on the histochemical side. At the present time there is no practical or academic incentive for making such comparisons.

As an illustration of parallel histological and chemical determinations of inorganic deposits in the lungs I report a study from experimental pathology. In ordinary histological sections of the lungs of some dogs, opaque masses were found in the thickened walls of the pulmonary alveoli. The masses stained intensively blue with hematoxylin and were to some extent, but not completely, soluble in hydrochloric acid. Hematoxylin-stained sections are illustrated in Fig. 55a, b. These properties suggested that the masses were, or contained, calcium salts. Additional staining reactions favored this interpretation, but definite proof could not be obtained by histochemical tests. Microincineration or spectrography of sparks could have proved the presence of calcium in a number of spots, but would not have allowed determination of calcium concentration in the lungs. Therefore large samples of the lungs preserved in 70% alcohol (in which calcium salts are not soluble) were dried to weight constancy and their calcium content was determined chemically. The same comparison was made with the lungs of a dog which showed histologically a minimum of amorphous masses, as shown in Fig. 56a, b. The lungs with the numerous deposits (Fig. 55a, b) contained 19 grams of calcium per kilogram of dry substance, whereas the lungs with the scarce deposits (Fig. 56a, b) contained only 0.55 grams of calcium per kilogram dry weight.[4] The large amount of calcium deposits in the lungs shown in Fig. 55a and b was probably the result of excessive ad-

[4] Chemical analysis under the direction of Dr. P. R. Averell, Dr. S. B. Davis, and Mr. R. P. Smith at the Stamford Research Laboratories of the American Cyanamid Company.

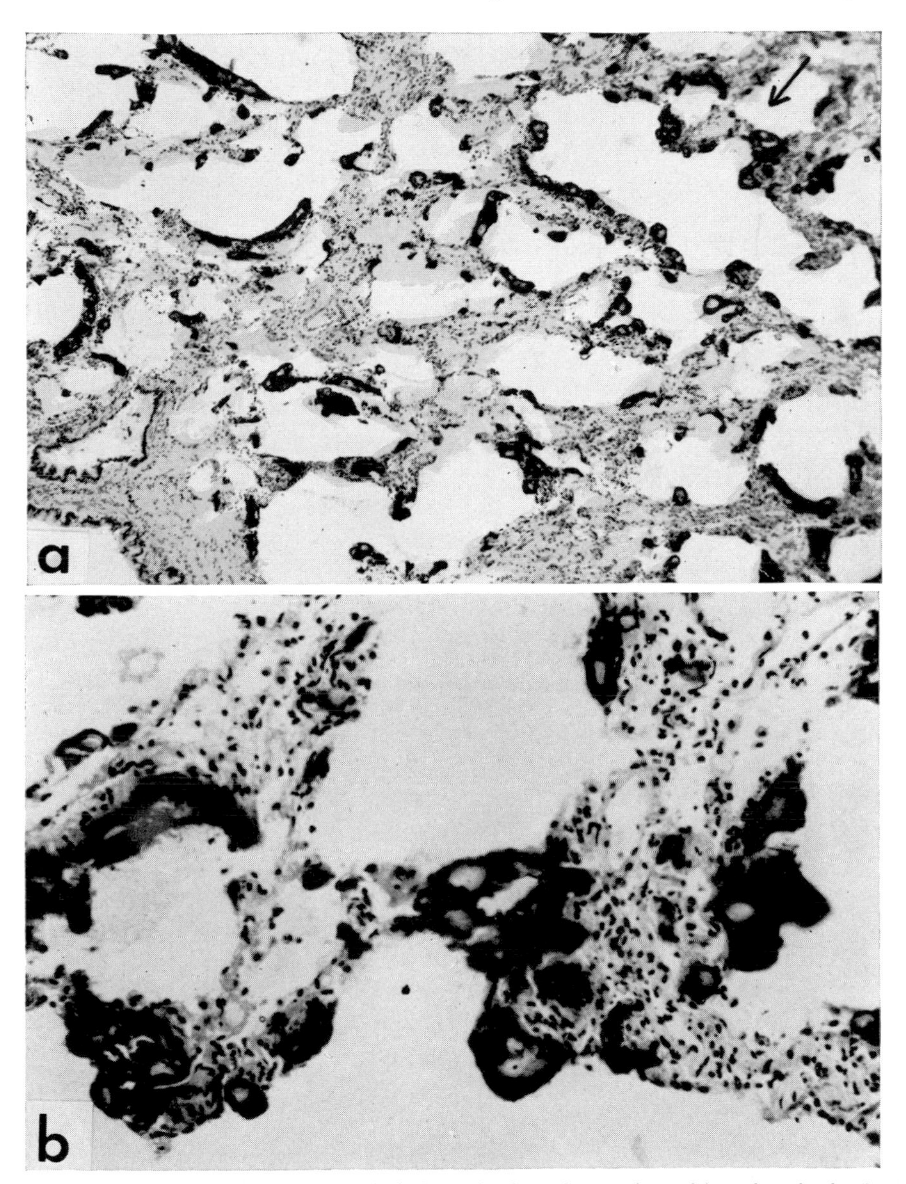

FIG. 55. Histochemical and chemical determination of excessive calcium deposits in the lungs of a dog. Lung section from 3-year-old male beagle (P 27) which as a puppy had received an overdose of vitamin D for 3 weeks. Black masses represent calcium-containing precipitates stained with hematoxylin. Alveolar walls thickened by connective tissue, alveolar spaces unusually wide, probably prevented from collapsing by the thick alveolar walls and the rigid precipitates. Fixation in 70% alcohol, paraffin section, 6 μ thick; stained with hematoxylin; no application of acid at any time. Chemically determined calcium concentration: 19 grams of calcium per kilogram of dry substance of lung. (a) Survey picture at magnification $\times 50$. (b) Area of (a) indicated by arrow at higher magnification $\times 200$.

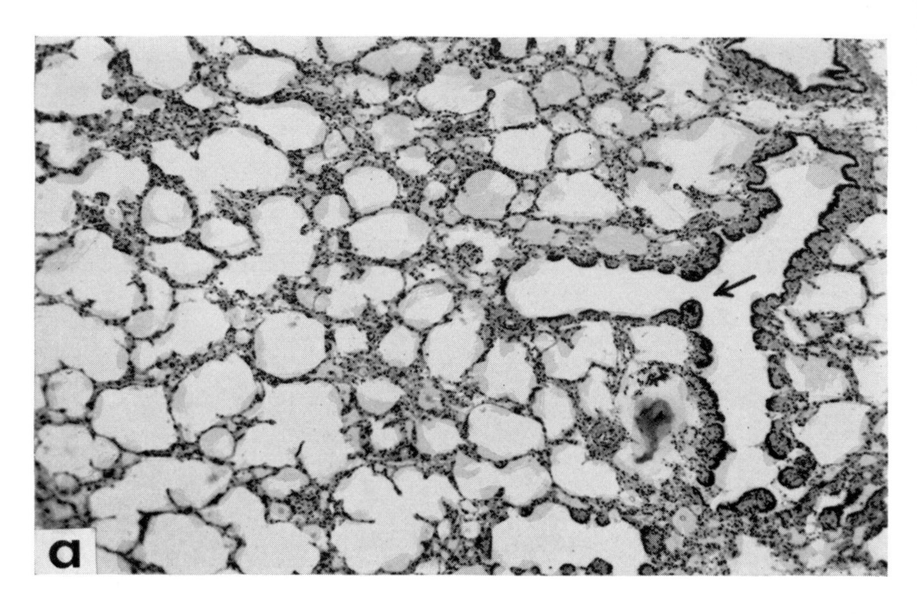

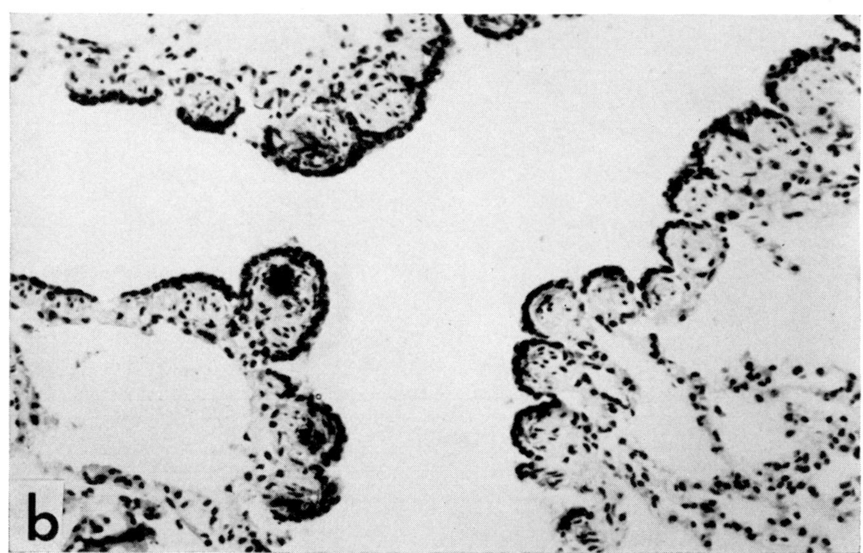

FIG. 56. Section from normal lung of 3-year-old male beagle (M 78) which as a puppy received the ordinary dose of vitamin D. Very few amorphous deposits were found in the lung which would possibly be interpreted as calcium salts stained with hematoxylin. Arrow points at one of the deposits. Alveolar walls are delicate (normal), alveolar spaces moderately wide as one would expect in a lung partially collapsed after opening of the chest. Histological technique as in Fig. 55. Chemically determined calcium concentration: 0.55 gram of calcium per kilogram of dry substance of lung. (a) Survey picture at magnification $\times 50$. (b) Area of (a) indicated by arrow at higher magnification $\times 200$.

ministration of vitamin D which the dog received as a puppy, two and one-half years prior to its death. This interpretation is supported by observations such as those of Hendricks, Morgan, and Freytag (1947) on irreversible retention of calcium in the lungs of dogs as part of hypervitaminosis D.

4. *Enzyme Determinations, Quantitative and Semiquantitative. Correlations with Physiological and Embryological Conditions*

The studies of Linderstrøm-Lang, Holter, and their associates on the *distribution of enzymes in the stomach* of the pig are the classic example of quantitative histochemistry of enzymes. It had been known for a long time that the mucous membrane of the stomach secretes digestive enzymes, particularly pepsin. From the surface of the mucous membrane tubular structures project into the deeper strata. These tubules are known as the gastric glands. On sections vertical to the surface they roughly resemble the crypts of the jejunum shown in my Fig. 33. The wall of each gastric gland consists of two morphologically distinct cell types, the chief cells and the parietal cells. In the so-called fundus area of the pig stomach the frequency of the parietal cells decreases from the neck of each gland, which is near the surface, to the bottom. Accordingly, the number of chief cells increases toward the bottom. The traditional assumption was that pepsin is produced by the chief cells. In order to prove or disprove this assumption, Linderstrøm-Lang, Holter, and Søeborg Ohlsen (1934) developed their cold microtome technique (see Part III, Chapter 2, B, 3, c).

Cylindrical blocks approximately 2 mm. in diameter were cut from the wall of the stomach so that the axis of the cylinder was perpendicular to the surface of the mucous membrane. In a cryostat, frozen sections 10 to 25 μ in thickness were made from the unfixed block. Alternately one section was fixed and stained to permit cell counts, and the next one left unstained to determine pepsin activity by measuring the rate of hydrolysis of an added protein substrate such as edestin. Two types of graphs were prepared. On the abscissa of each, the distance of the section from the surface of the mucous membrane was indicated. On the ordinate of one graph the changing frequency of chief cells was marked, while the ordinate of the other graph represented changing enzymatic activity. The shapes of the two resulting curves were strikingly similar. The conclusion was that the chief cells were indeed the producers of pepsin. A convenient summary of this work, with a histological diagram and representative graphs, is found in Linderstrøm-Lang's 1939 Harvey Lecture. As mentioned earlier O. H. Lowry (1953) and E. Robins (1957) adapted the methods of Linderstrøm-Lang and Holter to measurements of enzymes and other substances in single cells of the brain.

After this brief report on joint application of histochemical and chemical procedures, let us consider some investigations of the secretion of the stomach in which histochemical data on enzymes were correlated with physiological data obtained by stimulation of the vagus nerve or histamine administration. Bowie and Vineberg (1935) used anesthetized dogs for their study. A fistula in the stomach allowed the collecting of gastric juice. A sample for histological study of the mucous membrane was removed before experimental stimulation, and another sample at the termination of the experiment. A special fixation and staining procedure for the presentation of pepsinogen granules in sections was designed by the authors. During stimulations of the vagi by induction currents the pepsin concentration in the gastric juice increased. After prolonged rhythmic stimulation of the vagi the histological sections showed disappearance of the pepsinogen granules from the chief cells. In other experiments the stomach was stimulated by repeated subcutaneous administration of histamine. After this treatment there was no visible decrease in the amount of pepsinogen granules in the chief cells. The gastric juice produced upon histamine stimulation was copious and highly acid, but extremely low in pepsin. The physiological data were represented in graphs, while the histochemical differences were illustrated by photomicrographs of samples before and after experimental stimulation. It seems to me that Bowie and Vineberger have demonstrated a satisfactory parallelism between histochemical and physiological data, although only the physiological data were quantitative.

Enzyme development in the central nervous system was included in L. Flexner's (1950, 1955) timetable of morphological, functional, and chemical differentiation in the brain of guinea pigs (see Part II, Chapter 2, E). Special investigations were devoted to the *distribution of cholinesterase in the central nervous system.* Since acetylcholine is discharged by functioning synapses, the removal of its excess requires proper amounts of cholinesterase. The distribution of this enzyme in the adult central nervous system of amphibians was examined by Shen (1958), both with histochemical estimates (Koelle's 1951 technique) and quantitative determination on homogenates. Satisfactory correlation of the two methods was found. The cytochemical pictures revealed that the enzyme was concentrated in the so-called synaptic centers, in which interneuronal synapses are the main components. During the embryonic development of the brain, the first appearance of the enzyme is synchronous with the *morphological* differentiation of synapses in each area. If the morphological differentiation of a synaptic center is disturbed experimentally by cutting its nerve fibers at some distance from the center, no cholinesterase appears in the area of that particular center. Correlation between cholinesterase development and *functional* differentiation of the nervous system was demonstrated by Boell and Shen (1950) and summarized by Shen (1955). In the *Amblystoma* (axolotl) embryo the motor behavior shows that neural differentiation begins in the spinal cord and gradually progresses in a cranial direction to medulla, mesencephalon, and finally hemispheres. The appearance of cholinesterase proved to proceed in the same sequence. The first measurable amount of the enzyme was found in the spinal cord when the animals first become responsive to neural stimulations. Significant enzyme activity appeared in the hindbrain at the time when the animals first exhibit spontaneous swinging movements. At the time when cholinesterase became detectable in the midbrain, coordinated and sustained swimming started. Whereas cholinesterase showed a definite differential and sequential pattern of accumulation in different regions of the central nervous system, the respiratory activity and the respiratory enzymes, cytochrome and succinic oxidase, were found to be practically identical in all parts of the brain.

I mentioned previously Shen's statement that the precise distribution of enzymes on the cellular and subcellular level seems to be a more urgent question than absolute quantitative determinations which cannot be related to localizations on those levels. In his 1958 paper Shen followed the distribution pattern of cholinesterase in the chick retina during embryonic development. The first trace of cholinesterase was found on the surface of the ganglion cells which are the first neuronal elements which undergo differentiation. For details of the development in the retina and for interesting comparisons with cholinesterase localization during the development of the myoneural junctions I refer to the original paper.

E. Histochemistry and Laboratory Practice

It depends on the tasks of a service laboratory and on the special projects of a research laboratory what histochemical techniques should be used. Older methods which are still important are the Sudan stains for fatty substances, the iron reactions, the different glycogen stains, and the oxidase and peroxidase reactions for white blood cells. Among modern histochemical reactions which are widely used I mention the Feulgen stain for deoxyribonucleic acid, the periodic acid-Schiff (PAS) reaction for mucopolysaccharides, and the different reactions for acid and alkaline phosphatases. Fixatives, embedding media, and sectioning techniques must be planned so that the intended stains can be applied efficiently (Part III, Chapter 3, B). This is particularly important with respect to some histochemical and cytochemical procedures such as enzyme reactions. Again the reader is referred to Lillie's 1954 book. His Table 18 shows which enzymes require study in unfixed frozen sections and which tolerate fixation. The fixatives listed as necessary for the preservation of the different enzymes are acetone or alcohol or formalin, respectively. I also described the addition of specific substrates for enzyme studies in histological sections. It is important to remember that some very useful histological procedures do not allow any histo- or cytochemical interpretation. A good example is osmium tetroxide, which is an indispensable fixative in electron microscopy, but does not give much information on the chemical composition of the structures seen. Porter and Kallman's (1953) attempt at analyzing the mode of action of OsO_4 was reported earlier (Part III, Chapter 7, B, 1, c).

Before concluding this chapter it might be useful to point out a few practical aspects of histochemistry which are related not to recently developed techniques, but to older procedures.

1. *Histochemical Interpretations Derived from Unspecific Procedures*

A paraffin section from material fixed in formalin, Bouin's solution, or other routine fixative, stained with hematoxylin and eosin, and mounted in Clarite allows the following histochemical interpretations. Fatty substances were removed by the fat solvents used for dehydration, and glycogen does not stain with either hematoxylin or eosin. Consequently, practically all structures which are stained can be considered to be proteins or nucleoproteins. Unstained spaces in cells, so called vacuoles, are suggestive either of dissolved fat or glycogen, depending on the shape and location of the vacuoles and the known properties of the particular organ. In sections of the liver and adrenal cortex, fat vacuoles appear circular, with distinct outlines of cytoplasm around each vacuole, while the unstained spaces which correspond to glycogen deposits are not necessarily clearly separated. If cells are stuffed with glycogen, the outline of each cell becomes unusually distinct, which produces a resemblance to plant cells with cellulose walls (see Rich, Berthrong, and Bennet, 1950, Fig. 2).

Angular empty spaces in tissues or cells usually suggest crystalline deposits that were dissolved or removed mechanically in the course of the techniques applied. In the pathologically thickened inner layer of an artery (arteriosclerosis) the occurrence of narrow spaces with straight outlines indicates that there had been cholesterol crystals.

Estimates of relative concentrations of proteins can be derived indirectly from the morphological appearance of thyroid colloid in different functional conditions and after the use of different fixatives. This is illustrated in my Figs. 25, 26, and 37. Protein casts in renal tubules are interpreted in a similar way (my Fig. 38). In these studies staining properties of the proteins were utilized, but the staining procedures were not histochemically specific. The sections shown in Figs. 37 and 38 were stained with aniline blue and azocarmine, which are empirical dyes for various structures. In the thyroid colloid, even staining with hematoxylin and eosin can be indicative of differences in protein concentration.

2. *Comments on Iron-Containing Pigments*

Colored granules or droplets which occur in living tissues are known as pigments. They are frequently discovered in stained sections. For their identification, stains such as iron reactions and Sudan stain are helpful, but unstained sections are necessary in order to determine the natural color and solubility of pigment particles. The first tabular key for the identification of pigments was published by W. Hueck in 1921; it is reprinted in Romeis' "Mikroskopische Technik" (1948), and is still useful. Particles which give positive iron reactions are found in the tissues under two different conditions. They may result either from a transformation of hemoglobin after destruction of red blood corpuscles or from an occupational inhalation of iron dust. The two types are known as endogenous and exogenous iron pigmentation. When particles of either type are deposited in cells they are coated or possibly combined with proteins and, therefore, may not necessarily be different chemically. In most cases, however, the appearance of the particles as well as the history and condition of the individual in which they are found will tell whether their origin was endogenous or exogenous.

The present discussion will deal with certain iron-containing substances which are derived from hemoglobin or other sources and are stored in the tissues. One of these substances is known as hemosiderin, and the other as ferritin. Hemosiderin appears in the shape of amorphous brown granules which give direct iron reactions: it is an iron-positive pigment. Ferritin deposits are not directly visible under the light microscope. By treatment of ferritin-containing sections with a cadmium sulfate solution, characteristic brown crystals are produced which can be seen under the light microscope (Granick, 1943). Examination with the electron microscope shows no difference between the units which compose an amorphous granule of hemosiderin and those units which compose a ferritin crystal (Bessis, 1959, Fig. 2A).

There is an extensive literature on the relations between hemosiderin and ferritin. Many investigators assume that several types of hemosiderin exist. Two brief surveys may be mentioned here: Volland and Pribella's paper on the nature and origin of iron pigments (1955), and an article by Gubler (1956) on absorption and metabolism of iron. From a comprehensive review by Drabkin (1951) I summarize the following statements. Ferritin is a protein-colloidal ferric hydroxide-phosphate complex; besides the protein components, the iron-rich micelles have the approximate composition $(FeOOH)8(FeO\text{-}OPO_3H_2)$. Hemosiderin is a colloidal ferric hydroxide of unknown constitution, with as much as 55% of iron admixed with organic matter.[5] Hemosiderin most probably represents tissue iron in excess of the capacity of apoferritin to bind the iron (page 363 of Drabkin's review). Apoferritin is the name given to the protein to which the iron in ferritin is bound. I suggest that for practical purposes, hemosiderin should be characterized by the following two criteria: (1) it gives direct iron reactions, and (2) it is derived from hemoglobin. The first criterion is experimental whereas the second is a matter of interpretation. Ferritin is detected in tissue sections by its crystalline precipitation with cadmium sulfate. Ferritin can be a product of transformation of hemoglobin or a metabolic product of iron absorption through the intestines. It seems that the authors of manuals on histochemistry are reluctant to state how hemosiderin and ferritin are identified in sections.

[5] This value was evidently taken from S. F. Cook's 1929 paper.

In hemoglobin as well as in ferritin the iron is "masked," which means that positive iron reactions are obtained only after destruction of the organic components to which the iron is bound. There are two hemoglobin-derived particulate pigments which contain masked iron: malaria pigment and the so-called formalin pigment. These two pigments are histochemically indistinguishable, although their mode of formation is entirely different. Malaria pigment is a product of metabolic transformation of hemoglobin by malaria parasites in red blood corpuscles. For an early phase of malaria pigment-formation see Fig. 34 (B_1). Formalin pigment is a result of interaction of unbuffered formalin with hemoglobin. Formalin pigment is illustrated in my Fig. 35. Malaria and formalin pigment fail to give direct iron reactions, although their iron content can be shown analytically. Bleaching tests, solubility, and appearance of the two pigments are very similar; a possible difference in the degree of birefringence has been claimed by Hershberger and Lillie (1947).

Chapter 8

Presentation of Procedures, Observations, and Interpretations in Static and Dynamic Morphology

Introduction: Logical Interpretations and Visual Presentations. Demands for reproducibility and quantification of observations arose later in morphology than in physiology, chemistry, and physics for the following reasons. Traditionally it was expected that morphological observations would be presented by elaborate verbal descriptions and realistic drawings or photographs, with as much detail as possible. This is quite different from tabulation, since tabulation is limited by the number of selected criteria. Without tabulation of data, reproducibility cannot be tested and quantification cannot be attempted. Moreover, there is an inherent antagonism between the visual and the mathematical presentation of phenomena. At the Second Congress for Unity of Science (Copenhagen, 1936, transactions published in 1937), mathematicians, physicists, biologists, sociologists, and philosophers tried to establish common principles of investigation. Niels Bohr pointed out (p. 296 of transactions) that rigorous logical connections can be established only at the sacrifice of visualization. As an example he mentioned the imaginary number $\sqrt{-1}$, which is indispensable in equations of modern physics, but cannot be expressed in pictures or picturesque language.

A physicist, chemist, or biochemist may take a dim view of biological publications which contain large numbers of photomicrographs or drawings. He may conclude that morphologists compile only picture books. However, some collections or morphological pictures are evidently valuable for purposes of medical pathology, such as Custer's "Atlas of The Blood and Bone Marrow" (1949) and Papanicolau's "Atlas of Exfoliative Cytology" (1954). Atlases of sections of the central nervous system are essential tools of neurophysiology. Standard pictures made it possible to use number codes for identification of special structures or developmental stages; examples will be given later. Some phenomena of dynamic morphology seem to defy attempts at condensing numerous observations into a few standard pictures. In this class are the variations of capillaries in the nail fold of man which are recorded as collections of individual pictures (see for example, Walls and Buchanan, 1956, quoted in Part III, Chapter 2, A, 1).

It is interesting to notice an analogous situation in organic chemistry. Collections of infrared absorption spectra of numerous organic compounds have been compiled. These curves are consulted for structural analysis, since the infrared spectrum of a compound may indicate the presence or absence of certain groups or linkages. Sets of curves and their interpretations are found in publications such as Gore's 1950 paper on infrared spectra of organic thiophosphates. Some collections of infrared spectra are comprised of 20,000 or more curves. Large compilations of reference graphs have proved to be extremely useful.

A. Verbal and Graphic Techniques

Static biological structures are presented best by means of pictures and diagrams, although translation of pictures into verbal expressions is necessary for the interpretation and discussion of pictorial data. There was no way of describing the axes and planes of the body without diagrams (Fig. 47a and b; man and dog). In dynamic morphology structural changes cannot be discussed without pictures of different phases. It would hardly have been possible to convey to my readers the basic concepts of vertebrate embryology without diagrams of the development of the frog's egg (Fig. 3a–l).

Experimental embolic mouse tuberculosis was described by means of diagrams of the phenomena observed. Tabulation of the diagrams led to reconstruction of their chronological sequence (Fig. 31 and Table 7). It was not feasible to supply pictures for each item discussed in this book, but references are given to pictures available in the literature.

Verbal descriptions of biological structures involve problems of terminology. A brief comparison of principles of terminology in physics, chemistry, and biology will be useful. The preparation of pictures and diagrams in morphological work requires special techniques, depending on the nature of the object and the purpose of the illustration. Therefore a detailed discussion will be devoted to these procedures.

Two useful books appeared with contents closely related to the material of our present chapter: "Guide to Medical Writing" by Davidson (1957) and "The Preparation of Medical Literature" by Cross (1959). The two books and our presentation differ in emphasis, but agree in many technical recommendations.

1. *Verbal Descriptions*

In physics, ambiguity of terms is a minor problem since most terms are defined quantitatively as part of an equation. Mutual quantitative definitions were illustrated earlier by examples such as mass, acceleration, and force, or amount of heat (calories), temperature, and specific heat. An attempt to adapt the idea of mutual definitions to biology was made in Table 2, which shows the applicability of the concepts stimulus, response, and reactivity to many areas of biology (Part II, Chapter 2, C). In chemistry there are different ways of characterizing elements and compounds. Some principles of chemical classification will be discussed in Part V in connection with the problems of biological classifications. At this point let us examine the ways in which *organic* compounds are characterized. An organic insecticide known under the commercial name of DDT has the empirical formula $C_{14}H_9Cl_5$, which indicates how many atoms of each element are present in one molecule, irrespective of their arrangement. The arrangement of the atoms is shown in the structural formula:

$$Cl-C_6H_4-\underset{\displaystyle CCl_3}{\underset{|}{CH}}-C_6H_4-Cl$$

On the basis of the structural formula or "picture" the systematic chemical name is given. The traditional form of the systematic name is 1,1,1-trichloro-2,2-bis(*p*-chlorophenyl)ethane. For indexing purposes the index or "inverted" named is used: Ethane, 1,1,1-trichloro-2,2-bis(*p*-chlorophenyl)-. Finally there is an abbreviated designation known as the generic name: dichlorodiphenyltrichloroethane. Evidently, every effort is made in organic chemistry to characterize a compound in one or more unambiguous ways, and no one expects the abbreviated generic name to tell the full story. There are many compounds in organic chemistry that are much more complicated than DDT, and therefore carry systematic names that are several lines long. Brevity of a name becomes important only at the commercial end.

In biology, and particularly in medicine, the brevity of names seems to be more important than their clarity. Since biological units are supposed to be much more complicated than chemical units, the demand for brief and simple names in biology is certainly paradoxical. The trouble with short biological terms is that they are frequently based on an alleged but not proved correlation of different criteria. Two characteristic examples were mentioned earlier in the book: the term norm and the term growth. If the statistical and teleological norm were always correlated, the term norm alone would suffice. Since such correlation is present in some cases, but absent in others, the specifying adjectives are necessary (see discussion in Part II, Chapter 1, D). Similarly, the term growth would be self-explanatory and useful if increases of the following things were highly correlated: volume, cell number, cell size, intercellular substances, wet weight, dry weight, total protein, and nucleic acids. This subject will be treated in greater detail in connection with the variations of organs and their components (Part V, Chapter 3).

In some areas of biology, efforts have been made to arrive at unambiguous terminologies. The oldest attempts in this direction were devoted to taxonomy and will be discussed in the chapter on classification. International agreements have been reached with respect to the nomenclature of macroscopic anatomical structures of man. In genetics the use of symbols, in conjunction with chromosome maps, permits characterization of genes in a satisfactory way.

Dynamic interpretations of static pictures may be difficult, but it is certainly not easy to describe morphological phenomena without interpretations. In medicolegal pathology it is imperative that observations on a body which was possibly the victim of a crime or accident are given in strictly descriptive language. After the descriptive statements, interpretations are in order. A considerable number of terms in normal and pathological histology are not purely descriptive, but contain various histogenetic and functional admixtures. There is no objection to such combinations if all components are well founded. If some are proved and others hypothetical, confusion results. Descriptions and pictures of elementary biological structures will be given in Part IV, and different criteria for classification of normal and pathological structures will be discussed in Part V.

In various areas of static and dynamic morphology it has been possible to establish useful systems of letters and numbers in conjunction with standardized pictures. As examples I mention the mapping of the cerebral cortex (Figs. 48 and 49) and the labeling of pulmonary segments and bronchi (Fig. 23). Embryological stages of important species are characterized by numbers. Thus it is possible to refer to "Shumway stage 13" of *Rana pipiens* or to "Harrison stage 27" of *Amblystoma;* numbered stages of *Rana pipiens* and *Amblystoma* are pictured in Rugh's manual (1948). The seemingly simple architecture of fibroblast colonies cultivated *in vitro* was the source of considerable confusion until names such as "explant," "zone of growth," "periphery" were replaced by diagrams and numbers (Mayer, 1939).

Sometimes for the same object different systems of standardization have been established by two or more independent investigators. If each system of labeling is well defined, it is not difficult to translate one into the other. In my Fig. 23 the bronchi are labeled by letters and numbers according to Boyden's system. Two other authors have used anatomical names to identify the bronchi and pulmonary segments. A convenient key for translating the three terminologies (Boyden, Jackson-Huber, and Brock) was made by Lindskog and Liebow (1953, their Table 1).

2. *Pictures*

In any biological investigation, the documentary value of pictures depends on proper identification and distinct visibility. This statement may seem trivial. However, there are quite a few publications in which biophysical or biochemical data are quantified and well documented, while the accompanying morphological illustrations do not fulfill the minimum requirements of qualitative presentation. Frequently, morphological pictures are given without records of variability.

a. *Characterization of samples. Identification on macroscopic, light-microscopic and electron-microscopic levels. Insets. Indication of scale.* The importance of continuity between macroscopic and microscopic analysis has been emphasized in earlier chapters (Part III, Chapter 2, B, 1). Such continuity is customarily maintained in studies of the central nervous system. Microscopic pictures of other organs are not always characterized adequately with respect to their topographic origin. Yet, there can be marked differences between the different lobes of the lung or liver, between proximal and distal femoral marrow, or between the head and the tail of the pancreas. As an example of effective presentation of pulmonary conditions I mention Terplan's studies on human tuberculosis. He uses the same plate to show (1) outlines of both lungs and their lobes, with arrows indicating the areas of sampling for microscopy; (2) low magnification photomicrographs of the samples indicated in (1); and (3) high magnification insets from special areas of (2). See Plate 2 of Terplan's 1945 paper.

In many cases the interpretation of structures depends both on the analysis of their detail and on the environment in which they occur. Therefore the systematic identification of an unknown sample (diagnosis) requires a gradual transition from the naked eye inspection of a histological section to microscopic analysis first at low and then at higher magnifications. The continuity between different magnifications can be enhanced by *insets* similar to those used in geography for reference between maps on different scales. A stepwise identification of microscopic structures is given in my diagrams, Fig. 57, A–G.

Figure 57A represents an entire section from an organ of an animal the size of a dog or larger. Since the outline of the section is rectangular, it cannot be a cross section through a whole organ, but must be a piece which has been carved out from an organ. Figure 57A shows the section at a magnification of approximately 15 times. In an average hematoxylin-eosin section, nuclei cannot be identified with certainty at this low magnification. The picture resembles a town map with house blocks and streets. The "streets" are likely to be strands of connective tissue separating specific structures. The "house blocks" could be cross sections through bundles of skeletal muscle, or they could be sections through the acini of the pancreas or of the parotid gland. In the area (*p*) a circular structure is seen which could be a cross section through a duct or a blood-vessel. In the area (*q*) an indistinct oval mass seems to be different from surrounding structures.

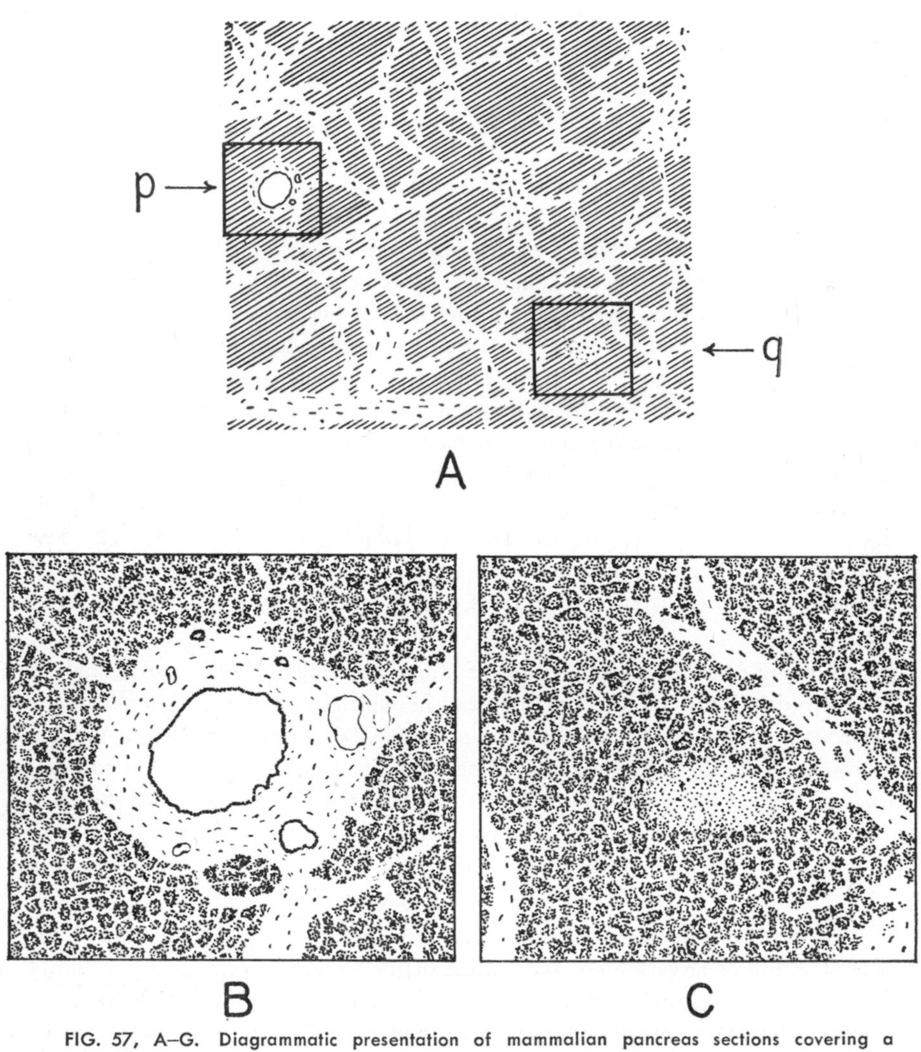

FIG. 57, A–G. Diagrammatic presentation of mammalian pancreas sections covering a wide range of magnification and resolution, from low power light microscopy (A) to electron microscopy (G). All magnifications are approximate.

FIG. 57 (A). Survey picture at magnification ×15. White strands suggest connective tissue, dividing the organ into lobules (represented as hatched areas.) Nuclei and acinar structures not recognizable. Insets *p* and *q* appear at higher magnifications in (B) and (C).

FIG. 57 (B and C). Insets of (A) at magnification ×50. Nuclei can be recognized, and their grouping suggests acinar pattern. (B) (inset *p*): ducts with thick lining and blood vessels with thin lining. (C) (inset *q*): an oval area without acinar pattern, suggesting an islet of Langerhans.

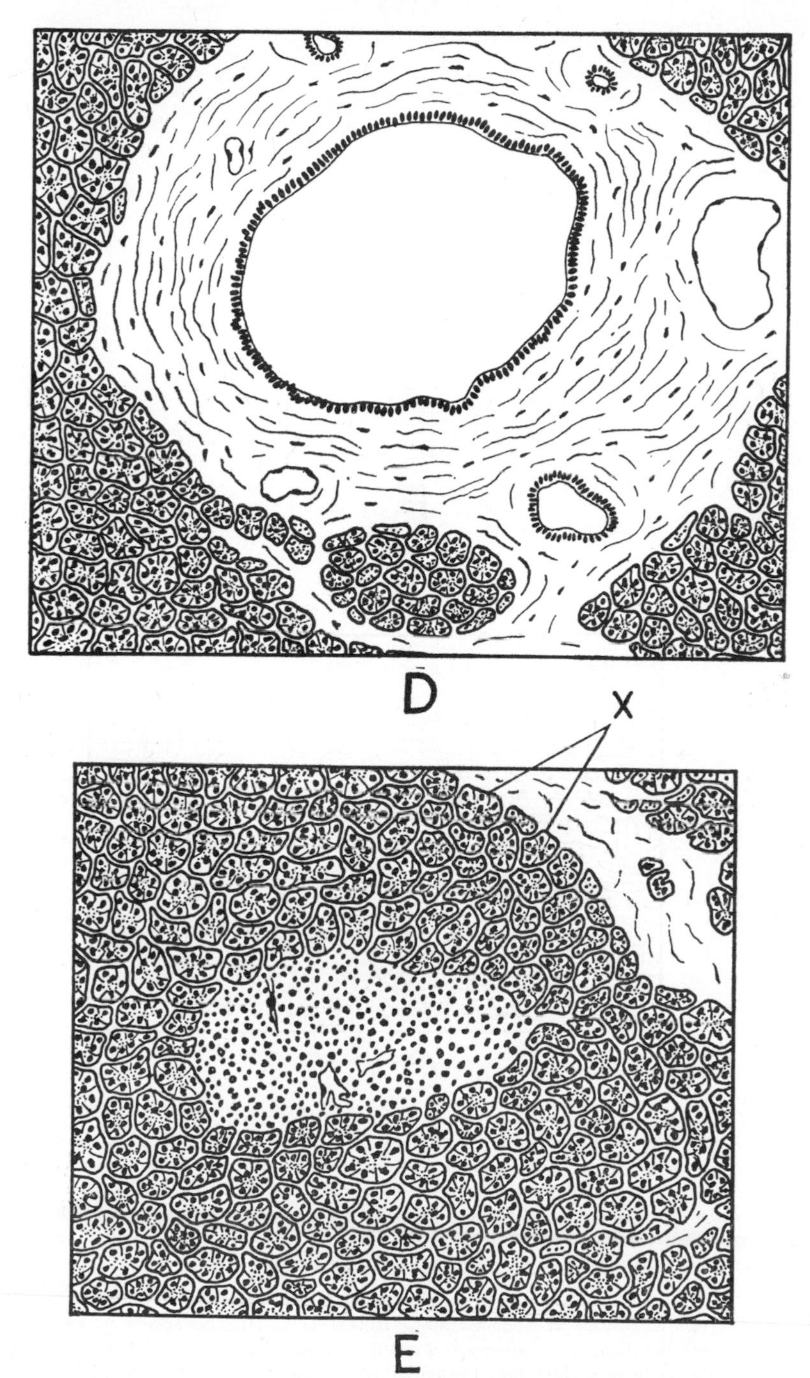

FIG. 57 (D and E). Central parts of (B) and (C) at magnification ×150. Nuclei and acini distinct; in (E), two acini are marked by X. Difference between lining of ducts and of blood vessels conspicuous in (D). Identification of islet of Langerhans confirmed in (E).

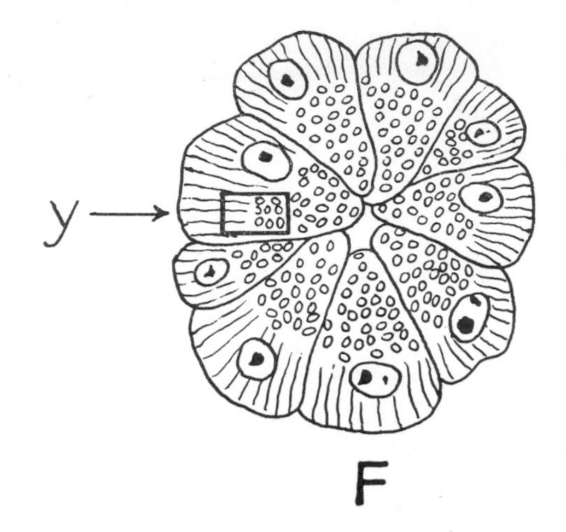

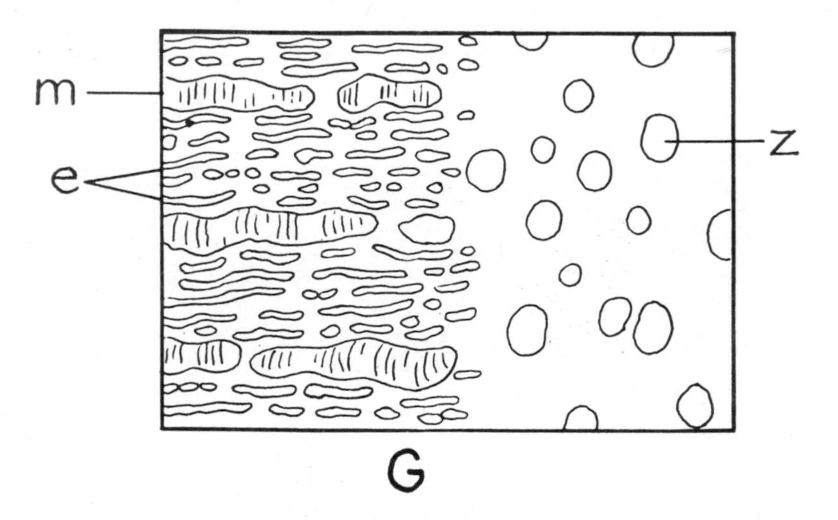

FIG. 57 (F). One acinus showing 9 cells in this plane of sectioning, at magnification ×1000. In this range is the highest magnification of light microscopy (oil immersion) which gives good resolution. Oval structures with dark dots are vesicular nuclei with nucleoli. Radial striation in the basal zone of each cell represents mitochondria. Small circles in the apical zone of each cell are zymogen granules. Inset *y* indicates area used for comparison with electron microscopic picture in an adjacent section (G).

FIG. 57 (G). Transition of basal and apical zone of a pancreatic exocrine (acinar) cell, seen under the electron microscope at magnification ×10,000. *m* = Mitochondria; e = endoplasmic reticulum (double lamellae); *z* = zymogen granules. These structures are not drawn to scale: *m* and *z* should be three or four times larger as compared to e. Comparison with inset *y* of adjacent section (F) shows that the e structures are not visible with light microscopy. *Note:* Structures shown in Fig. 57G were studied mostly in the guinea pig pancreas.

Figures 57B and C show the areas (*p*) and (*q*) of Fig. 57A at a magnification of approximately 50 times. Nuclei can be recognized as dots, and their grouping suggests acini of a gland such as the pancreas. In Fig. 57B the circular structure which was noticed in area (*p*) of Fig. 57A proves to have a ribbonlike lining with many nuclei, suggesting a duct. Similar structures of smaller caliber are visible, and also spaces with a shallow lining suggesting blood vessels. In Fig. 57C the vague oval mass which was seen in area (*q*) of Fig. 57A now appears distinctly different from the surrounding acinar structures: it suggests an islet of the pancreas. Figures 57D and E show parts of Figs. 57B and C at a magnification of approximately 150 times, confirming the tentative identifications. The lining of ducts and blood vessels are unmistakable in Fig. 57D. The oval structure in Fig. 57E is definitely a pancreatic islet, very different from the acini; two acini are marked by X.

Figure 57F is a diagram of one acinus, shown at a magnification of approximately 1000 times. The limit of light-microscopic resolution is usually at magnifications from 1200 to 1500 times. Enlargements made from negatives taken at such magnifications do not show finer structures, but can be convenient for printing and lettering. In Fig. 57F the radial lines in the basal part of each cell stand for mitochondria, while the little circles in the apical part of each cell indicate zymogen granules. The last picture, Fig. 57G, is a diagram of an electron micrograph which might have been obtained from a section adjacent to the section used in Fig. 57F. The area seen in Fig. 57G corresponds with the inset (*y*) of one of the acinar cells seen in Fig. 57F. The zymogen granules (*z*) appear very large now, and three of the delicate mitochondria of Fig. 57F form massive cross-striated structures (*m*). The spaces between the mitochondria show delicate structures with double outlines (*e*). These structures which are not visible under the light microscope were described originally as endoplasmic reticulum (Porter, 1953) and are also known as double lamellae or α-cytomembranes (see Sjöstrand, 1956). I decided to draw the diagram Fig. 57G on the basis of a magnification of not more than 10,000 times. As a consequence it was not possible to draw all structures in Fig. 57G to scale. The mitochondria and zymogen granules should be three to four times larger as compared to the endoplasmic reticulum. A compromise was made in order to ensure the continuity between all diagrams of Fig. 57. The original electron micrograph used for my diagram Fig. 57G shows the different structures of a small portion of an acinar cell of the guinea pig pancreas at a magnification of 26,000 times, occupying an area of 20 $\times$ 12 cm. in Plate 170 of Palade and Siekevitz (1956). Evidences that the endoplasmic reticulum or cytoplasmic membranes exist in the living cells were recounted previously (Part III, Chapter 3, C, 2, c).

Continuity between different magnifications is maintained if alternating sections are cut from a tissue block for light and electron microscopy so that adjacent planes of the same cell or nucleus are available for study with both methods. An instructive series of such paired pictures of nuclei is found in a paper by Bloch, Morgan, Godman, Howe, and Rose (1957) on intranuclear crystalline aggregates of adenovirus. The electron micrographs are magnifications of 9000 to 14,000 times, and the (enlarged) light photomicrographs approximately 3000 to 4000 times. The virus particles are visible only under the electron microscope. In a study of this type, light photomicrographs at lower magnifications are unnecessary.

In some investigations it is desirable to take electron micrographs from many or all areas of a structure and to maintain the topographic relations between the different electron micrographs. For instance, the identification of different cell types in gastric glands greatly depends on their location. To solve this problem the electron micrographs are pasted together, while light photomicrographs of adjacent sections serve as guides for the proper arrangement of the montage (see, Lillibridge, 1960). The montage technique was introduced into electron microscopy by Porter, Claude, and Fullam (1945). These investigators studied whole (not sectioned) connective tissue cells which were cultivated *in vitro.* They made a montage of five electron micrographs of one cell and compared this composite with a light photomicrograph of a similar cell.

The indication of scales and magnifications on pictures is essential. It is strange that this requirement is frequently neglected in *macroscopic* illustrations. In the endocrine literature one can find pictures of two men, who are very different in size, photographed side by side. The taller man could be normal and the shorter one a dwarf, or the shorter one could be normal and the taller one a giant. Usually the legends indicate which of the two alternatives is meant, but the pictures would be more convincing if a yardstick or at least a chair were part of the photograph. In histological courses it is not always easy to convey to the students the idea of orders of magnitude on microscopic levels. Beginners are inclined to interpret the outlines of vesicular nuclei as cell outlines and to assume that each nucleolus represents a nucleus. Such error would be avoided if it were customary for the students to use familiar structures as yardsticks which are present in most histological sections, e.g., red blood corpuscles.

b. *Diagrams versus realistic pictures.* In contrast to a realistic picture the diagram of a biological object is a selective presentation. Depending on the purpose of the diagram, some structures of the object are shown (or accentuated) and others are omitted (or toned down). Features of different objects may be combined in one diagram. Finally, interpretations or assumptions may also be expressed in a diagram.

Although there has never been any doubt concerning the place of diagrams in morphological research, the textbooks of morphology show interesting differences with respect to the use of diagrams. At all times diagrams have played a substantial role in textbooks of botany, zoology, embryology, and macroscopic human anatomy. In normal histology and pathology the early woodcut illustrations could not be very realistic for technical reasons. With the advent of the camera lucida which facilitated the drawing of microscopic preparations; with the development of better printing techniques and, finally, with the general availability of photomicrography, the textbooks of histology and pathology were illustrated predominantly by realistic pictures, sometimes to the complete exclusion of diagrams.

In recent years this trend has reversed itself. Increased emphasis on dynamic morphology made the use of diagrams imperative. Besides realistic pictures, numerous diagrams are found in modern textbooks such as Ham's "Textbook of Histology" (3rd ed., 1957), Hueck's "Morphologische Pathologie" (2nd ed., 1948), and Florey's "General Pathology" (2nd ed., 1958). In "Principles of Pathology" by Hopps (1959) *all* illustrations are diagrams. Three-dimensional semidiagrammatic pictures were introduced into textbooks of histology by Braus (1924) with reconstructions made by this artist, A. Vierling. Their picture of the relations between liver cord cells, blood vessels and bile capillaries has been reprinted in many textbooks.

In the present book three-dimensional diagrams are shown at various occasions, e.g., the layers of the abdominal wall of the rabbit (Fig. 24a) and the three-dimensional analysis of a multinucleated giant cell (Fig. 54b). The methods of three-dimensional analysis were discussed previously (Part III, Chapter 6, A, 4 and Chapter 6, B). I mentioned the areas in which three-dimensional analysis is of obvious importance, namely, embryology, regional and surgical anatomy, roentgenology, and the dynamic morphology of the central nervous system. Physical reconstructions were probably made first in the field of embryology. A description of the wax-plate reconstruction technique was given in an earlier chapter. Colored wax models called moulages are frequently made to present skin diseases, gangrena of extremities, and other changes of the surface of the body.

The usefulness and limitation of models were discussed extensively in an earlier chapter (Part III, Chapter 5). Both models and diagrams are simplified presentations of natural structures or processes. Similar to the standards for models, the value of a diagram can be appraised on the basis of three questions. (1) Is the diagram compatible with observed data? (2) Are observations and interpretations clearly separated in the diagram? (3) Does the diagram serve the purpose for which it was designed?

c. *Drawings versus photographs. Outline drawings as supplements to photographs.* It is usually assumed that important features are more readily emphasized in drawings and that photographs have greater documentary value. However, the documentary value of photographs as compared to drawings has been overrated. If an investigator decides to publish only those pictures which support what he wants to prove, he can select "favorable fields" for photomicrographs as well as for drawings. The possibility of retouching constitutes another element of uncertainty in photography.

Legitimate accentuation of some structures and attenuation of others can be achieved not only in drawings but also in photomicrographs. If, for instance, in a histological section the nuclei are stained red and the collagenous fibers blue, black-white photographs can be taken with different *filters:* the red nuclei can be accentuated by a green filter, and the blue fibers by a red filter. In publications, the author should indicate any accentuation of structures made by selective filters. In his book on photomicrography Shillaber (1944) illustrated how different *developers* were used with the same type of films in order to enhance specific effects (his Figs. 211 and 212).

My Fig. 57, A to G, stresses the importance of *different magnifications* for the identification of microscopic structures. In publications, the exclusive use of high magnifications frequently detracts from the documentary value of illustrations. If only high magnifications are shown, one cannot tell whether the structures in the particular picture were seen in many fields of vision, or only in the field chosen for illustration. Again, this holds for both photographs and drawings.

The main technical advantage of drawings over photomicrographs is connected with the problems of focal depth which were discussed earlier (Part III, Chapter 6). Diagrams of a mesentery, Figs. 51b and 52, indicate that, at high magnifications, the nuclei of the surface layers and the fibers of the intermediary layer are in different focal planes. Therefore photomicrographs would show either the fibers distinctly and the nuclei blurred, as seen in Fig. 51b, or the reverse. When making a drawing of the same preparation, one may change the focus continuously by using the micrometer. Thus a picture can be obtained which is a composite of different focal planes and, therefore, shows nuclei and fibers with equal distinctness. Whenever three-dimensional interpretations are involved, one may document some focal planes by photomicrographs and support the interpretation by a drawing which follows pertinent structures through different focal planes.

Mitotic divisions as seen in histological sections show some chromosomes of the same nucleus in different planes. Therefore it has been customary to combine photomicrographs of several planes in a composite drawing in which the chromosomes are shaded differently, indicating the planes. In many instances it is useful to present the same microscopic preparation in photographs which contain all details, and in outline drawings which emphasize the important structures. In his monograph on the mitotic cycle Hughes (1952) published 15 pairs of plates, one of each pair containing photomicrographs (mostly phase contrast) and the partner containing drawings of the important structures identified by lettering. In a book on the cytologic diagnosis of cancer (by the staff of the Vincent Memorial Laboratory, 1950) most preparations are represented three times: a black-white photomicrograph of one focal plane, a colored drawing combining several focal planes, and an outline drawing for lettering. Transparent sheets with outline drawings are very helpful supplements to photomicrographs. Examples are an atlas of the hypothalamic and hypophyseal region by Rioch, Wislocki, and O'Leary (1940) and a paper by Hamilton, Soley, and Eichhorn (1940) on autoradiographs of thyroid glands.

Certain morphological techniques depend on photographic emulsions. This is obviously the case in electron microscopy and radioautography. Resolution problems need consideration, even in radioautography where the original and the photographic emulsion are in direct contact. To increase resolution of small sources of radiation, the grain size of the emulsion must be reduced. On the other hand, increased contrast (more blackening) per unit of radiation is produced in coarse grain films. The resolving power decreases as the concentration of radiation or duration of exposure increase (after J. H. Taylor, 1956). It is possible to use the same section for autoradiography and histological staining. Therefore it is useful to have pictures of the radioactive deposits accompanied by pictures which show the important structures with conventional stains. I mentioned above an early study on the uptake of radioactive iodine in normal and pathological thyroid glands, by Hamilton *et al.* (1940). These authors supplied transparencies on cellophane which facilitated the superimposition of each autoradiograph picture with the corresponding hematoxylin-eosin picture.

As a rule, macroscopic conditions in pathology are presented effectively by photographic methods (for illustrations see Cross, 1959), but there are exceptional cases in which sketching is the best method. Some macroscopic biological objects produce so many highlights that they can be photographed only when submerged in water or saline, or with the use of several sources of light in adjusted positions. These techniques are not always compatible with the nature of the investigation. There can be other reasons that favor sketching rather than photographing. I give two examples from my own experience. In a study on variability in size and shape of the cecum in living rats under different conditions, the abdominal cavity of each anesthesized animal was opened repeatedly for inspection after certain intervals. It would have been difficult to design any photographic arrangement which would yield pictures of the complete outline of each cecum without much handling of the organ. Instead, an outline sketch of each cecum was made, supplemented by caliper measurements in several dimensions (E. Mayer and E. J. Robinson, 1948, unpublished data). In a study on the position of the chick embryo in the egg on different days after the start of incubation, I observed the embryos through mica windows in the shell. Since these conditions were very unfavorable for photography, I recorded the position of each embryo at a given time by a sketch which showed the outline of the embryo including wing bud and allantois, the heart, the main blood vessels, and the eye. These landmarks were sufficient to follow the changing or unchanging orientation of the embryo in the egg. Guided by unambiguous sketches one could recognize the same phenomena in hazy photographs (Mayer, 1942). The fact that camera lucida drawings still have a place in first-class investigations clearly appears from the publications of King and Briggs on transplantation of embryonic nuclei into enucleated eggs. In a 1956 paper they showed decreasing variability of development in three transplant generations which proved increasing homogeneity of clones of transferred nuclei. The different stages and abnormalities of frog embryos were documented by camera lucida drawings (their Figs. 2 and 3) and photographs.

d. *Zinc etchings versus halftone pictures. Black-white versus colored pictures.* After World War II a tendency developed toward the use of zinc etchings (line drawings) rather than halftones, and to replace colored pictures by black-white presentations. Probably both changes were caused by economic pressure. However, the results of this pressure were not altogether bad. The first edition of Hueck's "Morphologische Pathologie" which appeared in 1937 contained numerous colored pictures. In the 1948 edition all colored pictures were rendered in black and white, for stated reasons of economy. In most cases I did not notice any marked loss in clarity.

My discussion on the selection of illustration techniques will be limited to the efficiency aspect. Among the textbooks of macroscopic anatomy it is interesting to compare two extreme examples. The voluminous German textbook and atlas by Rauber-Kopsch contains a large number of beautiful colored plates; the 15th edition (1940) is somewhat condensed, but still consists of three volumes. Woodburne's "Essentials of Human Anatomy" (1957), to which I have referred repeatedly, is one volume. It has only one colored picture, the frontispiece, which seems to serve ornamental purposes. The black-white pictures of this book are zinc etchings, except for radiographs and few other items requiring halftones. Evidently, the different presentations by Rauber-Kopsch and Woodburne are expressions of different ideas on how to teach anatomy. The realistic colored plates of Rauber-Kopsch enable the student to learn every detail of systematic anatomy, while the more or less diagrammatic zine etchings of Woodburne attract the attention to the most important points, and at the same time achieve a remarkable unification of topographic and dynamic (functional) anatomy.

The choice between halftone pictures and zinc etchings presents an interesting problem. In Young's "The Life of Mammals" (1957), halftone pictures, which were taken from the literature, were transformed into line drawings and printed as zinc etchings. I have used this transformation in the present book at several occasions. Frequently the redrawn zinc etchings are clearer than the original halftone pictures.

Modern textbooks of *normal histology* have placed an increasing emphasis on line drawings and diagrams, e.g., Ham's textbook (1957) to which I have often referred, and the excellent German textbook by Bargmann (1959). The authors of these two books restricted colored pictures to a very small number used mainly for the presentation of blood and bone marrow cells.

In recent textbooks of *pathological morphology* different illustration techniques have been used. Photographs, reproduced in halftone technique, predominate in Moore's "Textbook of Pathology" (1951) and in Florey's "General Pathology" (1958). In Hopps' "Principles of Pathology" (1959) all pictures are diagrammatic zinc etchings.

Some publications are filled with color photographs which are not only less distinct than black-white photomicrographs, but are of no documentary value with respect to colors. It is remarkable how efficient black-white photos can be, both on the micro- and macroscopic level. Although the diagnosis of skin diseases in the living patient greatly depends on color differences, H. W. Siemens did not use any color photos in his "General Diagnosis and Therapy of Skin Diseases" (translated edition, 1958). It was rightly emphasized by a reviewer of Siemens' book that "each minute change in the gross pathology . . . has been illustrated by black and white photographs of a quality surpassing that of the average color photograph" (Warren, 1958).

Black-white photography is applied in most modern methods of histology and histochemistry such as dark-field, phase contrast, microscopy in ultraviolet light, and electron microscopy. Photographs or drawings of microincinerated material are reproduced effectively if the incinerated material is shown white on a black background (G. H. Scott, 1933; Horning, 1951).

e. *Reproduction of pictures.* In preparing illustrations for lectures or publications the author should not take it for granted that his readers are as familar with his subject as he is. This seemingly trivial remark will be appreciated by those who have to struggle with the present rate of new technical developments in biophysics and biochemistry, particularly on submicroscopic levels (see the preceding chapter). When pictures obtained by new methods are presented, it is useful to add related pictures derived from familiar procedures. Bessis in "Cytology of the Blood" (1956, Fig. 35) shows a shadowed film of red blood corpuscles, representing a new technique, side by side with a film of red blood corpuscles treated with conventional staining. This picture was mentioned earlier in connection with the three-dimensional interpretation of microscopic structures (Part III, Chapter 6, B).

The scientist should accept responsibility for the quality of drawings and photographs which appear in his publications. Many morphologists are competent in taking photomicrographs at high magnifications. For survey pictures at low magnifications usually the cooperation of a specialized photographer is needed; equipment for low power photomicrography is not available in all laboratories. As far as drawings are concerned, a number of scientists have enough artistic talent to produce pictures for publication. Well-known examples are the realistic and attractive drawings of radiolaria, siphonophora, and sponges by Ernst Haeckel (1899–1904) and A. Maximow's pictures of blood cells and connective tissue (1910); some of Maximow's drawings are reproduced in the textbook of Maximow and Bloom (see, for instance, 6th ed., 1952). Semidiagrammatic drawings of a high quality are those of Goodrich in his studies on the structure and development of vertebrates (1930) and of Patten in his "Human Embryology" (1953).

Even those morphologists who cannot prepare their own originals should be sufficiently familiar with the techniques to permit a fruitful collaboration with the artist or photographer.

In lectures the efficiency of illustrations depends on three points. (1) All material must be readily visible in the most distant part of the lecture room. (2) Lantern slides should not be crowded with an excessive amount of information. (3) The organization or plan of the lecture, large summarizing tables, and similar materials not suitable for lantern slides, can be presented profitably on blackboards or wall charts.

In publications the success of pictures depends on the quality of the original, on the grade of the screen used for reproduction, and on the type of paper used for printing in the particular periodical or book. The author should either verify that the quality of screen and paper will do justice to his originals, or he should make allowance for loss of detail in the process of reproduction. Structures which are poorly visible in the original cannot be expected to be distinct in the publication, even with the best techniques of reproduction. Some planning is necessary with respect to the final magnifications at which microscopic objects are to appear in the publication. Otherwise the picture may be too large for the page size of the particular periodical. If the author makes reasonable suggestions for the final size of his illustrations, it is not likely that the editor will interfere. A biconcave reducing hand lens is helpful in estimating the tolerated reduction. As a rule results are more satisfactory if the original is reduced 10% or 20% before reproduction; reductions do not require any additional step in the process of reproduction. In planning reduction of pictures, the size of the lettering requires attention. Particularly in the reproduction of graphs, frequently the lettering is too small. It is not enough that lettering is legible: it should be easily legible. The reader should be able to concentrate on the content of a scientific paper without wasting energy in deciphering letters and figures. Disproportionately heavy lettering can also be disturbing.

Zinc etchings can be prepared from originals without the use of screens, provided the originals are composed of lines and dots. Thickness and spacing of lines and dots in the original should be planned according to the reduction intended in the publication. A useful drawing guide for zinc etchings was published by A. J. Riker and Regina S. Riker (1936); it has been reprinted in various issues of the *American Journal of Botany* (for example see Vol. 47, April issue, 1960). This guide shows tolerated as well as excessive reduction of various line and dot drawings, and also covers the reduction of letters and numbers. Many of the suggestions which I have made here are to be found in an excellent article by Christman (1954).

f. *Motion pictures. Animated cartoons.* The contributions of motion pictures to biological research are obvious. Prior to the motion picture era, sequential short-exposure photographs were used by Stillman (1882) and his photographer Muybridge to reconstruct the different phases of motion in horses. Each exposure was 0.002 second, and up to 24 cameras worked in series. The traditional paintings and engravings showed racing horses with both forelegs stretched forward and both hindlegs stretched backward, but the sequential photographs revealed that such a phase does not occur except when leaping obstacles. The flight of birds and insects has been a particularly fruitful field for motion picture study. From the point of view of the present book it is interesting that Magnan applied the title "Morphologie Dynamique" to motion picture analysis of flight (1932). Activities of inner organs are recorded by Roentgen motion pictures. Finally I mention motion pictures on the microscopic level. Movements and multiplication of bacteria, protozoa, and cultivated tissue cells had been recorded by drawings with a certain amount of success. However, extensive observations became possible only with the use of motion pictures. Depending on the purpose of the studies, the frequency of exposures may vary over a wide range. To produce the impression of continuity in projected films, approximately 16 pictures per second are necessary. For numerical analysis of migration and multiplication of fibroblasts and monocytes *in vitro,* a simple and economical technique was designed which required not more than six pictures per minute (Willmer, 1933; Willmer and Jacoby, 1936).

When an investigator has been able to document his work by motion pictures, he may face difficulty in demonstrating his observations in a paper. In some cases it is possible to select stills which illustrate the main points satisfactorily, but in other cases it is not possible. In publishing his studies on the flight of birds, Magnan (1932 and 1933) included substantial portions of the films in each paper. Although the pictures are quite distinct and time scales are always supplied, it is difficult to derive from these pictures the conclusions which were probably quite obvious in the projected moving film. Sometimes a selection of stills from photomicrographic films provides valuable information, such as the pictures of mitotic phases given in A. Hughes' monograph, 1952.

Animated cartoons have been used successfully for research and teaching in biology. It seems to me that, at least in teaching, animated cartoons should be shown along with photographic motion pictures, a device similar to the supplementation of photographs by outline drawings in publications, as discussed before.

B. Tabulation of Qualitative and Semiquantitative Data in Morphology. Scoring and Ranking

Examples of the tabulation of qualitative data will be taken from pathological morphology. Let us start with a brief recapitulation of observations on experimental embolic mouse tuberculosis (Mayer *et al.*, 1954). It was possible to distinguish seven stages as seen in the diagrams, Fig. 31a–g. In Table 7 presence or absence of a stage was indicated by a plus or minus sign for each of the 71 mice, without attempts at grading. Inspection of the table shows that the changes from one stage to another were clearly time-associated and that the sequence was from stage a to stage b, from here to c, and so on, until stage g was reached. It is interesting to compare an earlier paper by Grün and Klinner (1952) with the presentation of Mayer *et al.* The photomicrographs of embolic clumps of tubercle bacilli and the descriptions of Grün and Klinner indicate that their observations were similar to those of Mayer *et al.* However, Grün and Klinner excluded the possibility of presenting individual records for each of 63 mice examined. They limited themselves to a summary of the conditions found in the majority of animals and to a chronological interpretation of the findings. There is not much difference in the results which the two laboratories obtained, but there is a great difference in the documentation of the results. Relations between morphological findings and survival statistics became more definite in the study by Mayer *et al.* than in the paper by Grün and Klinner.

The next example is a form of tabulation which proved particularly useful in experimental studies on the toxicity of new drugs administered to animals for periods of several weeks or months. At the end of the experimental period the animals, usually dogs and rats, are sacrificed and autopsied. Since the verbal record of each autopsy would be quite lengthy, summaries are necessary in which groups of animals can be compared with respect to important points. Table 10 shows the pattern of tables used in our laboratories. In the left column each animal is listed by number and sex. If animals listed in the same table received different drugs, different doses of the same drug, or were treated for different periods, some columns on the left side are made available. The bulk of the columns is provided for the different organs and systems. In these columns the entry + for normal indicates that the findings are within the statistical norm (see Part II, Chapter 1, D). If the findings are outside the statistical norm, they are interpreted either as harmful or as harmless variations. The harmful deviations are entered as p for pathological, and the harmless deviations as v for nonpathological variations. Organs which were not examined at all are marked by minus signs on the macro- and microscopic lines. If microscopic study was omitted, there is a minus sign on this line only. The entries +, p, and v cover morphological findings, including size or volume of organs. Absolute and relative weights of organs are given in separate tables. Each p or v entry is explained in a supplement to the table. Tables of this type are found in publications by Maren, Mayer, and Wadsworth (1954) and Roepke, Maren, and Mayer (1957).

This pattern of tabulation has a number of advantages. First of all, the pathologist commits himself to *interpretations* of his findings. Secondly, the *scope* of the morphological investigations is immediately apparent. The scope of so-called complete autopsies was described previously (Part III, Chapter 2, B, 1). In some studies, abbreviated autopsies are perfectly satisfactory; the selection of organs for study will depend on the particular project. The third advantage

TABLE 10

Summary of Scope and Results of Postmortem Morphological Studies: Macroscopic and Microscopic Findings[a]

Animal, number, sex	Age at death	Drug administration		Morphol. diagnostic procedure	Circ. system	Resp. system	Digest. tube	Liver	[c] ...	Nutrit. state
		Dose[b]	Duration							
Dog K 22 M	2 yr.	40 mg.	10 wk.	Macro. Micro.	+ +	+ +	+ —	+ +	...	Medium
Dog P 37 F	4 yr.	100 mg.	27 wk.	Macro. Micro.	+ +	+ p	+ —	+ v	...	Medium

[a] Pattern of table used particularly in safety tests of drugs (toxicity studies). Explanation of entries: + = normal; p = pathological; v = nonpathological variations; — not examined. Entries p and v are described in the supplement.

[b] Dose in milligrams drug per kilogram body weight per day.

[c] Dotted lines indicate provision for any number of organs included in the particular study.

Supplement to Table 10

Dog P 37, F Respiratory system: congestion and edema.
Liver: weight normal; granular bile pigment in liver cord cells; iron reaction in Kupffer cells positive, medium intensity.

of the pattern of Table 10 is that the *individual animals* are listed separately. This is not always done in publications which deal with groups of animals. Although, as a rule, the organs of each dog are described separately, there is a wide spread custom of pooling the findings on small animals such as rats and mice. Suppose that there was a group of 10 rats each of which received the same dose of a drug for the same period. Many investigators summarize the autopsy findings as follows: the livers were enlarged in 5 rats; the spleens were enlarged in 7 rats; small abscesses were observed in the kidneys of 3 rats. From such a report one cannot tell *how the different abnormalities were combined in individual rats.* Moreover, the failure to identify individual animals in a report makes it impossible to compare the results of postmortem studies of a particular animal with observations made during its life.

The two examples discussed here dealt with *qualitative* tabulation. The next example will illustrate procedures for grading morphological variations so that a *semiquantitative* scoring system can be established with the use of one or more plus signs. In experimental infections of monkeys with poliomyelitis virus it became desirable to tabulate the microscopic changes which occurred in the spinal cord at different intervals after infection. Systematic investigations on these lines were published by Bodian in 1948. The motor nerve cells in the anterior horn[1] of the spinal cord are the victims of the poliovirus (see my diagrams Fig. 58a–c). Bodian developed a standard for scoring the destruction of these motor cells. In a number of monkeys the motor nerves which innervate the left arm and leg were severed surgically. The segments which supply the arm are located in the cervical enlargement, and those which supply the leg in the lumbar enlargement, of the cord. The operation caused a gradual destruction and final disappearance of the motor neuron population in the respective segment of the left side of the spinal cord. It became possible to predict the number of large motor nerve cells occurring in intact or damaged segments. Ten to fifteen sections of 15 μ thickness proved representative of the four principle segments of either enlargement (totaling eight samples). After these standardization studies, 50 rhesus monkeys were infected with poliomyelitis virus. The period of observation varied from a few days to one year after infection. Bodian's quantitative tabulation of histopathological criteria became the basis for testing methods of poliomyelitis vaccines ("The Monkey Safety Test for Poliomyelitis Vaccine," edited by Bodian, 1956).

In the preparation of vaccines from attenuated poliomyelitis virus, the histopathological responses to the virus serve as a guiding principle. For scoring purposes two sets of criteria are used: the proportion of intact to destroyed nerve cells in the anterior horn, and the degree of inflammatory response in the anterior horn and other parts of the gray substance. The extensive work on vaccines made simplifications of scoring methods necessary. In a report by Cabasso *et al.* (1959) the authors describe two methods of neurohistopathological scoring. In their earlier method-A, scoring was based essentially on the estimated percentage of nerve cells lost from the anterior horns. No cell loss was indicated by O; no cell loss, but some inflammatory response by ±; cell loss of 10–20% by +; 21–40% by ++; 41–60% by +++; 61–80% by ++++; more than 80% loss by +++++. Method-B, which was recently adopted by Cabasso *et al.*, is the scoring according to Melnick and Brennan (1959). This method represents a further simplification. Estimates of destroyed nerve cells are made on a semiquantitative basis without the use of percentages, and the inflammatory response is given increased weight in the score. My diagrams illustrate the scoring system of Melnick and Brennan (1959). The normal condition, scored O, is shown in Fig. 58a. The lowest degree of changes consists of either a single focus

[1] In man and monkey, anterior is synonymous with ventral, and posterior is synonymous with dorsal (see Part III, Chapter 6, A, 1,).

of neuronal injury or a single inflammatory focus. This condition, scored + by Melnick and Brennan, is not represented in my diagrams. Medium or severe changes, scored ++ or +++, are shown in Fig. 58b; these changes consist of various degrees of motor cell destruction and inflammatory responses. The maximal changes, scored ++++, are represented in Fig. 58c: all motor nerve cells are destroyed, and inflammatory phenomena not only are intensive but also have spread to the dorsal horn, although the sensory nerve cells remained intact. The diagrams Fig. 58a–c were drawn on the basis of photomicrographs supplied by Dr. E. V. Orsi of the Virus and Rickettsia Section of our laboratories. The diagrams show similar conditions on the right and left side of each section. In reality the two sides are frequently different, and therefore are scored separately. The maximum score for one level is 8, based on ++++ on each side. The maximum for ten levels is 80.

Should greater precision be required it would be necessary to substitute a ranking method for the scoring by plus signs, and the destruction of nerve cells and the degrees of inflammation should be ranked separately. The present tests for monkey neurovirulence of attenuated poliovirus vaccines are much more complicated than one might conclude from my diagrams Fig. 58a–c. In addition to segments of the spinal cord, various areas of the brain are examined. Spinal as well as intracerebral injections require deposition of the inoculum in special areas. Therefore, histological check of the needle track becomes part of the evaluation of each test (see Jungherr, Cabasso, Moyer, and Cox, 1961).

An interesting attempt to score tuberculous changes in lungs of mice was made by Stewart (1950). Groups of mice were killed at different intervals after infection with finely dispersed suspensions of tubercle bacilli. Three criteria were scored separately: (A) macroscopic changes, scored 0 to 3; (B) histological changes, scored 0 to 4; (C) number and distribution of bacilli, scored 0 to 4. The combined or total scores for the lungs of each mouse could vary from 0 to 11. Therefore the group averages could also vary from 0 to 11. Stewart pooled the total scores of all mice killed at the same interval after infection and divided this figure by the number of mice. The average of the pooled groups served as an "index of severity" of pulmonary changes and was plotted against days after infection. The resulting curve rose from the 7th to the 20th day, but not much after that. All mice were dead by the 33rd day.

Let us compare Stewart's 1950 paper with the 1954 study of Mayer *et al.* on embolic mouse tuberculosis (see my Tables 7 and 8). In both investigations "morphology groups" and "survival groups" were handled independently. Stewart's presentation was semiquantitative. The correlation between changes in the lungs and time after infection was demonstrated in graphs, based on average group scores. Consequently, information on individual mice was not given. In the study of Mayer *et al.* no attempt was made at quantifying the morphological data, except for an estimate of consolidation of lung tissue. Each individual mouse was entered in a table with indication of time after infection and the presence or absence of bacterial clumps and of certain histopathological conditions in the lungs. This qualitative treatment of morphological data proved sufficient to demonstrate correlation with the time after infection.

Stewart's work on experimental mouse tuberculosis served as an example of scoring by assigned numbers. Scoring by plus signs was illustrated by the diagrams of spinal cords after experimental poliomyelitis infection in monkeys. Both types of scoring are similar since they present the data as *discrete classes*. As a rule the establishment of such classes is derived from some unsystematical ranking procedure. The number of useful classes is limited. Plus signs in excess of

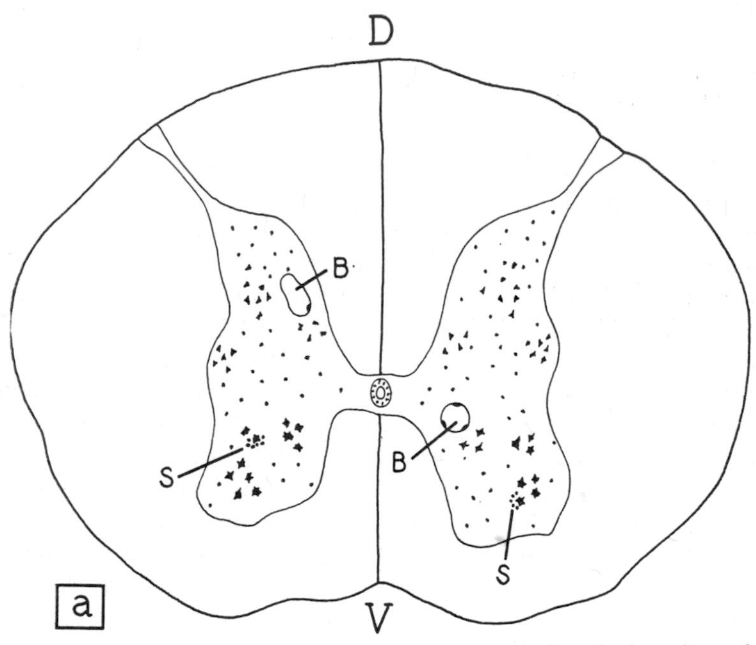

FIG. 58a–c. Scoring of morphological changes in the spinal cord of monkeys after experimental infection with poliomyelitis virus. Diagrams were made from the scoring system of Melnick and Brennan (1959). Cross sections showing "butterfly shape" of the gray substance with the ventral horns toward *V* and the dorsal horns toward *D*. The oval structure in the center is the central canal. *B*, blood vessels.

(a) Normal condition. Large black units in the ventral areas: motor nerve cells. Smaller black units in dorsal areas: sensory nerve cells. Dots, nuclei of neuroglia; *S*, small number of satellites close to nerve cells. Score 0. (b) Medium to severe changes after infection. Number of motor cells decreased. *N*, neuronophagia, i.e., numerous phagocytes surrounding destroyed nerve cells. Inflammatory response: increased number of neuroglia cells (dots) and formation of a "cuff" around blood vessels *B* in ventral horn. Score ++ or +++, depending on extent of nerve cell destruction and of inflammatory phenomena. (c) Maximal changes after infection. Motor nerve cells completely destroyed (sensory nerve cells intact). Inflammatory phenomena: intensity increased in ventral horn, some involvement of dorsal horn as indicated by "cuff" around blood vessel in dorsal horn. Score ++++. (Diagrams were drawn on the basis of photomicrographs supplied by Dr. E. V. Orsi. No diagram was made for score +.)

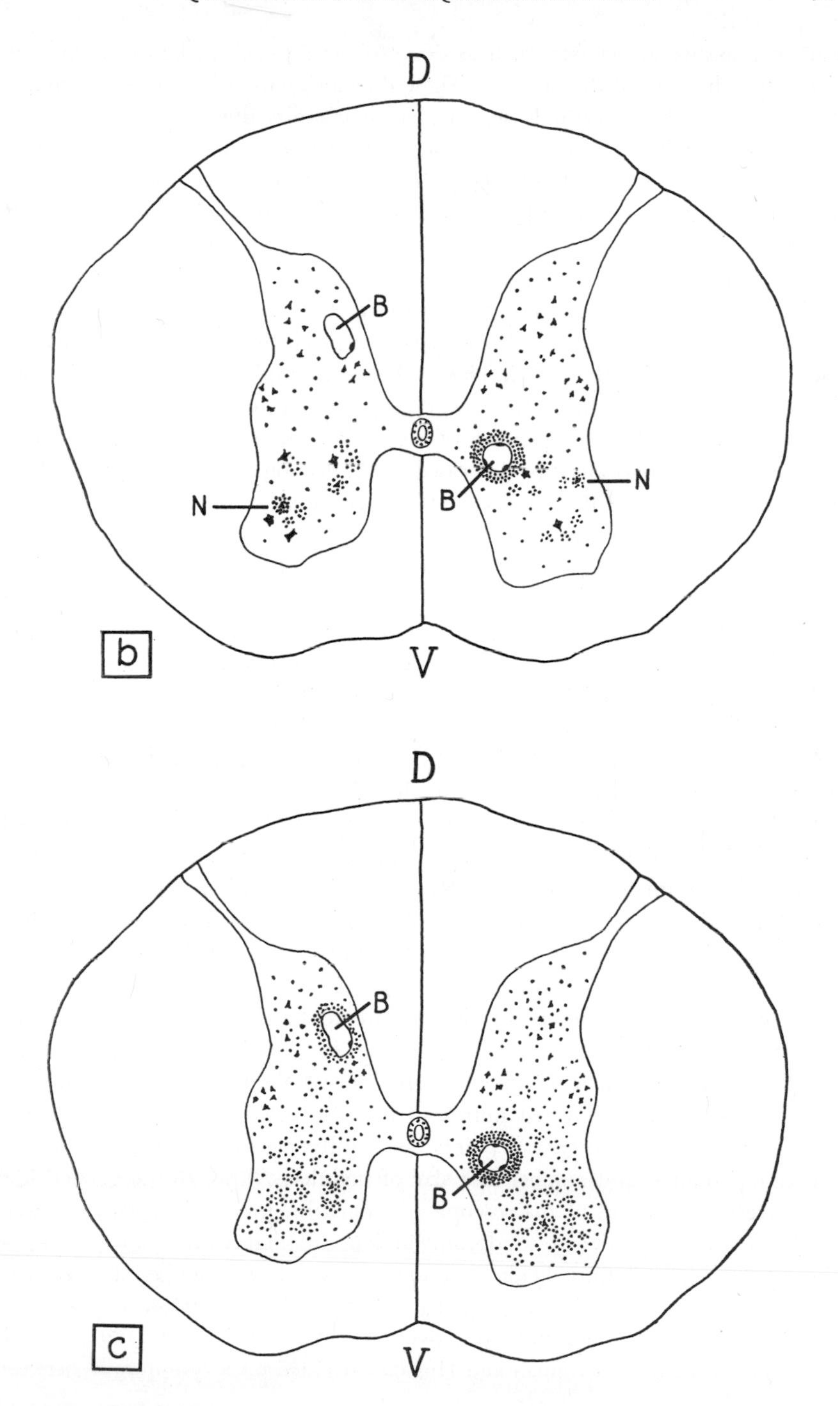
D
B
N
B
N
b
V
D
B
B
c
V

four, or intermediary classes, such as "two to three plus", make tabulations unwieldy and thus defeat the purpose. These difficulties are avoided by systematical *ranking* of data. Rank methods do not involve arbitrary division into classes, and there is no problem of intermediary steps since the number of rank places is unlimited. Suppose ten sections from spinal cords of monkeys infected with poliomyelitis are to be ranked for degrees of histopathological changes of the type shown in Fig. 58b and c. Code labels from *A* to *J* are given to the ten slides. Several observers are asked to arrange the sections according to increasing severity of changes (for instance, motor neuron destruction). If all observers arrive at the same ranking order, a result of marked reproducibility has been obtained. Then, one may divide the data in classes. In my example of ten histological sections, five sections will be in the lower half and five in the higher half of the rank numbers. Finally, the question of significant differences of rank data is to be determined statistically, as described in Wilcoxon's booklet "Some Rapid Approximate Statistical Procedures" (1949). An application of his *rank correlation* method will be shown in Table 13.

Sometimes the investigators feel that their morphological data do not lend themselves to division into classes and therefore are presented best as a series of pictures. If, for instance, the effect of a drug on experimental tumors has been studied, scale photographs of tumor nodes from treated and untreated animals will allow the readers of the publication to from their own opinion whether or not there is a marked difference. This procedure is useful if simple structures such as tumor nodes are the object. As an example I mention the charts in a paper by Suguira and Stock (1955) on the effect of phosphoramides on mouse and rat tumors. A small piece of tumor was implanted subcutaneously into each control and experimental animal. The progress of the tumors was observed by sacrificing animals 7, 14, or 21 days after implantation. The tumor nodes were dissected out and represented by silhouettes to scale. The silhouettes show convincingly how the tumors of untreated controls increased in size while those of experimental animals decreased or disappeared.

If complicated morphological changes are to be documented, photographs of each sample may or may not be helpful. For instance, this problem is encountered in the field of experimental tuberculosis. Youmans (1949) compared groups of lungs from infected untreated mice, infected and treated mice, and uninfected controls. The critical points were differences in volume, presence or absence of nodules, and number and size of nodules. Although the general idea was conveyed by groups of macroscopic photographs, many of the pictures were not distinct enough to serve as documentation. In their work on experimental tuberculosis in guinea pigs, Feldman and Hinshaw (1945) studied a number of organs besides the lungs. It was not feasible to publish photographs of the different organs of each guinea pig. Moreover, macroscopic as well as microscopic criteria had to be used. Feldman and Hinshaw decided on a compromise between photographic documentation and interpretive scoring (plus signs or numbers). They represented each animal by a diagram which showed the site of inoculation, local lymph nodes, lungs, spleen, and liver. The degree of tuberculous involvement of each organ was indicated by dotting of various density. Maximal involvement was shown as solid black, no involvement as white. This presentation would have been perfect if the rules by which the diagrams were made had been illustrated by some photographs. Other investigators, who were mentioned previously, published sample photographs to show the basis of their scoring or tabulation procedures (Stewart, 1950; Mayer *et al.*, 1954).

Various scoring systems have been suggested for purposes of practical medicine. The Gaffky score for the number of tubercle bacilli in the sputum may serve as an example here. The original (1884) Gaffky score comprised ten grades and was applied, with minor modifications, for many decades. This score carried much weight in deciding when a patient could leave the sanatorium. Eventually the precision of the Gaffky score proved to be excessive. At present, a score of no more than one zero and three plus grades is recommended (Diagnostic Standards, National Tuberculosis Association, 1955 ed., p. 50). Some investigators feel that even a division into "positive" and "negative" is quite satisfactory (Willis and Cummings, 1952, p. 89). Histological grading systems of malignant tumors for prognostic purposes will be discussed in connection with the medicopathological classification of diseases (Part V, Chapter 4, C, 2).

C. Presentation of Quantitative Data, Particularly on the Microscopic Level

Introduction: General Comments on Microscopic Samples. The quantification of morphological data consists of two steps: definition of the sample and measurements within the sample. Measurements within a biological sample are comparable, in many instances, to measurements in physics or chemistry. It is the *definition of samples* which is frequently difficult when morphological structures are concerned. Since these difficulties are greater on the microscopic than on the macroscopic level, subsequent discussions will be devoted mainly to microscopic structures.

Regional differences of structure and reactivity within one organ were discussed previously (Part III, Chapter 6). I mentioned that Heyl and Laqueur (1935) used the intermediary zone of the guinea pig thyroid for quantitative assays of thyroid-stimulating hormone, since the peripheral and central zones did not yield reproducible results. Regional differences in the bone marrow need consideration when cell counts of bone marrow samples are made. In animals the size of dogs and larger, irrespective of age, the spongy bones contain active bone marrow. The long bones of adult large animals contain inactive fat marrow in the distal part and active (red) marrow in the proximal part. The concentration of islets of Langerhans in the pancreas varies in different regions of the organ. In experiments which produce a selective destruction of the islets, such as administration of alloxan, the parts of the pancreas which are used for assay must be characterized properly. This is difficult when the diffuse pancreas of rodents is the subject of study; samples from many areas may be necessary in this case.

The occurrence of mitoses is usually expressed as the mitotic index, the number of mitotic nuclei per 1000 nuclei. If, in experimental animals, the organs which are studied have an extremely low mitotic index, one may consider the use of colchicine, which arrests mitoses in the metaphase and thereby produces a much higher count at a given moment. This introduces an interference which is not easily controlled (see Part III, Chapter 3, C, 2, a). Discrepancies between mitotic counts made by different authors result from the fact that some investigators include early stages of the prophase in their counts of mitotic nuclei, whereas others exclude them.

1. *Sampling, Counting, and Measuring*

a. *Relations between physical conditions of microscopic samples and methods of quantification.* Quantification of microscopic structures presents different possibilities and problems. The physical state of a sample ready for quantification may or may not be similar to its natural state. Red or white blood cells are counted in diluted blood for determination of their concentration, while a dry film of blood (smear) is used to estimate the proportion of the different types of white blood cells and of nucleated red cells. Details will be given later.

Counting of nuclei and other cytological structures in sections involves considerable physical alteration of the tissue sample. Sections which are thin enough for counting purposes require fixation, dehydration and embedding. In other words, the colloidal sample is transformed into a dry sample for quantification.

For our discussions, the material will be arranged according to its physical conditions. Definition of microscopic samples will be described first. Then counting and measuring procedures will be treated in this sequence: suspensions; smears; total mounts and sections.

b. *Definition of microscopic samples for quantitative study: fluids, smears, sections, and total mounts.* Whenever microscopic structures are measured or counted, one would expect the samples to be well defined. However, there is a wide range of definiteness of samples. As a rule, liquid samples are better defined than histological sections.

Counts of red and white blood cells are expressed as number per cubic millimeter of blood. Therefore the volume of the blood sample and of the counting chamber are integral parts of the data. So-called differential counts of blood cells are made by spreading a drop of blood on a slide and staining the dry film of blood (smear technique). The different types of white cells are recorded as per cent of all white cell types present, for example, 70% neutrophiles, 5% eosinophiles, and 25% lymphocytes, totaling 100%. Depending on the frequency of the different cell types, it may suffice to count 100 white cells, or it may be necessary to count several hundred. The total number counted should be stated, of course. In order to transform these relative figures into absolute concentrations, one uses the chamber counts which indicate the concentration of all white cells in a cubic millimeter. Although the thickness of the blood film and the density of cells vary in the marginal areas, the bulk of a good film presents one layer of cells which are not compressed excessively. This is the part used for counting.

In the diagnosis or study of malaria in patients or experimental animals, the search for malaria parasites can be made either in a thin film of blood or in the transparent margin of a thick drop of blood (both on glass slides). There are two tasks: first, to ascertain whether malaria parasites are present in the blood, and secondly, if they are present to estimate their number (for example, before and after treatment with an antimalarial drug).

The presence of bacteria in body fluids or tissue samples can be determined in different ways. A sample of sputum or of a crushed piece of tissue can be smeared on a slide and examined microscopically after staining; different stains are available for several classes of bacteria. In the case of tubercle bacilli, direct examination of a crushed preparation or smear may be negative upon acid-fast staining and fluorescence microscopy. After liquefaction of all other components of the sample by strongly alkaline solution of sodium hypochlorite and subsequent centrifugation, it might be possible to detect tubercle bacilli in the concentrated sediment. Besides these morphological procedures, culturing methods and animal inoculations are used to ascertain the presence or absence of bacteria, and to identify those which are present. I have described previously (Part III, Chapter 3, A) how the *number* of tubercle bacilli in the organs of experimentally infected mice has been estimated by plating out on semisolid media.

In the study of normal or pathological *histological sections,* characterization of samples and accounting for the material used are handled in many different ways depending on the purpose of the respective investigations. An earlier chapter dealt with the reconstruction of embryonic and other structures by means of serial sections. It is inherent to this procedure that the origin of the sample is clearly defined, and that the number of sections and their thickness is on record. Consequently, the dimensions of all structures which are present in the sample can be determined without difficulty. On the other hand, there are instances in which a small number of sections can be considered to be sufficient for quantitative estimates. Thyroid glands of different sizes require different numbers of sections for the quantification of dilute and concentrated colloid and determination of the height of follicular cells. In a beagle's oval thyroid which is approximately $2 \times 1 \times 0.5$ cm., one section through the main dimension of the organ may serve the purpose. The thyroid of a rat is so small that one section does not contain enough follicles for quantitative estimates. A human thyroid is too large to permit of preparing thin sections through the whole organ; in addition, regional variations may be overlooked if only one plane is examined.

If the presence or absence of certain structures in histological sections is to be determined, the area of the sections and the number of sections examined are of importance. Yet it is not customary in normal or pathological histology to give this kind of numerical information. In our own investigations on the protrusion of renal tubules into the space of Bowman's capsule (Mayer and Ottolenghi, 1947), we reported that 4 to 10 sections were examined for estimating the frequency of protrusions, and that the statements "absent" and "frequent" were always based on 3 or more sections whereas the diagnosis "occasional" was sometimes based on one section only. Obviously this was not a quantitative characterization of the samples examined. On the other hand, it was perhaps not necessary to apply greater precision since no quantitative claims were made. The important points of the paper were a complete description of the protrusions on the basis of *serial sections,* and a demonstration of the fact that the protrusions occurred not only in dogs which had been subject to drug experiments, but also in untreated control dogs.

Besides the thickness of sections and the number of sections examined, the area of the section may deserve recording. This is particularly important if the tissue block is taken from a large organ such as the liver of a dog. In small organs such as a rat kidney which can be sectioned as a whole, macroscopic dimensions and the orientation of the sections may give more information than areas of sections. General sampling procedures for histological purposes were discussed previously (Part III, Chapter 2, B, 3, a).

It is difficult to estimate the minimum of samples required for different investigations. The effect of poliomyelitis virus infection in monkeys has been estimated on the basis of histopathological criteria in the spinal cord (Fig. 58a–c) and brain. The number of sections which must be examined for testing of vaccines has been standardized by committees. Three-dimensional analysis of microscopic structures requires special techniques to provide representative samples. The use of serial sections and total mounts in chromosome counts and other cytological studies were discussed in Part III, Chapter 6, B, 1.

When tissue sections are studied with *electron microscopy,* the problems of representative samples is even greater than in studies with light microscopy since only a fraction of a cell is shown in an electron microscopic field. The joint mounting of adjacent electron microscopic fields and the use of insets for orientation were described previously. It is difficult to demonstrate that electron microscopic samples are representative of a large organ such as the human kidney. Farquhar, Vernier, and Good (1957) studied four children of the same family who suffered nephrotic disturbances. Biopsies of kidneys were examined with light and electron microscopy. In adjacent sections it was possible to study the same glomerulus by both methods. By light microscopy, changes in the glomeruli showed satisfactory parallelism with the clinical data, but the electron microscope revealed certain alterations of the glomerular capillaries which were similar in all the children regardless of the clinical phase or of the pathological conditions observed by light microscopy. The most conspicuous alteration was the loss of the characteristic foot processes which represent an elaborate organization of the normal epithelial cytoplasm. The authors concluded that further electron microscopic studies may supply a key to the functional disturbances which are observed in nephrosis.

It seems to me that some quantitative aspects should be considered here. Light microscopy enables one to estimate the ratio of those nephrons which are evidently unable to function over those which probably function normally. What is the proportion of glomeruli which show the electron microscopic changes described by Farquhar, Vernier, and Good? At least ten glomeruli were examined from each of three children who had been subjected to surgical biopsy, whereas only three glomeruli were available in the needle biopsy obtained from the fourth child. If all glomeruli examined showed the same alteration, one may conclude that probably the other glomeruli of each kidney also showed these alterations. However, such extrapolation may not be safe. Farquhar, Vernier, and Good did not state the number of sections examined or the number of sections in which the alterations of the glomerular wall were seen distinctly. Another question is: What proportion of each glomerulus showed the electron microscopic alterations? Because of its high technical quality I have selected the study of Farquhar, Vernier, and Good to demonstrate the tremendous difficulties in quantifying samples in electron microscopy.

c. *Counting and measuring procedures.* Isolated cells are usually counted in a *moist condition,* mostly in *suspensions.* Red or white blood cells are counted in known dilutions of blood samples with the use of chambers of calibrated volume (hemocytometers). Numerous studies have been published on the technical rules and statistical treatment of these counting procedures. A convenient presentation of the technique is found in Wintrobe's "Clinical Hematology" (3rd ed., 1951). For a brief discussion of the error of blood cell counts see J. Berkson's article in "Medical Physics" (Vol. 1, pp. 110-114, 1944). As an example Berkson uses an individual who has 5,000,000 red blood corpuscles per cubic millimeter of blood. Repeated counts are made by enumerating the cells in 80 squares of the hemocytometer chamber. According to the distribution curve shown one may expect 95% of the counts to fall between 4,220,000 and 5,780,-000. The expectation is that 50% fall between 4,740,000 and 5,260,000.

Nuclear counts of tissue cells can be made in suspensions which are prepared as follows. The organ sample is minced, a chemical is added to dissolve the cytoplasm, and the fluid preparation is centrifuged. The layer which contains the nuclei is stained, and samples of the stained nuclei are placed in a counting chamber of the type used for blood cells. In this way the concentration of nuclei in the organ sample can be determined. If only the peripheral zone of the cytoplasm is dissolved, nuclei in mitosis can be preserved. In a study by Andreasen and Christensen (1949) this method was used to estimate the percentage of mitotic nuclei in thymus and lymph nodes of young rats. The work of these authors was recounted previously (Part III, Chapter 3, B).

Instead of counting red blood corpuscles in a chamber, one can estimate the volume of packed blood corpuscles in a sample of centrifuged blood. This is known as the *hematocrit* method. For details I refer again to Wintrobe's book. If the volume of the individual red blood cells is normal, the hematocrit values can be readily converted into number of cells per cubic millimeter as obtained in counting chambers.

Recently speed and accuracy of blood cell counts have been greatly increased by the use of an electronic particle counter, known as the Coulter counter. The principle of the instrument has been summarized in a paper by Richar and Braekell (1959) as follows: "Briefly a suspension of cells in an electrically conductive medium is caused to flow through a minute aperture conducting an electric current between platinum electrodes. Each cell (being a relative nonconductor) causes a momentary increased impedance to the flow of electric current. The resulting pulses are recorded by a decade counter."

The Coulter counter can also be used for counting bacteria in suspensions. Until now, such counts were made conveniently by turbidimetry, which means the measuring of cloudiness produced by a bacterial population. In order to transform relative turbidity values into absolute concentrations, suspension samples were placed in counting chambers. All these counting methods do not differentiate between live and dead bacilli in a population. The proportion of viable bacilli is determined by plating out the sample on a semisolid medium and counting the colonies that develop. It is assumed that plating out of a sufficiently dilute suspension of bacteria will produce one colony from a very small number of viable bacteria, possibly from one bacterium (see Part III, Chapter 3, A).

Isolated cells obtained in a natural or artificial suspension can be *spread as a thin film* on a glass slide and *allowed to dry*. This so-called smear technique was mentioned at various occasions. As far as quantification is concerned it is mainly used for differential counts of white cells. I stated that each smear contains an optimal area where cells are packed neither too densely nor spread too widely. Mechanical stages are helpful in these counts in order to avoid repeated counting of the same fields. Although the different cell types are usually expressed as percentage of total white cells, it is advisable to count more than 100 white cells. Differential counts of cells obtained from the bone marrow can be useful, although they are not as reliable as those of blood samples.

Dry smears are used extensively in exfoliative cytology, which means cytological study of scrapings. This method originated with vaginal smears which permitted one to follow the estrous cycle in rodents. Subsequently this method proved useful in diagnosing human diseases (see the "Atlas of Exfoliative Cytology," by Papanicolau, 1954).

In order to count chromosomes in dividing isolated cells, crush or squash techniques can be used which I recapitulate here briefly. A small drop of fluid containing the cells in suspension is spread gently on a coverglass. The coverglass is inverted and placed on a glass slide. Adequate pressure is used to flatten the cells in the scanty fluid between coverglass and slide. Since the chromosomes must be counted in each individual cell, no mass counting procedures can be applied.

In any section (or total mount) the relative areas of two or more structures can be determined by direct *area measurements,* by *line sampling,* or *by point sampling.* Values obtained in several optical planes allow the calculation of relative volumes of structures, or of frequency of structures in the sample.

Direct measurement of areas is carried out by means of a planimeter. A sufficiently enlarged picture of the microscopic preparation must be obtained, either by photomicrography or by drawing the outlines with the help of a projectoscope. The paper picture is used for the measurements. Suppose there are two types of structures, A and B, whose areas are to be determined. Then two independent sets of area measurements are necessary, one for the A's and one for the B's. Start at a point X of structure A_1, guide the tracer needle of the instrument around A_1, and travel from here to A_2. Having outlined A_2, travel to A_3 and so on until all A-structures are covered. The round trip is completed by a return to point X of structure A_1. At this moment the planimeter has registered cumulatively all areas of type A. The same procedure is applied to the B-structures. If no planimeter is available, one may cut out the different areas on paper or cardboard and determine their ratios by weighing the pieces.

Line sampling and point sampling methods do not require photomicrographs or drawings, but can be used on the original microscopic slide placed on the stage of the microscope. The principle of these sampling methods is that a reference system is superimposed on the microscopic picture. Convenient for this purpose are eyepiece micrometers, which consist of a square-ruled network on a circular glass plate. When the glass plate rests on the diaphragm of the eyepiece, the lines of the squares and the microscopic preparation are in focus simultaneously. Either those squares are counted which are more than half filled with A-structures, or the intersection points which fall on A-structures are counted. Then the same procedures are applied to the B-structures. The total sums of squares, or of intersection points, give the relative areas of the two structures. Since it is frequently difficult to decide when half of a square is filled by an A- or B-structure, counting of intersection points is preferable. The results of the two methods are similar, as illustrated by Fig. 15 in Eränkö's "Quantitative Methods in Histology and Microscopic Histochemistry" (1955).

For *line sampling* some authors prefer reference systems which consist of a number of lines arranged at random (Richardson, 1953), while others prefer one or two single lines (Uotila and Kannas, 1952). In either case the length of the lines and the frequency of their intersection with structures A or B supply the values for the estimate of relative areas.

I mentioned one kind of *point sampling*, namely, the use of a square-ruled eyepiece micrometer for counting intersection points. Another point sampling method was designed by Chalkley (1943). Four short hairs are mounted radially on the diaphragm of an eyepiece. In a given position of the microscopic section, the tips of two hairs may hit *A*-structures, the tip of the third hair may hit a *B*-structure, and the tip of the fourth hair may hit neither *A*- nor *B*-structures. The first three observations are recorded as hits of *A*- and *B*-structures respectively, and the fourth is recorded as a miss. By moving the histological

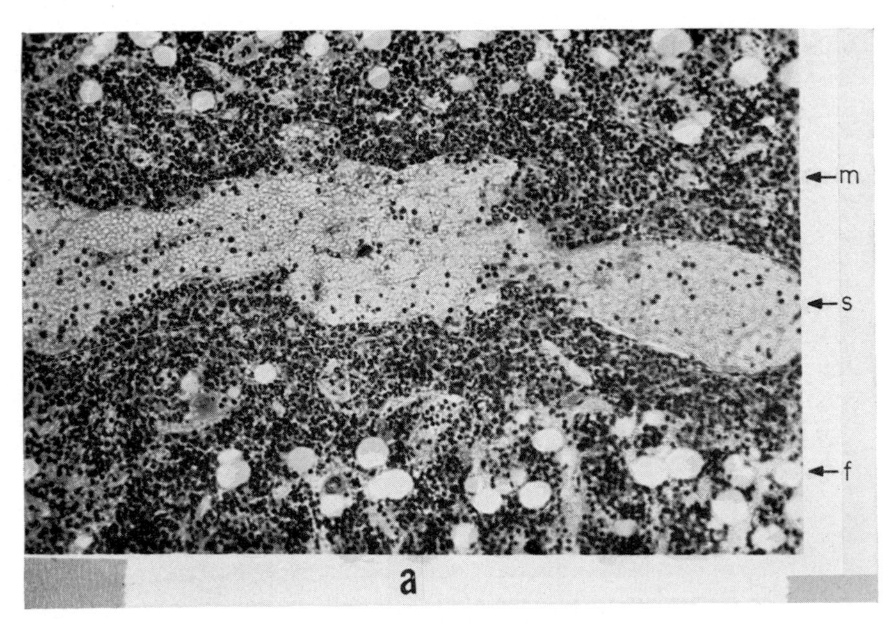

FIG. 59a–c. Comparison of two methods for measuring relative areas of components in histological sections. (a) Section of femoral marrow of a rat. Three components: m = erythromyeloid (hemopoietic) tissue; s = sinusoids; f = fat tissue. Magnification ×120. (b) Outline drawing of structures of (a) made for planimetric determination. Fat tissue = solid lines f. Sinusoids = interrupted lines s. The areas between fat tissue and sinusoids are erythromyeloid tissue m. Percentages obtained by planimetry: fat 5%; sinusoids 30%; erythromyeloid tissue 65%. (c) Grid superimposed on (b) for point sampling. For record of point sampling, see Table 11. Percentages obtained from Table 11: fat 5%; sinusoids 35%; erythromyeloid tissue 60%. (Figs. 59a and b from E. Mayer and A. Q. Ruzicka, 1945.)

section from field to field, cumulative counts of hits and misses are obtained which represent relative areas.

One may say that point sampling and planimetry represent the two extremes among methods of determining relative areas. Therefore a comparison between

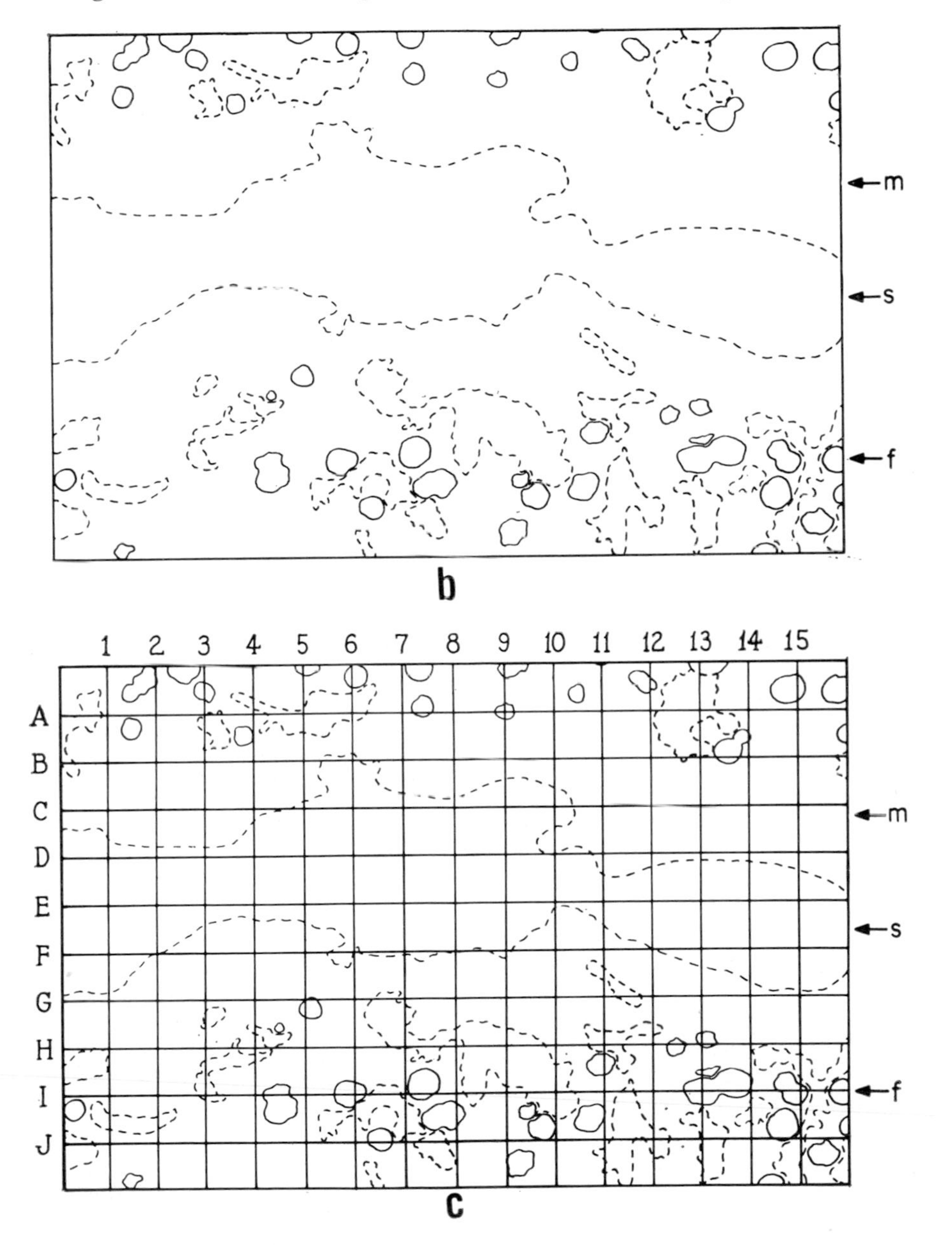

the results of planimetry and of point sampling in the same object was suggested by Dr. W. M. Layton of these laboratories. Pictures and planimetric determinations were available from a bone marrow study by Mayer and Ruzicka (1945). The point sampling was carried out by Dr. Layton. The task was to determine the relative areas of the three components of the bone marrow: (1) erythromyeloid tissue, representing the active marrow; (2) sinusoids, which are wide, thin-walled blood vessels; and (3) fat tissue, which occupies spherical spaces appearing as empty circles in paraffin sections. Figure 59a shows a photomicrograph of a section from the femoral marrow of a rat. In Fig. 59b, the outlines of two components of the photomicrograph are drawn, namely, the sinusoids and the fat tissue. The spaces between sinusoids and fat tissue are filled with the third component, the

TABLE 11

Record of Point Sampling for Determination of Relative Areas of Components in Femoral Marrow of Rat[a]

	1	2	3	4	5	6	7	8	9	10	11	12	13	14	15	m	s	f
A	m	m	m	m	s	s	m	m	f	m	m	m	s	m	m	11	3	1
B	m	m	m	m	m	s	m	m	m	m	m	m	s	m	m	13	2	0
C	m	m	m	m	s	s	s	s	s	s	m	m	m	m	m	9	6	0
D	s	s	s	s	s	s	s	s	s	s	m	m	m	m	m	5	10	0
E	s	s	s	s	s	s	s	s	s	s	s	s	s	s	s	0	15	0
F	s	s	m	m	m	s	s	s	s	m	m	m	s	s	s	6	9	0
G	m	m	m	m	f	m	m	m	m	m	m	m	m	m	m	14	0	1
H	m	m	m	m	m	m	s	m	s	m	s	m	f	m	m	11	3	1
I	m	m	s	m	m	f	m	m	m	s	m	m	f	f	f	9	2	4
J	m	m	m	m	m	m	m	m	m	m	m	s	s	m	s	12	3	0
															Grand totals	90	53	7
															Per cent	60	35	5

[a] Counts made from Fig. 59c; m = erythromyeloid tissue, s = sinusoids, f = fat tissue.

erythromyeloid tissue. Planimetric determination expressed as percentage of the whole picture gave the following results: erythromyeloid tissue 65%, sinusoids 30%, and fat 5%. In Fig. 59c, square-ruled tracing paper is superimposed on Fig. 59b. Each intersection of a vertical with a horizontal line hits either an erythromyeloid area (m), a sinusoid area (s), or a fat area (f). Table 11 shows these counts. Rounded to integrals, the results were as follows: erythromyeloid tissue 60%, sinusoids 35%, and fat 5%. This is in remarkable agreement with the planimetric values. If no pictures had been available, the point sampling could have been done using the original section and a square-ruled eyepiece micrometer. Assuming that the planimetric determinations have the highest degree of precision, the lower precision of the point sampling probably would be good enough for most questions concerning the composition of bone marrow. As a matter of fact, there are situations in which planimetry should not be used. When some of the histological structures are hollow spheres, ellipsoids, or tubules, oblique sectioning may introduce a marked variability of relative areas. In such instances point sampling may be adequate, while planimetry would be excessive in labor and precision. I give an example of a reasonable combination of planimetric and other measurement. Hartoch (1931) studied how partial thyroidectomy in dogs affects the remainder of the gland. The relative areas of concentrated and dilute colloid were determined planimetrically, and stage and eyepiece micrometers were used to measure diameters of follicles and heights of follicular cells.

Measurements of relative *areas* of different structures can be used for calculating relative *volumes*. Finally the absolute volume of each structure can be estimated if the sample which has been analyzed is a known fraction of an organ. In order to determine the relative volumes in the femoral marrow of a rat the section shown in Fig. 59a is not a sufficient basis for calculation. At least additional samples from the same plane would be needed, but a number of sections from several planes is preferable. In such study planimetry would consume too much time and produce false precision. Point sampling would be the method of choice.

Planimetry is particularly useful for determination of areas with irregular outlines (1) if the structures to be studied are plane, with a relatively small third dimension, and (2) if the object is uniform or contains only a small number of continuous (not scattered) components. Therefore, some experiments with tissue cell colonies *in vitro* are suitable for planimetric measurements (see Figs. 21 and 46).

Precautions are needed in certain determinations of relative volumes. For instance, in estimating the density of nuclear populations in any volume of tissue, corrections may be necessary for the relations between nuclear diameters and thickness of sections (Abercrombie, 1946).

2. *Time Factors That Affect Sampling*

Many time-associated factors need consideration in sampling. The estrous cycle modifies the morphology of female sex organs, and molting changes the structure of the integument. Seasonal factors are frequently interwoven with biological factors, particularly with endocrine mechanisms. In this category belong hibernation and migration phenomena. Hibernation of mammals is reflected in the morphology of the thyroid gland and of the fat tissue. In any new field the failure to obtain reproducible data should suggest the possibility of time-associated variability. Frequent sampling combined with planned variations of sampling times is apt to reveal time-associated mechanisms.

In human parasitology the effect of time factors on sampling results was recognized early. Well-known examples are the flooding of the blood stream with malaria parasites at intervals of 1, 2, or 3 days, and the strictly nocturnal appearance of a helminth, *Filaria nocturna,* in the circulating blood.

Mitotic activity is known to be time associated in different ways. Periodic regeneration is involved in molting (whole skin in reptiles, feathers in birds) and in the estrous cycle of the female reproductive system. Diurnal rhythms of mitotic activity were reported particularly in the skin and livers of mammals. In the liver of young rats there is a 10-hour period (from 8 P.M. to 6 A.M.) during which few mitoses occur (Jackson, 1959). Experimental inhibition of mitotic activity may be followed by a wave of increased mitotic activity with final stabilization on the normal level. Good illustrations of such "compensatory waves" are found in irradiation studies with tissue cell colonies *in vitro* (see Canti and Spear, 1929).

The frequency of mitoses seen in a fixed and stained preparation is expressed as the number of mitotic nuclei in a total of 1000 nuclei counted, or the mitotic

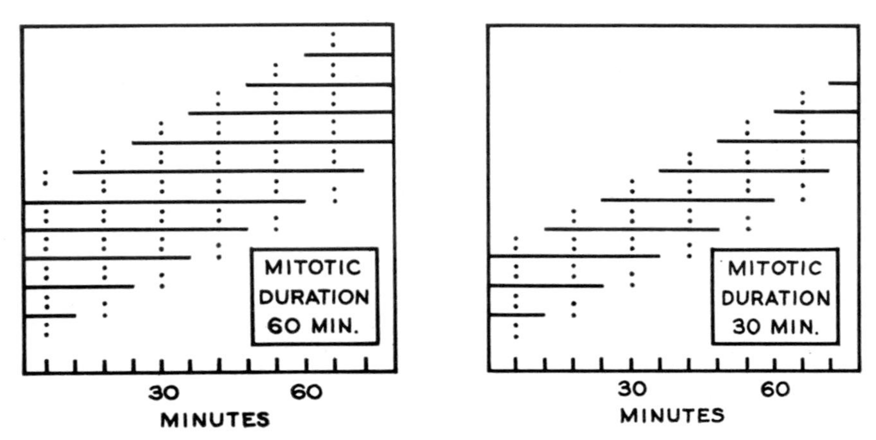

FIG. 60. Effect of mitotic duration on mitotic frequency as seen in static preparations. The diagrams represent two groups of histological samples; for instance, liver samples from rats and from mice. It is assumed that samples of equal size are examined and that, in both groups, mitoses start at 12-minute intervals. In one group the mitotic duration is 60 minutes with the result that at any given moment 5 mitoses are present. In the other group the mitotic duration is 30 minutes with the result that only 3 mitoses are present at any time. The number of mitoses present appears from the intersections of dotted (vertical) lines with continuous (horizontal) lines. (Modified after Hansemann, 1893.)

index (see above). Evidently this ratio of mitotic to intermitotic nuclei depends (1) on the time required by a cell to complete a mitosis and (2) on the interval between the end of the mitosis, which one may call the mother-cell mitosis, and the start of mitotic activity in the daughter cells. The diagram Fig. 60 (after Hansemann, 1893) illustrates the effect of mitotic duration on mitotic frequency in static preparations. I have mentioned pretreatment of animals with colchicine in order to change a very low mitotic index to a high one. Since this change is produced by arresting all mitoses in the metaphase, colchicine obscures chronological aspects of mitotic activity (for other interferences by colchicine, see Part III, Chapter 3, C, 2, a).

The duration of mitoses is of technical importance in microcinematography.

In tissue cell colonies *in vitro,* with an approximate mitotic duration of 30 minutes, exposures at intervals of 6 minutes proved to be sufficient for obtaining precise records of mitotic activity under different conditions (Willmer and Jacoby, 1936). This technique was included in the general discussion of motion pictures in biological research (Part III, Chapter 6, A, 2, f).

Indirect effects of time-associated factors can lead to surprising results in animal experiments. In the early period of sulfonamide research, the administration of the drug in the diet of mice or birds seemed to produce erratic concentrations of the drug in the blood. Although it was well known that small birds, such as hummingbirds, feed only during periods of light, special studies were needed to demonstrate a cyclic food intake in mice and ducklings (Litchfield, White, and Marshall, 1939; Marshall, Litchfield, and White, 1942). Frequency and spacing of feeding periods were investigated by recording the movements of a food cup according to the method of Richter (1927). In this way it was verified that under laboratory conditions ducklings fed during periods of light, whereas mice fed during periods of darkness. The cyclic food intake proved to be the main factor which controlled the concentration of the drug in the blood of the experimental animals.

The glycogen content of liver cord cells in young rats decreases with increasing intervals after the last feeding (see Table 12). On the other hand the mitotic activity of these cells seems to depend on available glycogen (Part III, Chapter 7, D, 1). These examples may suffice to indicate the innumerable ways in which morphological samples can be affected by time-associated factors.

3. *Statistical Treatment of Morphological Data*

In the present book quantitative relations between stimulus, response, and reactivity were presented statistically. In order to illustrate the difference between the statistical and the teleological (desirable) norm, a fictive example was used of three groups of ten scouts exposed to drenching rain for three different periods of time, with three different responses as result (Table 3). Quantitative studies of interaction between mice and disease-producing bacteria were described with the use of two different schemes (Table 4); bacteria were the stimulus, and the survival rates of the infected mice were the expression of the response. In one of the schemes, containing three variates, the results were presented by three-dimensional graphs (Fig. 6a–c). In the other scheme, which contained two variates, the observed cumulative frequencies were plotted arithmetically as well as on logarithmic probability paper (Fig. 7a, b).

In connection with S-shaped cumulative frequency curves, I discussed the deeply-rooted tendency to interpret formal similarity as an indication of similar mechanisms. The classic example of this misconception was the idea that the increase in number of a *Drosophila* population, the increase in weight of a cucumber, and monomolecular autocatalytic reactions were related in some way since all of them produced S-shaped curves. A useful brief discussion of curves of growth and related subjects is found in an article by Berrill (1955). I give a simple example of merely formal resemblance. Euler's theorem of polyhedra is expressed by the formula $s + v = e + 2$, wherein s is the number of faces (or sides), v is the number of vertexes, and e the number of edges of a polyhedron. The phase rule of Gibbs is expressed by the formula $f + p = c + 2$, wherein f is the number of freedoms, p is the number of phases, and c the number of components. No one would try to extract any meaning from the similarity of the two formulas.

Morphological data are to be treated according to the same statistical rules as other data. I repeat here some of the most elementary rules. Averages alone are useless. Averages with indication of ranges are acceptable in many cases. Standard deviations give satisfactory information, provided the number of items (size of population) is also indicated. Sometimes raw data are desirable in addition to their statistical summary.

The question whether statistical treatment is needed in a morphological study must be decided by the same criteria as in other studies. For quantitative as well as semiquantitative data, the rule holds that differences which can be seen on inspection do not need statistical analysis. One cannot decide by inspection whether differences between derivatives, such as ratios and logarithms, are significant or not. If two classes of measurements overlap markedly, statistical determination of significant differences between the two classes is necessary. The rank method of determination of significant differences was mentioned previously, in connection with the general treatment of semiquantitative data in morphology.

Two examples will illustrate what data make it possible to draw valid conclusions from mere inspection, and what data require statistical treatment. Both examples refer to a subject that was discussed earlier, namely, the relation between histochemical estimates of glycogen in liver sections and the quantity of glycogen determined chemically in the same liver (Part III, Chapter 7, D, 1). I recounted the work of Deane *et al.* (1946), who found a satisfactory correlation between histochemical and chemical determinations, and I mentioned that Grafflin *et al.* (1941) had not noticed any marked correlation of their data. For the purpose of the present discussion, the data of Deane *et al,* are summarized in my Table 12. I distributed the histochemical values (photometric densities) in four classes with the symbol *Y,* and the chemical values (grams glycogen per 100 grams liver) in four classes with the symbol *Z.* Inspection shows a perfect parallelism between the *Y* classes and the *Z* classes in spite of a slight overlapping of ranges between two neighbor classes in *Y.* Therefore statistical analysis is not necessary. However, it is interesting to apply the rank correlation method of Litchfield and Wilcoxon (1955) to the raw data of Deane *et al.* The rank correlation coefficient is 0.91 with a probability of chance result <0.01. This is close to the maximum correlation possible, thus confirming the estimate derived from inspection of the grouped data in my Table 12. (Rank correlation calculated by Dr. B. Jackson; for procedure see Table 13.)

Grafflin *et al.* (1941) used 14 samples of guinea pig livers for parallel histochemical and chemical determinations of glycogen. Best's carmine was used as histochemical stain, and the Pflüger method for chemical determination (see Part III, Chapter 7, D, 1). In my Table 13 the original data of Grafflin *et al.* are reprinted in columns (a) and (b). In column (a) the liver samples, with

TABLE 12

Comparison of Histochemical (Y) and Chemical (Z) Estimates of Glycogen in Livers of 20 Rats, Sacrificed in Groups of Four at Different Intervals after Last Feeding[a,b]

Hours after last feeding	Numbers of rats	(Y) Ranges of histochemical determinations in average optical densities[c]	(Z) Ranges of chemical determinations in grams glycogen per 100 grams liver[d]
6 and 12	4 + 4	23 –34	4.0 –5.3
18	4	16 –22	2.2 –3.1
24	4	1.5–11	0.5 –1.7
48	4	0 – 1.5	0.02–0.08

[a] Condensed from Table I of Deane, Nesbett, and Hastings (1946).

[b] Ranges of values were arranged in four classes (6- and 12-hour data pooled). High correlation of histochemical and chemical values is apparent from inspection of data. Treatment of original values by the method of Litchfield and Wilcoxon (1955) gave a rank correlation coefficient 0.91, with a probability of chance result <0.01.

[c] From column 5 of original; averages of left and right lobe, rounded to nearest integer.

[d] From column 7 of original.

laboratory numbers of the animals, are arranged in order of increasing histochemical glycogen content, and the corresponding chemical values are entered in column (c) as per cent of liver wet weight.

From the inspection of their data, Grafflin *et al.* concluded that the histological estimation of liver glycogen content is practically valueless in experimental studies. The authors arrived at this sceptical interpretation because of the "numerous individual discrepancies." They were particularly impressed by the fact that the sample from animal No. 19, which was the third lowest histochemically, was in the group of the six highest chemical values, i.e., above 7.0%. Dr. B. Jackson of our laboratories analyzed the data of Grafflin *et al.* with the rank correlation method of Litchfield and Wilcoxon (1955). For this purpose the histochemical ranks of the 14 samples were entered in column (b), while the chemical rank values were entered in column (d); when ties occurred in column (d), averages were entered: 1.5 instead of 1 and 2; and 12.5 instead of 12 and 13. Column (e) gives the rank differences (RD) between values in column (b) and their partners in column (d), always subtracting the smaller from the larger value. Column (f) contains the squares (RD^2) of the rank differences. The sum of the squares of the rank differences is 103.50. This value, derived from 14 samples, gives the rank correlation coefficient $\rho = 0.78$

TABLE 13

Comparison of Histochemical (Y') and Chemical (Z') Estimates of Glycogen in Livers of 12 Rats[a,b]

(Y') Histochemical estimates of glycogen arranged in order of increasing content		(Z') Chemically determined glycogen			
Lab. nos. of animals[c] a	Rank b	Per cent of liver wet weight[d] c	Rank d	Rank differences (RD) e	Squares of rank differences (RD^2) f
22	1	4.0	1.5	0.50	0.25
15	2	4.0	1.5	0.50	0.25
19	3	7.1	9.5	6.50	42.25
17	4	4.7	3	1.00	1.00
13	5	6.2	6	1.00	1.00
20	6	7.0	8	2.00	4.00
16	7	5.4	5	2.00	4.00
18	8	5.3	4	4.00	16.00
14	9	7.7	11	2.00	4.00
21A	10	6.6	7	3.00	9.00
23	11	7.8	12.5	0.50	0.25
24A	12	7.8	12.5	0.50	0.25
24B	13	8.2	14	1.00	1.00
21B	14	7.1	9.5	4.50	20.25
				Sum of (RD^2)	103.50
				Rank correlation coefficient	0.78
				Probability of chance result	<0.01

[a] Data from Grafflin, Marble, and Smith (1941).

[b] Histochemical estimates by ranking only; chemical determinations in per cent of liver wet weight, with subsequent ranking. Rank correlation of (Y') and (Z') determined according to Litchfield and Wilcoxon, 1955. There is a moderate degree of correlation between the two sets of estimates.

[c] A and B indicate two liver samples from the same animal.

[d] Two decimal places given by the authors were rounded to one decimal place.

according to the nomogram in the Litchfield and Wilcoxon paper. On the right scale of this nomogram, the coefficient 0.78 is approximately halfway between 0.0 (no correlation) and 0.95 (highest correlation). There is a probability of less than 0.01 that the rank correlation coefficient 0.78 is due to chance.

Statistical treatment revealed a marked correlation between the histochemical and chemical data of Grafflin *et al.* contrary to the interpretation of the authors.

The correlation coefficients obtained from the data of Grafflin *et al.*, and from the data of Deane *et al.*, cannot be compared because of differences in the material and methods. The histochemical estimates were made by eye in the study of Grafflin *et al.*, whereas Deane *et al.* determined photometric densities. The chemical glycogen values of Deane *et al.* covered a more than eightfold variation, whereas the values of Grafflin *et al.* were within a twofold variation. Finally, the sampling of Deane *et al,* always included a right and left lobe of the liver of each of 20 rats, but in the guinea pig study by Grafflin *et al.* 10 samples were taken from 10 different animals, and two pairs of samples from two additional animals. One cannot agree with the generalization of Grafflin *et al.* that "histological estimation of liver glycogen content . . . is entirely unreliable where an accurate knowledge of relative glycogen content is required in experimental work." I summarize this discussion as follows: (1) histochemical estimates of glycogen based on liver sections from *one animal,* do not necessarily warrant conclusions with respect to the chemical glycogen content of the liver, (2) if *groups of animals* are studied, one can expect a marked correlation between the results of the two procedures.

D. Comments on Publication Techniques Other Than Illustrations

The rules for organizing a paper in dynamic morphology are not different from those in other areas of natural science. Some aspects of publishing which are frequently neglected will be discussed here.

E. Bright Wilson, in his "Introduction to Scientific Research" (1952) gives a brief, but excellent, summary of the points which are important in organizing a publication. One may add two points to Wilson's list. (1) In any paper the conclusions should refer to the questions which have been raised in the introduction; it should be stated which questions have been answered and which have not. (2) The author should express his opinion whether the type of work presented in the paper is worth continuing, and if so, in what direction. I mention two books in which the authors made useful recommendations for future work. In de Beer's "Development of the Vertebrate Skull" (1937, Chapter "Agenda") 45 problems were formulated which relate to the morphology of the chondocranium, and references are given to the places of the book where the particular question originated. At the end of his comprehensive book on "Bird Malaria" (1940), Hewitt listed 85 questions which should be investigated. Dr. Hewitt informed me that approximately 50 of these questions had been answered by 1959. The presentation of programmatic papers in conferences should be encouraged. The probable future of tissue culture work was discussed by invited speakers at the Decennial Review Conference held in 1956 at Woodstock (see Part III, Chapter 5, C, 2).

Clear-cut conclusions, hypothetical interpretations, and recommendations for future work add to the value of a publication. Yet many authors are reluctant to commit themselves: they fear future work may show that they were wrong. The answer to this fear is that honorable and constructive errors are unavoidable in the development of any science. If the observed material leads to certain conclusions one should not hesitate to express them. Suppose some interpretations of an author become untenable in the light of new data. His work should still be considered meritorious if his data were good at that time, and if he has discussed adequately the pros and cons of his interpretations. All of us admire J. de Meyer for having committed himself by coining the term insulin in 1909. At that time the internal secretion of the pancreas was merely a postulate. The existence of insulin was not demonstrated before 1922; the main steps in its discovery were described in Part III, Chapter 1, A.

An author can expect a fruitful continuation of his work by other investigators only if his data are reproducible. It is difficult to estimate the amount of technical detail necessary in a publication so that another laboratory can duplicate the work. A remarkable solution of this problem was found by the first editors of "Organic Syntheses" (R. Adams, J. R. Conant, *et al,* Vol. 1, 1921). They established the following policy. When a laboratory has submitted a paper, it is sent to another laboratory for checking of procedures and yields. The checker attempts to follow the directions as closely as possible. Discrepancies are straightened out by communication between the two laboratories. If satisfactory confirmation is obtained, the paper is published. The names of the submitting and of the checking authors are indicated separately. This example has been followed by other publications such as "Biochemical Preparations" (H. E. Carter, editor-in-chief, Vol. 1, 1949). It might not be a bad idea to have a similar policy adopted for publications in experimental biology, including dynamic morphology. However, this becomes a vicious circle. In many areas of biology the standardization procedures are less developed than in chemistry. Consequently no "checking laboratory" may be in the position to repeat, right away, the special work done in the "submitting laboratory." Observations of rare conditions, particularly in human beings, may deserve immediate publication irrespective of the prospects of confirmatory observations.

Standardization of procedures in dynamic morphology will be enhanced if the authors indicate the degree of reliability of their data. Various discussions in earlier parts of this book centered around the reliability of data. The importance of separating observations and interpretations was emphasized particularly with reference to dynamic interpretations of static data (Part III, Chapter 3, A and B).

The presentation of human genetics is another area which calls for "grading of reliability." If the inheritance of some abnormality is to be analyzed, the members of the family tree are characterized customarily by different symbols for male or female, and for presence or absence of the abnormality. In addition, symbols for each person should indicate reliability of information which might be graded as follows: (1) individuals examined by the author, (2) individuals examined by another competent investigator who supplied a written report (or photographs), (3) individuals reported by members of the family, with documentation, and (4) individuals reported by the family, without documentation. Most medical pedigrees fail to indicate these differences. As an example of satisfactory presentation, I mention a paper by Fergusson (1949) on familial polycystic disease of the kidney. Different symbols were used for proved polycystic renal disease; suspicious, though unproved; no evidence of disease clinically or pyelographically; and finally, not examined. Autopsied cases were stated. One pedigree of four generations contained five proved cases, all limited to the second generation. Another pedigree of four generations had one proved case in the first generation and four in the second generation.

It is easier to document the inheritance of exterior abnormalities such as six fingers on each hand than the inheritance of an abnormal condition of viscera such as polycystic kidneys. With respect to internal abnormalities, it is necessary to indicate the procedures applied in examining the individual members of the family tree; this was done in Fergusson's paper. Omission of any members can lead to genetical misinterpretations. A clear statement of these rules is found in Czellitzer's 1928 paper on errors and difficulties in human genetics.

In animal experiments complete accounting for all animals mentioned in the paper adds to the reliability of the report. Unfortunately, one can find papers which allegedly cover observations on 200 mice, while experiments on only eight mice are reported. The customary statement that the published experiment is "representative of many others" is usually open to question, although acceptable if properly documented. Similarly it is desirable that any procedures listed in "Material and Methods," also appear somewhere in the remainder of the paper. This rule which seems self-evident is not always heeded. An author might state that "the dye was administered by means of intravenous, intramuscular, subarachnoid and intraventricular injections," but never report any intramuscular injection in the experimental part.

For a long time medicobiological investigators did not like to write papers with so-called *negative results,* and editors were reluctant to publish them. Gradually it has been recognized that the publication of negative data can be of great importance; for instance, with respect to screening results in experimental chemotherapy. I mention a report edited by C. C. Stock (1953) on "Negative Data from Experimental Cancer Chemotherapy Studies."

Much has been written on the desirable language in scientific publications. I conclude the present chapter with two remarks on *intellectual style.* (1) In serial papers a certain amount of recapitulation is necessary. The terse statement "for techniques, see preceding papers" can be most discouraging. At least the main procedural points should be given in each paper of a series. (2) If an author discusses the reports and interpretations of another investigator, his own statements should be clearly distinguishable from those of the other investigator. To achieve this, stereotype formulas are unavoidable; for example "it seems to me," "the present author feels." I repeat that which I have expressed earlier: in scientific writing, clarity has priority over elegance.

Part III: References

Abel, J. J., Rowntree, L. G., and Turner, B. B. (1913). *J. Pharmacol.* **5**, 275.

Abercrombie, M. (1946). *Anat. Record* **94**, 238.

Adams, Roger, Conant, J. R., Clarke, Hans T., and Kamm, Oliver. (1921). *Org. Syntheses* **1**, Preface.

Algire, G. H. (1943). *J. Natl. Cancer Inst.* **4**, 1.

Algire, G. H., and Legallais, F. Y. (1949). *J. Natl. Cancer Inst.* **10**, 225.

Alpatov, V. V., and Nastyukova, O. K. (1948). *Doklady Akad. Nauk. S. S. S. R.* **59**, 1221.

Ambrus, C. M., Ambrus, J. L., Johnson, G. C., Packman, E. W., Chernick, W. S., Back, N., and Harrison, J. W. E. (1954). *Am. J. Physiol.* **178**, 33.

Anderson, N. G. (1956). *In* "Physical Techniques in Biological Research" (G. Oster and A. W. Pollister, eds.), Vol. 3, p. 300. Academic Press, New York.

Andreasen, E., and Christensen, S. (1949). *Anat. Record* **103**, 401.

Andreasen, E., and Ottesen, J. (1945). *Acta. Physiol. Scand.* **10**, 258.

Andresen, N. (1942). *Compt. rend. trav. lab. Carlsberg, Sér. Chim.* **24**, 139.

Andrews, J. R. (1944). *In* "Medical Physics" (O. Glasser, ed.) Vol. 1, p. 1264. Yearbook Publ., Chicago, Illinois.

Anson, M. L. (1938). *In* "The Chemistry of the Amino Acids and Proteins" (Carl L. A. Schmidt, ed.), p. 407. C. C Thomas, Springfield, Illinois.

Ariëns Kappers, C. U., Huber, G. C., and Crosby, E. C. (1936). "The Comparative Anatomy of the Nervous System of Vertebrates Including Man." Macmillan, New York.

Ashby, W. (1919). *J. Exptl. Med.* **29**, 267.

Atwood, W. H. (1947). "A Concise Comparative Anatomy." Mosby, St. Louis, Missouri.

Bailey, P. and Cushing, H. (1926). "A Classification of the Tumors of the Glioma Group on a Histogenetic Basis with a Correlated Study of Prognosis." Lippincott, Philadelphia, Pennsylvania.

Bainbridge, F. A., and Evans, C. L. (1914). *J. Physiol. (London)* **48**, 278.

Banting, F. G., and Best, C. H. (1922). *J. Lab. Clin. Med.* **7**, 251.

Barcroft, J., Harris, H. A., Orahovats, D., and Weiss, R. (1925). *J. Physiol. (London)* **60**, 443.

Barcroft, S., and Stephens, J. G. (1927). *J. Physiol. (London)* **64**, 1.

Bargmann, W. (1959). "Histologie und Mikroskopische Anatomie des Menschen," 3rd ed. Thieme, Stuttgart.

Barlow, C. F., Schoolar, J. C., and Roth, L. J. (1957). *Neurology* **7**, 820.

Bartelmez, G. W. (1940). *Anat. Record* **77**, 509.

Bartone, N. F., and Grieco, R. V. (1955). *Am. J. Surg.* **89**, 170.

Bassett, D. L. (1943). *Anat. Record* **73**, 251.

Bataillon, E. (1910). *Compt. rend. acad. sci.* **150**, 996.

Baumberger, J. P., Suntzeff, V., and Cowdry, E. V. (1942). *J. Natl. Cancer Inst.* **2**, 413.

Bayliss, W. M. (1915). "Principles of General Physiology," 1st ed. Longmans, Green, London.

Bear, R. S. (1952). *Advances in Protein Chem.* **7**, 69.

Beard, J. W., and Rous, P. (1934). *J. Exptl. Med.* **59**, 593.

Beckner, M. (1959). "The Biological Way of Thought." Columbia Univ. Press, New York.

Bělař, K. (1927). Diagram of germinal tract and somatic differentiation, Fig. 283 in M. Hartmann's "Allgemeine Biologie," 1st ed. Fischer, Jena.

Bělař, K. (1930). *Protoplasma* **9**, 209.

Belling, J. (1930). "The Use of the Microscope." McGraw-Hill, New York.

Belt, A. E., Smith, H. P., and Whipple, G. H. (1920). *Am. J. Physiol.* **52**, 101.

Bennett, H. S. (1950). *In* "McClung's Handbook of Microscopical Technique," (Ruth McClung Jones, ed.), 3rd ed., p. 591. Hoeber, New York.

Berkson, J. (1944). *In* "Medical Physics" (O. Glasser, ed.), Vol. 1, p. 110. Yearbook Publishers, Chicago, Illinois.

Bernard, Claude (1865). "Introduction à l'Étude de la Médecine expérimentale," New ed. (1952), Flammarion, Paris.

Bernstein, J. (1894). "Lehrbuch der Physiologie des tierischen Organismus, im speciellen des Menschen." Enke, Stuttgart.

Berrill, N. J. (1955). *In* "Analysis of Development" (B. H. Willier, P. Weiss, and V. Hamburger, eds.), p. 620. Saunders, Philadelphia, Pennsylvania.

Bessis, M. (1956). "Cytology of the Blood and Blood-forming Organs" (E. Ponder, transl. and ed.). Grune & Stratton, New York.

Bessis, M. (1959). *In* "Eisenstoffwechsel" (W. Kreiderling, ed.), p. 11. Thieme, Stuttgart.

Best, C. H., and Taylor, N. B. (1955). "The Physiological Basis of Medical Practice," 6th ed. Williams & Wilkins, Baltimore, Maryland.

Biedermann, W. (1911). *In* "Handbuch der vergleichenden Physiologie" (H. Winterstein, ed.), Vol. 2, Part 2, p. 273, Fischer, Jena.

Biesele, J. J., Berger, R. E., Wilson, A. Y., Hitchings, G. H., and Elion, G. B. (1951). *Cancer* **4**, 186.

Blank, H., McCarthy, P. L., and DeLamater, E. D. (1951). *Stain Technol.* **26**, 193.

Bloch, D. P., Morgan, C., Godman, G. C., Howe, C., and Rose, H. M. (1957). *J. Biophys. Biochem. Cytol.* **3**, 1.

Bloch, H. (1950). *J. Exptl. Med.* **91**, 197.

Blumenthal, H. T., Handler, F. P., and Blache, J. O. (1954). *Am. J. Med.* **17**, 337.

Bodian, D. (1948). *Bull. Johns Hopkins Hosp.* **83**, 1.

Bodian, D. editor (1956). *Am. J. Hyg.* **64**, 104.

Boell, E. J., and Shen, S. C. (1950). *J. Exptl. Zool.* **113**, 583.

Bohr, Niels (1936). *In 2nd Intern. Congr. for Unity of Sci. Copenhagen* (R. Carnap and H. Reichenbach, eds.), pp. 293-303. F. Meiner, Leipzig.

Bok, S. T. (1936). *Verhandel. Koninkl. Akad. Wetenschap. Sect. II,* **35**, 1.

Boll, I. (1958). *Folia Haematol.* [N. F.] **3**, 78.

Bonner, J. T. (1947). *J. Exptl. Zool.* **106**, 1.

Born, G. (1883). *Arch. mikroskop. Anat.* **22**, 584.

Bostick, W. H. (1957). *Trans. N. Y. Acad. Sci.* **20**, 79.

Bowie, D. J., and Vineberg, A. M. (1935). *Quart. J. Exptl. Physiol.* **25**, 247.

Boycott, A. E. (1929). *Proc. Roy. Soc. Med.* **23**, 15.

Boycott, A. E., and Diver, C. (1923). *Proc. Roy. Soc.* **B95**, 207.

Boycott, A. E., Diver, C., Hardy, S. L., and Turner, F. M. (1929). *Proc. Roy. Soc.* **B104**.

Boyden, E. A. (1955). "Segmental Anatomy of the Lungs." McGraw-Hill (Blakiston), New York.
Bozler, E. (1941). *Biol. Symposia* **3**, 95.
Brachet, J. (1947). *In* "Nucleic Acid" (J. F. Danielli and R. Brown, eds.), p. 207. Cambridge Univ. Press., London and New York.
Brachet, J. (1955). *In* "The Nucleic Acids" (E. Chargaff and J. N. Davidson, eds.), Vol. 2, pp. 475-519. Academic Press, New York.
Brachet, J. (1957). "Biochemical Cytology." Academic Press, New York.
Bradley, O. C. (1943). "Topographic Anatomy of the Dog," 4th ed. Macmillan, New York.
Braus, Hermann (1924). "Anatomie des Menschen." Springer, Berlin.
Bremer, J. L., rewritten by Weatherford, H. L. (1944). "Textbook of Histology," 6th ed. Blakiston, Philadelphia, Pennsylvania.
Brickner, R. M. (1927). *Bull. Johns Hopkins Hosp.* **40**, 90.
Briggs, R. and King, T. (1953). *J. Exptl. Zool.* **122**, 485.
Briggs, R. and King, T. J. (1955). *In* "Biological Specificity and Growth" (E. G. Butler, ed.), p. 207. Princeton Univ. Press, Princeton, New Jersey.
Brobeck, J. R. (1946). *Physiol. Revs.* **26**, 54.
Brown, F. A. (1950). *In* "Comparative Animal Physiology" (C. Ladd Prosser, ed.), Chapter 22, p. 725. Saunders, Philadelphia, Pennsylvania.
Brumpt, Émile (1949). "Précis de parasitologie, Collection de précis médicaux" 6th ed. Masson, Paris.
Bryant, J. C., Earle, W. R., and Peppers, E. V. (1953). *J. Natl. Cancer Inst.* **14**, 189.
Bryant, J. C., Evans, V. J., Schilling, E. L., and Earle, W. R. (1961). *J. Natl. Cancer Inst.* **26**, 239.
Bucher, Nancy L. R., Scott, Jesse F., and Aub, Joseph C. (1951). *Cancer Research* **11**, 457.
Buchner, P. (1921). "Tier und Pflanze in intracellulärer Symbiose." Borntraeger, Berlin.
Buchsbaum, R. (1937). *Physiol. Zoöl.* **10**, 373.
Buchsbaum, R. (1948). *Anat. Record* **102**, 19.
Buchsbaum, R., and Buchsbaum, M. (1957). "Basic Ecology." Boxwood Press, Pittsburgh, Pennsylvania.
Buchsbaum, R., and Williamson, R. R. (1943). *Physiol. Zoöl.* **16**, 162.
Bullough, W. S. (1949). *J. Exptl. Biol.* **26**, 83.
Bullough, W. S. (1952). *Biol. Revs. Cambridge Phil. Soc.* **27**, 133.
Burns, B. D. (1950). *J. Physiol. (London)* **111**, 50.
Burns, R. K. (1949). *Survey Biol. Progr.* **1**, 233-266.
Burton, C., Barnes, R. B., and Rochow, T. G. (1942). *Ind. Eng. Chem.,* **34**, 1429.
Cabasso, V. J., Jervis, G. A., Moyer, A. W., Roca-Garcia, M., Orsi, E. V., and Cox, H. R. (1959). *1st Intern. Conf. on Live Poliovirus Vaccines Washington, D. C., Sci. Publ. No.* **44**, *Pan Am. Sanit. Bur. Washington, D. C.*
Callander, C. Latimer (1939). "Surgical Anatomy", 2nd ed. Saunders, Philadelphia, Pennsylvania.
Canti, R. G., and Spear, F. G. (1929). *Proc. Roy. Soc.* **B105**, 93.
Carleton, H. M. (1957). "Histological Technique," 3rd ed. Oxford Univ. Press, London and New York.
Carr, J. G. (1945). *Nature* **156**, 143.
Carrel, A. (1912). *J. Exptl. Med.* **15**, 516.
Carrel, A. (1913). *J. Exptl. Med.* **18**, 155.
Carrel, A. (1923). *J. Exptl. Med.* **38**, 521.
Carrel, A., and Lindbergh, C. A. (1938). "The Culture of Organs." Hoeber, New York.

Carter, Herbert E. (1949). "Biochemical Preparations," Vol. 1. Wiley, New York.

Caspersson, T. (1941). *Naturwissenschaften* **29**, 33.

Caspersson, T. (1950). "Cell Growth and Cell Function." Norton, New York.

Cavendish, Henry (1776). *In* Maxwell's edition (1879) of "Cavendish's Electrical Researches," pp. 194-215. Cambridge Univ. Press, London.

Chain, E. B. Larsson, S., and Pocchiara, F. (1960). *Proc. Roy. Soc.* **B152**, 283.

Chalkley, H. W. (1943). *J. Natl. Cancer Inst.* **4**, 47.

Chambers, R. W., and Kopac, M. J. (1950). *In* "McClung's Handbook of Microscopical Technique" (Ruth McClung Jones, ed.) 3rd ed., p. 492. Hoeber, New York.

Champy, C. (1922). *Arch. morphol. gén. exptl.* **4**, 1.

Chatton, E., and Lwoff, A. (1936). *Bull. soc. franç. microscop.* **5**, 25.

Chaves, A. D., and Abeles, H. (1952). *Am. Rev. Tuberc.* **65**, 128.

Cheronis, N. D. (1954). *In* "Techniques of Organic Chemistry" (A. Weissberger, ed.), Vol. 6, p. 180. Interscience, New York.

Child, C. M. (1941). "Patterns and Problems of Development." Univ. of Chicago Press, Chicago, Illinois.

Christeller, E. (1927). "Atlas der Histotopographie gesunder und erkrankter Organe." Thieme, Leipzig.

Christeller, E., and Eisner, G. (1929). *Beitr. pathol. Anat. u. allgem. Pathol.* **81**, 524.

Christman, Ruth C. (1954). *Science* **119**, 534.

Clara, Max (1956). "Die arterio-venösen Anastomosen," 2nd ed. Springer, Vienna.

Clark, E. R., and Clark, E. L. (1943). *Am. J. Anat.* **73**, 215.

Clark, E. R., Kirby-Smith, H. T., Rex, R. O., and Williams, R. G. (1930). *Anat. Record* **47**, 187.

Clarke, G. L. (1954). "Elements of Ecology." Wiley, New York.

Cohnheim, Julius. (1872). Cited in Sir Howard Florey, "General Pathology" 2nd ed. (1958), pp. 47 and 67. Saunders, Philadelphia.

Cohrs, P., Jaffé, R., and Meessen, H. (1958). "Pathologie der Laboratoriumstiere." Springer, Berlin.

Congdon, E. D. (1915). *Anat. Record* **9**, 343.

Conn, H. J. (1946). "Biological Stains," 5th ed. Biotech. Public. Geneva, New York.

Cook, S. F. (1929). *J. Biol. Chem.* **82**, 595.

Coons, A. H., Creech, H. J., Jones, R. N., and Berliner, E. (1942). *J. Immunol.* **45**, 159.

Costello, Donald P. (1955). *In* "Analysis of Development" (B. H. Willier, P. Weiss, and V. Hamburger, eds.), p. 213. Saunders, Philadelphia, Pennsylvania.

Couch, N. P., Cassie, G. F., and Murray, J. E. (1958). *Surgery* **44**, 666.

Cowdry, E. V. (1952). "Laboratory Technique in Biology and Medicine," 3rd ed. Williams & Wilkins, Baltimore, Maryland.

Crick, F. H. C., and Kendrew, J. C. (1957). *Advances in Protein Chem.* **12**, 133.

Crile, G., Telkes, M., and Rowland, A. F. (1932). *Protoplasma* **15**, 339.

Crosby, William H. (1957). *Blood* **12**, 165.

Cross, Louise Montgomery (1959). "The Preparation of Medical Literature." Lippincott, Philadelphia, Pennsylvania.

Custer, R. P. (1949). "Atlas of the Blood and Bone Marrow." Saunders, Philadelphia, Pennsylvania.

Czellitzer, A. (1928). *Deut. med. Wochschr.* **54**, 1629.

Danielli, J. F. (1946). *J. Exptl. Biol.* **22**, 110.

Davenport, C. B., and Davenport, G. C. (1910). *Am. Naturalist* **44**, 641.

Davidson, Henry A. (1957). "Guide to Medical Writing." Ronald, New York.

Deane, H .W., Nesbett, F. B., and Hastings, A. B. (1946). *Proc. Soc. Exptl. Biol. Med.* **63**, 401.
de Beer, G. R. (1937). "The Development of the Vertebrate Skull." Oxford Univ. Press (Clarendon), London and New York.
DeBruyn, P. P. H., Robertson, R. C., and Farr, R. S. (1950). *Anat. Record,* **108**, 279.
Deichmann, W. B., Kitzmiller, K. V., Witherup, S., and Johansman, R. (1944). *Ann. Internal Med.* **21**, 803.
Deitch, A. D. and Murray, M. R. (1956). *J. Biophys. Biochem. Cytol.* **2**, 433.
Deitch, A. D., and Moses, M. J. (1957). *J. Biophys. Biochem. Cytol.* **3**, 449.
de Kiriline, Louise (1946). *Audubon Mag.* **48**, 284.
Delgado, J. M. R. (1952). *Yale J. Biol. and Med.* **24**, 351.
de Meyer, J. (1909). *Arch. fisiol.* **7**, 96.
Dempsey, E. W., and Singer, M. (1946). *Endocrinology* **38**, 270.
Dempster, W. T., and Williams, R. C. (1952). *In* "Laboratory Technique in Biology and Medicine" (E. V. Cowdry, ed.), 3rd ed., p. 312. Williams & Wilkins, Baltimore, Maryland.
Denny-Brown, D. (1929). *Proc. Roy. Soc.* **B104**, 252.
Deutsche Pathologische Gesellschaft (1925). *Zentr. allgem. Pathol.* **36**, Suppl.
"Diagnostic Standards and Classification of Tuberculosis" (1955), 1st ed. National Tuberculosis Association, New York.
Doan, Charles A. and Ralph, Paul H. (1950). *In* "McClung's Handbook of Microscopical Technique" (Ruth McClung Jones, ed.), 3rd ed., p. 571. Hoeber, New York.
Dobell, C. Clifford (1911). *Arch. Protistenk.* **23**, 269.
Dodson, Edward O. (1946). *Stain Technol.* **21**, 103.
Dorris, Frances (1938). *Wilhelm Roux' Arch. Entwicklungsmech. Organ.* **138**, 323.
Dounce, A. L. (1943). *J. Biol. Chem.* **147**, 685.
Drabkin, David L. (1951). *Physiol. Revs.* **31**, 345.
Dubos, René J. (1945). "The Bacterial Cell." Harvard Univ. Press, Cambridge, Massachusetts.
Duke-Elder, W. S. (1938). "Textbook of Ophthalmology." Mosby, St. Louis, Missouri.
Earle, W. R. (1944). A.A.A.S. Research Conf. on Cancer, pp. 139-153. American Association for the Advancement of Science, Washington, D. C.
Earle, W. R., and Nettleship, A. (1943). *J. Natl. Cancer Inst.* **4**, 213.
Ebeling, Albert H. (1913). *J. Exptl. Med.* **17**, 273.
Eccles, J. C. (1957). "The Physiology of Nerve Cells." Johns Hopkins Press, Baltimore, Maryland.
Edinger, Tilly (1948). "Evolution of the Horse Brain." Geological Society of America, Memoir 25.
Edwards, E. A., and Duntley, S. Q. (1939). *Am. J. Anat.* **65**, 1.
Ehrlich, Paul (1879). *Arch. Anat. u. Physiol., Physiol. Abt.,* p. 571.
Ehrlich, Paul (1900). *Proc. Roy. Soc.* **B66**, 424.
Eickhoff, W. (1949). "Schilddrüse und Basedow." Thieme, Stuttgart.
Elias, H. (1949). *Am. J. Anat.* **84**, 311.
Ellinger, P. (1940). *Biol. Revs. Cambridge Phil. Soc.* **15**, 323.
Ellinger, P., and Hirt, A. (1929). Z. *Anat. Entwicklungsgeschichte* **90**, 791.
Enders, J. F., Weller, T. H., and Robbins, F. C. (1949). *Science* **109**, 85.
Engström, A. (1956). *In* "Physical Techniques in Biological Research" (G. Oster and A. W. Pollister, eds.), Vol. 3, p. 489. Academic Press, New York.
Engström, A., and Finean, J. B. (1958). "Biological Ultrastructure." Academic Press, New York.

Ephrussi, B. (1933). *Arch. anat. microscop.* **29**, 95.

Ephrussi, B. (1935). Phénomènes d'intégration dans les cultures des tissues. *Actualités sci. et ind. No.* **240**, Hermann, Paris.

Ephrussi, B. (1956). *In* "Enzymes: Units of Biological Structure and Function" (O. H. Gaebler, ed.), p. 29. Academic Press, New York.

Eränkö, O. (1955). "Quantitative Methods in Histology and Microscopic Histochemistry." Little, Brown, Boston.

Escher, H. (1919). *Correspondenz-Blatt für Schweizer Ärzte* **49**, 1609.

Evans, V. J. (1957). *J. Natl. Cancer Inst.* **19**, 539.

Evans, V. J., Earle, W. R., Sanford, K. K., Shannon, J. E., and Waltz, H. K. (1951). *J. Natl. Cancer Soc.* **11**, 907.

Farquhar, M., Vernier, R., and Good, R. (1957). *Am. J. Pathol.* **33**, 791.

Farr, R. S. (1951). *Anat. Record* **109**, 515.

Fawcett, D. W., and Ito, S. (1958). *J. Biophys. Biochem. Cytol.* **4**, 135.

Feldman, W. H., and Hinshaw, H. C. (1945). *Am. Rev. Tuberc.* **51**, 582.

Fell, H. B., and Robison, R. (1929). *Biochem. J.* **23**, 767.

Fenn, W. O. (1941). *Biol. Symposia* **3**, 1.

Fenn, W. O. (1945). *In* "Physical Chemistry of Cells and Tissues" (R. Höber, ed.), p. 453. Blakiston, Philadelphia, Pennsylvania.

Fenner, F., Martin, S. P., and Pierce, C. H. (1949). *Ann. N. Y. Acad. Sci.* **52**, 751.

Fergusson, J. D. (1949). *Proc. Roy. Soc. Med.* **42**, 806.

Fernández-Morán, H. (1953). *Exptl. Cell Research* **5**, 255.

Finerty, John C. (1952). *Physiol. Revs.* **32**, 277.

Finerty, John C., and Panos, Thomas C. (1951). *Proc. Soc. Exptl. Biol. Med.* **76**, 833.

Firor, W. M. (1931). *Am. J. Physiol.* **96**, 146.

Firor, W. M., and Gey, G. O. (1945). *Ann. Surg.* **121**, 700.

Fischer, A. (1930). *Arch. pathol. Anat. u. Physiol. Virchow's* **279**, 94.

Fischer, A. (1941). *Naturwissenschaften* **29**, 650.

Fischer, E. (1894). *Ber. Deut. Chem. Ges.* **27**, 2985.

Fitzpatrick, T. B., Larner, J., and Landing, B. H. (1948). *Bull. Intern. Assoc. Med. Museum* **28**, 96.

Flesch, P. (1954). *In* "Physiology and Biochemistry of the Skin" (S. Rothman, ed.), p. 601. Univ. of Chicago Press, Chicago, Illinois.

Flexner, L. B. (1950). *In* "Genetic Neurology" (P. Weiss, ed.), p. 194. Univ. of Chicago Press, Chicago, Illinois.

Flexner, L. B. (1955). *In* "Biochemistry of the Developing Nervous System" (H. Waelsch, ed.), p. 281. Academic Press, New York.

Florey, Sir Howard (1958). "General Pathology," 2nd ed. Saunders, Philadelphia, Pennsylvania.

Forbus, W. D. (1952). "Reaction to Injury," Vol. 2. Williams & Wilkins, Baltimore, Maryland.

Friedenwald, Jonas S., and Buschke, Wilhelm (1944). *J. Cellular Comp. Physiol.* **23**, 95.

Fulton, John F. (1949). "A Textbook of Physiology," 16th ed. Saunders, Philadelphia, Pennsylvania.

Gaffky, G. T. A. (1884). *Mitt. Kaiserl. Gesundheitsamte* (*Berlin*). **2**, 126.

Gaillard, P. J. (1948). *In* "Growth in Relation to Differentiation and Morphogenesis," *Symposia Soc. Exptl. Biol. No.* **2**, 139.

Galileo, G. (1638). "Unterredungen und mathematische Demonstrationen über zwei neue Wissenszweige." German translation by A. von Oettingen (1917). Engelmann, Leipzig.

Gatenby, J. B., and Beams, H. W. (1950). "The Microtomist's Vademecum", 11th ed. Blakiston, Philadelphia, Pennsylvania.

Geiger, A. (1958). *Physiol. Revs.* **38**, 1.

Geigy, R. (1931). *Rev. suisse zool.* **58**, 187.

Gerard, R. W. (1950). *In* "Genetic Neurology" (P. Weiss, ed.), p. 199. Chicago Univ. Press, Chicago, Illinois.

Gersh, I. (1932). *Anat. Record* **53**, 307.

Gersh, I., and Bodian, D. (1943). *J. Cellular Comp. Physiol.* **21**, 253.

Gessler, A. E., and Fullam, E. F. (1946). *Am. J. Anat.* **78**, 245.

Gettner, M. E., and Ornstein, L. (1956). *In* "Physical Techniques in Biological Research" (G. Oster and A. W. Pollister, eds.) Vol. 3, p. 627. Academic Press, New York.

Gey, G. O. (1933). *Am. J. Cancer* **17**, 752.

Glasser, O., Editor (1944). "Medical Physics," Vol. 1. Yearbook Publishers, Chicago, Illinois.

Glasser, O., Editor (1950). "Medical Physics," Vol. 2. Yearbook Publishers, Chicago, Illinois.

Glick, D., Engström, A., and Malmström, B. G. (1951). *Science* **114**, 253.

Glimstedt, G. (1932). *Verhandl. deut. anat. Ges.* **41**. *Anat. Auz. Suppl.* **75**.

Glimstedt, G., and Håkansson, R. (1951). *Nature* **167**, 397.

Goldstein, Kurt (1939). "The Organism — a Holistic Approach to Biology, Derived from Pathological Data in Man," American Psychology Series. American Book Company, New York.

Goldstein, L., and Plaut, W. (1955). *Proc. Natl. Acad. Sci. U. S.* **41**, 874.

Gomori, George (1952). "Microscopic Histochemistry." Univ. of Chicago Press, Chicago, Illinois.

Gomori, George (1957). *In* "Methods in Enzymology" (S. P. Colowick and N. O. Kaplan, eds.), Vol. 4, p. 381. Academic Press, New York.

Gonzales, Th. A., Vance, M. Helpern, M., and Umberger, Ch. J. (1954). "Legal Medicine," 2nd ed. Appleton-Century-Crofts, New York.

Goodrich, E. S. (1958). "Studies on the Structure and Development of Vertebrates," Dover edition. Dover Publications, New York.

Gordon, G., and Phillips, C. G. (1953). *Quart. J. Exptl. Physiol.* **38**, 35.

Gore, R. C. (1950). *Discussions Faraday Soc. No.* **9**, 138.

Gough, J., and Wentworth, J. E. (1948). *Proc. 9th Intern. Congr. Ind. Med., London,* pp. 661-665.

Gould, S. E. (1953). "Pathology of the Heart." C. C Thomas, Springfield, Illinois.

Grafflin, A. L. (1947). *Am. J. Anat.* **81**, 63.

Grafflin, A. L., Marble, A., and Smith, R. M. (1941). *Anat. Record* **81**, 495.

Granick, S. (1943). *J. Biol. Chem.* **149**, 157.

Gray, D. F. (1959). *J. Hyg.* **57**, 473.

Gray, P. (1954). "The Microtomist's Formulary and Guide." McGraw-Hill (Blakiston), New York.

Gray, P., Cox, G., Worthing, H., Everhart, W. H., and MacDonald, E. (1942). *Anat. Record* **82**, 416.

Greene, Harry S. N. (1941). *J. Exptl. Med.* **73**, 475.

Gregg, John R., and Løvtrup, Soren (1950). *Compt. rend. trav. Carlsberg, Sér. chim.* **27**, 307.

Grieco, R. V. (1939). "On meristematic cells in stratified epithelia of mammals." M.S. Thesis, Dept. of Pathology, American Univ. of Beirut, Lebanon.

Gross, Jerome (1956). *J. Biophys. Biochem. Cytol.* **2**, (Suppl.), 261.

Gross, Sidney W. (1944). *In* "Medical Physics," (O. Glasser, ed.), Vol. 1, p. 1262. Yearbook Publishers, Chicago, Illinois.

Grün, H., and Klinner, W. (1952). *Arch. pathol. Anat. u. Physiol. Virchow's* **332**, 311.

Gubler, Clark J. (1956). *Science* **123**, 87.

György, Paul, and Staff Editors (1958). "Germ Free Animal Studies." Univ. of Pennsylvania project. Walter Reed Army Institute of Research, Washington, D. C.

Haeckel, Ernst (1899-1904). "Kunstformen der Natur." Leipzig Bibliographisches Institut.

Halasz, N. A., Halloran, K. H., and Liebow, A. A. (1946). *Circulation* **15**, 826.

Hallauer, C. (1931). *Z. Hyg. Infektionskrankh.* **113**, 61.

Ham, A. W. (1957). "Histology," 3rd ed. Lippincott, Philadelphia, Pennsylvania.

Hamburger, V. (1942). "Manual of Experimental Embryology." Univ. of Chicago Press, Chicago, Illinois.

Hamilton, J. G., Solely, M. H., and Eichhorn, K. B. (1940). *Univ. Calif. (Berkeley) Publs. Pharmacol.* **1**, 339.

Hamilton, W. F., Brewer, G., and Brotman, I. (1934). *Am. J. Physiol.* **107**, 427.

Hämmerling, J. (1934). *Naturwissenschaften* **22**, 829.

Hansemann, D. (1893). "Studien über die Spezificität, den Altruismus und die Anaplasie der Zellen." Hirschwald, Berlin.

Hardy, A. C. (1956). "The Open Sea." Houghton Mifflin, Boston, Massachusetts.

Hardy, J. D., Wolff, H. G., and Goodell, H. (1950). *J. Clin. Invest.* **29**, 115.

Harris, R. J. C., Editor. (1954). "Biological Applications of Freezing and Drying." Academic Press, New York.

Harrison, R. G. (1921). *J. Exptl. Zool.* **32**, 1.

Harrison, R. G. (1928). *Arch. exptl. Zellforsch.* **6**, 4.

Harrison, R. G., Astbury, W. T., and Rudall, K. M. (1940). *J. Exptl.* Zool. **85**, 339.

Hartoch, W. (1931). *Arch. pathol. Anat. u. Physiol. Virchow's* **281**, 507.

Hartoch, W. (1933). *Klin. Wochschr.* **12**, Jahrgang **24**, 942.

Hartroft, W. Stanley (1954). *Anat. Record* **119**, 71.

Harvey, E. B. (1950). *In* "McClung's Handbook of Microscopical Technique" (Ruth McClung Jones, ed.) 3rd ed., p. 586. Hoeber, New York.

Harvey, E. N., and Loomis, A. L. (1930). *Science* **72**, 42.

Hauduroy, P. (1950). *In* "Bacilles Tuberculeux et Paratuberculeux" (P. Hauduroy, ed.), p. 31. Masson, Paris.

Hayashi, Teru, Editor (1959). "Subcellular Particles" Symposium. Ronald, New York.

Haythorn, Samuel R., and Taylor, Fred A. (1945). *Am. J. Pathol.* **21**, 123.

Hegner, R., and Hewitt, R. (1937). *Science* **85**, 568.

Heidenhain, R. (1875). *Arch. ges. Physiol.* **10**, 557.

Hendricks, J., Morgan, A., and Freytag, R. (1947). *Am. J. Physiol.* **149**, 319.

Herrera, Alfonso L. (1932). *Protoplasma* **15**, 361.

Hershberger, L. R., and Lillie, R. D. (1947). *J. Tech. Methods* **26**, 162.

Hesse, R. (1924). "Tiergeographie auf Ökologischer Grundlage." Fischer, Jena.

Heuser, C. H., and Streeter, G. L. (1929). *Carnegie Inst. Wash. Contribs. Embryol.* **20**, 1.

Hewitt, R. I. (1940). "Bird Malaria." Johns Hopkins Press, Baltimore, Maryland.

Heyl, J. G., and Laqueur, E. (1935). *Arch. intern. pharmacodynamie* **49**, 338.

Hiebert, C. A. (1959). *Cancer* **12**, 663.

Himmelweit, F. (1938). *Brit. J. Exptl. Pathol.* **19**, 108.

Hirschberg, E. (1958). *Cancer Research* **18**, 869.

His, Wilhelm (1874). "Unsere Körperform und das Physiologische Problem ihrer Entstehung." F. C. W. Vogel, Leipzig.

His, Wilhelm (1901). *Arch. Anat. u. Entwicklungsgeschichte,* p. 307.

Höber, Rudolf (1945). "Physical Chemistry of Cells and Tissues." Blakiston, Philadelphia.
Hodgkin, A. L. (1951). *Biol. Revs. Cambridge Phil. Soc.* **26**, 339.
Hodgkin, A. L., and Huxley, A. F. (1939). *Nature* **144**, 710.
Hoerr, Normand (1944). *In* "Medical Physics" (O. Glasser, ed.), Vol. 1, p. 625. Yearbook Publishers, Chicago, Illinois.
Hokin, L. E. (1955). *Biochim. et Biophys. acta* **18**, 379.
Holmes, Oliver Wendell (1861). "Elsie Venner," 35th ed. (1889). Houghton Mifflin, Boston.
Holt, S. J. (1959). *In* "The Structure and Function of Subcellular Components," *Biochem. Soc. Symposium (Cambridge, Engl.) No.* **16**, 44.
Holtfreter, J. (1934). *Arch. exptl. Zellforsch.* **15**, 281.
Holtfreter, J. (1948). *Ann. N. Y. Acad. Sci.* **49**, 709.
Holtfreter, J., and Hamburger, V. (1955). *In* "Analysis of Development" (B. H. Willier, P. Weiss, and V. Hamburger, eds.), p. 230. Saunders, Philadelphia, Pennsylvania.
Hooghwinkel, G. J. M., Smits, G., and Kroon, D. B. (1954). *Biochim. et Biophys. Acta* **15**, 78.
Hopps, H. C. (1959). "Principles of Pathology." Appleton-Century-Crofts, New York.
Horning, E. S. (1951). *In* "Cytology and Cell Physiology," (G. H. Bourne, ed.), 2nd ed., p. 287. Oxford Univ. Press (Clarendon), London and New York.
Horsley, V., and Clarke, R. H. (1908). *Brain* **31**, 45.
Hörstadius, Sven (1950). *In* "McClung's Handbook of Microscopical Technique," (Ruth McClung Jones, ed.), 3rd ed., p. 555. Hoeber, New York.
Houssay, Bernardo A. (1955). "Human Physiology," 2nd ed. McGraw-Hill, New York.
Hubbert, M. King (1937). *Bull. Geol. Soc. Am.* **48**, 1459.
Hueck, W. (1921). *In* "Handbuch der allgemeinen Pathologie" (Krehl and Marchand, eds.), Vol. 3, p. 298. Hirzel, Leipzig.
Hueck, W. (1937). "Morphologische Pathologie," 1st ed. Thieme, Leipzig.
Hueck, W. (1948). "Morphologische Pathologie," 2nd ed. Thieme, Leipzig.
Huettner, A. F. (1949). "Fundamentals of Comparative Embryology of the Vertebrates." Macmillan, New York.
Hughes, A. (1952). "The Mitotic Cycle." Academic Press, New York.
Huxley, H. E. (1957). *J. Biophys. Biochem. Cytol.* **3**, 631.
Huxley, H. E. (1958). *Sci. Am.* **199** (5), 67.
Huxley, Julian S., and de Beer, G. R. (1934). "Elements of Embryology." Cambridge Univ. Press, London and New York.
Hydén, H. (1943). *Acta Physiol. Scand.* **6**, Suppl. 17, 1.
Hyman, Libby H. (1951). "The Invertebrates," Vol. 3. McGraw-Hill, New York.
Islami, A. H., Pack, G. T., and Hubbard, J. C. (1959). *Surg. Gynecol. Obstet.* **108**, 549.
Jackson, B. (1959). *Anat. Record* **134**, 365.
Jackson, C., and Jackson, C. L. (1944). *In* "Medical Physics" (O. Glasser, ed.), Vol. 1, p. 125. Yearbook Publishers, Chicago, Illinois.
Jacoby, F. (1937). *Arch. exptl. Zellforsch. Gewebezücht.* **19**, 241.
Jacoby, F., and Marks, J. (1953). *J. Hyg.* **51**, 541.
Jacoby, F., Trowell, O. A., and Willmer, E. N. (1937). *J. Exptl. Biol.* **14**, 255.
Janssen, W. A., Fukui, G. T., and Surgalla, M. J. (1958). *J. Infectious Diseases* **103**, 183.
Japp, F. R. (1898-1899). *Nature* **59**, 29, 55, and 101.
Jensen, C. O. (1903). *Zentr. Bakteriol.* **34**, 28.
Jones, F. W. (1920). "The Principles of Anatomy As Seen in the Hand." Blakiston, Philadelphia.

Jungherr, E. L., Cabasso, V. J., Moyer, A. W., and Cox, H. R. (1961). *J. Infectious Diseases* **108**, 247.

Katsch, G., and Borchers, E. (1913). *Z. exptl. Pathol. u. Therap.* **12**, 225.

Kaufmann, Carl, and Lehmann, Erich (1926). *Arch. pathol. Anat. u. Physiol. Virchow's* **261**, 623.

Kazeeff, W.-N. (1939). *Compt. rend. soc. biol.* **130**, 722.

Kemp, T., and Engelbreth-Holm, T. (1931). *Arch. Exptl. Zellforsch. Gewebezücht.* **10**, 117.

Kerr, S. E. (1938). *J. Biol. Chem.* **123**, 443.

Kerr, S. E., and Ghantus, M. (1936). *J. Biol. Chem.* **115**, 9.

Kerr, S. E., and Ghantus, M. (1937). J. Biol. Chem. **117**, 217.

Kersten, H., Brosene, W. G., Jr., Ablondi, F., and SubbaRow, Y. (1947). *J. Lab. Clin. Med.* **32**, 3.

Kiaer, S. (1927). *Arch. klin. Chi. Langenbecks* **149**, 146.

Kiernan, F. (1833). *Phil. Trans. Roy. Soc. London* **B123**, 711. Quoted by F. P. Mall (1906).

King, T. J., and Briggs, R. (1956). *Cold Spring Harbor Symposia Quant. Biol.* **21**, 271.

Knaysi, Georges (1951). "Elements of Bacterial Cytology," 2nd ed. Cornell Univ. Press (Comstock), Ithaca, New York.

Knisely, Melvin H. (1950). *In* "McClung's Handbook of Microscopical Technique" (Ruth McClung Jones, ed.), 3rd ed., p. 477. Hoeber, New York.

Knower, H. McE. (1950). *In* "McClung's Handbook of Microscopical Technique" (Ruth McClung Jones, ed.). 3rd ed., p. 544. P. Hoeber, Inc., New York.

Knowlton, F. P., and Starling, E. H. (1912). *J. Physiol. (London)* **44**, 206.

Koehler, Wilhelm (1932). *Z. Morphol. u. Ökol. Tiere* **24**, 582.

Koelle, G. B. (1951). *J. Pharmacol. Exptl. Therap.* **103**, 153.

Kölliker, A. (1867). "Handbuch der Gewebelehre des Menschen," 5th ed. Engelmann, Leipzig.

Kopac, M. J. (1956). *Ann. N. Y. Acad. Sci.* **63**, 1219.

Kracht, J., and Spaethe, M. (1953). *Arch. pathol. Anat. u. Physiol. Virchow's* **324**, 83.

Krieg, Wendell, J. S. (1946). *Quart. Bull. Northwestern Univ. Med. School* **20**, 199.

Krogh, A. (1922). "Anatomy and Physiology of Capillaries." Yale Univ. Press, New Haven, Connecticut.

Kudo, Richard R. (1954). "Protozoology," 4th ed. C. C Thomas, Springfield, Illinois.

Kugler, J. H., and Wilkinson, W. J. C. (1959). *J. Histochem. and Cytochem.* **7**, 398.

Landström, H., Caspersson, T., and Wohlfart, G. (1941). *Z. mikroskop.-anat. Forsch. (Abt. 2, Jahrb. Morphol. mikroskop. Anat.)* **49**, 534.

Langeron, M. (1949). "Précis de Microscopie," 7th ed. Masson, Paris.

Lanman, J. T., Bierman, H. R., and Byron, R. L., Jr. (1950). *Blood* **5**, 1099.

Larner, J., Illingworth, B., Cori, G. T., and Cori, C. F. (1952). *J. Biol. Chem.* **199**, 641.

Lasnitzki, I. (1951). *Brit. J. Cancer* **5**, 345.

Latta, H., and Hartmann, J. F. (1950). *Proc. Soc. Exptl. Biol. Med.* **74**, 436.

Lawrence, A. S. C., Miall, M., Needham, J., and Shen, S. C. (1944). *J. Gen. Physiol.* **27**, 233.

Lawrence, J. S., Ervin, D. M., and Wetrich, R. W. (1945). *Am. J. Physiol.* **144**, 284.

Laws, R. M. (1957). *Nature* **179**, 1011.

Leblond, C. P., and Bogoroch, R. (1952). *In* "Laboratory Technique in Biology and Medicine" (E. V. Cowdry, ed.), 3rd ed., p. 297. Williams & Wilkins, Baltimore, Maryland.

Leblond, C. P., and Messier, B. (1958). *Anat. Record* **132**, 247.

Leblond, C. P., and Stevens, C. E. (1948). *Anat. Record* **100**, 357.

Leblond, C. P., Stevens, C. E., and Bogoroch, R. (1948). *Science* **108**, 531.

Leblond, C. P., and Walker, B. E. (1956). *Physiol. Revs.* **36**, 255.
Lederberg, J. (1952). *Physiol. Revs.* **32**, 403.
Leduc, S. (1911). "The Mechanism of Life," transl. by W. D. Butcher, Rebman, London.
Lehmann, Otto, (1910). "Flüssige Kristalle." Barth, Leipzig.
Leopold, I. H. (1952). *Trans. Am. Ophthalmol. Soc.* **49**, 625.
Levi-Montalcini, R. (1950). *J. Morphol.* **86**, 253.
Levine, M. (1951). *Ann. N. Y. Acad. Sci.* **51**, 1365.
Levy, F. (1920). *Sitzber. preuss. Akad. Wiss.* **24**, 417.
Lewis, F. T. (1925). *Proc. Am. Acad. Arts Sci.* **61**, 1.
Lewis, F. T. (1943). *Am. J. Botany* **30**, 74.
Lewis, M. R., and Geiling, E. M. K. (1935). *Am. J. Physiol.* **113**, 529.
Lewis, T. (1927). "Blood Vessels of the Human Skin." Shaw, London.
Liebow, A. A. (1953). *Am. J. Pathol.* **29**, 251.
Liebow, A. A., Hales, M. R., Lindskog, G. E., and Bloomer, W. E. (1947). *J. Tech. Methods* **27**, 116.
Lillibridge, C. (1960). *Anat. Record* **136**, 234.
Lillie, R. D. (1954). "Histopathologic Technic and Practical Histochemistry," 2nd ed. McGraw-Hill (Blakiston), New York.
Lillie, Ralph S. (1918). *Science* **48**, 51.
Lillie, Ralph S. (1925). *J. Gen. Physiol.* **7**, 473.
Linderstrøm-Lang, K. (1939). Harvey Lecture, *Bull. N. Y. Acad. Med.* **15**, 719.
Linderstrøm-Lang, K. Holter, H., and Søeborg Ohlsen, A. (1934). *Z. physiol. Chem. Hoppe-Seyler's* **227**, 1.
Lindskog, G. E., and Liebow, A. A. (1953). "Thoracic Surgery and Related Pathology." Appleton-Century-Crofts, New York.
Lison, L. (1953). "Histochimie et Cytochimie Animales," 2nd ed. Gauthier-Villars, Paris.
Litchfield, J. T., Jr. (1949). *J. Pharmacol. Exptl. Therap.* **97**, 399.
Litchfield, J. T., Jr., White, H. J., and Marshall, E. K., Jr. (1939). *J. Pharmacol. Exptl. Therap.* **67**, 437.
Litchfield, J. T., Jr., and Wilcoxon, F. (1955). *Anal. Chem.* **27**, 299.
Lloyd, D. P. C. (1949). *In* "Textbook of Physiology" (J. F. Fulton, ed.), 16th ed., p. 6. Saunders, Philadelphia, Pennsylvania.
Loeb, L. (1945). "The Biological Basis of Individuality." C. C Thomas, Springfield, Illinois.
Loewi, O. (1921). *Arch. ges. Physiol. Pflüger's* **189**, 239.
Looss, A. (1905). "Die Wanderung der Ankylostomum-und Strongiloideslarven von der Haut nach dem Darm." *Compt. rend. congr. intern. zool. Geneva.*
Lotka, A. J. (1956). "Elements of Mathematical Biology." Dover Publications, New York.
Low, Frank N. (1953). *Anat. Record* **117**, 241.
Lowry, O. H. (1953). *J. Histochem. and Cytochem.* **1**, 420.
Lowry, O. H., Robert, N. R., and Chang, M-L. W. (1956). *J. Biol. Chem.* **222**, 97.
Lucké, B., and Mallory, T. (1946). *Am. J. Pathol.* **22**, 867.
Ludford, R. J. (1935). *Arch. exptl. Zellforsch. Gewebezücht.* **17**, 339.
Ludwig, Wilhelm (1932). "Das Rechts-Links-Problem im Tierreich und beim Menschen." Springer, Berlin.
Luft, John, and Hechter, Oscar (1957). *J. Biophys. Biochem. Cytol.* **3**, 615.
Luyet, B. J. (1951). *In* "Freezing and Drying" (R. J. C. Harris, ed.), p. 77. The Institute of Biology, London.
Macdougald, T. G. (1937). *Arch. exptl. Zellforsch. Gewebezücht.* **20**, 35.

Macdougald, T. J., and Gatenby, J. B. (1935). *Arch. exptl. Zellforsch. Gewebezücht.* **17**, 325.
McClung Jones, Ruth, Editor (1950). "McClung's Handbook of Microscopical Technique," 3rd ed. Hoeber, New York.
McCune, R. M., Jr., and Tompsett, R. (1956). *J. Exptl. Med.* **104**, 737.
McCune, R. M., Jr., Tompsett, R., and McDermott, W. (1956). *J. Exptl. Med.* **104**, 763.
McIlwain, Henry (1955). "Biochemistry and the Central Nervous System," Little, Brown, Boston, Massachusetts.
McQuilkin, W. T., Evans, V. J., and Earle, W. R. (1957). *J. Natl. Cancer Inst.* **19**, 885.
Magnan, A. (1932). "Exposés de morphologie dynamique (mécanique de mouvement)," *Actualités sci. et ind.* Nos. **35** and **46**. Hermann, Paris.
Mall, F. P. (1906). *Am. J. Anat.* **5**, 227.
Malpighi, M. (1661). Quoted in Encyclopedia Britannica (1953), **14**, 731.
Mann, Gustav (1902). "Physiological Histology. Methods and Theory." Clarendon Press, Oxford.
Maren, T. H., Mayer, E., and Wadsworth, B. C. (1954). *Bull. Johns Hopkins Hosp.* **59**, 199.
Markee, J. E. (1929). *Am. J. Obstet. Gynecol.* **17**, 205.
Markowitz, J., Archibald, J., and Downie, H. G. (1954). "Experimental Surgery," 3rd ed. Williams & Wilkins, Baltimore, Maryland.
Markowitz, J., and Essex, H. E. (1930). *Am. J. Physiol.* **92**, 205.
Marks, J., and James, D. M. (1953). *J. Hyg.* **51**, 340.
Marshall, E. K., Jr., Litchfield, J. T., Jr., and White, H. J. (1942). *J. Pharmacol. Exptl. Therap.* **75**, 89.
Marshall, I. M. (1954). *Exptl. Cell Research* **6**, 240.
Martin, B. F., and Jacoby, F. (1949). *J. Anat.* **83**, 351.
Martini, E. (1912). *Z. wiss. Zool.* **102**, 425.
Martinovitch, P. N. (1950). *Nature* **165**, 33.
Marvin, J. W. (1939). *Am. J. Botany* **26**, 280.
Matzke, E. B. (1950). *Bull. Torrey Botan. Club* **77**, 222.
Maximow, A. (1906). *Arch. mikroskop. Anat.* **67**, 680.
Maximow, A. (1910). *Arch. mikroskop. Anat.* **76**, 1.
Maximow, A., and Bloom, W. (1952). "A Textbook of Histology," 6th ed. Saunders, Philadelphia, Pennsylvania.
Mayer, E. (1922). *Arch. pathol. Anat. u. Physiol. Virchow's* **236**, 279.
Mayer, E. (1933). *Wilhelm Roux' Arch. Entwicklungsmech. Organ.* **130**, 382.
Mayer, E. (1934). *Arch. exptl. Zellforsch. Gewebezücht.* **16**, 23.
Mayer, E. (1935). *Skand. Arch. Physiol.* **72**, 249.
Mayer, E. (1937). *Scientia* **61**, 101.
Mayer, E. (1939). *In "Tabulae Biologicae" Vol.* **19**, pp. 65-275. W. Junk, The Hague.
Mayer, E. (1942). *Anat. Record* **84**, 359.
Mayer, E. (1947). *Endocrinology* **40**, 165.
Mayer, E. (1949). *Anat. Record* **103**, 71.
Mayer, E., and Fürstenheim, A. (1930). *Arch. pathol. Anat. u. Physiol. Virchow's* **278**, 391.
Mayer, E., Jackson, E., Whiteside, E., and Alverson, C. (1954). *Am. Rev. Tuberc.* **69**, 419.
Mayer, E., and Ottolenghi, L. (1947). *Anat. Record* **99**, 477.
Mayer, E., and Ruzicka, A. Q. (1945). *Anat. Record* **93**, 213.
Mayer, E., and Schreiber, H. (1934). *Protoplasma* **21**, 34.
Medawar, P. B. (1941). *Nature* **148**, 783.
Meessen, Hubert (1952). "Experimentelle Histopathologie." Thieme, Stuttgart.
Meier, R., Schuler, W., and Desaulles, P. (1950). *Experientia* **6**, 469.

Mellors, R. C. (1959). *In* "Analytical Cytology" (R. C. Mellors, ed.), 2nd ed., p. 1. McGraw-Hill, New York.
Mellors, R. C., Kupfer, A. and Hollender, A. (1953). *Cancer* **6**, 372.
Melnick, J. L., and Brennan, J. C. (1959). *1st Intern. Conf. on Live Poliovirus Vaccines Washington, D. C., Sci. Publ. No.* **44**, Pan Am. Sanit. Bur. Washington, D. C.
Melnick, J. L., and Godman, G. C. (1951). *J. Exptl. Med.* **93**, 247.
Merrill, J. P. (1960). *A.M.A. Arch. Internal Med.* **106**, 143.
Meyer, K., and Loewenthal, H. (1928). *Z. Immunitätsforsch.* **54**, 420.
Michaelis, Leonor (1920). *Arch. mikroskop. Anat.* **94**, 580.
Miescher, Friedrich (1871). *Hoppe-Seyler's med.-chem. Untersuch.* 441.
Miller, T. G., and Abbott, W. O. (1934). *Am. J. Med. Sci.* **187**, 595.
Minami, S. (1923). *Biochem. Z.* **142**, 334.
Minot, C. S. (1910). "Laboratory Textbook of Embryology," 2nd ed. Blakiston, Philadelphia, Pennsylvania.
Mitchison, J. M., and Swann, M. M. (1954). *J. Exptl. Biol.* **31**, 443.
Montagna, W. (1949). *Anat. Record* **103**, 77.
Montagna, W., Chase, H. B., and Hamilton, J. B. (1951). *J. Invest. Dermatol.* **17**, 147.
Moolten, S. E. (1935). *Arch. Pathol.* **20**, 77.
Moore, R. A. (1951). "Textbook of Pathology," 2nd ed. Saunders, Philadelphia, Pennsylvania.
Morau, Henry (1894). *Arch. méd. exptl.* **6**, 677.
More, Robert H., and Duff, G. Lyman (1951). *Am. J. Pathol.* **27**, 95.
Morgan, J. F., Morton, H. J., and Parker, R. C. (1950). *Proc. Soc. Exptl. Biol. Med.* **73**, 1.
Morton, Avery Adrian (1938). "Laboratory Technique in Organic Chemistry," 1st ed. McGraw-Hill, New York.
Murray, M. R., and Kopech, G. (1953). "A Bibliography of the Research in Tissue Culture 1884-1950," 2 vols. Academic Press, New York.
Murray, M. R., Peterson, E. R., Hirschberg, E., and Pool, J. L. (1954). *Ann. N. Y. Acad. Sci.* **58**, 1147.
Murray, M. R., and Stout, A. P. (1954). *Texas Repts. Biol. and Med.* **12**, 898.
Murray, M. R., and Stout, A. P. (1958). *In* "Treatment of Cancer and Allied Diseases" (G. T. Pack and I. M. Ariel, eds.), 2nd ed., Vol. 1, p. 124. Hoeber, New York.
Naegeli, O., de Quervain, F., and Stalder, W. (1930). *Klin. Wochschr.* **9**, 924.
Needham, J. (1931). "Chemical Embryology," Vol. 1. Cambridge Univ. Press, London and New York.
Needham, J. (1942). "Biochemistry and Morphogenesis." Cambridge Univ. Press, London and New York.
Neurath, H., Greenstein, J. P., Putnam, F. W., and Erickson, J. O. (1944). *Chem. Revs.* **34**, 157.
Newell, R. R. (1944). *In* "Medical Physics" (O. Glasser, ed.), Vol. 1, p. 1269. Yearbook Publishers, Chicago, Illinois.
Newman, H. H. (1936). "Outlines of General Zoology," 3rd ed. Macmillan, New York.
Newman, S. B., Boryska, E., and Swerdlow, M. (1949). *Science* **110**, 66.
Nielsen, N., Okkels, H., and Stochholm-Borresen, C. Chs. (1932). *Acta Pathol. Microbiol. Scand.* **9**, 258.
Nissl, F. (1892). *Allgem. Z. Psychiat.* **48**, 197.
Nissl, F. (1910). *In* "Enzyklopädie der mikroskopischen Technik" (P. Ehrlich *et al.*, eds.), 2nd ed., Vol. 2, p. 252. Urban u. Schwarzenberg, Berlin and Vienna.
Noll, H., Bloch, H., Asselineau, J., and Lederer, E. (1956). *Biochim. et Biophys. Acta* **20**, 299.

Novikoff, A. B. (1951). *Science* **113**, 320.

Oliver, J. R. (1954). *In* "Ciba Foundation Symposium on the Kidney" (A. A. G. Lewis *et al.*, eds.), p. 1. Little, Brown, Boston, Massachusetts.

Olszewski, J. (1952). "The Thalamus of the Macaca Mulatta." Karger, Basel.

Osterhout, W. J. V. (1927-1928). *J. Gen. Physiol.* **11**, 83.

Osterhout, W. J. V., and Harris, E. S. (1927–1928). *J. Gen. Physiol.* **11**, 391.

Ostwald, Wolfgang (1908). "Über die zeitlichen Eigenschaften der Entwicklungsvorgänge." Engelmann, Leipzig.

Pack, G. T., and Ariel, I. M., Editors. (1958). "Treatment of Cancer and Allied Diseases," 2nd ed., Vol. 1, Hoeber, New York.

Palade, G. E. (1952). *J. Exptl. Med.* **95**, 285.

Palade, G. E. (1959). *In* "Subcellular Particles" (T. Hayashi, ed.), p. 64. Ronald, New York.

Palade, G. E., and Siekevitz, P. (1956). *J. Biophys. Biochem. Cytol.* **2**, 671.

Palay, Sanford L. (1958). *Exptl. Cell Research, Suppl.* **5**, 275.

Papanicolau, George N. (1954). "Atlas of Exfoliative Cytology." Harvard Univ. Press, Cambridge, Massachusetts.

Parker, R. A. (1958). *Stanford Med. Bull.* **16**, 87.

Parker, R. C. (1929). *Arch. exptl. Zellforsch. Gewebezücht.* **8**, 340.

Parker, R. C. (1950). "Methods of Tissue Culture," 2nd ed. Hoeber, New York.

Parker, R. C., Castor, L. N., and McCulloch, E. A. (1957). *In* "Cellular Biology, Nucleic Acids and Viruses," *Spec. Publ. N. Y. Acad. Sci. No.* **5**, 303.

Parker, T. J., and Haswell, W. A. (1940). "Textbook of Zoology," 6th ed., revised by O. Lowenstein, Macmillan, New York and London.

Pasieka, A. E., Morton, H. J., and Morgan, J. F. (1956). *J. Natl. Cancer Inst.* **16**, 995.

Pasteur, L. (1860). "Recherches sur la dissymétrie moléculaire des produits organiques naturels," Oeuvres de Pasteur réunies. Masson, Paris, 1922.

Paterson, R. (1948). "The Treatment of Malignant Disease by Radium and X-Rays," Arnold, London.

Patten, B. M. (1953). "Human Embryology," 2nd ed. McGraw-Hill (Blakiston), New York.

Pearse, A. G. E. (1960). "Histochemistry," 2nd ed., Little, Brown, Boston, Massachusetts.

Perutz, M. F. (1958). *Endeavour* **17**, 190.

Peterson, E. R., and Murray, M. R. (1955). *Am. J. Anat.* **96**, 319.

Peterson, W. H., and Strong, F. M. (1953). "General Biochemistry." Prentice-Hall, Englewood Cliffs, New Jersey.

Petrunkevitch, A. (1937). *Anat. Record* **68**, 267.

Pfeffer, W. (1897). "Pflanzenphysiologie," 2nd ed. Engelmann, Leipzig.

Pflüger, E. F. W. (1905). "Das Glykogen und seine Beziehungen zur Zuckerkrankheit." Hager, Bonn.

Pickering, G. W. (1943). *Brit. Med. J. I,* 1, *II,* 31.

Politzer, G. (1931). *Z. Zellforsch. u. Mikroskop. Anat.* **13**, 334.

Pomerat, C. M., Drayer, G. A., and Painter, L. T. (1946). *Proc. Soc. Exptl. Biol. Med.* **63**, 322.

Popper, H., and Schaffner, F. (1957). "Liver: Structure and Function." McGraw-Hill (Blakiston), New York.

Porter, K. R. (1953). *J. Exptl. Med.* **97**, 727.

Porter, K. R., and Blum, J. (1953). *Anat. Record* **117**, 685.

Porter, K. R., Claude, A., and Fullam, E. F. (1945). *J. Exptl. Med.* **81**, 233.

Porter, K. R., and Kallman, F. (1953). *Exptl. Cell Research* **4**, 127.

Pratt, H. S. (1935). "A Manual of the Common Invertebrate Animals (Exclusive of Insects)." Blakiston, Philadelphia.
Prinzmetal, M., Ornitz, E. M., Jr., Simkin, B., and Bergman, H. C. (1948). *Am. J. Physiol.* **152**, 48.
Prosser, C. L. (1950). *In* "Comparative Animal Physiology" (C. L. Prosser, ed.), p. 776. Saunders, Philadelphia, Pennsylvania.
Puck, T. T., Marcus, P. I., and Cieciura, S. J. (1956). *J. Exptl. Med.* **103**, 273.
Raczkowski, H. A., Kloos, K., and Opitz, E. (1953). *Z. Krebsforsch.* **59**, 261.
Raleigh, G. W., and Youmans, G. P. (1948). *J. Infectious Diseases* **82**, 197.
Ranvier, L. (1889). "Traité technique d'histologie," 2nd ed. Savy, Paris.
Rappaport, A. M., Borowy, Z. J., Lougheed, W. M., and Lotto, W. N. (1954). *Anat. Record* **119**, 11.
Rashevsky, N. (1928). *Z. Physik* **46**, 568.
Rashevsky, N. (1940). "Advances and Applications of Mathematical Biology." Univ. of Chicago Press, Chicago, Illinois.
Rashevsky, N. (1948). "Mathematical Biophysics." Univ. of Chicago Press, Chicago, Illinois.
Rashevsky, N. (1949). *Bull. Math. Biophys.* **11**, 1.
Rasmussen, A. T. (1921). *Endocrinology* **5**, 760.
Rauber-Kopsch (1940). "Lehrbuch und Atlas der Anatomie des Menschen" (F. Kopsch, ed.), 15th ed., 3 Vols. Thieme, Leipzig.
Rawles, M. E. (1955). *In* "Analysis of Development" (B. H. Willier, P. Weiss, and V. Hamburger, eds.), p. 499. Saunders, Philadelphia, Pennsylvania.
Regaud, Cl., and Policard, A. (1913). *Compt. rend. soc. biol.* **74**, 449.
Reyniers, J. A., Editor (1943). "Micrurgical and Germ Free Techniques." C. C Thomas, Springfield, Illinois.
Rhumbler, L. (1923). *In* "Handbuch der biologischen Arbeitsmethoden" (E. Abderhalden, ed.) Abt. V, 5. Teil, 3A, pp. 219-440. Urban und Schwarzenberg, Berlin and Vienna.
Rich, A. R. (1951). "The Pathogenesis of Tuberculosis," 2nd ed. C. C Thomas, Springfield, Illinois.
Rich, A. R., Berthrong, M., and Bennet, I. L., Jr. (1950). *Bull. Johns Hopkins Hosp.* **87**, 549.
Rich, A. R., and Lewis, M. R. (1932). *Bull. Johns Hopkins Hosp.* **50**, 115.
Richar, W. J., and Breakell, E. S. (1959). *Am. J. Clin. Pathol.* **31**, 384.
Richards, O. W. (1944). *In* "Medical Physics" (O. Glasser, ed.), Vol. 1, p. 750. Yearbook Publishers, Chicago, Illinois.
Richardson, K. C. (1953). *J. Endocrinol.* **9**, 170.
Richter, C. P. (1927). *Quart. Rev. Biol.* **2**, 315.
Richter, C. P. (1950). *Science* **112**, 20.
Riker, A. J., and Riker, R. S. (1936). "Introduction to Research on Plant Diseases." Dept. of Plant Physiology, Univ. of Wisconsin, Madison, Wisconsin.
Rioch, D. McK., Wislocki, G. B., and O'Leary, J. L. (1940). *Research Publs. Assoc. Research Nervous Mental Disease* **20**, 3.
Ritchie, P. D. (1933). "Asymmetric Synthesis and Asymmetric Induction." Oxford Univ. Press, London and New York.
Rivers, D., and Wise, M. E. (1960). *Brit. J. Ind. Med.* **17**, 93.
Rivers, D., Wise, M. E., King, E. J., and Nagelschmidt, G. (1960). *Brit. J. Ind. Med.* **17**, 87.
Robertson, T. B. (1908). *Arch. Entwicklungsmech. Organ.* **25**, 581.
Robbins, F. C., Weller, T. H., and Enders, J. F. (1952). *J. Immunol.* **69**, 673.

Robinow, Carl (1936). *Protoplasma* **27**, 86.
Robinow, C. F. (1956). *J. Biophys. Biochem. Cytol.* **2**, 233.
Robins, Eli (1957). *Exptl. Cell Research, Suppl.* **4**, 241.
Roepke, R. R., Maren, T. H., and Mayer, E. (1957). *Ann. N. Y. Acad .Sci.* **69**, 457.
Rogers, D. E. (1958). *In* "Pasteur Fermentation Centennial," Symposium, p. 61. Pfizer, New York.
Romeis, B. (1948). "Mikroskopische Technik," 15th ed. Leibniz, Munich.
Rösch, G. A. (1925). *Z. vergleich. Physiol.* **2** ,571.
Rösch, G. A. (1927). *Z. vergleich. Physiol.* **6**, 264.
Rosenblueth, A. (1941). *Biol. Symposia No.* **3**, 111.
Rosenblueth, A., and Wiener, N. (1945). *Phil. Sci.* **12**, 316.
Rosenthal, T. B. (1952). *In* "Laboratory Technique in Biology and Medicine." (E. V. Cowdry, ed.), 3rd ed., p. 203. Williams & Wilkins, Baltimore, Maryland.
Rothman, S. (1954). "Physiology and Biochemistry of the Skin." Univ. of Chicago Press, Chicago, Illinois.
Roulet, F. (1948). "Methoden der Pathologischen Histologie." Springer, Vienna.
Rous, P., and Beard, J. W. (1934). *J. exptl. Med.* **59**, 577.
Royer, G. L., and Maresh, C. (1947). *Calco Tech. Bull. No.* **796**. American Cyanamid Co., Bound Brook, New Jersey.
Royer, G. L., Maresh, C., and Harding, A. M. (1945). *Calco Tech. Bull. No.* **770**, American Cyanamid Co., Bound Brook, New Jersey.
Ruch, T. C. (1949). *In* "A Textbook of Physiology" (John F. Fulton, ed.), 16th ed., p. 292. Saunders, Philadelphia, Pennsylvania.
Rugh, R. (1948). "Experimental Embryology." Burgess, Minneapolis, Minnesota.
Sacks, J. (1956). *In* "Physical Techniques in Biological Research" (G. Oster and A. W. Pollister, eds.), Vol. 2, p. 1. Academic Press, New York.
Salomonsen, C. J. (1919). "Erindringsord for Deltagere i de experimental-pathologiske Øvelser ved Københavns Universitet." 2nd ed. Nordisk Forlag, Copenhagen and Oslo.
Sandison, J. C. (1924). *Anat. Record* **28**, 281.
Sanford, K. K., Earle, W. R., and Likely, G. D. (1948). *J. Natl. Cancer Inst.* **9**, 229.
Sanford, K. K., Likely, G. D., and Earle, W. R. (1954). *J. Natl. Cancer Inst.* **15**, 215.
Sarnoff, S. J., Case, R. B., Welch, G. H., Jr., Braunwald, E., and Stainsby, W. N. (1958). *Am. J. Physiol.* **192**, 141.
Sauerbruch, F., and Heyde, M. (1908). *Münch. med. Wochschr.* **I**, 153.
Schade, H., Claussen, F., and Birne, M. (1928). *Z. klin. Med.* **108**, 581.
Schaper, A., and Cohen, C. (1905). *Arch. Entwicklungsmech. Organ.* **19**, 348.
Schleip, W. (1923). *Arch. Zellforschg.* **17**, 289.
Schönheimer, R. (1929). *Z. physiol. Chem. Hoppe-Seyler's* **182**, 148.
Schultz, Julius (1929). "Die Maschinentheorie des Lebens." Bornträger, Berlin.
Schultz-Brauns, O. (1929). *Arch. pathol. Anat. u. Physiol. Virchow's* **273**, 1.
Schulze, Walter (1900). *Arch. mikroskop. Anat.* **56**, 491.
Schumacher, F. X., and Eschmeyer, R. W. (1942). *J. Tennessee Acad. Sci.* **17**, 253.
Scott, G. H. (1933). *Am. J. Anat.* **53**, 243.
Seifter, S., Dayton, S., Novic, B., and Muntwyler, E. (1950). *Arch. Biochem.* **25**, 191.
Sellers, A. L., Roberts, S., Rask, I., Smith, S., Marmorston, J., and Goodman, H. C. (1952). *J. Exptl. Med.* **95**, 465.
Selye, H. (1949). "Textbook of Endocrinology," 2nd ed. Acta Endocrinologica, Montreal, Canada.
Selye, H., and Jasmin, G. (1956). *Ann. N. Y. Acad. Sci.* **64**, 481.
Sharp, L. W. (1926). "Introduction to Cytology," 2nd ed. McGraw-Hill, New York.

Shen, S. C. (1955). *In* "Biological Specificity and Growth" (E. G. Butler, ed.), p. 73. Princeton Univ. Press, Princeton, New Jersey.
Shen, S. C. (1958). *In* "Symposium on Chemical Basis of Development" (W. D. McElroy and B. Glass, eds.), p. 416. Johns Hopkins Press, Baltimore, Maryland.
Shepherd, R. G., Howard, K. S., Bell, P. H., Cacciola, A. R., Child, R. G., Davies, M. C., English, J. P., Finn, B. M., Meisenhelder, J. H., Moyer, A. W., and van der Scheer, J. (1956). *J. Am. Chem. Soc.* **78**, 5051.
Shillaber, C. P. (1944). "Photomicrography in Theory and Practice." Wiley, New York.
Shull, A. F. (1918). *J. Morphol.* **30**, 455.
Siekevitz, P., and Palade, P. (1958). *J. Biophys. Biochem. Cytol.* **4**, 203.
Siemens, H. W. (1958). "General Diagnosis and Therapy of Skin Diseases." Univ. of Chicago Press, Chicago, Illinois.
Singer, M. (1952). *Intern. Rev. Cytol.* **1**, 211.
Singer, M., and Morrison, P. R. (1948). *J. Biol. chem.* **175**, 133.
Sjöstrand, F. S. (1956). *In* "Physical Techniques in Biological Research" (G. Oster and A. W. Pollister, eds.), Vol. 3, p. 241. Academic Press, New York.
Sjöstrand, F. S., and Hanzon, V. (1954). *Exptl. Cell Research* **7**, 393.
Sollmann, T. H., and Hanzlik, P. J. (1939). "Fundamentals of Experimental Pharmacology," 2nd ed. Stacey, San Francisco.
Sonneborn, T. M. (1959). *Advances in Virus Research* **6**, 229.
Spemann, H., and Falkenberg, H. (1919). *Wilhelm Roux' Arch. Entwicklungsmech. Organ.* **45**, 371.
Sprent, J. F. A. (1956). *Parasitology* **46**, 54.
Ssobolew, L. W. (1900). *Centr. allgem. Pathol. u. pathol. Anat.* **11**, 202.
Starling, E. H. (1930). "Principles of Human Physiology," 5th ed. Lea & Febiger, Philadelphia, Pennsylvania.
Starling, E. H., and Verney, E. B. (1925). *Proc. Roy. Soc.* **B97**, 321.
Stearns, M. L. (1940). *Am. J. Anat.* **67**, 55.
Sternberg, J., and Podoski, M.-O. (1953). *Ann. inst. Pasteur* **84**, 853.
Stetten, D., Jr., and Stetten, M. R. (1956). *Science* **124**, 241.
Stetten, M. R., Katzen, H. M., and Stetten, D., Jr. (1956). *J. Biol. Chem.* **222**, 587.
Stetten, M. R., and Stetten, D., Jr. (1954). *J. Biol. Chem.* **207**, 331.
Stevens, C. E., and Leblond, C. P. (1953). *Anat. Record* **115**, 231.
Stewart, G. T. (1950). *Brit. J. Exptl. Pathol.* **31**, 5.
Stillman, J. D. B. (1882). "The Horse in Motion As Shown by Instantaneous Photography." G. A. R. Osgood, Boston, Massachusetts.
Stock, C. C. (1953). *Cancer Research Suppl.* **1**, 1.
Storer, T. I., and Usinger, R. L. (1957). "General Zoology," 3rd. ed. McGraw-Hill, New York.
Straub, Walter (1901). *Arch. exptl. Pathol. Pharmakol. Naunyn-Schmiedeberg's* **45**, 346.
Sucquet, J.-P. (1862). Quoted by M. Clara (1956).
Suguira, K., and Stock, C. C. (1955). *Cancer Research* **15**, 38.
Swann, M. M., and Mitchison, J. M. (1958). *Biol. Revs. Cambridge Phil. Soc.* **33**, 103.
Syverton, J. T., and Scherer, W. F. (1954). *Ann. N. Y. Acad. Sci.* **58**, 1056.
Sze, L. C. (1953). *Physiol. Zool.* **26**, 212.
Tasaki, I. (1939). *Am. J. Physiol.* **127**, 211.
Taylor, D. W. (1948). *Science* **107**, 164.
Taylor, J. H. (1956). *In* "Physical Techniques in Biological Research" (G. Oster and A. W. Pollister, eds.). Vol. 3, p. 545. Academic Press, New York.
Terplan, K. (1945). *Am. Rev. Tuber.* **51**, 91.

Thimann, K. (1960). *In* "Fundamental Aspects of Normal and Malignant Growth" (W. W. Nowinski, ed.), p. 748. Elsevier, New York.

Thoma, R. (1878). Quoted by H. Florey (1958).

Thompson, D'Arcy W. (1917). "On Growth and Form," 1st ed. Cambridge Univ. Press, London.

Thompson, D'Arcy W. (1942). "On Growth and Form," 2nd ed., Cambridge Univ. Press, London and New York.

Thornton, N. C. (1953). *In* "Growth and Differentiation in Plants" (W. E. Loomis, ed.), p. 137. Iowa State College Press, Ames, Iowa.

Thygeson, P. (1958). *Bull. World Health Organization* **19**, 128.

Tolansky, S. (1954). *Sci. Am.* **191** (2), 54.

Topley, W. W. C., Greenwood, M., and Wilson, Joice (1931). *J. Pathol. Bacteriol.* **34**, 523.

Tsui, C. L. (1935). *Zool. Jahrb., Abt. Anat. u. Ontog. Tiere* **60**, 37.

Tyler, A. (1955). *In* "Analysis of Development" (B. H. Willier, P. Weiss, and V. Hamburger, eds.), p. 170. Saunders, Philadelphia, Pennsylvania.

Underhill, B. M. L. (1955). *Brit. Med. J.* **4950**, 1243.

Uotila, U., and Kannas, O. (1952). *Acta. Endocrinol.* **11**, 49.

Uretz, R. B., Bloom, W, and Zirkle, R. E. (1954). *Science* **120**, 197.

Vallance-Owen, J. (1948). *J. Pathol. Bacteriol.* **60**, 325.

Van Cleave, H. J. (1922). *Biol. Bull.* **42**, 85.

Van Cleave, H. J. (1932). *Quart. Rev. Biol.* **7**, (1), 59.

Van Creveld, S. (1939). *Medicine* **18**, 1.

Van der Reis, V., and Schembra, Fr. W. (1924). *Z. exptl. Med.* **43**, 94.

Van Dyke, D. C., Huff, R. L., and Evans, H. M. (1948). *Stanford Med. Bull.* **6**, 271.

Vincent Memorial Laboratory (1950). "The Cytologic Diagnosis of Cancer." Saunders, Philadelphia, Pennsylvania.

Vlès, F., and Gex, M. (1928). *Compt. rend. soc. biol.* **98**, 853.

Vogt, W. (1925). *Wilhelm Roux' Arch. Entwicklungsmech. Organ.* **106**, 542.

Vogt, W. (1934). *Arch. exptl. Zellforsch. Gewebezücht.* **15**, 269.

Volland, W., and Pribilla, W. (1955). *Klin. Wochschr.* **33**, 145.

von Frisch, K. (1931). "Aus dem Leben der Bienen," Verständliche Wissenschaft, Springer, Berlin.

von Gierke, E. (1929). *Beitr. pathol. Anat. u. allgem. Pathol.* **82**, 497.

von Gutfeld, F., and Mayer, E. (1932). *Zentr. Bakteriol. Parisitenk. Abt. I, Orig.* **124**, 122.

von Hansemann, D. *See* Hansemann.

von Mering, J., and Minkowski, O. (1889). *Arch. exptl. Path. Pharmakol. Naunyn-Schmiedeberg's* **26**, 371.

von Recklinghausen (1863). *Arch. pathol. Anat. u. Pathol. Virchow's* **27**, 419.

von Uexküll, J. (1912). "Umwelt und Innenwelt der Tiere," 1st ed., quoted by W. M. Bayliss (1915).

von Uexhüll, J. (1921). "Umwelt und Innenwelt der Tiere," 2nd ed. Springer, Berlin.

Wagoner, G., and Custer, R. P. (1932). "A Handbook of Experimental Pathology." C. C Thomas, Springfield, Illinois.

Wajda, S. H. (1954). *Cuad. Cient. Histol., Mendoza Arg.* **1**, 11.

Waldeyer, W. (1863). *Henle's und von Pfeufer's Z. rationelle Med.* [3] **20**, 193.

Walker, C. E., and Allen, M. (1927). *Proc. Roy. Soc.* **B101**(712), 468.

Walker, P. M. B. (1956). *In* "Physical Techniques in Biological Research," (G. Oster and A. W. Pollister, eds.), Vol. 3, p. 401. Academic Press, New York.

Walls, E. W., and Buchanan, T. J. (1956). *J. Anat.* **90**, 329.

Waltz, H. K., Tullner, W. W., Evans, V. J., Hertz, R., and Earle, W. R. (1954). *J. Natl. Cancer Inst.* **14**, 1173.
Warburg, O. (1923). *Biochem. Z.* **142**, 317.
Warren, L. H. (1958). *Science* **128**, 24.
Watson, J. D., and Crick, F. H. C. (1953). *Nature* **171**, 964.
Wearn, J. T., and Richards, A. N. (1924). *Am. J. Physiol.* **71**, 209.
Weiss, P. (1947). *Yale J. Biol. and Med.* **19**, 235.
Weiss, P. (1950). *Quart. Rev. Biol.* **25**, (2), 177.
Weiss, P. (1955). *In* "Analysis of Development," (B. H. Willier, P. Weiss and V. Hamburger, eds.), p. 346. Saunders, Philadelphia, Pennsylvania.
Weiss, P., and Andres, G. (1952). *J. Exptl. Zool.* **121**, 449.
Weiss, P., and Kavanau, J. L. (1957). *J. Gen. Physiol.* **41**, 1.
Weiss, S., and Parker, F., Jr. (1939). *Medicine* **18**, 221.
Weismann, A. (1904). "Vorträge über Deszendenztheorie," 2nd ed., Vol. 1, Fischer, Jena.
West, E. S., and Todd, W. R. (1955). "Textbook of Biochemistry," 2nd ed. Macmillan, New York.
Westwood, J. C. N., Macpherson, I. A., and Titmuss, D. H. J. (1957). *Brit. J. Exptl. Pathol.* **38**, 138.
Wheland, G. W. (1955). "Resonance in Organic Chemistry." Wiley and Sons, New York; Chapman and Hall, London.
White, P. R., Editor (1957). "Decennial Review Conference on Tissue Culture" [Woodstock, Vermont (1956)]. *J. Natl. Cancer Inst.* **19**, 467.
Wigand, R. (1958). "Morphologische, biologische und serologische Eigenschaften der Bartonellen." Thieme, Stuttgart.
Wigglesworth, V. (1940). *J. Exptl. Biol.* **17**, 201. Quoted by F. A. Brown (1950).
Wilcke, J. (1935). *Acta Brev. Neerl.* **5**, (5/6), 99.
Wilcoxon, F. (1949). "Some Rapid Approximate Statistical Procedures," rev. ed. American Cyanamid Co., New York.
Williams, R. G. (1944). *Am. J. Anat.* **75**, 95.
Williams, R. C., and Kallman, F. (1955). *J. Biophys. Biochem. Cytol.* **1**, 301.
Willier, B. H. (1939). *In* "Sex and Internal Secretions" (Allen, Danford, and Doisy, eds.), 2nd ed., p. 64. Williams & Wilkins, Baltimore, Maryland.
Willier, B. H. (1948). *In* "Biology of Melanomas," *Spec. Publ. N. Y. Acad. Sci. No.* **4**, 321.
Willis, H. S., and Cummings, M. M. (1952). "Diagnostic and Experimental Methods in Tuberculosis," 2nd ed. C. C Thomas, Springfield, Illinois.
Willmer, E. N. (1933). *J. Exptl. Biol.* **10**, 340.
Willmer, E. N. (1954). "Tissue Culture," 2nd ed. Wiley, New York.
Willmer, E. N., and Jacoby, F. (1936). *J. Exptl. Biol.* **13**, 237.
Wilson, C. W., Hughes, A. F., Glücksmann, A., and Spear, F. G. (1935). *Strahlentherapie* **52**, 519.
Wilson, Edmund B. (1934). "The Cell in Development and Heredity," 3rd ed. Macmillan, New York.
Wilson, E. Bright, Jr. (1952). "An Introduction to Scientific Research." McGraw-Hill, New York.
Wilson, H. V. (1907). *J. Exptl. Zool.* **5**, 245.
Wilson, J. W. (1948). *Anat. Record* **101**, 672.
Winton, F. R., and Bayliss, L. E. (1955). "Human Physiology," 4th ed. Little, Brown, Boston, Massachusetts.

Wintrobe, M. M. (1951). "Clinical Hematology," 3rd ed. Lea & Febiger, Philadelphia, Pennsylvania.

Wittaker, S. R. F., and Winton, F. R. (1933). *J. Physiol.* (*London*) **78**, 339.

Wolff, Et., and Dubois, F. (1948). Rev. suisse zool. **55**, 218.

Woodburne, R. T. (1957). "Essentials of Human Anatomy." Oxford Univ. Press, London and New York.

Woodstock Conference (1956). *See* P. R. White (1957).

Yegian, D., and Porter, K. (1944). *J. Bacteriol.* **48**, 83.

Youmans, G. P. (1949). *Ann. N. Y. Acad. Sci.* **52**, 662.

Young, J. Z. (1936). *Cold Spring Harbor Symposia Quant. Biol.* **4**, 1.

Young, J. Z. (1957). "The Life of Mammals." Oxford Univ. Press, London and New York.

Zeiger, K. (1938). "Physikochemische Grundlagen der histologischen Methodik." Steinkopff, Dresden and Leipzig.

Zworykin, V. K., and Hillier, J. (1950). *In* "Medical Physics" (O. Glasser, ed.), Vol. 2, p. 511. Yearbook Publishers, Chicago, Illinois.

Zworykin, V. K., Hillier, J., and Vance, A. W. (1944). *In* "Medical Physics" (O. Glasser. ed.), Vol. 1, p. 387. Yearbook Publishers, Chicago, Illinois.

Part IV

ELEMENTARY GENERAL STRUCTURES

The role of the cell as an important unit of organization was discussed previously (Part III, Chapter 4, C). Cellular and non-cellular organisms were described. The present part deals with general properties and components of cells, the association of tissue cells, and intercellular substances and spaces. A brief chapter will be devoted to the subject of potential spaces on both the macro- and microscopic level.

Chapter 1

General Properties and Components of Cells

As a result of cytochemical and electron microscopic discoveries the variety of intracellular structures not only has increased, but more and more differences between tissue cells of different types have been discovered. With these restrictions in mind, the exocrine cell of the pancreas acinus illustrated in Fig. 57F may serve as a representative tissue cell. Light microscopy at high magnification shows the following structures: mitochondria in the basal part and zymogen granules in the apical part of the cytoplasm, and a nucleus with one or more nucleoli in each cell. The electron microscopic picture (Fig. 57G) reveals additional structures such as the endoplasmic reticulum. The picture does not include infoldings of the cell membrane visible with the electron microscope. The polar differentiation of cells of this type is, at least to some extent, an expression of their aggregation pattern. Each exocrine glandular cell has a base bordering the blood capillaries which carry the raw material to the cell, and an apex from which the specific product of the cell is secreted into the duct system (Fig. 87A).

Many authors subsume cytoplasm and nucleus under the common concept of protoplasm, whereas others use cytoplasm and protoplasm as synonyms. Since there are no operational criteria of protoplasm, the term is avoided in the present book (see also Hardin's criticisms, 1956).

A. Nuclear Structures and Nuclear Material

Figure 61a–d illustrates differences in intermitotic nuclei as seen with hematoxylin or other conventional nuclear stains. Intranuclear material demonstrated with conventional nuclear stains is known as chromatin. Nucleoli stain with cytoplasmic stains such as eosin. Opposite types of nuclei may be compatible with the same function: the vesicular nucleus of the egg cell and the opaque nucleus of the spermatozoon are equally capable of transmitting hereditary properties. Diagrams of mitotic phases are shown in Fig. 61, M1 to M4.

The characteristic component of nuclear chromatin is deoxyribonucleic acid (DNA), and the Feulgen reaction provides a specific stain for this compound. According to Leuchtenberger (1950) the Feulgen reaction increases in intensity as nuclei change from the vesicular to the condensed (pycnotic) type. Microspectrophotometric determinations show that the increased intensity of the Feulgen reaction indicates an increased concentration of DNA in the nucleus. It is not

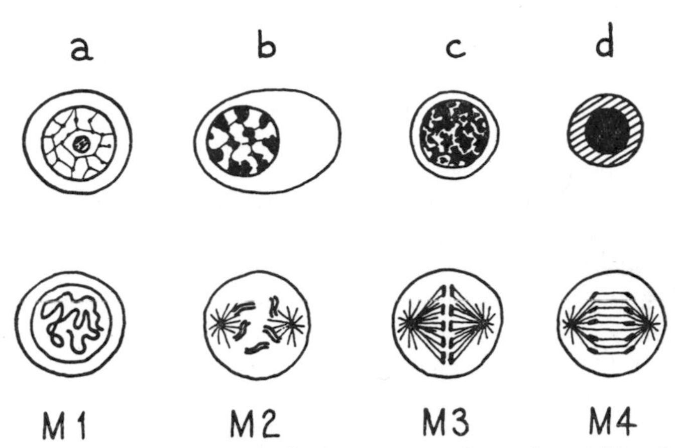

FIG. 61. Diagram of chromatin distribution, as seen after nuclear staining (hematoxylin, carmine, etc.).

(a) to (d) Various spherical intermitotic nuclei. (a) Chromatin finely distributed. Nucleus transparent, with distinct membrane ("vesicular nucleus"); nucleolus shaded. Most frequent type of nuclei, occurring in all embryonic and many postembryonic tissues. (b) Margination of chromatin clumps. Nucleus less transparent than in (a). Typical of "plasma cells" but also occurring in other cells. (c) Large amount of chromatin particles, more or less evenly distributed, with small spaces between them. Nuclear membrane obscured. Typical of lymphocytes, thymocytes, and microglia cells. (d) Maximal density of chromatin, producing an entirely opaque, so-called pycnotic, nucleus. The picture shows an immature nucleated red blood corpuscle of a mammal; shading of cytoplasm indicates hemoglobin.

(M 1) to (M 4) Mitotic phases. (M 1) Early prophase, transformation of (a). Chromatin concentrated in a long filament (spireme). (M 2) Later prophase. Nuclear membrane has disappeared. Distinct paired chromosomes. (M 3) Metaphase. Twin chromosomes lined up in equatorial plate. (Stellate structures, centrosomes; fine lines, spindle fibers. (M 4) Anaphase. Daughter chromosomes are separated and moving to opposite poles.

Note: After stage (M 4) the chromatin is redistributed in two daughter nuclei and the cytoplasm divides. The resulting daughter cells with their nuclei resemble the mother cell (a).

the absolute amount of DNA which changes but the amount per unit of volume.

Chemical determinations of the average amount of deoxyribonucleic acid per nucleus were made by Boivin, Vendrely, and Vendrely (1948). Liver, pancreas, and thymus of calf were homogenized with citric acid and the isolated nuclei were counted in blood cell chambers (see Part III, Chapter 3, B). The average DNA per nucleus was similar in these three organs, but only half the value was found in spermatozoa of bulls. This seemed to be in good agreement with the fact that reproductive cells contain half the chromosome number of somatic cells. The authors pointed out that ova should be studied in the same way. Since spermatozoa have pycnotic nuclei and ova have particularly transparent vesicular nuclei, a comparison of their DNA content would be interesting with respect to my Fig. 61a–d. Investigations of this question have yielded contradictory results, even in closely related forms of sea urchins. In *Arbacia,* the ratio of sperm DNA: egg DNA was found to be 1: 290 by Vendrely and Vendrely (1949), and in *Paracentrotus* the ratio of 1: 28 was determined by Elson and Chargaff (1952). Vendrely and Vendrely felt that the high value in the egg did not make sense. They rejected the interpretation of reserve DNA in the cytoplasm of the egg and rather ascribed the excessively high value in the whole egg to technical errors. Elson and Chargaff were satisfied that the discrepancy of their values was "less dramatic."

Astonishing values were reported in frogs. Sze's (1953) data on *Rana pipiens* indicate a ratio of DNA in sperms : DNA in unfertilized eggs at an order of magnitude of 1: 3000, if we assume that the DNA content which he determined in diploid nuclei of blastulas and gastrulas was twice that of haploid sperm nuclei. Sze pointed out that his enormous values of DNA in the nucleus of the fertilized ovum are incompatible with cytological observations, which do not indicate any extraordinary amount of chromatin. In a paper by Pollister, Swift, and Alfert (1951), dealing with the DNA content of somatic and reproductive cells of mice, spectrophotometric comparisons between the DNA content of mature (haploid) spermatozoa and mature (haploid) oocytes proved difficult. As the mature oocyte nucleus swells, the DNA is diluted so that it cannot be estimated reliably. However, an indirect answer was obtained as follows. In fertilized mouse eggs it was possible to determine the DNA separately for the female and male pronuclei. It seems that in any one fertilized ovum the DNA values were similar for the male and female pronucleus.

In studies of this kind one is faced with a number of problems which have been summarized by Dr. W. M. Layton of these laboratories (personal communication, 1961). (1) In chemical determinations of DNA, questions of specificity have not been entirely settled. (2) Quantitative estimates of DNA by spectrophotometry depend on the completeness of the individual nuclei. Therefore determinations in sections are less reliable than those in suspensions (see earlier discussion of three-dimensional analysis, Part III, Chapter 6, B). (3) Spectrophotometry of Feulgen-stained material gives different values depending on the fixative used (Di Stefano, 1948; Swift, 1950). Ultraviolet spectrophotometry without Feulgen stain is beset with optical difficulties and too tedious for handling many samples. (4) The distribution of DNA in nuclei is not random (for margination of chromatin see Fig. 61b). One or more nucleoli containing some DNA are part of the material to be studied (Fig. 61a). Through the analysis of such factors Iversen (1960) arrived at the conclusion that simple measurement of the geometric dimensions of nuclei will yield as much quantitative information on DNA as spectrophotometry does. (5) Besides the technical problems mentioned above, the biological interpretations require the following considerations. It would not be justifiable to assume that the chromosome number and DNA content of nuclei are parallel. Chromosomes must duplicate their DNA content prior to increasing their number. The phase in the mitotic cycle, the ploidy, and the DNA content are not connected in any simple or general way.

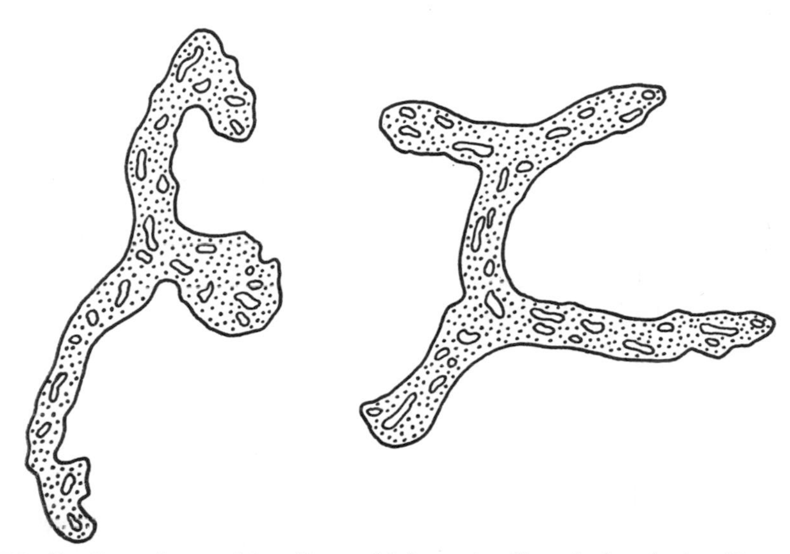

FIG. 62. Unusual type of branching nuclei from caterpillar spinning glands, with granular dispersion of chromatin (dots), and numerous nucleoli (outlined spaces). Diagrams were made by combining pictures from Brachet and Meves. Magnification approximately ×500.

Brachet (1957, Fig. 32): glands from *Bombyx mori,* preparation stained with Feulgen reaction; used for drawing nuclear shapes. Meves (1897, Fig. 6): glands from *Pieris rapae,* section stained with orange-G, acid fuchsin, and methyl green; used for drawing chromatin granules and nucleoli.

In an earlier part of the book it was mentioned that hematoxylin stain does not allow one to differentiate between dispersed nuclear material, bacteria, or calcium particles (Part III, Chapter 2, B, 3, d). If nuclear structures are not clearly recognizable, the Feulgen reaction is used to decide whether the material in question is nuclear or not. Figure 62 shows an unusual type of branching nuclei which occurs in spinning glands of caterpillars. Meves (1897) interpreted as chromatin the masses of fine dots which stained with methyl green, and as nucleoli the large particles which stained red with fuchsin. Korschelt (1896) had stated that the fine granules are not identical with "the usual chromatin." It seems to me that this question has been decided in favor of Meves's opinion. In Brachet's "Biochemical Cytology" (1957) a Feulgen preparation of silk gland nuclei shows that most of the content of the branched nuclei is chromatin. My diagram Fig. 62 was made by combining pictures of Meves and Brachet. Pictures at low magnification showing a maze of ramified nuclei in spinning glands were published by Helm (1876).

Chemical and biological investigations on nucleic acids and nucleoproteins were discussed previously at various occasions. As an example of combined morphological and cytochemical studies of the nucleus the work of Mazia and Prescott (1954) was recounted, which provided evidences of synthesis of deoxy-

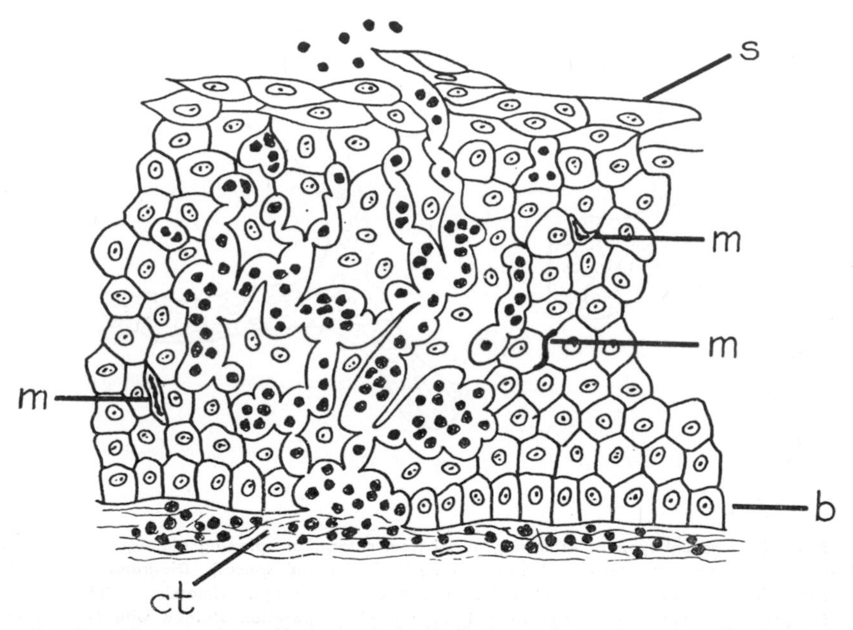

FIG. 63. Diagram of stratified mosaic (epithelium) covering a human tonsil, with normal invasion by lymphocytes; lymphocyte nuclei appear dark. Local disruption and passive transformation of mosaic pattern into cytoplasmic network. *ct*, Connective tissue layer separating epithelial mosaic from lymphatic tissue of the tonsil; *b*, basal layer of mosaic; *s*, surface layer of mosaic; *m*, lymphocytes greatly elongated as they migrate through narrow crevices of mosaic.

ribonucleic acid by the intermitotic structures of the nuclei (Part II, Chapter 2, H, 2). Other aspects of mitosis will be discussed in connection with phases of the life cycle (Part V, Chapter 2, B, 2).

The presence of DNA in bacteria had been demonstrated by chemical analysis of homogenized cultures (bulk chemistry), but it was controversial whether the DNA could be localized in individual bacteria. With improved techniques, visible accumulations of nuclear material were demonstrated in bacteria of many types. The nuclear masses in bacteria resemble the dense types of nuclei shown in Fig. 61c and d: they are opaque rather than "vesicular." Once the DNA aggregates were verified in bacteria cytochemically through the Feulgen reaction, aniline dyes such as Giemsa's stain were used for the analysis of visible structures of the nuclear material (Robinow, 1945). It has not yet been decided to what extent the visible nuclear material in bacteria can be compared to chromosomes. One should take into account that some protozoa contain several nuclei with different functions. The small nuclei serve reproduction while the large ones are metabolic organs.

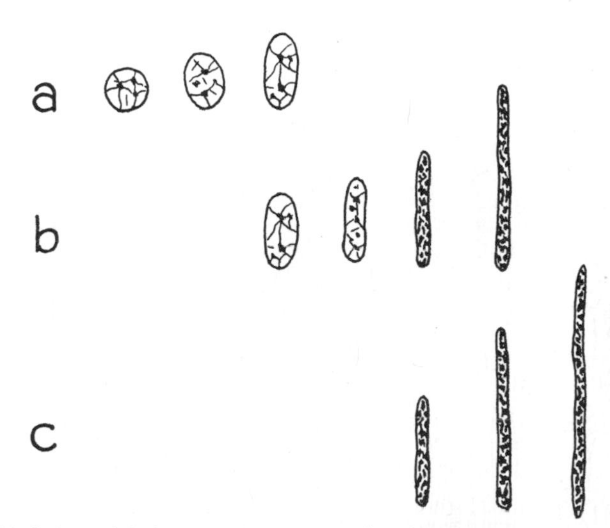

FIG. 64. Variations of isolated nuclei of connective tissue cells and smooth muscle cells. Changes associated with maturation result in overlap of different types (for aggregation patterns of these nuclei, see Fig. 65). In each horizontal row maturity progresses from left to right. (a) Reticulum cell nuclei: slight changes with maturation. (b) Fibroblast nuclei: the youngest on the left is indistinguishable from a mature reticulum cell nucleus; mature forms of (b) resemble nuclei of smooth muscle (c). (c) Smooth muscle nuclei: young vesicular forms have been omitted since they can hardly be identified; mature dark forms similar to mature (b) nuclei. Note: Vesicular nuclei as seen in row (a) and on the left of row (b) are not limited to the cells mentioned, but are the most frequent type occurring in nerve cells, in cells of exocrine and endocrine glands, and in cells covering the external and internal surfaces. The dark type, third from the left in row (b), also occurs in cells lining the interior of vessels (endothelium).

The morphological appearance of nuclei can be modified by environmental factors. In a hypotonic fluid, nuclei swell and become more and more transparent. Upon entering a narrow space, spherical, oval or segmented nuclei assume the shape of elongated rods. Figure 63 shows the invasion of tonsillar epithelium of lymphocytes, which is a normal process. It would not be possible to identify individual distorted lymphocytes as they appear in the crevices between epithelial cells: their identity is concluded from their association with the population of migrating lymphocytes. Under pathological conditions the cornea of the eye is infiltrated with white blood cells. Since many of their nuclei are compressed between the lamallae of the cornea they are indistinguishable from the nuclei of the connective tissue (see B. Rones, 1941, p. 118 and Fig. 9).

Nuclei with similar morphological characteristics occur in different cells and tissues. In addition, nuclei undergo marked changes with the aging of the cell. Figure 64 illustrates the maturation of nuclei in connective tissue and smooth muscle. A young fibroblast nucleus is similar to a mature reticulum cell nucleus; and a moderately aged smooth muscle nucleus resembles the most mature phase of a fibroblast nucleus. Obviously the identification of *isolated*

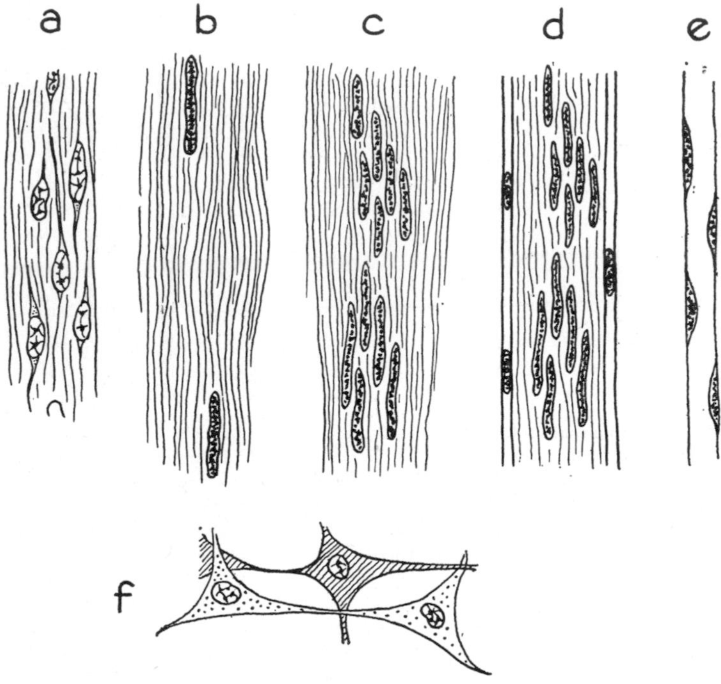

FIG. 65. Distribution of nuclei in fibrous tissues, inner linings of vessels, and cytoplasmic nets. (a) to (e) are diagrams of longitudinal sections. (a) Strand of young connective tissue, for instance, from embryonic tendon. (b) Strand of mature connective tissue, for instance, from adult tendon. Nuclei sparser, darker, and longer than in (a). (c) Bundle of smooth muscle fibers. Individual nuclei similar to (b), but assembled in groups. Nuclei would appear wavy in (c), if the muscle tissue had contracted before fixation. (d) Bundle of nerve fibers from a peripheral nerve encased by a sheath of connective tissue fibers (epineurium); the nerve fibers are associated with delicate connective tissue (endoneurium), of which nuclei only are visible. Nuclei of endoneurium are grouped like those of (c) and also would appear wavy, if the nerve was not stretched during fixation (see the "snake-fence" picture in Ham's Histology, 1957, Fig. 319). (e) Two parallel rows of nuclei usually allow identification of a longitudinally cut blood or lymph capillary; blood corpuscles in the bore would serve as confirmation. In fixed sections, nuclei tend to bulge into the bore as pictured here (for three-dimensional diagram, see Fig. 71e). During life the blood pressure seems to cause outward bulging of nuclei. (f) Cytoplasmic net with nuclei similar to those of (a). For other forms of the reticular pattern see Fig. 73a and b.

nuclei is uncertain in most cases. The distribution of nuclei and the pattern of the tissue in which they occur supply the criteria for identification. Figure 65 shows the distribution of nuclei in bundles of connective tissue, both young (a) and mature (b), in bundles of smooth muscle fibers (c), and in the connective tissue which forms sheaths around nerve fibers in peripheral nerves (d). Bundles or nets of fibers with a relatively high concentration of nuclei are known as "cellular connective tissue" (Figs. 65a and 73b), in contrast to "acellular connective tissue" with widely scattered nuclei (Fig. 65b). The cellular connective tissue represents the young phase as seen in embryonic tendon or developing scar tissue, while the acellular type is found in mature tendons or scars. (A photomicrograph of a mature tendon sectioned longitudinally is shown in Ham's "Histology," 1957, Fig. 84.) Oval nuclei lined up in parallel rows suggest a blood or lymph capillary sectioned longitudinally (e). Nuclei in this situation are called endothelial nuclei since the cells which line vessels are known as endothelium. Nuclei of cells which form a cytoplasmic network or reticulum (f) can be identified as reticulum cell nuclei if the cytoplasmic pattern is distinct. Unfortunately, some authors give the impression that they can identify isolated nuclei and cells whereas, in my opinion, everything depends on association patterns. This subject will be resumed in Part V, which deals with classification procedures.

Electron microscopy has not contributed much to the finer analysis of nuclear structures, in contrast to the remarkable achievements of this new technique in the field of cytoplasmic structures.

The characteristics of nuclei play an important role in human pathology. Human autopsy material, as a rule, is not as well preserved as the samples in experimental pathology or zoology (see Part III, Chapter 2, B, 1). Cytoplasmic structures suffer from postmortem autolysis more than nuclear structures. Therefore pathologists learned to rely on a system of *nuclear diagnosis.* At the same time, they developed the habit of referring to round cells, spindle cells, or cylindric cells, on the basis of nuclei irrespective of preservation of cytoplasm. Nuclear diagnosis is a very useful tool in pathology, but the principles of this procedure are not easily found in textbooks or other literature.

B. Cytoplasm and Its Components. Factors Controlling Shapes of Cells

Procedures for studying the properties of cytoplasm were discussed in many parts of this book. Separation of cytoplasmic components by microcentrifugation was described in connection with micrurgical techniques (Part III, Chapter 2, A, 4). Mitochondria, the Golgi apparatus, and the Nissl bodies were treated in the discussion on so-called artifacts (Part III, Chapter 3, C, 2, a and c). The chapter on models included a recent analysis of the mechanical properties of the surface of sea urchin eggs and early approaches to the colloidal chemistry of cytoplasm (Part III, Chapter 5, B). Intracellular pigment, fat drops, glycogen deposits, and zymogen granules served as examples for cytochemical procedures (Part III, Chapter 7). It may be mentioned that the distinction between microscopic granules, drops, and vacuoles (or vesicles) is not always rational. Particularly in stained and fixed preparations the appearance of structures does not necessarily indicate whether they were solid, semisolid, or liquid in their original state. The term vacuole is used both for microscopic spaces with unknown content and for spaces whose known content has been dissolved (fat vacuoles).

A number of visible structures in the cytoplasm have been interpreted as parasites or symbionts by some authors, and as components of the cytoplasm by others. Mitochondria in tissue cells of metazoa and kappa particles in paramecia posed the question, "symbiont or cytoplasmic component?" The fact that a structure is regularly present and is transmitted from one generation to another does not preclude the symbiotic or parasitic nature of the structure. The reader is referred to P. Buchner's classic book, "Tier und Pflanze in intracellulärer Symbiose" (1921); to Lederberg's review on "Cell Genetics and Hereditary Symbiosis" (1952); and to Sonneborn's article "Kappa and Related Particles in Paramecium" (1959). At this writing mitochondria are no longer considered as possible symbionts.

Three items may suffice here to illustrate subcellular and supercellular aspects of the cytoplasm: the *ergastoplasm,* the *shapes of stacked cells,* and the *sickle cell variation of red blood corpuscles.*

The concept of *ergastoplasm* was introduced jointly by M. and P. Bouin and by C. Garnier in a number of papers which appeared in the "Bibliographie anatomique" from 1897 to 1899. These investigators observed that the cells of plants as well as of vertebrate animals contained filaments in the cytoplasm. The filaments showed an affinity to basic dyes, comparable to that of the chromatin of nuclei. Appearance and position of the filaments varied with different functional phases of the cells. In cells of the embryonic sac of Liliaceae one stage showed the filaments running concentrically around the nucleus, whereas they were arranged in radial bundles at another stage (M. and P. Bouin, 1898, Figs. 1 and 2). In the serous cells of salivary glands of rats the filaments changed their position and appearance conspicuously after injection of pilocarpine, a drug that stimulates salivation (Garnier, 1899, Figs. 1 and 2). The authors assumed that the cytoplasmic filaments are connected with activities which take place in the "chemical" or intermitotic phase, such as general cell metabolism and specific secretions. The authors proposed for the basophilic cytoplasmic filaments the name of *ergastoplasm* meaning "plasma which elaborates while transforming." This term was created as opposed to other components of the cell, and particularly as opposed to *kinoplasm* which represented the total of mitotic structures. M. and P. Bouin and Garnier referred to the mitotic phase as the "mechanical period" in the life cycle of the cell. The current concept of ergastoplasm no longer emphasizes the contrast to kinoplasm. Ergastoplasm now refers to the basophilic components of the cytoplasm, and to the endoplasmic reticulum which seems to be the main carrier of the basophilic material. The basophilic properties are ascribed to the ribonucleic acid content of the cytoplasm. The endoplasmic reticulum proved to be a system of parallel membranes. Bennett (1956a) proposed the hypothesis of membrane flow which assigns special transport functions to the endoplasmic reticulum. The particular activity of the ergastoplasm in the modern sense seems to be protein synthesis. Other structural components of the cytoplasm, particularly the mitochondria, are connected with special enzymatic activities, as verified by microcentrifugation and chemical analysis. Finally there are structures and materials in the cytoplasm which cannot as yet be interpreted. Some investigators refer to them as the ground substance of the cytoplasm (Dempsey and Wislocki, 1955, p. 249). This is not related to intercellular ground substances, which will be discussed later (Part IV, Chapter 3, B). According to Frey-Wyssling's classification (1955), the ground plasma (*Grundplasma*) of the cytoplasm offers four different *aspects,* depending on the object investigated and the technique used: the reticular (or spongy), the granular, the fibrillar (chains of beads) and the lamellar aspect. The latter refers to the double lamellae discussed above (see my Fig. 57F).

Let us now turn from the subcellular level to an interesting supercellular problem, namely, the *shapes of cells stacked in several layers.* This subject is mentioned in the textbook of Bremer-Weatherford (1944) and in its rewritten edition by Greep (1954), but is neglected in other textbooks. If polyhedric cells form a one-layer mosaic, usually their shapes are verified without much difficulty (my Fig. 71a–c). If cells are stacked in several closely packed layers, reconstruction in three dimensions is needed to analyze the shape of the individual cells. In the human adult the vaginal epithelium contains up to thirty cell layers (Bargmann, 1959, Fig. 65). My example Fig. 66, taken from one of the earlier papers by F. T. Lewis (1925), shows a section through a thick layer of stacked cells, obtained from the lining of the oral cavity of a human fetus. Two groups of cells were selected for reconstruction in wax: one group near the surface, and one from deeper layers. Pictures of the reconstructed models are reproduced in Fig. 67a and b. Of the three cells in Fig. 67a, two had thirteen faces and one had fifteen faces. Of the five cells in Fig 67b, one cell had twelve faces, three had fourteen, and one had fifteen faces. Determinations of this type supported Lewis's claim that a close variation around an average of fourteen faces is characteristic of stacked cells in animals and plants. Polyheders with an average of fourteen faces were also obtained in model experiments with compressed buckshot (Marvin, 1939), and with soap bubbles in a froth (Matzke, 1950). These studies were recounted previously in the review of models (Part III, Chapter 5). It was mentioned at that occasion that observations were at variance with the mathematical expectation that molding under pressure should produce units with 12 faces.

It is evident that the shapes of cells are controlled by compression and surface tension, but not much is known concerning relations between the molecular composition of cells and their shapes on the level of light microscopy. The only known instance of such relations is the *sickle-cell variation of red blood corpuscles.* In 1949 Pauling, Itano, Singer, and Wells reported that the abnormal red

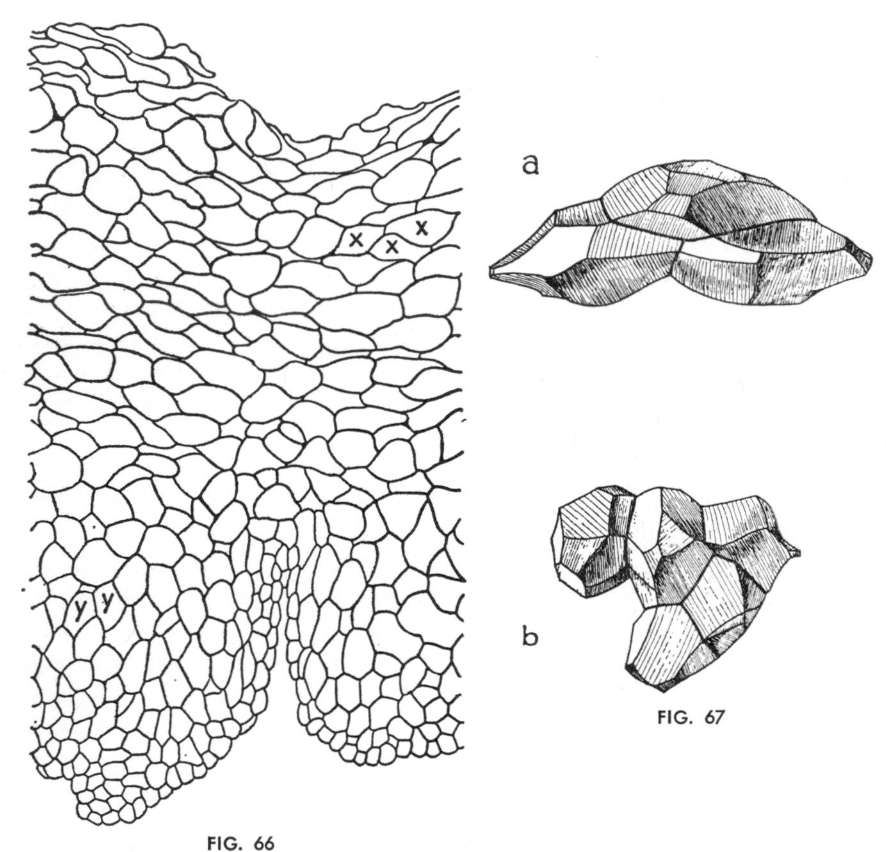

FIG. 67

FIG. 66

FIG. 66. Section through an exceptionally thick layer of stacked cells, from the lining of the oral cavity of a human fetus. Plane of section vertical to the surface. Magnification ×220. Three cells near the surface are marked *X*; their reconstruction in wax is shown in Fig. 67a. Two cells in a deep layer are marked *Y*; they are included in the wax reconstruction of five cells seen in Fig. 67b. (From F. T. Lewis, 1925; his Fig. 19.)

FIG. 67. Wax reconstructions of cell groups indicated in Fig. 66. Magnification ×600. (a) Three cells marked *X* in Fig. 66, from a layer near the surface. Two cells have 13 faces, and one cell has 15 faces. (b) Five cells from a deeper layer, including the two cells marked *Y* in Fig. 66. One cell has 12 faces, three cells have 14 faces, and one cell has 15 faces. Redrawn and reduced from F. T. Lewis, 1925; his Figs. 27 and 28 on Plate III.

blood corpuscles in sickle cell anemia contain an abnormal hemoglobin, which was called hemoglobin-S. The hemoglobin-S molecule is more positive than the ordinary hemoglobin molecule by about three charges. Deoxygenated hemoglobin-S causes a reversible change in the red cell shape from a biconcave disc to a sickle-shaped form. The speed and intensity of the sickling process depends on the proportion of hemoglobin-S present (see Itano's review 1957).

As a result of studies with lipid models, Holtfreter (1948) concluded that the cell membrane controls the shape of the cell. His statement refers to isolated cells and to the loosely aggregated cells of early amphibian embryos.

C. Interfaces, Visible and Invisible

Cell membranes are interfaces on the outside of a cell which separate its cytoplasm from neighboring cells or body fluids. The different morphological patterns in which neighboring cells (or nuclear territories) can be associated will be described in the next chapter (Part IV, Chapter 2), and the relations between intercellular spaces and body fluids will be the subject of a special analysis (Part IV, Chapter 3, C). The interfaces to be discussed here are the boundaries between a cell and its environment, and the partitions within the cytoplasm of one cell.

Morphological properties of cell membranes were discovered microscopically in those tissues which consist of a closely packed mosaic of cells. Many plant cells form a distinct box of cellulose which is coated on the inside by a layer of cytoplasm (cytoplasmic sac). The cytoplasm of adjacent cells is connected by strands (plasmodesmata) which pass through pores in the cellulose wall. Although the existence of plasmodesmata had been accepted on the basis of light microscopy, the pores were finally demonstrated by electron microscopy (Meeuse, 1957). The cellulose wall of plant cells is the most conspicuous type of cell membrane. Cell membranes in animals (metazoa) have been studied most extensively in red blood corpuscles. The presence of a membrane in red blood corpuscles is postulated on the basis of biochemical observations. It has been shown, for instance, that cations cannot enter red blood corpuscles. With the light microscope no membrane is visible on the outside of intact red blood corpuscles. However, if the hemoglobin is extracted (hemolysis) by hypotonic solutions or other means, the outline of the corpuscle becomes distinct, and the picture suggests an empty bag. The existence of a membrane is supported by the observation that the contents of the corpuscle flow out upon puncturing with a needle. Frey-Wyssling (1955) mentions a mathematical objection to the idea of a discrete membrane. The living (hydrated) membrane is supposed to be 0.5 μ thick. The two membranes would be 1.0 μ thick, which is approximately the thickness of the whole corpuscle. Therefore a gradual transition between the external and internal zone of the red blood corpuscle must be assumed. For details of the inner structure of the red blood corpuscle see Bessis (1956).

In white blood corpuscles, a direct demonstration of cell membranes has not been possible. One can tell where the leucocyte ends and where its environment begins, but this optical line of demarcation is not associated with any definite layer of cytoplasm. In other words, the cell membrane of the leucocyte cannot be seen microscopically, but has been inferred from observations of phagocytosis under different physicochemical conditions. I do not know whether the theoretical assumption of a monomolecular surface layer in leucocytes has been helpful. Most illuminating were theoretical and experimental studies by Fenn (1922a,b) on adhesiveness or stickiness of leucocytes. Fenn emphasized that a discussion of stickiness of cells should be introduced by the question, "sticking to what?" The adhesiveness of leucocytes (from peritoneal exudate of rats) was measured as follows. A suspension of the cells was allowed to settle on a glass slide for a certain period. Then a standard stream of liquid was used to wash off those of the sedimented cells which did not stick to the slide. One of the factors responsible for their failure to stick is of interest here: the cell may be so fluid that it tears away even though a thin invisible film of the cell, widely spread out, may continue to adhere. It seems to me that this observation clearly illustrates the difficulty of defining a cell membrane in leucocytes. In order to determine mutual stickiness of leucocytes, Fenn first counted the unclumped cells. Then the suspension was allowed to settle for 6 hours, was resuspended, and clumps of four or more cells were counted. Depending on various factors, leucocytes proved to be sticky to glass, but not to other leucocytes, or vice versa.

Protoplasts of plant cells were liberated from their cellulose boxes by Pfeiffer (1934) and tested for adhesiveness to various surfaces. The adhesiveness was greatest to glass, less to mica, and least to paraffin-coated glass. The mutual adhesiveness of tissue cells from adult human beings has been measured by Coman (1944) with the use of a blunt and a pointed glass needle guided by a micromanipulator. In a pair of cells isolated from a mosaic pattern and placed on a coverglass, one cell was held by the blunt needle, while the other cell was pulled by the pointed needle. The bend of the pointed needle was gauged in milligrams for measuring the force needed to separate the two cells. Coman observed that cells from squamous cell carcinomas have a decreased mutual adhesiveness as compared to that of squamous cells from normal epidermis.

In experimental embryology the mutual adhesiveness of cells has received much attention. I mentioned earlier Holtfreter's (1948) observation that there is a low degree of adhesiveness between the cells of early amphibian embryos. This is interpreted by Holtfreter as a possible prerequisite of the morphogenetic movements which occur during this period (see my Fig. 32). Tyler (1955) reviewed the ontogeny of immunological properties and their relation to specific adhesion between cells. Details will be given in the discussion of classification problems (Part V, Chapter 2, B, 1).

Absence of cell boundaries may be the natural and healthy condition, as in the surface layer of chorionic villi (Fig. 54a); or the absence of visible cell boundaries may be due to a pathologic process, to postmortem autolysis, or to the technique used. Presence of cell boundaries usually represents a natural condition; in other words, visible cell boundaries are rarely produced by autolysis or uncontrolled techniques. In the discussion on natural or artificial units (Part III, Chapter 4, C, 2), some small plants and animals without visible cell boundaries were described. Impregnation with $AgNO_3$ for demonstration of cell boundaries was mentioned in connection with three-dimensional analysis (Fig. 51a), and will occupy us again in the discussion of nucleocytoplasmic patterns (Fig. 72c). When individual cells are surrounded by a dense network of fibers, it may be difficult to tell whether the fibers are part of the cell membrane or not. The bases of most gland cells rest on a special structure known as the *basement membrane*. Although this membrane stains with connective tissue stains, it has been controversial whether it is part of the cell surface or belongs to the adjacent connective tissue. Figure 40 shows that a segment of renal tubules can be displaced by detaching itself from the basement membrane. This observation speaks in favor of the interpretation that the basement membrane is not part of the cell wall, but is a sheath of connective tissue. It is still possible that in other locations the cell mosaic has produced the basement membrane.

On the free surface of cell mosaics certain reinforcements of cell outlines occur which are known as *terminal bars*. They are shown in Fig. 68 in a section through a human thyroid in a highly activated phase (comparable to Fig. 26D from a dog thyroid). The terminal bars form hexagons, and therefore appear as dots, short lines, or long lines, depending on the angle at which they are cut. The terminal bars have physical properties different from those of the rest of the cell surface. This can be concluded from the behavior of the dilute colloid at fixation. The colloid remained anchored to the terminal bars, while it retracted from the rest of the cell surface. G. E. Wilson (1929) observed that in some cases the terminal bars were pulled away from the tops of the thyroid cells, probably by particularly forceful retraction of the thyroid colloid at fixation.

In very shallow cell mosaics the terminal bars may be the only demonstrable part of the cell boundaries. This was illustrated in Fig. 52 (vertical section through a rat's mesentery).

The outlines of cells in a mosaic pattern, which show particularly with silver impregnation (Fig. 72c), are interpreted by some authors as a *cement substance* which fills the gaps between cells. If this interpretation is accepted, one should be interested in the consistency of the intercellular cement. It seems to me that a particular firmness can be assumed at least for superficial parts of the intercellular cement, namely, the terminal bars. This interpretation is supported by phenomena seen in the thyroid gland: colloid arches in my Fig. 68, and Wilson's picture of detached terminal bars.

The *continuity and stability of cell membranes* deserves a brief discussion here. Perforations in the cellulose wall of plant cells forming a closely packed mosaic were mentioned previously. In some animal tissues with mosaic pattern, secondary silver impregnation produces rows of dots on the free surface of the cells where one would expect continuous outlines. This was observed by Grafflin and Foote (1938) in renal tubules of the cat (pars convoluta of proximal tubule).

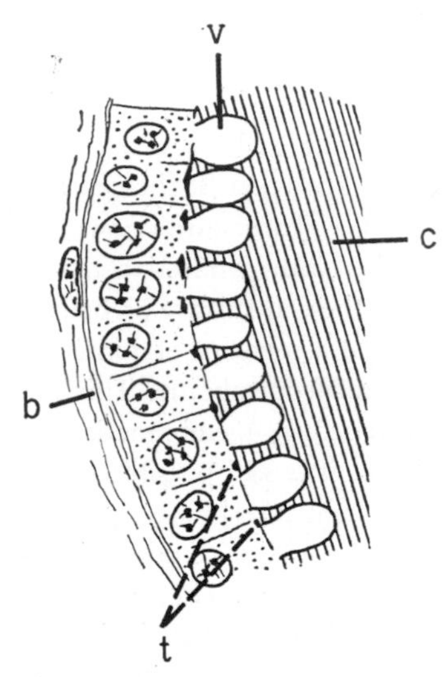

FIG. 68. Terminal bars. Diagram of part of a follicle of a human thyroid in activated phase. Colloid (c) remained anchored to the terminal bars (*t*) while it retracted from the rest of the cell surfaces, thus forming marginal vacuoles (*v*). Terminal bars are cut transversely in (*t*), skew or lengthwise in other places. (*b*) Basal membrane of follicle cells. Diagram drawn from a photomicrograph taken at a magnification of approximately ×1000. Note: For lower magnifications of thyroid in morphological activation see Figs. 26C and D.

In cultures of isolated macrophages W. H. Lewis (1931) studied the behavior of surface membranes in living cells. He observed that wavy sheetlike protusions or thin membranous pseudopodia projected from the cell margin, engulfed small drops of the medium, and finally retracted so that the outline of the cell was reconstructed. In this way the drops were transported into the interior of the cell, similarly to the manner in which milk is lapped up by the cat. Lewis called the phenomenon pinocytosis, which means "drinking action of cells," in analogy to phagocytosis of solid or semisolid particles, or "eating action of cells." Remarkable instability of the cell membrane was observed by Robinow (1936) in epithelial colonies cultivated from rabbit kidney. The most impressive of Robinow's pictures was published in Fischer's "Biology of Tissue Cells" (1946, Fig. 1). This picture shows marginal cells of the colony which are incompletely impregnated with silver. From one of the cells a tonguelike piece of cytoplasm protrudes into the medium. As mentioned previously, some of Robinow's pictures show cytoplasmic margins outside the official silver-impregnated boundaries. The reasons why Robinow's observations cannot be considered as artifacts were discussed in connection with technical interferences (Part III, Chapter 3, C, 2, c).

Electron microscopy has revealed marked corrugations in cell membranes which appear smooth under the light microscope. Fawcett (1955) described foldlike structures which occur along the whole surface of liver cells where they make contact with neighboring cells. On the surface of high-prismatic cells in the mouse jejunum, Zetterqvist (1956) observed a pattern of folds. It seems that the grooves between the folds have pores from which narrow winding spaces extend into the depth of the cytoplasm. In sections, the membranes which border these spaces appear as two parallel lines. From a discussion at the Edinburgh Symposium on "Biological Organization, Cellular and Subcellular" (1959) I summarize Sjöstrand's definitions of three types of cytomembranes. The α-cytomembranes are double membranes covered on one side with a great number of uniformly distributed particles which consist of ribonucleic acid. These membranes frequently occur close to each other in the ergastoplasmic regions of the cytoplasm (see my Fig. 57G). Whenever this pattern is found, the coating granules are responsible for the basophilia of that region. The α-cytomembranes were originally interpreted as an endoplasmic reticulum or a system of channels. The β-cytomembranes are infoldings of the cell membrane, due either to invagination or to a complicated wrinkling of the interface between two adjacent cells. The γ-cytomembranes are, according to Sjöstrand, the electron microscopic equivalent of the Golgi apparatus.

Assuming that a cell membrane is a steady state rather than a stable structure, Bennett (1956a) proposed a hypothesis that the flowing membrane may be part of a transport mechanism which carries ions or particles into and out of cells and also within the cells. "Membrane flow" is derived from the idea that a membrane may be formed or synthesized in one region and broken down or enzymatically destroyed in another region. As a second mechanism by which particles can be transported, Bennett suggested a transient vesiculation of cell membranes. The hypotheses of Bennett are derived mostly from well-documented observations of Palade (for references see Bennett's paper).

In many instances the absence of light microscopic cell membranes or intracellular partitions has been puzzling. In rotifers with constant nuclear numbers, cell boundaries are present in some areas but not visible in adjacent areas (see Part III, Chapter 4, C, 2). Possibly the electron microscope would reveal cell boundaries around each nuclear territory. The cleavage of insect eggs goes through the following phases: a centrally located single nucleus migrates to the periphery; the nucleus divides into a number of nuclei which are aligned near the surface of the egg; finally partitions appear between the nuclei (Bodenstein, 1955). It would be interesting to use the electron microscope in order to determine the earliest appearance of these partitions. Unfortunately most embryological material is lacking differential absorption of electrons and, at the same time, is inaccessible to shadowing techniques. I mention an embryological observation in which the absence of cell boundaries occurs as an abnormality. For unknown reasons, developing eggs of ascidians, such as *Ciona intestinalis,* undergo a regression of the cleavage furrows, thus returning to the external appearance of an undivided egg. Andresen *et al.* (1944) studied the cytology and oxygen consumption of *Ciona* eggs in which reversal of cleavage had taken place. Regular mitotic divisions of nuclei were not seen. However, the nuclei increased in number, probably by fragmentation. Surprisingly, the oxygen consumption in larvae without cell boundaries was only slightly less (approximately 15%) than that of normal larvae of corresponding age. It seems to me that in this study the loss of light microscopic cell boundaries is biologically significant, since it is accompanied by disturbances of the normal nuclear divisions.

On the basis of light microscopy it had been assumed that the myocardium is not divided into cellular units, but forms a continuous mass (syncytium). The intercalated discs which transverse the muscle fibers[1] were interpreted in various ways, but not as cell boundaries. Electron microscopic studies of various investigators led to the opposite interpretation. Cardiac muscle is no longer considered a syncytium since the intercalated discs appear to be specialized junctions between cellular units of the myocardium (Fawcett and Selby, 1958). At another occasion Fawcett *et al.* (1959) found that cell membranes, which were believed to form complete partitions on the basis of light microscopy, exhibited large gaps when studied with the electron microscope. These observations were made in spermatids which are cells in an early phase of spermatogenesis. They occur in clusters of four cells which progress synchronously to the next spermatogenetic stage. In traditional pictures of histological sections each cell is surrounded by a distinct and complete membrane. Electron microscopic studies reveal that the members of each four-cell group are connected by broad cytoplasmic bridges. Although these bridges are particularly conspicuous in OsO_4-fixed sections examined under the electron microscope, they can also be seen in groups of living cells with the use of phase contrast light microscopy. A diagram of this condition is given in my Fig. 70c.

Invisible partitions in the interior of amebas have been postulated by Harvey and Marsland (1932) on the basis of experiments with the centrifuge microscope (for description of apparatus, see Part III, Chapter 2, A, 4). By micro-

[1] If applied to muscle, the term fiber refers to specialized cytoplasm rather than to intercellular structures.

manipulation drops of oil or cream were injected into amebas, which contain crystals as natural components. Centrifugation showed that the crystals are heavy since they accumulated at the end opposite to the fat drops. When the amebas were observed under the microscope during centrifugation, the heavy crystals fell in jerks even when moving through a completely clear field, as though they had met invisible obstructions. This peculiar "move and stop and move and stop" was imitated in a model consisting of a gelatin drop mixed with a drop of milk placed on the centrifuge microscope slide. The authors interpreted the invisible obstacle to be cytoplasm in such a condition that a slight alteration could change it from the sol to a gel state and vice versa.

Cell boundaries on the level of light microscopy can vary widely in distinctness, depending on biological and technical conditions. As far as I know, these relations have not been analyzed systematically. Liver cord cells are an instructive example. In an average nutritional condition, liver cell boundaries are usually quite distinct. If the liver cord cells are stuffed with glycogen, the cell membranes become very conspicuous, comparable in appearance to the cellulose walls of plant cells. Liver sections mounted in Clarite may not show cell boundaries, or only indistinct boundaries, whereas sections of the same block mounted in aqueous media exhibit distinct cell outlines. In livers of living animals injected with fluorescein and observed *in situ,* cell boundaries were not visible in ultraviolet light (Grafflin, 1947).

A special type of interface is involved in the synapses of the central nervous system. The concept of synapse refers to the site of transmission of stimuli from one neuron to another. The synaptic mechanisms were mentioned previously (Part III, Chapter 1, A). A neuron consists of the nerve cell (perikaryon) with several ramifications (dendrites) and one fibrous extension, the axon. The end of the axon splits into a large number of fine branches, each terminating with a small bulb (end bulbs, *boutons terminaux,* synaptic knobs). The end bulbs of the axon of each neuron associate with the next neuron in two places, namely, with the terminations of dendrites and with the surface membrane of the cell body. The cleft between an end bulb of the first neuron and the cell surface of the next neuron is of the order of magnitude of 200 Å. This value is based on electron micrographs of sections and electrochemical determinations in the living animal (for reference see Part III, Chapter 1, A). Since neurons are associated in this peculiar way, they form "continuous as well as separate entities" (Ham, 1957, p. 437). Hundreds or thousands of end bulbs may be associated with the surface of one nerve cell. As much as 38% of the surface of a nerve cell may be studded with end bulbs (Bodian, 1942, p. 159).

D. Cilia, Filaments, and Flagella

Cilia, filaments, and flagella are related structures. They serve locomotion if they are part of independent organisms such as paramecia and spermatozoa. On the free surface of tissue cells, cilia move particles or fluids in a definite direction. Electron microscopy revealed that the unit structures of cilia, filaments, and flagella contain nine peripheral small filaments plus two in the center. Frequently the peripheral filaments consist of two units arranged tangentially or radially, as shown in diagrams of cross sections (Fig. 69). This

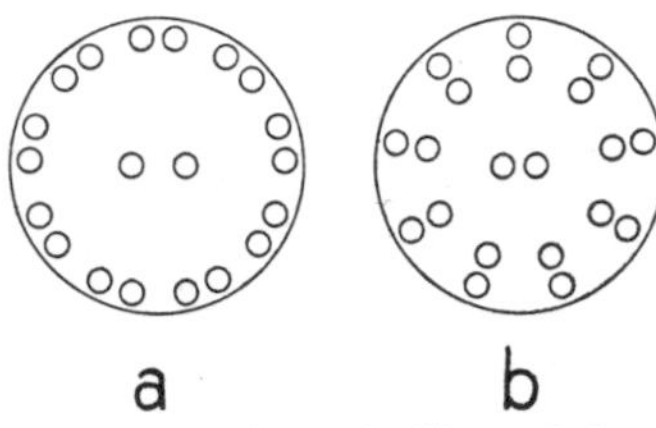

FIG. 69. Diagrams of transverse sections of cilia and flagella, showing internal pattern of nine peripheral pairs of small filaments plus two central filaments as revealed by electron microscopy. (a) Tangential arrangement of peripheral paired filaments, typical of most cilia and flagella. (b) Radial arrangement of peripheral paired filaments, occurring in tail piece of mammalian sperm. (Redrawn from Engström and Finean, 1958, Fig. IV, 25.)

same pattern has been found in all organisms that have been examined. Fawcett and Porter (1954) observed the 9 + 2 pattern in ciliated epithelia of various mollusks, amphibians, and mammals. Bradfield (1955) reported the presence of the pattern in plants such as algae, mosses, and ferns; in sperm tails of sea urchins, slugs, cockroaches, spider beetles; and in vertebrates such as fish, cocks, frogs, and human beings. I mentioned previously (Part III, Chapter 7, C) that there is no bridge between the general 9 + 2 pattern of filaments and the data of present protein chemistry.

The ciliary apparatus of paramecia has been investigated mostly by silver nitrate impregnation of dry or moist preparations (Chatton and Lwoff, 1936). Paramecia treated with these procedures were studied in electron micrographs by Ehret and Powers (1959). Three-dimensional reconstructions revealed a remarkable variety of structures in the "pellicle surface" and its junction with the cilia.

Astonishingly complicated structures extend from the surface of the epithelial cells which line the inner surface of the nictitating membrane of the eye of many birds and reptiles. An illustration of this feather epithelium is found in Andrew's "Textbook of Comparative Histology" (1959, Fig. 16). Possibly these structures have the function of cleaning the eyeball, like "feather-dusters on the histological level," as Andrew puts it.

E. Interaction of Nucleus and Cytoplasm

The relations between nuclei and cytoplasm have topographic, biochemical, genetic, and embryological aspects which are interwoven in various ways. Some topographic distributions of nuclei in the cytoplasm were illustrated and discussed earlier (syncytium of chorionic villi, Fig. 54a; multinucleated giant cell of Langhans type, Fig. 54b). This subject will be treated more fully in the chapter on nucleocytoplasmic patterns (Part IV, Chapter 2). The biochemical relations between nucleus and cytoplasm center mainly on the distribution of nucleic acids. As an example of studies in this field I reported the experiments of Goldstein and Plaut (1955), who showed that most of the ribonucleic acid of the cytoplasm is synthesized in the nucleus (Part III, Chapter 7, B, 3). It is now generally assumed that the nucleolus is the site of this activity. Exchange of material between nucleus and cytoplasm may include particles, since pores in the nuclear membrane were revealed by electron microscopy. Hughes (1952) called attention to indirect evidence of action across the nuclear membrane in the so-called bouquet stage of the meiotic prophase. The ends of the chromosomes are bunched together inside the nuclear membrane in the place closest to the centrosome which is located in the cytoplasm. Morphological aspects of the interaction between nucleus and cytoplasm were discussed in a survey article by Altmann (1955).

The interaction of nuclei and cytoplasm is one of the most important points in analyzing the relation between genetics and embryologic differentiation. The main problems of this relation can be summarized as follows. (1) How is the differentiation of cytoplasm controlled by the genes which are located in the chromosomes? (2) What effect do metabolic processes in the cytoplasm exert on the genes? These problems suffered additional complication when more and more instances of cytoplasmic inheritance (plasma genes) were discovered. I mentioned earlier the so-called killer factor, kappa, in *Paramecium aurelia.* The kappa factor is one of the best-studied cytoplasmic components with genetic autonomy (Lederberg, 1952; Sonneborn, 1959).

Since embryonic cells differentiate in various ways, although they are endowed with the same assortment of chromosomal genes, one is forced to assume unequal reactivities of the cytoplasm of different cells. It became possible to study the different reactivities of cytoplasma in embryonic cells when Briggs and King (1953) developed their technique of transplanting nuclei as described previously (Part III, Chapter 3, C, 2, a). Within the same species, transfer of nuclei from differentiated cells of the blastula or gastrula stage into enucleated eggs caused cleavage and subsequent development into normal tadpoles. If different species (*Rana catesbeiana and Rana pipiens*) were used, the early cleavage and formation of blastulas proceeded normally. However, at the beginning of gastrulation a specificity of nucleocytoplasmic interaction emerged, and the development suddenly stopped in the late blastula or early gastrula stage. The same principle was manifested when different orders were used for nuclear transfer. Cleavage and blastula formation occurred quite frequently when nuclei from blastulas of newts (*Triturus pyrrhogaster*) were transferred into enucleated eggs of frogs (*Rana pipiens*), but later developments did not take place. This is a brief summary of experiments of Briggs and King (1955).

Chapter 2

Associations of Tissue Cells. Nucleocytoplasmic Patterns

Nuclei are distributed in the cytoplasm in various ways. It may or may not be possible to allot a definite mass of cytoplasm to each nucleus, and there are nuclei with very little or no cytoplasm. When associations of tissue cells are mentioned in the literature, frequently descriptions and interpretations are mixed. Concepts such as "epithelium" and "mesenchymal tissues" are supposed to cover visible patterns as well as origins and potentialities. The present chapter will be devoted to a *description of the different nucleocytoplasmic patterns* and to some *observed transformations* of one pattern into another. The diagrams and descriptions given here will be utilized, together with other criteria, for the classification and identification of normal and pathologic tissues (Part V, Chapters 2 and 4). Diagrams of nucleocytoplasmic patterns were published by Mayer (1925) and his collaborator Cohn (1926). Similar descriptive diagrams are found in Sharp's "Fundamentals of Cytology," 1943 (his Fig. 5).

A. Aggregates of Single Cells

Figure 70a–c shows single cells, each with one nucleus. In Fig. 70a, the cells have smooth, round, or polygonal outlines. Groups of round cells tend, under pressure, to assume polyhedric shapes, appearing as polygons in sections. Myelocytes in active bone marrow are an example of pattern Fig. 70a. If the cells are very closely packed, it may be difficult to tell whether or not they form a mosaic (see Fig. 72a and b). If single cells float freely in a fluid, they may appear spherical. Isolated motile cells in colloidal media or on interfaces between solid and liquid media show highly variable so-called ameboid forms with pseudopodia, as seen in Fig. 70b. This behavior is common in white blood cells. Cytoplasmic bridges between single cells were usually ascribed to poor technique, causing separate cells to be glued together. However, broad cytoplasmic bridges between adjacent cells occur as regular natural structures in spermatids (Fig. 70c). Spermatids form groups of four cells, and the whole group may be seen in a favorable section. The discovery of this interesting phenomenon by Fawcett *et al.* (1959) was recounted previously in the discussion of interfaces. Because of incomplete cell membranes, the pattern of Fig. 70c can also be considered a variation of the cytoplasmic network in Fig. 73a, or of the unpartitioned cytoplasm in Fig. 74.

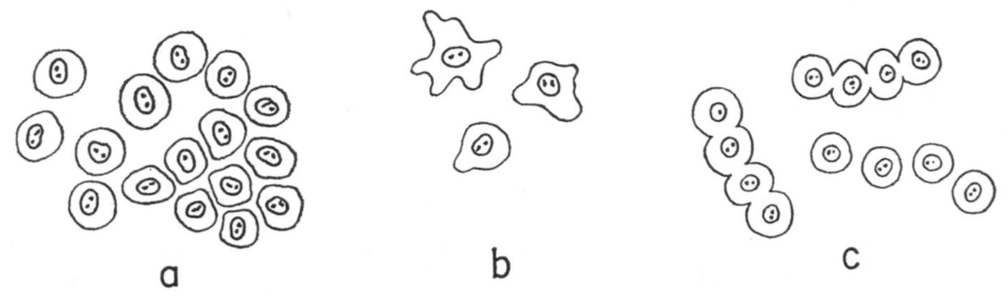

FIG. 70. Nucleocytoplasmic patterns: single cells. (a) Free cells, rounded or slightly flattened against each other depending on space available. (b) Single cells with pseudopodia suggesting motility. (c) Groups of four spermatids, each group derived from one ancestral cell (spermatogonium). The phase with broad cytoplasmic bridges between individual cells is intermediary between the single cell pattern and various patterns of aggregation. The four spermatids remain joined during their synchronous maturation, but finally separate. (Diagram made from photo- and electron micrographs of Fawcett, Ito, and Slautterback 1959.)

B. Mosaic Pattern

In art, the term mosaic refers to a picture or pattern composed of one layer of closely packed units which may or may not vary in color and shape. I borrow the term mosaic with one modification for the description of nucleocytoplasmic patterns: it will be applied not only to one layer of units, but also to units stacked in several layers. The term mosaic pattern is used here in contrast to the single-cell pattern, the network patterns, and the pattern of multinucleated unpartitioned cytoplasm.

Figure 71a–e shows three-dimensional models of one-layer mosaics. High-prismatic cells (Fig. 71a) form the lining of the mucous membranes in trachea, bronchi, stomach, intestines, and uterus. This is illustrated in Fig. 33 showing the jejunum of a mouse. A modification of high-prismatic cells is found in the acini of glands (Figs. 57 F and 87A). Since the secretory cells are arranged around a central opening, the prisms become truncated pyramids, appearing triangular with rounded tips in sections. The traditional name for the pattern Fig. 71a is "columnar epithelium."

In Fig. 71b, the prismatic cells are of medium height. Their traditional name is "cuboidal epithelium," since in sections vertical to the surface of the layer, the cells appear as squares. Ducts of glands are frequently lined by cells of this shape (Fig. 87A and C). The pattern Fig. 71c represents very shallow prisms. This occurs in small caliber ducts. The traditional name is "low cuboidal epithelium."

The patterns Fig. 71 a–c are found in different segments of renal tubules (Bailey, 1953, Fig. 313). In the thyroid, the wall of the follicles may be formed by either of the three patterns. My diagrammatic Figs. 25 and 26 show that

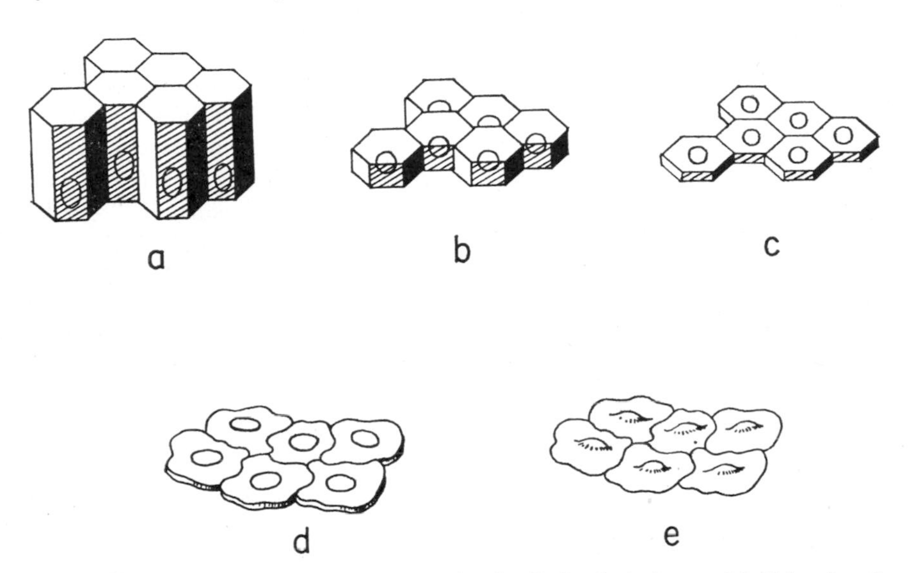

FIG. 71. Nucleocytoplasmic patterns: mosaic of cells in single layers. (a) High prismatic cells (columnar epithelium). (b) Prismatic cells of medium height (cuboidal epithelium). (c) Low prismatic cells (low cuboidal epithelium). (d) Low cells with wavy outlines. (e) Extremely thin cells with thickenings produced by nuclei (endothelial cells). A sheet of such cells forming a tube is shown in Fig. 65e (tube sectioned lengthwise). Note: In these three-dimensional diagrams, outlines of nuclei are drawn to indicate that the nuclei are enclosed in cytoplasm.

the height of the follicular cells in the thyroid depends on the functional phase of the gland and on the fixative applied. The height of cells in a one-layer mosaic may be inferred from the shape or position of the nuclei, even though cell boundaries may not be visible (Fig. 25B and D).

In Fig. 71d the cells resemble in height those of Fig. 71c, but their outlines are wavy instead of polygonal. The pattern Fig. 71d commonly occurs on the lower surface of leaves (for a three-dimensional diagram see Eames and MacDaniels (1947, Fig. 149). It is probable that the polygonal pattern (c) and the wavy pattern (d) are expressions of different mechanical conditions during the development of the tissues. Once established, the two patterns seem to be independent of external mechanical factors. If the mesentery of a small animal is fixed while stretched on a frame, the outlines of the covering cells still remain wavy (Fig. 51a). A botanical example shows that the difference between the two patterns can be hereditary. Karpechenko (1928, pp. 35 and 36) observed that epidermal cells of the petals were wavy in *Raphanus sativa* (radish), but polygonal in *Brassica oleracea* (cabbage). The wavy outlines of *Raphanus sativa* proved to be dominant in hybrids of the two species. Karpechenko took care to select branches of equal nutrition and light conditions for these comparisons (for genetic aspects see Part V, Chapter 1, A).

In animals, one-layer mosaics of cells with wavy outlines are usually so shallow that only the places of the nuclei can be represented in three dimensions (Fig. 71e). Seen on edge, these structures appear as thin lines, with thickenings at intervals, produced by the nuclei (Fig. 65e). Very thin cells with wavy outlines form the walls of blood and lymph capillaries, the inner layer of larger vessels, and the lining of the thoracic, peritoneal, and other cavities (Fig. 51a). Traditional names for this pattern are "simple squamous epithelium," "endothelium," or "mesothelium," depending on the location or probable origin.

Under pathologic conditions the thin cells with wavy outlines may swell to such an extent that structures resembling Fig. 71d are produced. As a normal occurrence, relatively thick cells with wavy outlines have been described by Grafflin and Foote (1938, 1939) in certain segments of renal tubules.

Relations between the three-dimensional shapes of cells shown in Fig. 71a–c, and their appearance in sections are illustrated in various textbooks of histology (Bailey, 1953, Fig. 22; Ham, 1957, Fig. 118). Sections through *several layers* of the mosaic pattern are also illustrated in most textbooks. Three-dimensional reconstructions (by F. T. Lewis) of cells in stacked mosaics were described previously (Figs. 66 and 67). Different types of stacked mosaics occur in different organs. In the parathyroid gland the mosaic of chief cells forms clumps and irregular cords. In sections, the liver cord cells appear either as single rows or as double rows as seen in the diagrams Fig. 42 and 87C. Three-dimensional reconstructions of liver tissue by Elias (1949) showed a maze of perforated plates consisting of a mosaic of cells.

How can one distinguish between an organized cell mosaic and mere crowding of single cells? The pictures seen in Fig. 71a–c could represent an organized mosaic of fused cells, or a mass of isolated cells so closely packed that their shapes became polygonal or polyhedral. Since this problem can hardly be studied in histological sections, let us start the discussion with observations on tissue cells cultivated *in vitro*. Figure 72a represents cultured cells from monkey kidney, and Fig. 72b cultured cells which are derived from the bone marrow of a monkey.

The diagrams Fig. 72a and b, were made from photographs kindly supplied by Dr. R. C. Parker together with his interpretations. Reference to the published material is made in the legends. It is not surprising that upon explantation of kidney fragments the renal tubules, which are mosaic pattern structures, give rise to sheets of cells consisting of a so-called epithelial mosaic. By contrast the bone marrow does not contain structures with a mosaic pattern (except in its vascular lining). Therefore, explanted bone marrow usually produces populations of single cells (Fig. 70a and b) or aggregates of the network pattern (Fig. 76a and b). The cells in Fig. 72b represent an unusual type, probably altered by conditions of cultivation. Their pattern resembles that of Fig. 72a, from monkey kidney. However, in view of the absence of mosaic patterns in

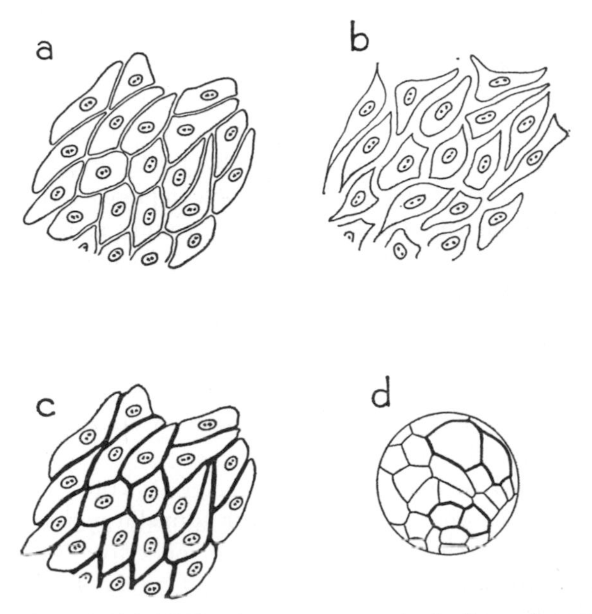

FIG. 72. Problem of distinguishing between an organized cell mosaic and mere crowding of single cells. Diagrams (a), (b) and (c) represent cell aggregates cultivated *in vitro*. (a) Cells derived from the tubules of explanted monkey kidney; usual type. Diagram made from photo interpreted by Dr. R. C. Parker as epithelial mosaic (see Fig. 11 of Parker, Castor and McCulloch, 1957). (b) Cells derived from connective tissue cells of explanted monkey bone marrow; unusual (altered) type. Diagram made from photo interpreted by Dr. R. C. Parker as closely packed single cells (see Fig. 17 of Parker, Castor and McCulloch, 1957). (c) Identical with (a), except for filling of gaps with black lines representing the picture which would result from primary impregnation with $AgNO_3$. Probably this technique has never been applied to preparations of type (b). (d) Heterogeneous impregnation of cell boundaries seen in late blastula of *Teredo norvegica* (shipworm). Silver impregnation was preceded by brief fixation with OsO_4. Cells with strongly impregnated boundaries are primordia of special organs such as shell glands. (After Fauré-Frémiet and Mugard, 1948.)

the bone marrow, one would be inclined to interpret the pattern of the bone marrow derivatives as crowded single cells rather than as epithelial mosaic.

To the tissue culturist, the shape of the colonies and other characteristics suggest that the cells of Fig. 72a are probably derived from mosaic pattern ancestors (epithelial tissue) and that those of Fig. 72b are probably derived from blood-forming cells or connective tissue cells (mesenchymal tissue). One may consider the possibility that the spaces are filled by extensions of the cytoplasm, which are too delicate to be seen in living cells or to be stained effectively after fixation. Electron microscopy has been used to study the boundaries between adjacent cells in sections. What was known as single lines in light microscopy, appeared as double lines in electron microscopy. I do not know whether total mounts of cultivated cells lend themselves to this type of electron microscopic analysis.

What tests on the light microscopic level are available to distinguish an organized mosaic from a mere aggregate of crowded cells? The first test which comes to mind is treatment of the fresh (unfixed) cell colony with $AgNO_3$. With this technique distinct cell boundaries were demonstrated by Robinow (1936) in cultures from rabbit kidneys, and by Jacoby and Marks (1953) in cultures from guinea pig kidneys. My Fig. 72c represents an experiment of thought: let us assume that the kidney-derived culture seen in Fig. 72a was treated with $AgNO_3$ with the result that the gaps between the individual cells were filled by thick black lines. The fictive Fig. 72c looks like an enlargement of Fig. 5 of Jacoby and Marks. Let us further assume that $AgNO_3$ treatment of the bone marrow-derived culture seen in Fig. 72b, did not result in black outlines of individual cells. The conclusion would be that, with silver impregnation as criterion, the cells in Fig. 72a form an organized mosaic pattern, while those in Fig. 72b do not.

The next question is: how reliable is the $AgNO_3$ criterion? As shown in Fig. 51a and b, of a mesentery, prefixation with formalin prevents $AgNO_3$ from staining of cell outlines, but produces staining of connective tissue fibers instead. This, however, is not a general rule. In the early blastula of a mollusk, *Teredo norvegica,* Fauré-Frémiet and Mugard (1948) observed the formation of distinct cell outlines as the result of $AgNO_3$ treatment preceded by brief fixation with OsO_4 (see my Fig. 72d). Neurons can be impregnated by a variety of silver nitrate procedures. The original Golgi silver method consists of immersing fresh pieces of nervous tissue first in a solution containing potassium dichromate (and usually OsO_4), and then in $AgNO_3$ (see McClung's Handbook, 1950, p. 356). In a modification of the Golgi stain by Porter and Davenport (1949) the sequence is reversed: the sample is placed first in a mixture of $AgNO_3$, formalin, and pyridine for 48 hours and then transferred into a potassium dichromate solution.

The cultures represented in my Fig. 72a–c contain neither connective tissue nor nerve fibers. Therefore positive impregnation with $AgNO_3$ would support the interpretation that the black lines in Fig. 72c represent cell outlines. Yet this criterion would not be entirely decisive. Narrow crevices or channels can also appear as solid black lines after treatment with $AgNO_3$. In this way, bile capillaries are presented with the use of Golgi methods. On the surface of paramecia an intricate system of lines is demonstrated by $AgNO_3$ treatment of dry or moist preparations (see Part IV, Chapter 1, D). These lines are certainly not cell boundaries, since the organisms are noncellular (unicellular). I also mentioned the possibility that the cells in Fig. 72b might be connected by cytoplasmic films too delicate to be seen with light microscopy. Because of these uncertainties, the silver impregnation test should be supplemented by other studies. In the discussion of interfaces, it was described how the mutual adhesiveness of cells can be determined (Fenn, 1922a, b; Coman, 1944). In view of the availability of these techniques, the following experimental plan is suggested in order to define the relations between closely packed cells and to decide whether cells are in contact or fused. Isolated cells obtained from suspensions should be placed together as closely as possible, either on a solid or on a semisolid surface. They should be compressed to such an extent that polygonal outlines form. Having been kept in this condition for different periods of time, comparable preparations should be divided into three groups: one group for measuring mutual adhesiveness, by Fenn's or Coman's method; a second group for testing the exchange of dyes between individual cells; and a third group for $AgNO_3$ impregnation. If these experiments give evidence in support of mosaic organization, one may study the reversibility of such a pattern by creating conditions which facilitate the dissociation of the cells.

C. Cytoplasmic Network, Fibrous Network

Figure 73a presents a network of cytoplasm containing a number of nuclei. One may assume that a certain area of cytoplasm is controlled by each nucleus. Such hypothetical areas are known as nuclear territories, but adjacent territories are not separated by visible boundaries. The bridges between the nuclear territories vary in width. They may be so thin that they become indistinguishable from fibrils, as seen in Fig. 73b. A reticular structure intermediary between Figs. 73a and b was illustrated in Fig. 65f. If the proportion of cytoplasmic mass to spaces or holes increases in favor of the cytoplasm, a condition shown in Fig. 74a results. Traditional names for structures shown in both Fig. 73a and Fig. 74a are "syncytium" and "plasmodium." In zoology and botany, plasmodium refers to structures resulting from repeated nuclear divisions without divisions of cytoplasm, whereas the term syncytium refers to the product of fusion of a number of cells. In human histology the term syncytium is used preferentially. The question of "fusion or contact" which was discussed with reference to mosaic patterns has also been raised with respect to cytoplasmic networks (see Lewis, "Is Mesenchyme a Syncytium?," 1922).

The multinuclear cytoplasmic network (Fig. 73a) occurs as an important embryological tissue, the mesenchyme. The interaction of mosaic patterns (epithelium) and cytoplasmic networks (mesenchyme) is characteristic not only of embryonic scaffolding (Fig. 19C), but also of regenerative processes in the adult organism, including wound healing. Lymphatic tissue consists of a cytoplasmic network with meshes of variable caliber, filled with lymphocytes. Transitions between cytoplasmic and fibrillar networks are common both in embryonic

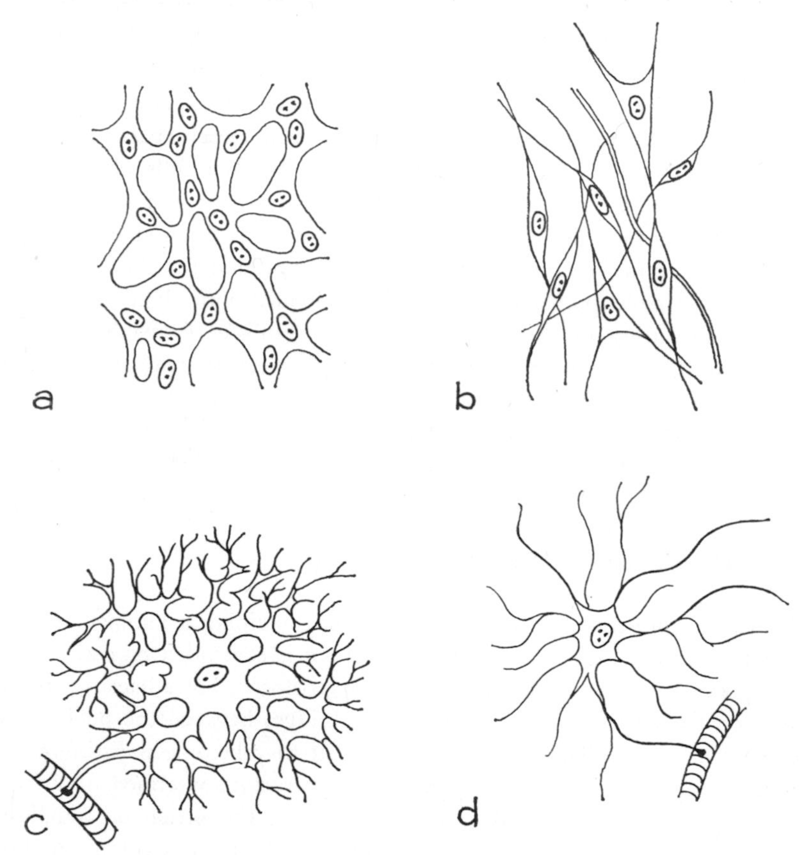

FIG. 73. Nucleocytoplasmic patterns: network pattern (reticulum). (a) Multinucleated cytoplasmic network. Functional nuclear territories may be assumed but are not separated by visible boundaries. Conventional name: syncytium (or plasmodium). (b) Thin bridges between cells can be interpreted as fibrils (fibrillar network) or as narrow strands of cytoplasm. A form intermediary between type (a) and type (b) is given in Fig. 65f. For similar types see also Fig. 76b and d (cultivated cells) and Fig. 75 (stellate cell in hepatic sinusoid). (c) Neuroglia cell forming a cytoplasmic network (diagram of a cytoplasmic astrocyte from gray matter). (d) Neuroglia cell forming fibrils (diagram of a fibrous astrocyte from white matter). Note: In (c) and (d) some processes are attached by perivascular feet to a blood capillary.

and adult organs. Three-dimensional presentations of mesenchymal cytoplasmic nets are found in Bargmann's textbook (1959, Figs. 85 and 86).

In the neuroglia, which is the supporting tissue of the central nervous system, one differentiates between cytoplasmic and fibrous astrocytes. Diagrams of these two types of astrocytes are given in Fig. 73c and d. An interesting photomicrograph of a living astrocyte of the cytoplasmic type, cultivated *in vitro*, is given by Finerty and Cowdry (1960, Fig. 106). Processes of the cells which terminate on the wall of capillaries are represented in my diagrams. Anastomoses between adjacent astrocytes are observed, but not indicated in the diagrams. The richest arborization of nervous elements occurs in the Purkinjé cells of the cerebellum.

Fibrils are considered as modified cytoplasm, products of the cytoplasm, or intercellular substances. Fibrillar structures are characterized by mechanical properties, tensile strength, elasticity, staining properties, chemical analysis, and

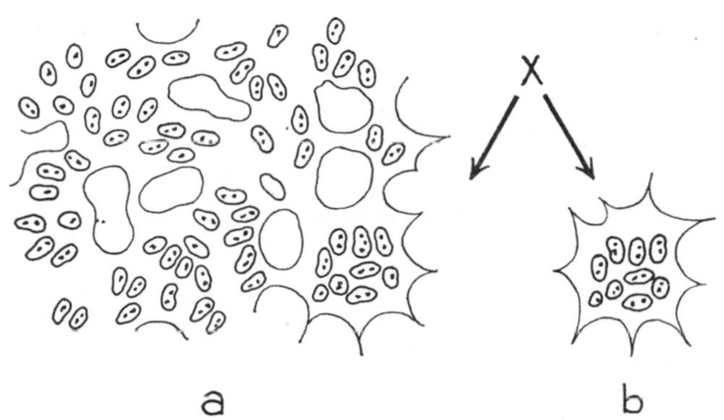

FIG. 74. Nucleocytoplasmic patterns: unpartitioned multinucleated cytoplasm. (a) Large mass of cytoplasm showing fewer holes and greater crowding of nuclei than the cytoplasmic network in Fig. 73a, but otherwise similar. Conventional name: syncytium (or plasmodium). (b) Small mass of cytoplasm separated from the bulk of diagram (a), either by a process during life or by histological sectioning. Note resemblance with Fig. 54a, C, and Fig. 54b, C″. Conventional name: multinucleated giant cell.

by their appearance in polarized light, electron microscopy, and Roentgen diffraction. However, there is no hard and fast borderline between fibrils and long cytoplasmic threads of the type shown in Fig. 77b. Fibers and fibrils will be discussed later in the chapter on intercellular substances.

D. Unpartitioned Multinuclear Cytoplasm

The characteristic feature of unpartitioned multinuclear cytoplasm is the impossibility of allotting a cytoplasmic area to each nucleus.[1] Otherwise the difference between this pattern and a cytoplasmic reticulum is only quantitative, namely different ratios of cytoplasmic mass to holes. This was mentioned above in comparing Figs. 73a and 74a.

[1] Partitions on the *electron microscopic* level (cytomembranes, endoplasmic reticulum) are discussed elsewhere.

The spatial relations of unpartitioned multinuclear cytoplasm were analyzed in the chapter on topographic and three-dimensional morphology. Figure 54a illustrates the syncytial trophoblast forming the outer layer of a chorionic villus. Different sections through this layer show clearly that no nuclear territories can be recognized in the cytoplasm. If the right lower corner of Fig. 74a is separated from the bulk of the cytoplasmic mass, either by a biological process or by sectioning, the isolated structure of Fig. 74b is the result. This is comparable to structure *C* in Fig. 54a. Such patches of unpartitioned multinuclear cytoplasm are named conventionally "multinucleated giant cells," or briefly "giant cells." The giant cell of the Langhans type illustrated in Fig. 54b is characterized by a peculiar distribution of the nuclei, but is still a modification of the general pattern of Fig. 74a. Foreign bodies, such as splinters or suture threads, stimulate the production of unpartitioned multinuclear cytoplasm, with the nuclei clustering around the foreign body (foreign body giant cells).

E. Transformations of Nucleocytoplasmic Patterns

Transformations of nucleocytoplasmic patterns are an important part of the development of tissues. An example was given previously in Fig. 70c, showing a phase of maturation of spermatids. The development of the central nervous system in vertebrate embryos offers a particularly impressive illustration of the transformation of nucleocytoplasmic patterns. The early blastula is a solid ball of closely packed cells. In the next stages the dorsal cell masses arrange themselves into one sheet of cells which forms the roof of a cavity (blastocoel). By a process of infolding the single layer becomes a double layer (Fig. 3e′ and f′, and Fig. 32). Further infolding processes produce the neural groove, the neural tube, the brain, and the eye vesicles (Fig. 3g–l). Up to this point the structures mentioned represent a mosaic pattern of prismatic cells. The subsequent developments can be divided into two phases: (1) loosening of the mosaic resulting in populations of more or less independent migratory cells (Fig. 30), and (2) differentiation into neurons and neuroglia, which means the establishment of the final network pattern of the central nervous system. A small fraction of the nervous tissue either maintains the original one-layer mosaic pattern or returns to it. This is the origin of the ependyma which lines the central canal of the spinal cord and the ventricles of the brain, and covers the choroid plexuses (Fig. 79a). Transformations of mosaic patterns into cytoplasmic networks are illustrated in Hardesty's 1904 paper on the development of neuroglia in the embryonic brain of the rabbit. Several of Hardesty's pictures are reproduced in Ham's "Histology" (1957, Fig. 283).

I mentioned previously the early development of insect eggs in which an unpartitioned multinucleated mass of cytoplasm first acquires incomplete partitions between the nuclei in the surface layer, and finally transforms into a mosaic pattern of well-defined cells (Part IV, Chapter 1, C).

In the adult organism, transformations of nucleocytoplasmic patterns occur under various normal and pathological conditions. Changes from one mosaic pattern to another are not too surprising in places where different mosaic patterns border each other. The uterine cavity is lined by a one-layer mosaic of high prismatic cells (epithelium of endometrium), but the vaginal portion of the cervix uteri is lined with a stacked mosaic (stratified squamous epithelium) similar to that of the vagina. The place of transition is unstable. Under pathological conditions either of the two types may occupy more than its normal area. If stratified squamous epithelium of the vaginal portion is replaced by high prismatic cells, the normally opaque whitish layer assumes a transparent red appearance simulating an ulcer (hence the term erosion).

If the structures of the human pelvis which hold the uterus in its proper position lose their strength, gravity causes the uterus to drop into the vagina and finally protrude through the vulva. In this condition which is known as prolapse, the cervix of the uterus is turned inside out. Thus the endometrium is exposed to friction and to air. In response to this abnormal environment the one-layer mosaic is transformed into a stacked mosaic of several layers, and the moist surface of the mucous membrane is changed to a dry condition resembling that of the epidermis.

In experiments with rats, Wolbach and Howe (1925) observed that lack of vitamin A transformed the moist one-layer mosaic of mucous membranes into a stacked mosaic with keratinization of the surface layers. This transformation occurred in the mucous membranes of the respiratory and genitourinary tracts, and also in the pancreatic ducts, but not in the stomach, intestines, liver or kidneys. The mode of transformation was the following. First the original lining of high prismatic cells became very shallow or disappeared (atrophy), then a new lining formed with increased mitotic activity and piling up of cells so that a stacked mosaic resulted. It is remarkable that upon adequate supply of vitamin A, a recovery took place in which the transformed epithelium of each region returned to its normal type (Wolbach and Howe, 1933).

Let us now consider transformation of network patterns into mosaics, of mosaics into network patterns, and similar dramatic changes. In the mucous membrane of the human uterus (endometrium) the layer known as tunica propria consists of a cytoplasmic or fibrillar network during some phases of the estrus, while it forms a mosaic (decidua cells) during other phases. In this mosaic, delicate fibers may be visible between individual cells. Another transformation is illustrated in Fig. 75, a diagrammatic cross section through a sinusoid[2] of the liver. The thin wall (*v*) of the vessel is coated on its inside with a shallow

[2] Liver sinusoids are the blood vessels between and along the trabeculae (rows of liver cord cells); see Figs. 42 and 87C.

mosaic. A nucleus (*e*) of this so-called endothelium bulges into the bore. This is the pattern shown in Fig. 65e and Fig. 71d and e. Under certain conditions some of the endothelial cells phagocytose red blood corpuscles. This is possible only when the very shallow cytoplasm swells at least to the thickness of a red blood cell. When more red blood cells are engulfed, the cell increases in volume and loses its contact with the neighbor cells. Instead of being part of a mosaic, the cell is transformed into a polygonal structure with cytoplasmic processes moored to the wall of the vessel. In diagram Fig. 75 the cell which has phago-

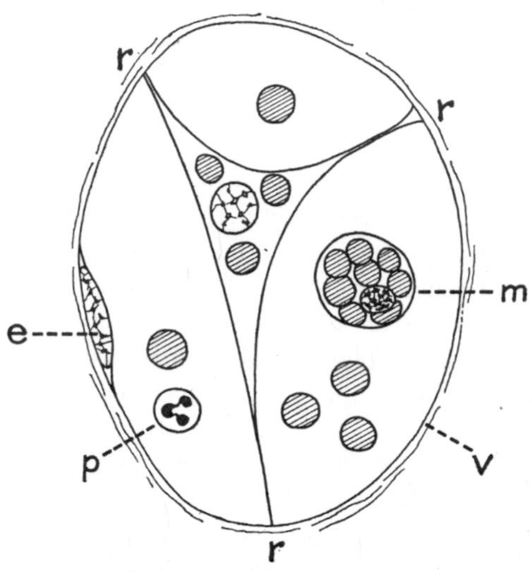

FIG. 75. Transformation of nucleocytoplasmic patterns *in situ*, from mosaic to cytoplasmic network and eventually to single cell. Diagrammatic cross section through sinusoid of mammalian liver. Free and phagocytosed red blood corpuscles shaded. *v*, Wall of sinusoid. *e*, Nucleus of endothelial lining seen on edge, similar to nuclei in Fig. 65e; for surface view of endothelial mosaic, see Fig. 71e. *r*, Three places of attachment of cytoplasm of a stellate cell (Kupffer cell). This is a transformation of type e, resulting from swelling and uptake of some red blood corpuscles and loss of contact with neighbor cells. *m*, Macrophage stuffed with red blood corpuscles; probably a transformation from the stellate form. The term macrophage as used here means large phagocyte. The fate of *m* in the circulating blood is controversial. *p*, Polymorphonuclear leucocyte.

cytosed three red blood corpuscles has assumed a triangular shape with thin attachments (*r*) to the wall. Such transformed phagocytic cells in liver sinusoids are known as Kupffer cells or stellate cells. When filled with more and more red cells, the cell rounds off as indicated by (*m*) in the picture. These transitions are illustrated convincingly in an old drawing by Alexander Maximow showing the effect of India ink injection in a rabbit liver (reprinted by Maximow and Bloom, 1952, Fig. 385). It is controversial whether the rounded isolated cells become the monocytes of the circulating blood. As described earlier (Part III, Chapter 3, B) Rous and Beard (1934) attempted to verify this transformation in rabbit experiments by loading the Kupffer cells with iron particles, collecting the loaded cells with a magnet, and cultivating them *in vitro*. The experiments did not support the hypothesis that Kupffer cells become monocytes of the blood (Beard and Rous, 1934).

The examples described represented active transformations, in other words, responses to various stimuli. The human tonsils are a good example of passive

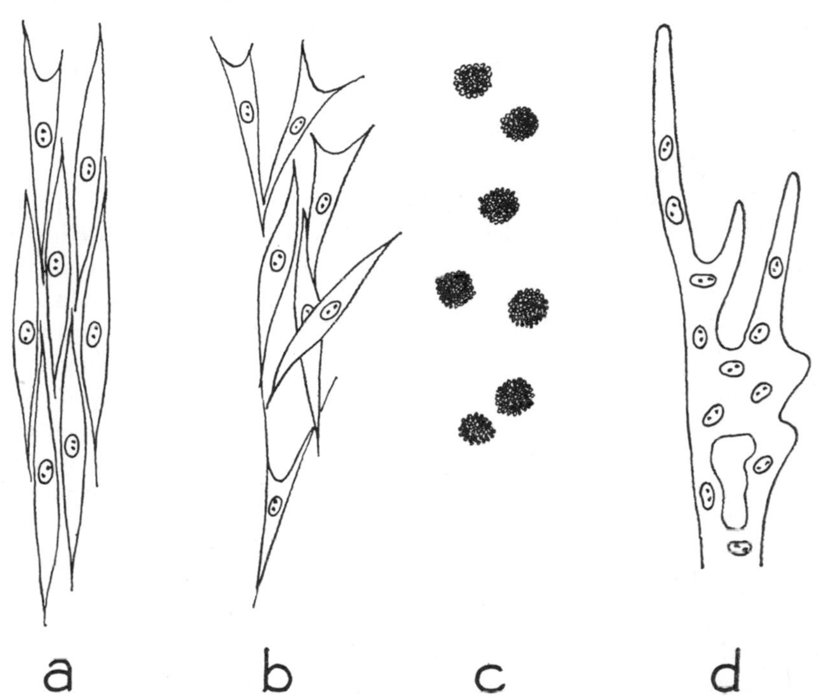

FIG. 76. Transformation of nucleocytoplasmic patterns and cell aggregates cultivated *in vitro*; from network to single cells or to unpartitioned multinucleated cytoplasm. (a) Closely packed spindle-shaped cells, connections not visible. (b) Looser spacing of same cell strain as in diagram (a), resulting in visible network pattern. (c) Transformation of (a) or (b) into isolated spherical cells; a result of injury. Accumulation of fat droplets indicated by black masses in the picture. (d) Transformation of (a) or (b) into unpartitioned multinucleated cytoplasm. Note: Cells shown here are cultivated from connective tissue, cartilage, bone, or muscle. They are briefly termed fibroblasts in tissue culture papers. Diagrams made from photomicrographs of living cells. (a), (b), and (d) from E. Mayer, 1930; and (c) from Mayer and Schreiber, 1934).

mechanical transformation. The stacked mosaic (stratified epithelium) which covers the human tonsils is regularly loosened by lymphocytes invading this layer. If the migration is intensive, the epithelial mosaic is disrupted by clusters of lymphocytes. As a result of the disruption the picture of a cytoplasmic network is produced, sometimes with loss of visible cell boundaries (Fig. 63).

Many of the criteria used for identification and classification of tissue cells in their natural situation are not applicable to cells cultivated *in vitro* (see Part V, Chapter 2, B, 1). Yet the main nucleocytoplasmic patterns of tissue cells *in situ* also occur in cultures. Since tissue cultures are particularly suitable for the study of certain factors which produce changes of patterns, some transformations *in vitro* will be described here.

Figure 76 shows cell forms which were observed in cultures from chick embryo hearts. The technique was that applied in the cultivation of fragments of bone, as illustrated in Fig. 21. In their most active phases the colonies consist of spindle-shaped cells which are so closely packed that they resemble a mosaic; their delicate cytoplasmic connections are not visible in that condition (Fig. 76a). When the multiplication and migration rate is below maximum, the cells are spaced more loosely so that a cytoplasmic network becomes apparent (Fig. 76b). Severe injury to the cells by failure to renew media, overheating, or ultraviolet irradiation leads to a withdrawal of the cytoplasmic connections between cells: the network breaks down and the cells become rounded. The rounding-off is accompanied by the appearance of numerous fat droplets in the cytoplasm (Fig. 76c). This is usually an irreversible condition leading to the death of the cells. However, cells may also round off prior to mitotic division, as part of their regular life cycle. Figure 76d is a relatively rare occurrence. The broad ribbons of unpartitioned multinucleated cytoplasm were observed after passages in media containing proteoses with high nonproteinic nitrogen (Mayer, 1930).

Figure 77a and b shows cells derived from mouse fibroblasts. The strain was established by Earle in 1940 (published 1943), had developed malignant properties in some substrains, as tested by grafting into animals, and has been cultivated since in various laboratories (Earle's L cells). After several years of uniform treatment Parker and Healy (unpublished data) prepared a number of subcultures from the same suspension. Some of the subcultures were cultivated in media containing B vitamins, while others were cultured without B vitamins. As a result, the cells formed the two different patterns seen in Fig. 77a and b (photomicrographs and data supplied by Dr. R. C. Parker, 1957). If the cells in Fig. 77a were a little more crowded, they would resemble those seen in Fig. 72b. The pattern in Fig. 77b resembles a fibrillar network, although there is no evidence of fusion between the long processes projecting from different cells.

Mechanisms which control the transformation of nucleocytoplasmic patterns were studied by Bonner (1947) in the slime mold *Dictyostelium discoideum*. Individual uninuclear myxamebas (single cells) terminate the so-called vegetative stage by entering the aggregation stage. Having ceased to feed and to multiply, the individuals stream in together to form a coherent mass of unpartitioned multinuclear cytoplasm (pseudoplasmodium). This mass then crawls as a body for variable distances during the migration stage. Finally this mass differen-

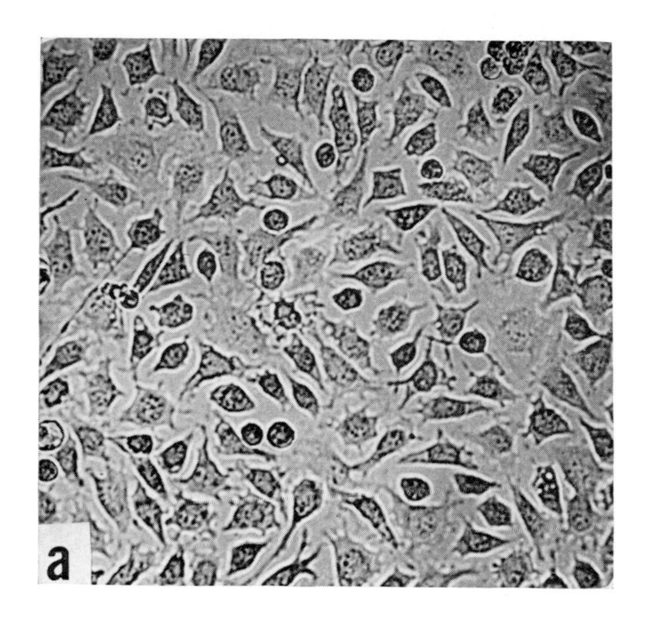

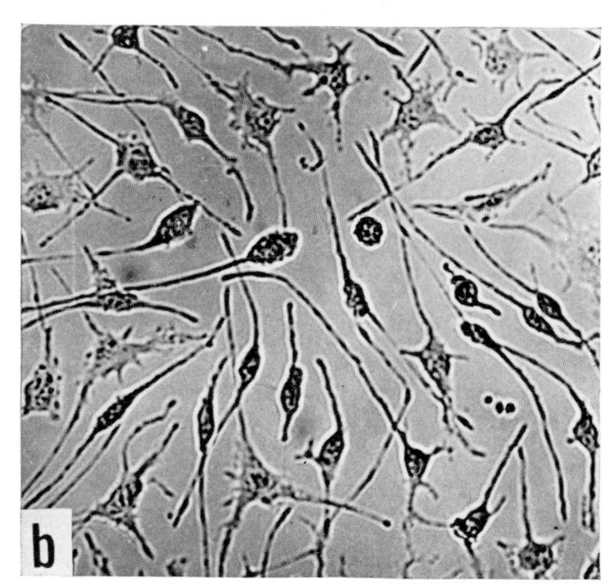

FIG. 77. Variation of pattern of cells *in vitro* depending on culture media. W. R. Earle's L cells originally derived from explanted mouse connective tissue. (a) Subcultures in media containing B vitamins showing polygonal cells with short cytoplasmic processes, moderately crowded. (b) Subcultures in media without B vitamins showing oval cells with long fibril-like processes. (Photomicrographs and data supplied by Dr. R. C. Parker, 1957.)

tiates into a spore-forming organism (culmination stage). Bonner posed the question how a number of independent myxamebas can be drawn together to form one organism. From his ingenious experiments I recount the following results. When placed in a thin film of water on an inverted coverglass, myxamebas showed very distinct attraction up to 200 μ distance (13 myxameba diameters). At four times this distance a weak but definite orientation was observed. No directing effects were produced by electric currents, magnetic fields, fibers, or grooves. The aggregation stimulus was not transmitted through glass, mica, or quartz. The effect was demonstrable across a water-filled gap of the substrate, and over the edge of a coverglass held under water. These observations suggested chemotaxis as the most probable mechanism to draw the single cells together. Bonner postulates that an active substance, "acrasin," diffuses from the single cells, but as yet this substance has not been identified (Bonner, 1959).

Chapter 3

Intercellular Substances and Spaces

Intercellular substances and the materials which fill tissue spaces can be defined by morphological, mechanical, or chemical criteria. In some instances these criteria are correlated, as in bone substance where calcium salts are deposited in the spaces between the bone cells. In other cases either the morphological or the chemical aspect is in the foreground. Papers on the production of collagen in tissue cultures may refer to production of collagenous fibers or to an increase in hydroxyproline (Gerarde and Jones, 1953), an amino acid which is characteristic of gelatin and collagen. The staining behavior of fibers may characterize their morphological as well as their chemical properties.

On the basis of morphological criteria intercellular substances and intercellular spaces are easily defined in some instances, but in other instances it is controversial where the cells end and the intercellular substances or spaces begin. In the mature cartilage one or two cells are encapsulated in a distinct space known as a lacuna. Between the lacunae are more or less homogeneous cell-free masses called the matrix. This is a distinct separation of cells and intercellular substance. In contrast to cartilage, the cells of bone tissue are closely interwoven with the intercellular substance. The scattered cells of immature bone tissue are connected by cytoplasmic processes, thus forming a cytoplasmic network. Gradually the meshes of the network are filled with intercellular substance, the matrix of the bone. In very thin sheets of mature bone canaliculi between cells can be demonstrated (Bremer-Weatherford, 1944, Fig. 97), but it is difficult to tell how far the cytoplasm extends into the canaliculi.

There is no general agreement on the concepts of intercellular substance, matrix, ground substance, interstitial fluid, and tissue fluid. Some authors distinguish between cells and cell-free matrix so that organized fibers as well as amorphous substances are included in their concept of matrix. Other authors divide intercellular substances into fibrous and amorphous (jellylike) types (Ham, 1957, p. 120). Fibers consist of proteins, whereas amorphous intercellular substances contain special carbohydrates, namely, mucopolysaccharides; one of them is hyaluronic acid and the other chondroitin sulfonic acid. Dilute jellies are classified as tissue fluids if they are rich in proteins (globulin and albumin), and as ground substance, if they are rich in mucopolysaccharides.

The same compound may appear fibrous or amorphous on the light microscopic level as illustrated by the coagulation of fibrinogen, a protein of the blood plasma. Fibrinogen from blood plasma of fowl coagulates with the formation of fibrin fibers if blood platelets are present, but in the absence of platelets a fibrin jelly forms which appears structureless under the light microscope (Mayer, 1935). Electron microscopic studies of clots from bovine fibrinogen showed a meshwork of single and compound fibers, with an increased tendency for lateral association of unit fibers as the pH decreased (Hawn and Porter, 1947). Production of fibrin fibers from elongated molecules may pass through an intermediary phase, termed profibrin by Mommaerts (1945). Mommaerts suggested a parallel action of electrostatic and hydrogen bonds in the aggregation of micelles of profibrin, myosin, gelatin, and other proteins. These examples illustrate how a fibrous state depends on physicochemical conditions. The interpretation of the observations will depend on the order of magnitude of units studied: light microscopic, electron microscopic, or molecular. For recent physicochemical studies of the fibrinogen-fibrin transition see publications of K. Laki and his associates (1953, 1960).

A. Fibers

A viscous thread of hot liquid glass is not considered a fiber. However, after cooling, the solid structure may be called a fiber. Evidently the concept of fiber is a mechanical one. It refers to long thin structures in a relatively stable state. Usually a certain degree of flexibility is implied. Originally biological fibers were examined mechanically, mainly by teasing techniques, and tested chemically for swelling and solubility in different agents. With the increased use of fixed and stained sections some structures were interpreted as fibers on the basis of their optical appearance, although no mechanical or chemical tests were applied. In diagrams Figs. 65f and 73b, the bodies of the cells are shown to taper gradually into thin structures. If the thin structures are relatively short (Fig. 65f) they usually stain like the cytoplasm and, therefore, are considered cytoplasmic strands. As mentioned previously, there are no absolute criteria for differentiating between fibrils and thin cytoplasmic threads. Periodical structures seen with electron microscopy and X-ray diffraction are not necessary characteristics of connective tissue fibers since they are present in collagen and reticular fibers, but not in elastic fibers. In fibrillar networks the relations between cell bodies and fibrils are not always as distinct as shown in the diagrams Fig. 73b and d. If a cell extends along a fibril, it is hardly possible to tell whether the cytoplasm is fused with the fibril or only in contact with it. For an illustration of these relations see Bargmann's textbook (1959, Fig. 88). In fibers which extend over long distances and form bundles, relations to individual cells or nuclei can no longer be recognized. As mentioned before, this condition is known as "acellular connective tissue" (Fig. 65b), in contrast to its immature phase, the "cellular connective tissue" (Fig. 65a).

A particularly intimate relation prevails between the nerve fiber (axon) and the nerve cell with which it is connected. The cytoplasm of the nerve cell contains delicate fibrils (neurofibrils) which are continuous with the fibrils of which the axon is composed. Fibrils in the nerve cell and in the axon stain with the same techniques. Nerve cell plus axon are considered a unit, the neuron. In order to emphasize this unity, many authors have replaced the term nerve cell by perikaryon, which means "the part surrounding the nucleus." Nerve fibers form a loose distinct network in coelenterates; for a picture of such network in *Hydra* see Parker and Haswell (1940, Vol. 1, Fig. 111). In vertebrates including man, a particularly intricate maze of nerve fibers is found in the retina. The closely packed mass of axons and dendrites in the cerebral cortex is known as neuropil. Some investigators assumed that the neurofibrils pass continuously from the axon of one neuron to the dendrites and perikaryon of the next neuron. It is now recognized that there is no continuity between the neurofibrils of different neurons. However, since the synaptic cleft between motor neurons is on the electron microscopic level of 200Å (see Part III, Chapter 1, A), it is still justifiable to refer to the neurons as a network on the light microscopic level. I quote again Ham's statement that neurons are "continuous as well as separate entities" (Ham, 1957, p. 437). Electron microscopic studies have shown that neurofibrils consist of small units (neuroprotofibrils) parallel with the longitudinal axis of the fibril. No cross striation or other periodical structures were observed (DeRobertis, 1954). I do not know of any Roentgen-ray diffraction studies in neurofibrils. Such studies would be complicated by the fact that, in peripheral nerves, the nerve fibrils are always closely associated with reticular fibers of the connective tissue.

A word may be said here concerning the functional role of neurofibrils. Local currents which are produced by potential differences between the interior fluid of an axon and its fluid environment proved to be the mechanism by which nervous impulses are conducted (see Part III, Chapter 5, B and C). Therefore, the role of neurofibrils has now been restricted to mechanical support or nutritional activities. It is interesting that in the meantime the nodes of Ranvier became important functional structures when Tasaki (1939) demonstrated the electrosaltatory transmission of the nerve impulse. Previously Ranvier's nodes, which are fairly regularly spaced interruptions of the myelin sheath, had been considered functionless or possibly an aid in the nutrition of the myelinated nerve. For subsequent developments of the saltatory conduction theory see Frankenhaeuser (1952; with discussion by Gasser). Nerves without myelin sheath have no Ranvier nodes of course. Nervous impulses are conducted at the highest speed in some myelinated fibers and at the lowest rate is some unmyelinated fibers, but the presence of myelin is not the only factor determining the rate of conduction (Lloyd, 1949, Table 1, p. 45.)

Muscle fibers have hardly any resemblance to connective tissue or nerve fibers. A smooth muscle fiber is an elongated cell tapering on both ends and having a maximal thickness of approximately 10 μ. The smooth muscle cell or fiber contains a number of smaller units, myofibrils, less than 1 μ in width. In most sections aggregates of smooth muscle cells or fibers do not show outlines of the units. They may be identified by the distribution of nuclei (Fig. 65c). Striated muscle consists of special structures which form parallel bundles in skeletal muscle while they are ramified in cardiac muscle. Striated muscle fibers measure from 1 to 40 mm. in length and from 10 to 40 μ in width, exceeding by far the dimensions of smooth muscle (Ham, 1957). Some of the cross striations are on the light microscopic level. They show birefringence which shifts its place depending on the state of contraction. Electron microscopic analyses of myofibrils are found in the extensive literature (for instance, DeRobertis' brief 1954 review). I referred previously to probable structures on the molecular level in connection with the "sliding filament" model of H. E. Huxley (1957). Muscle fibers contain a second component, the sarcoplasm, which is not fibrillar. An instructive three-dimensional model by Bennett (1956b) shows relations between the different components of striated muscle.

Finally, let us consider chromosomes which may form long fibers (Fig. 61, M 1) or may be so short that the concept of fiber is hardly applicable. In contrast to other biological fibers which are products of cytoplasmic differentiation, chromosomes are integral structures of the nucleus. A chromosome consists of unit filaments (chromonemata). Some 1000 filaments may be associated in one giant chromosome. In such polytene chromosomes the alternation of dark bands with clear zones produces a superficial resemblance to striated muscle fibers, as mentioned by Saez (1954). In *Drosophila* one of the giant chromosomes may exceed 400 μ in length; this is 250 times the length of the ordinary somatic chromosome. While in general the function of chromosomes is to carry the genes through the mitotic cycle, the polytene giant chromosomes occur in abortive mitoses which do not proceed beyond the prophase. The functional role of these giant chromosomes is unknown.

FIG. 78. The various orders of magnitudes ("hierarchies") of structural elements of collagen. Microscopic level (top diagram) refers to bright light microscopy. Ultramicroscopic level (second diagram from top) refers to dark-field microscopy. Electron optical level is equivalent to electron microscopic level. Protofibrils (bottom diagram) carry a periodic pattern of chemical content with a period of approximately 640 Å. indicated by the repetition of letter a. Under usual conditions a collagen molecule is approximately 640Å. long. (From Bear, 1952, Fig. 1 and pp. 152–153).

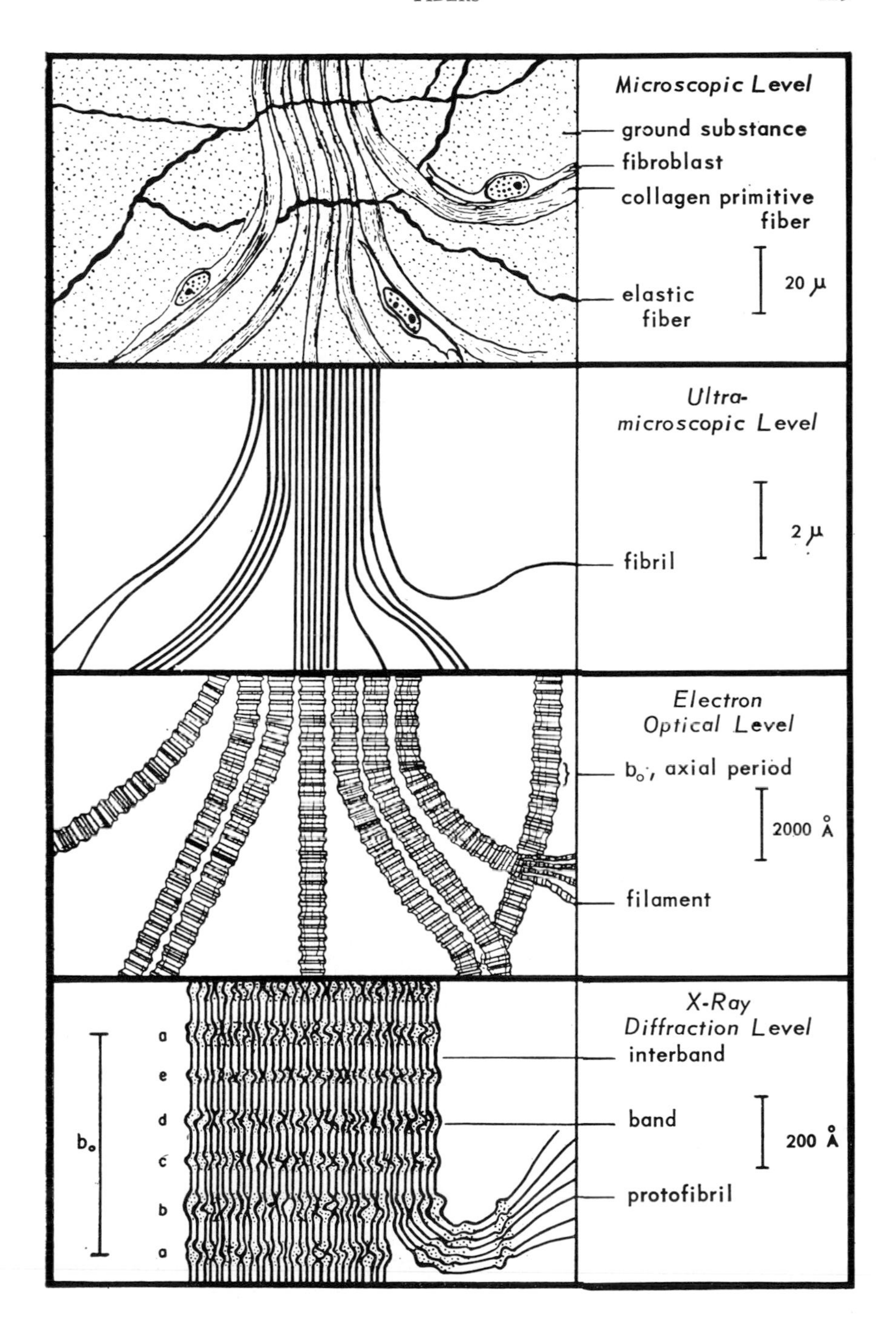
Microscopic Level
ground substance
fibroblast
collagen primitive fiber
elastic fiber
20 μ
Ultra-microscopic Level
2 μ
fibril
Electron Optical Level
b₀', axial period
2000 Å
filament
X-Ray Diffraction Level
interband
band
200 Å
protofibril
b₀
a
e
d
c
b
a

I stated in the preceding chapter that it is not always possible to distinguish between cell boundaries and a basketlike enclosure of cells formed by connective tissue fibers. The decidua cells of the endometrium present this alternative. In a mosaic pattern the gaps between individual cells may be crossed by narrow structures, called intercellular bridges. Fibrils which run inside these bridges and also through the cell bodies are known as tonofibrils. They have been observed mainly in the deeper layers of the epidermis (see Bailey, 1953, Fig. 13). In view of the fact that the terminal bars can be detached from the rest of the cells, one could hardly object if someone wanted to consider the network of terminal bars equivalent to a fiber network. The basement membrane is interpreted by most investigators as being a connective tissue structure, but some maintain that it is part of the epithelial cells.

Modern methods such as Roentgen-ray diffraction and electron microscopy, in conjunction with protein-chemical analysis, are the basis of the well-known diagram by Bear (1952), which is reproduced here (Fig. 78). This diagram was mentioned previously in the discussion of natural and artificial units (Part III, Chapter 4, E). With electron microscopy, fairly uniform spacing was discovered in collagen fibers from the tendon of the rat's tail: the average period was 640 Å. (Schmitt *et al.,* 1942). This value was confirmed subsequently in collagen from various sources. Using improved X-ray diffraction techniques Bear (1942) found a similar periodicity, namely, 640 Å. with a range from 638 to 648 Å., in airdried collagen of tendons, skin, and cornea from different species.

In comparative studies of collagen fibers and reticular fibers, light microscopic techniques including staining have been used as well as chemical analysis, electron microscopy, or X-ray diffraction. Collagen fibers do not branch, but run as single fibers or in bundles of parallel fibers. Reticular fibers branch and anastomose, forming networks, as their name indicates. If certain aniline dyes are applied in conjunction with proper pretreatment and differentiation, they stain collagen fibers almost selectively. Reticular fibers may stain with the same procedures, but not as regularly. After secondary silver impregnation, reticular fibers appear as distinct black structures (Fig. 51b), while collagen fibers exhibit a variable brownish stain. Under the polarizing microscope collagen fibers are birefringent, but reticular fibers are not. All these observations led to the question whether reticular fibers are merely thin collagen fibers or differ with respect to their chemical composition.

Kramer and Little (1953) studied reticular fibrils from the renal cortex, which is rich in this material, but hardly contains collagen fibers. Under the electron microscope the reticular fibers varied in thickness from 600 Å. down to less than 100 Å. They showed cross striation with a period of the order of 650 Å., similar to that of collagen. X-Ray diffraction photographs of reticulin revealed spacings identical with those of collagen, with additional refraction rings interpreted by the authors as an amorphous component. The volume of this amorphous matrix, which seems to be rich in carbohydrates, is not less than the volume occupied by reticular fibers. If such matrix is associated at all with collagen fibers, the amount is negligible. Kramer and Little compared collagen to rope and reticulin to linoleum: the composition of fibers may be the same, but the arrangement is different, and the matrix is conspicuous only in the case of reticular fibers. The presence of the matrix may modify the appearance, pattern, and stainability of fibers, as Jacobson (1953) pointed out. He suggested that the relative resistance of collagen and reticular fibers to trypsin digestion may be due to some protective polysaccharide coat. It seems to me that some optical properties of the reticular fibers are obscured by the matrix so that, for instance, birefringence of the fibers cannot be detected.

It is still controversial whether reticular fibers yield gelatin when boiled in water as collagen fibers do (Kramer and Little, 1953). A characteristic amino acid, hydroxyproline, is present in collagen and gelatin. Using Kramer and Little's 1953 technique, Windrum *et al.* (1955) isolated reticulin. They found its hydroxyproline content to be similar to that of collagen.

Elastic fibers differ from collagen and reticular fibers by mechanical properties, as indicated by their name and by their staining characteristics. Similar to collagen and reticular fibers, the elastic fibers consist of fibrous proteins, but the hydroxyproline content is lower in elastic fibers (W. H. Stein and E. G. Miller, 1938). It is quite certain that elastic fibers do not yield gelatin upon boiling. I gather from the literature that neither electron microscopy nor X-ray diffraction reveal any distinct pattern in elastic fibers (Gross, 1949; Martin, 1953, p. 87).

B. Nonfibrous Intercellular Substances, Ground Substance

A close association between fibers and nonfibrous ground substance occurs in reticular fibers, as mentioned above. In seemingly homogeneous ground substance of cartilage, a considerable number of fibers can be shown with adequate methods. The present discussion will deal with the nonfibrous intercellular substance or ground substance. As stated in the introduction to this chapter, different authors do not agree on the concepts of intercellular substance, matrix, ground substance, and tissue fluid.

It is interesting to follow the changing ideas concerning ground substance in the central nervous system. Some authors took it for granted that ground substance is present in the central nervous system, but did not mention any characteristics by which it might be recognized (for instance, Weed, 1922, p. 199). Taft (1938) and other investigators felt that the existence of ground substance was evidenced by granules which were seen with dark-field microscopy in the areas between nerve cells. Spatial relations between the ground substance and the nerve or neuroglia fibers remained problematic.[1] According to Angevine (1951, p. 28) no ground substance has been demonstrated in the central nervous system, although a cement substance might be present between the fibrils. A successful histochemical approach to the question of ground substance in the central nervous system was initiated by Hess (1953). He observed a homogeneous positive periodic-acid Schiff (PAS) reaction in the gray matter of the adult central nervous system. Since this reaction can be ascribed to a neutral mucopolysaccharide which is characteristic of ground substance, Hess concluded that such ground substance is present in the gray matter of the central nervous system. Hess (1955) studied the probable origin of the PAS-positive material in the embryonic central nervous system. His pictures of different stages show increasing amounts of stainable ground substance between unstained cell bodies. The spatial relations between the PAS-positive ground substance and the nerve fibers or neuroglia fibers have been discussed by Dempsey and Wislocki (1955). In electron micrographs of pericapillary areas of rat brains they found the endothelium, the ramifications of neuroglia cells, and the axons so closely packed that there seemed to be no room for ground substance. Special analysis, however, revealed narrow spaces between these structures, filled with an amorphous substance of moderate electron density. I mention the additional possibility that axons and neuroglia may be soaked with the PAS-positive ground substance of Hess. The distribution of this substance in submicroscopic spaces between axons and neuroglia, and possibly inside these structures, is supported by an argument of Davson (1956). He questioned how the presence of ground substance, as demonstrated by Hess, could be reconciled with the large proportion of tissue fluid in the brain which has been determined chemically (15 to 30% of the brain weight). Davson's answer is that the mucopolysaccharide of Hess must have a very high water content, which also is a prerequisite for ready diffusion of dissolved material throughout the brain substance.

What is the relation of unorganized ground substance to tissue fluid? Yoffey and Courtice (1956) assume that the water in tissue spaces is always in a bound state. On this basis tissue fluid and jellylike ground substance appear to be similar. However, modern histochemistry has established a new well-defined difference: tissue fluids are dilute protein solutions, whereas ground substance mainly consists of mucopolysaccharides. This was mentioned previously in the introduction to this chapter.

Some comments on the physical properties of ground substance may be appropriate here. Ground substances can be gel-like, semisolid, or solid. The hard ground substance in bone consists of collagenous fibers embedded in amor-

[1] Bouton's (1940) concept of ground substance is difficult to understand. He states that "the amorphic colloidal ground substance . . . constitutes the bulk of the cerebral cortex." In my view, neurons and neuroglia form the bulk of the cerebral cortex with relatively little space for amorphous ground substance.

phous material: both components are impregnated with calcium salts (for instructive diagrams see Ham, 1957, Fig. 163). The vitreous body of the eye, although not without some fibrillar components, may be considered a macroscopic mass of ground substance. The viscosity and elasticity of the vitreous body were determined by Robertson and Duke-Elder (1933). Microscopic nickel particles were pulled through the jelly by means of an electromagnet, and the displacement of the particles was measured.

Chemical extraction and enzyme reactions have indicated some physical properties of ground substance. Hyaluronic acid, one of the mucopolysaccharides in the ground substance of connective tissue, possesses marked viscosity. In the body, this property prevents the spreading of particles such as India ink and bacteria. Hyaluronic acid is hydrolyzed by an enzyme, hyaluronidase, which is present in spermatozoa, snake venom, and certain tumors. If India ink is injected subcutaneously, its spreading is greatly enhanced by simultaneous injection of hyaluronidase. Therefore this enzyme became known as the *spreading factor.* Not only the spreading of particles, but also the spreading of finely dispersed suspensions, is controlled to some extent by local concentrations of hyaluronic acid and hyaluronidase. This is part of the mechanisms which are summarized in the literature as local tissue permeability.

Some authors divide the amorphous intercellular substances into "soft ground substances" and "firm cement substances" (Ham, 1957, p. 121). Others seem to use the term cement substance for finely distributed ground substance without implying special physical properties. Evidently it was in this sense that Angevine (1951) referred to the possible presence of a cement substance between the fibrils of the cerebral cortex, as quoted above. The *cohesion* between cells in a mosaic pattern has been studied experimentally, but I do not know of any direct test to examine the *consistency* of the so-called cement substance between the cells. However, a relatively firm consistency can be ascribed to the terminal bars, the superficial portion of the cement substance. The reasons were discussed previously (Part IV, Chapter 1, C; Fig. 68).

The term ground substance is also applied to components or areas of the cytoplasm (see Part IV, Chapter 1, B). This has nothing to do with the intercellular ground substances.

C. Intercellular Spaces, Extracellular Fluid

Intercellular spaces which are filled with fluid are studied with biochemical and morphological procedures. The aim of the biochemical procedures is to obtain a balance sheet of fluid distribution for the whole body or for special organs. The techniques are known as dilution and clearance tests. The goal of morphological analysis is the demonstration of spaces on the macroscopic, light microscopic or electron microscopic level.

1. *The Three Compartments of Body Fluids*

The biochemical balance sheet of body fluids is composed of three fluid compartments: blood and lymph plasma in vessels, interstitial fluid, and intracellular fluid. Interstitial fluid is also known as intercellular or tissue fluid. A system of tracers has been developed for dilution tests. For instance, a known amount of Evans blue is injected into a vein. At intervals blood samples are obtained in which the amount of Evans blue per unit of plasma is determined. When the concentration values have stabilized, the total volume of plasma in the vascular system can be calculated. Inulin and heavy water (D_2O) are tracers for other dilution tests. Table 14 shows the use of these tracers and the values

TABLE 14

Determination of the Three Fluid Compartments of the Body: Blood Plasma in Vessels, Interstitial Fluid (Intercellular Fluid) and Intracellular Fluid

Tracer used for dilution test	Fluid measured	Components of fluid measured	Amount of fluid measured in per cent of body weight[a]
Evans blue	Blood plasma in vessels	Blood plasma in vessels } A	5%
Inulin	Extracellular fluid	Blood plasma in vessels + interstitial fluid } B	20%
Heavy water	Total water	Blood plasma in vessels + interstitial fluid + intracellular fluid } C	70%

Values for the three fluid compartments:

A =	Blood plasma in vessels	5%
B − A =	Interstitial fluid	15%
C − B =	Intracellular fluid	50%

[a] Approximate values in the human adult.

which are obtained for the different fluid compartments. Without going into biochemical detail, I mention the fact that the extracellular fluid contains the greater part of the sodium and chloride in the body, while most of the potassium is contained in the intracellular fluid. These differences, which are expressed in terms such as sodium or potassium compartment, indicate a control of electrolyte transport by the cell membranes. The possible mechanisms, including the so-called electrolyte pumps, have been studied extensively. For a brief and lucid presentation I refer to Yoffey and Courtice (1956).

The three fluid compartments are characterized by the biochemical operations indicated in Table 14. To what extent can corresponding spaces be demonstrated with morphological methods? Two of the biochemical compartments have well-defined morphological equivalents: the vessels, which are filled with blood plasma;

and the cells which contain intracellular fluid. For the third biochemical fluid compartment, the tissue fluid, there is no distinct morphological container. Biochemical determinations of interstitial fluid (tissue fluid) in the kidney are based on the assumption that the tubules, the renal corpuscles, and the blood vessels are "bathed" in the interstitial fluid. The term "bathed" can mean different things. The object may be soaked by the bath like a sponge, or the contact with the fluid may be limited to an external surface. Many surfaces of individual cells or of cell aggregates appear smooth under the light microscope, but corrugated or deeply folded under the electron microscope. If a thin film of fluid covers an object, or fills a narrow crevice between several objects, this is still considered a form of bathing in biochemical language. Inspection of stained sections does not always help in deciding whether or not in the morphological sense, intercellular spaces are present. This decision may be difficult even under the favorable conditions of cultivated cell colonies which consist of a one-layer mosaic of well-displayed cells (see previous discussion of Fig. 72a–c). Whole living or dead animals have been used in various ways for the study of interstitial spaces. The injection of dyes into the vascular system of an isolated organ seems to be an obvious technique for the presentation of interstitial spaces. However, too much pressure will burst the capillaries, but too little pressure prevents the dye from filling the interstitial spaces adequately. Direct injection into interstitial spaces is feasible with micromanipulation, but puncturing with micropipettes may distort the natural conditions and always involves problems of representative sampling.

There is little doubt that the lymph is formed from interstitial fluid. It was believed that structures such as cytoplasmic networks (Figs. 65f and 73a) and loose fibrillar networks (Fig. 73b) were filled with interstitial fluid, as a sponge is filled with water. However, there is much evidence that the water in such tissues is not free, but is associated with the colloid gel which makes up the unorganized ground substance of connective tissue. I have followed here the presentation of Yoffey and Courtice (1956, pp. 57–59). These authors refer to supporting observations by Clark and Clark (1933). In the ear chamber of the rabbit, the behavior of injected dyes and the absence of Brownian movement in the tissue between the blood vessels indicated that there was no free fluid in the morphological spaces.

Cytoplasmic and fibrillar networks occur mainly in lymphatic tissues, including the tunica propria of mucous membranes, and in the connective tissue layers of the skin. How does the lymph which seems to form in these places find its way into the tubular lymph vessels, the lymphatics? How do the lymphatics originate in the tissues? Answers to these questions were obtained through the study of the lacteals which are the lymphatics in the tunica propria of intestinal villi. After a fat-containing meal these vessels proved to be filled with milky fluid, and in this phase it was possible to demonstrate that they have blind ends. It was also possible artificially to fill lacteals with dyes. Early illustrations of lacteals are reproduced in the monograph of Yoffey and Courtice (1956, Figs. 45 and 46) and in Maximow and Bloom's textbook (1952, Fig. 229). Most investigators agree that fluid from the surrounding connective tissue can enter the lacteals only by diffusion or through submicroscopic pores. It seems to me that other lymphatics may not necessarily have blind ends. The alternative is that spongelike structures of reticular or loose connective tissue gradually change to perforated membranes and finally to tubules which form the origin of lymph vessels.

In Ham's instructive diagrams (1957, Figs. 92–97) the normal amount of interstitial fluid in connective tissue of the skin is compared to pathological increases of the fluid (edema) produced by various disturbances. The pictures illustrate the exchange of colloids and crystalloids through the wall of a lymph capillary and through the wall of the arterial and venous end of a blood capillary. All lymphatics in Ham's diagrams have blind ends.

2. *The Cerebrospinal Fluid. Sites of Potential Exchange of Materials*

The cerebrospinal fluid is a body fluid which does not readily fit into the scheme of three fluid compartments as represented in Table 14. The cerebrospinal fluid fills macroscopic cavities of the central nervous system, such as the ventricles of the brain and the central canal of the spinal cord. The largest amount of the cerebrospinal fluid is contained in the meshes of the subarachnoid space, which forms a liquid-filled jacket around the spinal cord and the brain (Fig. 79a). The tissues of brain and spinal cord, though not containing any lymph vessels, are bathed in fluid. Therefore three potential exchanges of material can be visualized: between nervous tissue and blood, between nervous tissue and cerebrospinal fluid, and between blood and cerebrospinal fluid (Walter, 1933). If each exchange can occur in two directions, the total of potential types of exchange is six. There are morphological prerequisites for biochemical exchanges. If, for instance, morphological studies should show that certain areas of nervous tissue are always separated from their blood capillaries by a space with cerebrospinal fluid, a direct exchange between blood and nervous tissue would be excluded in that area. On the other hand, visible direct contact between two structures does not necessarily mean permeability of their interface. Permeability at an interface may be greater for some substances than for others. The mechanisms controlling exchanges of material in the central nervous system are known as "blood-brain barrier," "blood-cerebrospinal fluid barrier," and "cerebrospinal fluid-brain barrier." The barriers may be absolute or relative for any particular substance. Therefore I shall use the term "border" interchangeably with "barriers." The procedures by which the barriers have been discovered will be described later. It seems useful to start with a survey of the *morphological sites* of potential exchanges of material. A well organized presentation of this subject is found in Bakay's 1956 book.

a. *Morphological sites of potential exchange of material.* Diagram Fig. 79a, represents a frontal section through a hemisphere of the human brain. The picture shows arteries which bring the blood to the brain, veins which carry

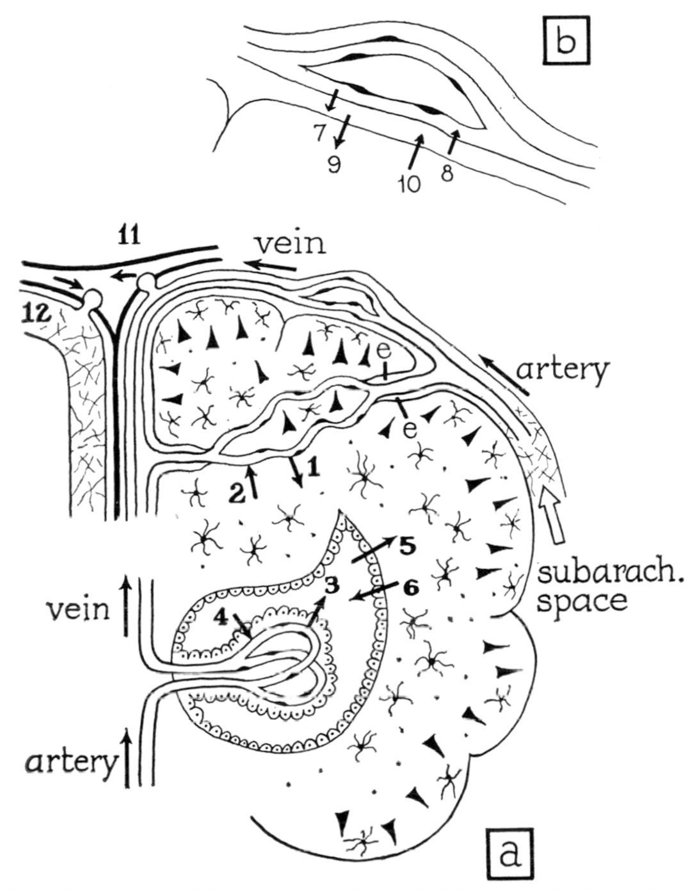

FIG. 79. Sites of potential exchanges of material between blood, cerebrospinal fluid, and brain substance. Diagram of a frontal section through a hemisphere of a human brain (a). The artery in the subarachnoid space divides into two branches, one entering the brain substance and one continuing on the surface of the brain. Both branches form capillary networks represented as simple loops in the diagram. The rich vascularization of the cerebral cortex is not indicated. Subarachnoid capillaries are shown enlarged in (b). In reality most subarachnoid arteries and veins are cut transversally in a frontal brain section. Nerve cells are represented by black triangles, and neuroglia cells by stellate structures. In the lower part of (a), a lateral ventricle and choroid plexus (with capillary loops) are shown lined with a continuous layer of ependymal cells. The letter (e) indicates a periarterial space (concept A of Fig. 80). Numbered arrows indicate theoretical possibilities of exchange. For observed exchanges see text. Blood-brain border: arrows *1* and 2. Cerebrospinal fluid-brain border: arrows *5* and *6*; *9* and *10*. Blood-cerebrospinal fluid border: arrows *3* and *4*; *7* and *8*; *11* and *12*. Arrows *11* and *12* show potential exchange between arachnoid villi and superior sagittal sinus.

the blood away,[2] capillary nets which connect the arteries with the veins, and places where fluid can be exchanged between the capillaries and their environment. Capillary nets are presented as single loops. In the upper margin of Fig. 79a, a capillary loop is shown inside the subarachnoid space. An enlarged picture of this loop is given in Fig. 79b. Topographic relations between scalp, skull, meninges, brain, and the blood vessels of these structures are well illustrated in Clara's diagram of a frontal section (Clara, 1953, Fig. 444). The cerebrospinal fluid flows from the lateral ventricles to the third and fourth ventricles and from here through one median and two lateral openings into the subarachnoid space. The median opening is seen in mid-sagittal diagrams (Ranson, 1942, Fig. 368).

In my Fig. 79a and b, the morphological sites of potential passage of material are indicated by arrows and numbers. These sites are *borders between the blood, the cerebrospinal fluid, and the tissue fluid of the nervous substance.* The arrows represent three different phases of investigation: passage in the direction indicated has been demonstrated; passage in the direction indicated has been excluded; the particular type of passage has not yet been investigated, or if so, no decison has been reached.

Is there a direct contact between blood capillaries and brain substance, as indicated by the arrows *1* and *2,* or is the brain substance separated from the capillaries by a pericapillary space? The two alternatives are shown diagrammatically in Fig. 80A and B. The existence of a definite space between each small artery, (or vein) and the brain substance is generally recognized (Virchow-Robin space). The question is whether the space ends at the level at which the artery branches into capillaries, as shown in Fig. 80A (*e*), or extends along the capillaries, as seen in Fig. 80B (*pc*) Investigators convinced of the latter condition assumed that the pericapillary space extends to the nerve cells, thus communicating with a perineuronal space (*pn* in Fig. 80B). My Fig. 80B is based on Weed's well-known diagram (Fig. 2 of his 1923 paper), which has been reprinted in most textbooks.

The *absence of a pericapillary space* in the normal spinal cord was demonstrated by electron microscopic studies of Dempsey and Wislocki (1955). Their most convincing electron micrograph (Fig. 4) is reproduced in my Fig. 81a. It shows a double line where the wall of a blood capillary borders the adjacent structures of the central nervous system. My tracing (Fig. 81b) of the electron micrograph will facilitate identification of the different structures. I agree with Dempsey and Wislocki that, with a high degree of probability, their findings in the spinal cord also exclude the existence of pericapillary space in the brain.

---

[2] Diagrams shown in most publications, including textbooks, give the impression that the subarachnoid space contains only veins. A good picture with arteries and veins is found in Patek's 1944 paper (Plate 6, Fig. 19).

For this reason I drew the spaces marked (*e*) in Fig. 79a similar to those in Fig. 80A, not extending them beyond the level of arteries or veins. The decision that Fig. 80A represents the normal condition is supported not only by the electron microscopic studies of Dempsey and Wislocki, but also by light microscopic studies of Woollam and Millen which were published during the same year. Woollam and Millen (1955) observed that India ink, injected into the subarachnoid space of rats, was found in spaces around intracerebral arteries

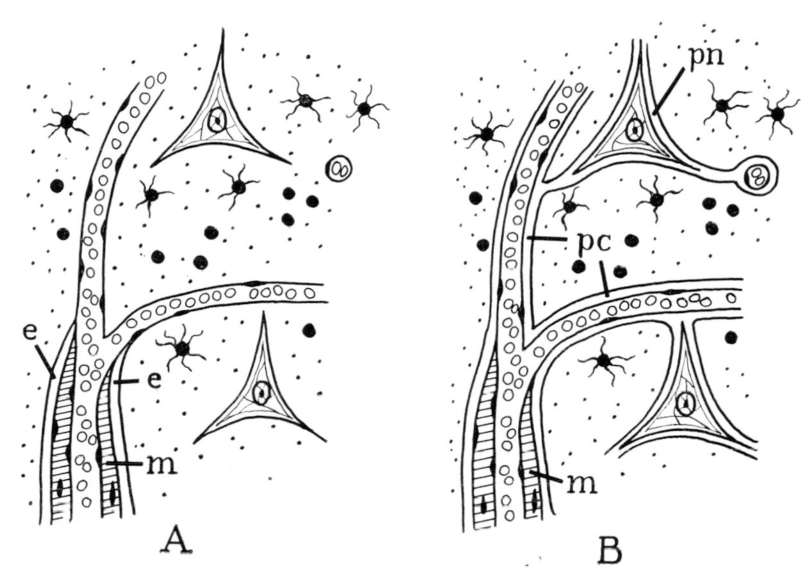

Fig. 80. Diagrams illustrating alternative concepts, (A) and (B), of perivascular spaces in the central nervous system. In both pictures *m* points to the muscular wall of a small artery or vein, and the small open circles indicate blood corpuscles. (A) Perivascular spaces are limited to the level of arteries and veins, with blind end of space shown in (e). There are neither pericapillary nor perineuronal spaces. (B) Perivascular spaces continue along the capillaries and extend into the brain tissue with open communication between pericapillary spaces (pc) and perineuronal spaces (*pn*). Note: For evidence in support of the concept (A) as representing normal conditions, see text and Fig. 81.

and veins, but did not reach the level of capillaries or nerve cells. Pericapillary and perineural spaces can be produced by pathological conditions and also by certain histological techniques. Therefore these spaces belong to the *potential spaces,* which will be the subject of the next chapter.

The subarachnoid capillary loop seen at the upper margin of Fig. 79a is enlarged in Fig. 79b. Arrows *7* and *8* show sites of potential exchange between the capillary and the subarachnoid space. Arrows *9* and *10* indicate potential exchange between this space and the brain substance. The lower half of Fig. 79a, contains a ventricular space and a part of the choroid plexus. The ependyma lining of the ventricle is continuous with that of the choroid plexus. The following sites of potential exchange of material are indicated: *3* and *4* between the capillaries of the choroid plexus and the fluid in the ventricular space; *5* and *6* between ventricular fluid and brain substance. The arrows *3, 4, 5,* and *6* involve passage through the ependyma, while there is a direct contact between cerebrospinal fluid and brain in the sites of arrows *9* and *10*. The arrows *11* and *12* point to arachnoid villi which dip into the venous sinuses of the dura mater. My diagram shows the traditional picture of macroscopic "Pacchioni's granulations."

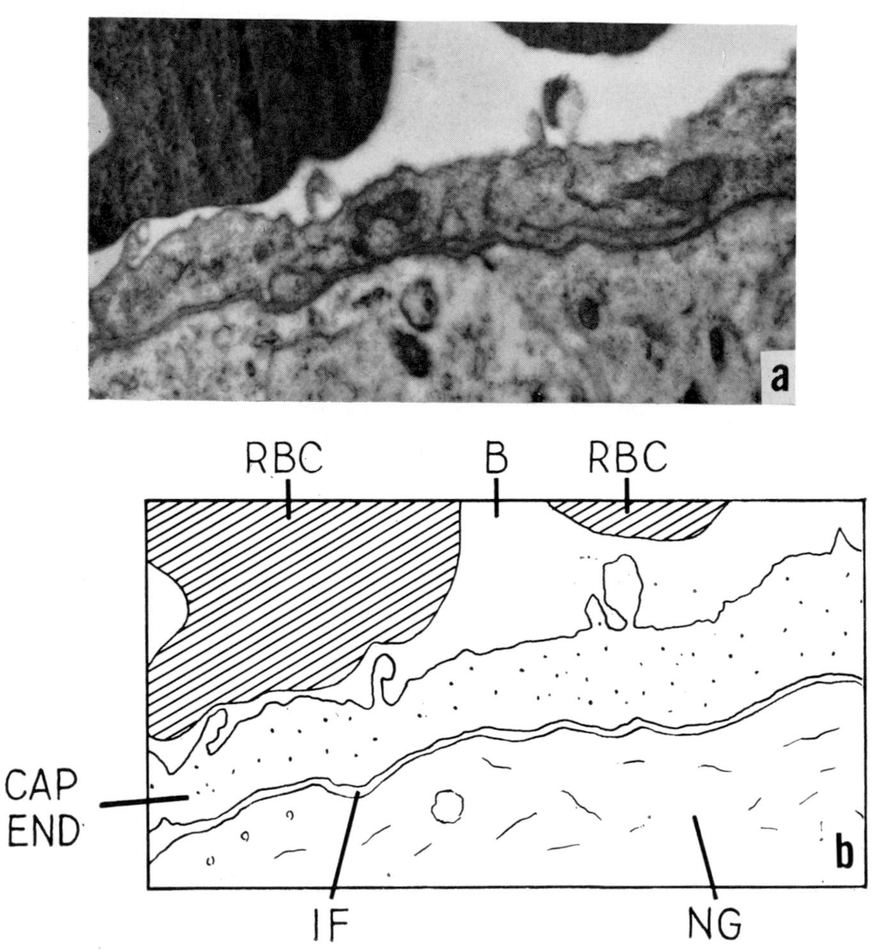

FIG. 81. Demonstration of contact between capillary endothelium and neuroglia. (a) Electron micrograph of a section through spinal cord of a mouse. Magnification $\times$25,000. (From Dempsey and Wislocki, 1955, Fig. 4.) (b) Tracing of Fig. 81a, with labels of structures. *B*, bore of capillary; *RBC*, red blood corpuscle; *CAP END*, endothelial wall of capillary; *NG*, neuroglia; *IF*, double-lined interface representing place of contact between capillary endothelium (basement membrane) and neuroglia.

b. *Procedures for determining presence or absence of exchange (barriers, permeability).* Procedures for determining exchange of material, between blood, cerebrospinal fluid, and brain tissue fall roughly into four classes: (1) observation of results of mechanical blocking, natural or experimental; (2) biochemical comparisons between blood or lymph plasma and cerebrospinal fluid; (3) injection of visible tracers into the circulating blood or into the cerebrospinal fluid; and (4) various studies on the distribution of drugs, hormones, bacterial toxins, antibodies, etc. Some data obtained with these procedures will be selected for applications of the diagram Fig. 79.

Present knowledge of the origin and flow of the cerebrospinal fluid is based on observations of *mechanical blocking.* The choroid plexuses produce cerebrospinal fluid which is secreted through the ependyma into the ventricular space, as indicated by arrow *3* in Fig. 79a. Reabsorption of fluid by the plexus does not occur; in other words, arrow *4* is ruled out. The lining ependyma of the ventricle seems to absorb fluid from the ventricular space, according to arrow *5;* this ependyma does not contribute to the ventricular fluid, a fact that rules out arrow *6.* The particular observations from which these conclusions were derived are summarized by Best and Taylor (1955, p. 1085) as follows. Occlusion of the so-called aqueduct which connects the third and fourth ventricle, leads to dilation of the lateral and the third ventricles. During an operation on a patient, clear fluid was seen exuding from a choroid plexus. In animals the sustained outflow of fluid from the aqueduct was determined and found similar to that from the subarachnoid space. If the exit of a lateral ventricle is blocked experimentally the ventricle will be dilated by accumulated fluid; this does not take place when the choroid plexus is removed from the ventricle before blockage.

The *selective exchange* of material between blood plasma and cerebrospinal fluid may be illustrated by some *biochemical data.* The composition of the cerebrospinal fluid is qualitatively similar to that of blood and lymph plasma, except for the absence of cholesterol in the cerebrospinal fluid. The greatest quantitative difference is the protein concentration which is 0.02% in the cerebrospinal fluid, as compared to 6% in the blood plasma and 3% in the lymph plasma. Ratios of concentrations in cerebrospinal fluid over concentrations in blood plasma vary with different substances. The ratio is 0.5 for calcium, 0.6 for potassium, 0.6 for urea (average), 0.96 for sodium, 1.2 for magnesium, 1.15 for chloride, and 2.4 for glucose (average). Evidently magnesium and chloride ions are actively secreted from the blood plasma into the cerebrospinal fluid, whereas potassium and urea ions are kept out (data mostly from Winton and Bayliss, 1955, p. 285).

Injection of visible tracers by different routes proved to be very informative. Whether or not a particulate dye, injected intravenously, passes through the capillary walls depends on properties of the dye such as particle size and electric charge, concentration of the dye in the blood, binding of the dye by the blood plasma, local pressure and flow of blood. If the tissue which is supplied by blood capillaries fails to retain a certain dye, even for a short period of time, this test cannot be used to determine permeability of the capillaries. The properties of the capillary wall itself vary with location, age, and animal species. By and large, the capillaries in brain and spinal cord proved to differ in permeability from those in other locations. However, the capillaries in the caudal part of the fourth ventrical (area postrema) behave like those outside the central nervous system. In newborn animals the brain capillaries do not possess the selective permeability found in older animals. Newborn guinea pigs are an exception: their brain capillaries have properties of those of adult guinea pigs. Keeping in mind the possible effects of numerous factors on the distribution of visible tracers, let us resume our discussion of the sites of the different barriers in the central nervous system.

Injection with trypan blue, a particulate vital stain, was among the first experiments to show the distribution of a visible tracer in the animal body. Goldmann (1913) observed that intravenous injection of this dye stained all organs except the brain. The connective tissue of the choroid plexus was stained but no dye appeared in the fluid of the cerebral ventricles. The conclusion was that trypan blue is held back by a blood-cerebrospinal fluid barrier. Injection of trypan blue directly into the subarachnoid space caused extensive staining of the brain substance. These data, in terms of my Fig. 79, mean that the transmission of trypan blue is blocked in the sites of arrows *1, 3,* and *7,* but is not blocked in the site of arrow *9.* There is a blood-brain barrier and a blood-cerebrospinal fluid barrier for trypan blue, but not a cerebrospinal fluid-brain barrier. To what extent the transmission of vital dyes may be determined by their electrochemical properties is controversial. Concerning particle size, Bakay (1951) reported that dyes penetrated the brain from the cerebrospinal fluid if their particle radius was less than 10 Å.

As mentioned before, the cerebrospinal fluid may be in direct contact with the superficial nervous tissue, arrows *9* and *10,* or may be separated from the nervous tissue by a layer of ependymal cells, arrows *5* and *6.* This morphological difference can lead to marked differences in the passage of some substances from the cerebrospinal fluid into the brain. Dandy and Blackfan (1913) studied the absorption of phenolsulfonphthalein in a patient with internal hydrocephalus. When the dye was injected into the dilated lateral ventricles, 0.25 to 1.0% of it appeared in the urine within 2 hours. If the same patient received the dye injection into the subarachnoid space, 35 to 60% was excreted in the urine within 2 hours. Evidently, in either case the injected dye had penetrated into the brain substance and from there into the blood capillaries. The passage of phenolsulfonphthalein at the site of arrow *9* in my Fig. 79 was much faster than at the site of arrow 5. The ependyma of the ventricle represents a *relative barrier* to this dye. Transmission of the dye from the brain to the blood capillaries is indicated by arrow *2*.

Radioactive phosphorus (P^{32}) was injected intravenously into rabbits by Hevesy and Hahn (1940). The deposition in the brain of not more than 0.02% of the total P^{32} administered was ascribed by the authors to a slow turnover of phosphorus in the nervous tissue. However, when Bakay and Lindberg (1949) injected P^{32} into the subarachnoid space (cisterna magna) of rabbits, they found large accumulations of the isotope in the brain. Similar to findings with trypan blue, barriers for P^{32} were present in the sites of the arrows *1, 3,* and *7,* but not of arrow *9* in Fig. 79. According to subsequent investigations by Bakay (1951) there is *no absolute barrier* to the transmission of P^{32} from the blood to the brain, but the amount and rate of transmission is less than from the cerebrospinal fluid to the brain. The maximum concentration of the isotope was reached 30 to 60 minutes after subarachnoid administration. The maximum concentration after intravenous injection was attained after 12 to 24 hours, and then did not exceed 2 to 5% of the maximum concentration after subarachnoid injection. All values were expressed as counts per minute per milligram of tissue.

The emphasis in Bakay's study was on quantitative determinations of radioactivity in various parts of the central nervous system after intravenous or subarachnoid injection of P^{32}. Radioautographs obtained from frontal sections 2.5 mm. in thickness served as crude illustrations of the quantitative data. No attempt was made to study the distribution of radioactive material on a microscopic level. Trypan blue injection into the subarachnoid space did not yield much information on the microscopic level. Trypan blue has a great affinity for connective tissue, but the scarcity of connective tissue in the central nervous system makes it difficult to estimate histochemically how much dye has been bound. Nerve cells are not stained distinctly with trypan blue. The binding of trypan blue by the blood plasma represents another complication.

The doses of trypan blue necessary for such studies are toxic when injected in the subarachnoid space. Recognizing the imperfections of trypan blue, Rodriguez (1955) applied nontoxic concentrations of aminoacridine dyes for the study of the histological sites of exchanges between blood, cerebrospinal fluid, and brain tissue. Aminoacridine dyes were introduced into histology by DeBruyn, Robertson, and Farr (1950) and used by Farr (1951) for the tagging of lymphocytes (see Part III, Chapter 3, B). Aminoacridine dyes produce selective fluorescence of living nuclei which is well preserved in frozen-dried sections. Intravenous injections of these dyes stain all nuclei except those of the central nervous tissue. Testicular nuclei stain erratically, for unknown reasons. The failure of staining in the central nervous system proved to be caused by a blood-brain barrier. This became evident when Rodriguez succeeded in staining the nuclei of the brain by subarachnoid and intraventricular injections of proflavine hydrochloride (3,6-diaminoacridine hydrochloride). This particular dye was selected because of low toxicity and bright fluorescence. Intravenous injections of the dye showed a barrier in the direction of the arrows *1* and *3* of my Fig. 79, since neither the brain substance nor the ventricular fluid received the dye. After subarachnoid injection not only the brain substance, but also tissue outside the central nervous system, was reached rapidly by the dye. This proved penetration in the direction of arrows *2* and *8*. Intraventricular injections showed transmission in the direction of arrow *5*, but no decision was reached with respect to arrow *4*. From the fact that there is a barrier in *1* but not in *9*, one concludes that no transmission occurred in the direction of arrow *7*.

Rodriguez's microscopic analysis produced some remarkable results. Upon intravenous injection the proflavine hydrochloride did not enter the vascular endothelium of the cerebral blood vessels, but it entered the endothelium elsewhere. Rodriguez concluded that the blood-brain barrier for this dye is located in the membrane of the endothelial cells which borders the capillary space. After subarachnoid injection the dye not only penetrates the nervous tissue, but also the endothelium of blood capillaries inside the brain. This histological observation agrees with the fact that, after subarachnoid injection, the dye appears rapidly in tissues outside the central nervous system.

Chemical agents which produce characteristic localized responses have supplied indirect evidence of the presence or absence of barriers. I mention the classic study of Lewandowsky (1900). When sodium ferrocyanide solutions were administered via the blood circulation (subcutaneous or intravenous injections in rabbits, cats, or dogs), no marked effects were observed, whereas injections into the cerebrospinal fluid caused severe responses of the central nervous system including convulsions. From these differences in response, Lewandowsky concluded that blood capillaries of the brain contained a barrier to the passage of ferrocyanide. The presence of a *relative* blood-brain barrier was also demonstrated by Lewandowsky. There was no doubt that strychnine passed the capillaries of the central nervous system when injected intravenously, but Lewandowsky found that it was ten times more toxic when injected into the cerebrospinal fluid. In Friedemann's (1942) article, the blood-brain barriers are reviewed mainly with respect to bacterial toxins, such as tetanus, diptheria, and botulinus toxin, to cobra venom, neurotropic viruses, and antibodies.

c. *The permeability of blood capillaries in the brain under various conditions.* As mentioned previously, the use of visible tracers for permeability studies in the brain requires consideration of numerous factors. For mechanisms of permeability of capillaries in general and of brain capillaries in particular the reader is referred to comprehensive presentations such as Broman's 1949 monograph and Davson's 1956 book. A few examples may suffice here to illustrate the variety of aspects involved.

In a study of hormonal factors Friedemann and Elkeles (1932) showed that adrenaline increases the permeability of brain capillaries for substances such as ethyl alcohol, paraldehyde, strychnine, and alizarin blue-S, which can pass through these capillaries under ordinary conditions. The impermeability of brain capillaries to trypan blue and arsphenamine was not altered by adrenaline. If cerebral capillaries are deprived of oxygen for a short time one would expect them to be sufficiently damaged to show an increase in permeability. However, studies of Broman (summarized in his 1949 monograph) demonstrated that the permeability for trypan blue did not change in brains of animals and humans for 2 or more hours after death. On the other hand, Becker and Quadbeck (1952) reported that the brain was stained 15 to 30 minutes after death of experimental animals if the dyestuff Astra violet was injected into the veins shortly before death; the brain capillaries are not permeable to this dye during life.

Exposure of the human brain, as produced by surgical opening of the skull, leads to edema of the brain. This observation prompted several investigators to test the effect of exposure on permeability of cerebral capillaries in experimental animals. Samorajski and Moody (1957) applied the microscopic aminoacridine method of Rodriguez (1955). Openings one-half inch in diameter were made in the skulls of rabbits. After 45 minutes of exposure of the brain, the opening was closed. At different intervals after operation (one-half hour to 21 days) the dye was injected intravenously, and one-half hour or one hour after injection, brain samples for histological study were removed under anesthesia. Fluorescence of nuclei in capillary endothelium and nervous tissue appeared in or near the exposed area after 2 hours, reached a maximum on the third day, and disappeared 2 or 3 weeks after exposure. Histologically clear indications of an increase in tissue fluid were observed, also with a peak around the third day and disappearing after 2 or 3 weeks. It is the synchronous correlation with tissue fluid increase which makes the aminoacridine test a reliable method for this type of study. No rupture of capillaries was observed. Evidently the capillaries of the brain, impermeable to aminoacridine in control animals, became permeable after exposure but returned to the original state after a few weeks: a relatively mild abnormal stimulus had changed the reactivity of the capillaries in a reversible way.

According to Hess (1955) stabbing of the brain with a hot needle causes the ground substance to disappear in the area of the wound. As described previously this ground substance is characterized by positive PAS reaction (periodic acid-Schiff reaction). Three days after the injury the ground substance starts to reappear, and after 10 days its regeneration is complete. Hess pointed to probable relations between presence or absence of the blood-brain barrier and presence or absence of the ground substance. The conclusions of Hess were based on earlier trypan blue experiments of other investigators. Subsequently his ideas received additional support from the aminoacridine study of Samorajski and Moody (1957).

Do blood capillaries in *brain tumors* differ in permeability from capillaries of normal brain tissue? This question was discussed in a study by Tocus (1959). Not only I^{131}-labeled octoiodofluorescein, but also other isotopes and dyes, including trypan blue and aminoacridine, showed a strong tendency to localize in brain tumors. Tocus found a small amount of labeled octoiodofluorescein in normal brain substance, but a large amount in the choroid plexus. According to Tocus, this indicator enters the brain tumor tissue readily, because the tumor capillaries contain either no barrier mechanism, or a damaged mechanism.

Chapter 4

Potential Spaces on Macro- and Microscopic Levels

In an earlier chapter comparisons were made between living and dead material and between natural conditions and artificial conditions produced by the techniques used (Part III, Chapter 3, C). Problems of visible spaces were not discussed in that chapter, and will be treated now. Let us start with a macroscopic example, the mammalian pleural cavity. The surface of the lung (visceral pleura) is separated by a thin film of fluid both from the inner lining of the thoracic cavity (parietal pleura) and from the diaphragm (diaphragmatic pleura). A small amount of fluid fills the angular space between thoracic wall, lung, and diaphragm. Under normal conditions there is no air space between the lung and the thoracic wall (or diaphragm). If a hypodermic needle, connected with a manometer, is thrust into the fluid-filled angular space between thoracic wall and diaphragm, pressure below that of the atmosphere is registered. A similar pressure is observed if the manometer needle is introduced through other places of the chest wall, thus separating the parietal from the visceral layer of the pleura. Evidently a suction effect is produced by the elastic tissue of the lung. The close contact between the surface of the normal lung and the parietal pleura is maintained by two factors: (1) the difference between intrapulmonary pressure, which is one atmosphere, and the intrapleural pressure with a manometer reading 4 to 5 mm. of mercury lower, and (2) the hydraulic traction exerted by the film of fluid between visceral and parietal pleura. Hydraulic traction becomes apparent if one attempts to pull two opposed moist glass plates apart without sliding movements. The 1887 paper by West in which these factors were analyzed makes fascinating reading. West's emphasis on the hydraulic traction rather than on differences in atmospheric pressure is accepted by Best and Taylor (1955, pp. 347-348). Various pathological processes involving perforations of the thoracic wall or the lung allow air to accumulate between visceral and parietal pleura. For these reasons, the normal pleural cavity is referred to as a potential space. A large space between lung and thoracic wall can also be produced by pathological fluids. If air or fluid is present in the pleural cavity, the lungs are found partially or completely collapsed. For therapeutic reasons one produces collapse of a diseased lung by filling the pleural cavity with air or nitrogen; the lung expands again as the gas is gradually absorbed. Pathological adhesions between visceral and parietal pleura may lead to irreversible obliteration of the potential pleural space. This renders collapse of the lungs impossible, irrespective of changing pressures.

The concept of *potential spaces* proved very useful with reference to all macroscopic or microscopic spaces which are not present during life under normal conditions but are produced by pathological processes, by postmortem changes, or by technical interferences during life or after death.

An example of microscopic potential space is illustrated in Fig. 80A and B. Although under normal conditions the perivascular spaces in the central nervous system do not extend to the level of capillaries (Fig. 80A), pericapillary spaces can be produced during life by inflammatory processes and by various injection techniques (Fig. 80B). After death histological procedures involving shrinkage of nervous tissue are apt to create not only pericapillary but also perineuronal spaces. Bargmann (1959, p. 790) refers to pericapillary spaces as "potential Virchow-Robin spaces," in contrast to the normal Virchow-Robin space around arteries or veins.

A potential microscopic space in the liver is known as the perisinusoidal space or the space of Disse (see my diagram Fig. 87C). Under ordinary conditions the liver cell cords are lined with the endothelial wall of the sinusoids, but under exceptional conditions not yet understood a distinct space appears between the endothelial cells and the liver cord cells. A human liver showing this space is illustrated in Bremer-Weatherford (1944, Fig. 333). I observed a similar condition in the liver of a rhesus monkey. This liver appeared otherwise macro- and microscopically normal. The perisinusoidal spaces were very distinct in formalin-fixed frozen sections when the sections were dehydrated and mounted in Clarite. If sections from the same block were not dehydrated and were mounted in glychrogel no such spaces were visible (Mayer, 1945, unpublished data). Popper and Schaffner (1957) stated that in human livers perisinusoidal spaces are not visible after instantaneous death, but are visible after an agony period of only a few minutes. They ascribed this difference to an increase in permeability of the sinusoidal wall for blood plasma proteins resulting in leakage of fluid and opening of the potential spaces. In electron microscopic studies of rat livers Fawcett (1955) found that the surfaces of liver cord cells facing the sinusoids are often covered with short processes (microvilli) which cannot be resolved with light microscopy. Since these processes vary in length, they control the distance between liver cell surface and wall of sinusoids. According to Fawcett, this variation of microvilli may account for the presence or absence of perisinusoidal spaces under the light microscope.

In transluminated livers of living frogs Knisely *et al.* (1945) were not able to demonstrate lymphatics or perisinusoidal spaces, but the authors assumed the presence of these structures because of functional reasons. The textbook of Maximow and Bloom (1952) also favors a possible role of the space of Disse

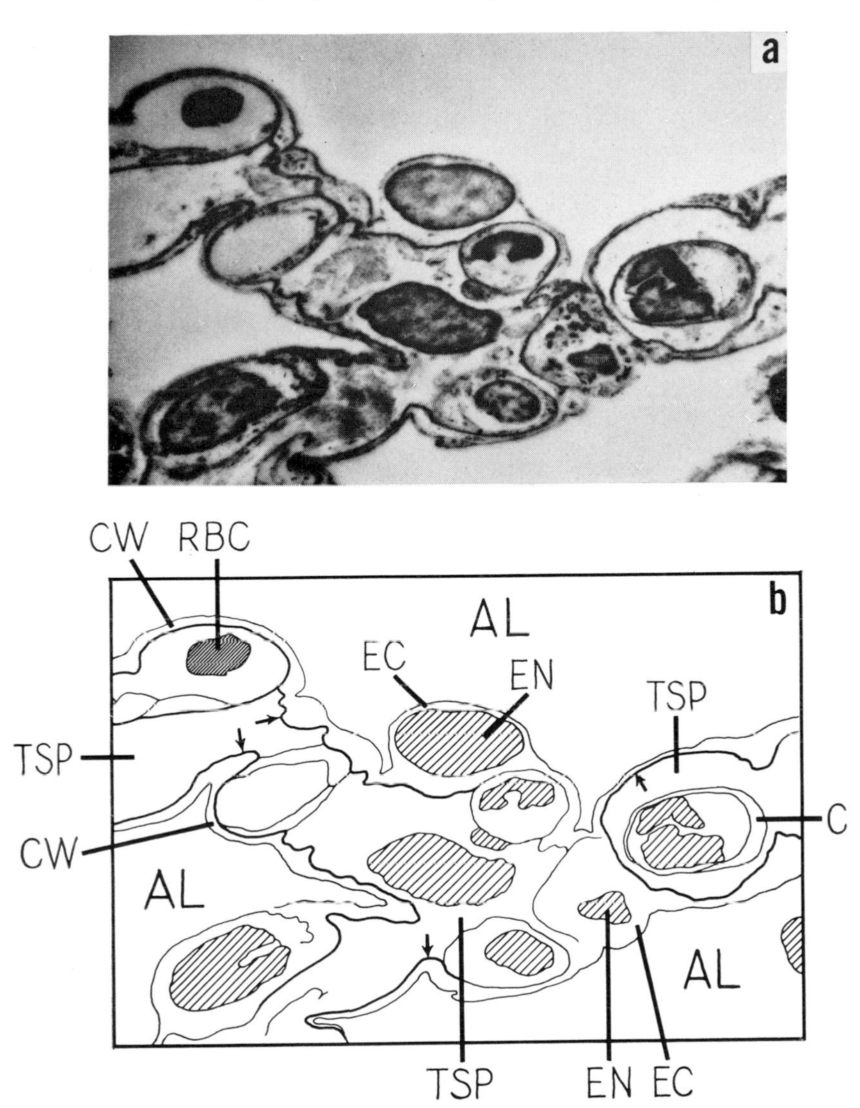

FIG. 82. Demonstration of potential tissue space between alveolar epithelium and capillary wall in the lung. (a) Electron micrograph of a section from the alveolar wall of the normal lung of a rabbit. Magnification ×3000. (From Low, 1953, Fig. 2.) (b) Tracing of (a), with labels of structures. *AL,* Alveolar spaces; *TSP,* tissue spaces; *CW,* capillary walls; *RBC,* red blood corpuscle; *EC,* epithelial cytoplasm; *EN,* epithelial nucleus. Arrows point to basement membranes of epithelium.

in the production of lymph. Approximately one-third to one-half of the body lymph comes from the liver, but lymphatic capillaries are not visible in the liver lobules. Therefore Maximow and Bloom postulate a potential lymphatic space between the sinusoidal endothelium and the surface of the liver cord cells, although this space cannot be demonstrated by injection procedures. Hemodynamic arguments against any role of the space of Disse in the production of hepatic lymph have been raised by Ham (1957, pp. 618–619). It seems to me that the question of perisinusoidal spaces deserves a coordinated attack by physiological experiments, microscopic observations of the liver during life, and light and electron microscopy of sections.

The alveoli of the lungs may serve as a third example of microscopic potential spaces. The partitions between the alveoli of the lungs consist of a network of blood capillaries with small meshes and a variable number of elastic and reticular fibers. The fact that blood capillaries form the bulk of the alveolar wall is well illustrated by Miller (1947, Fig. 62). The nature of the lining of the alveoli of the lungs had been controversial for a long period. Electron microscopic studies confirmed the presence of a continuous layer of extremely shallow cells (alveolar epithelium). Another question was whether tissue spaces existed between the alveolar epithelium and the wall of the blood capillaries. No such space was demonstrable in normal lungs, and all attempts failed to separate the epithelial layer from the capillaries by mechanical or chemical dissection. However, in human pneumonia, Miller observed continuous sheets of epithelial lining being separated from the capillary walls (see Fig. 49 of his book). I agree with Miller's interpretation that this separation was produced by the inflammatory exudate, and I assume that the space was somewhat enlarged by the histological technique (shrinkage).

Electron microscopic studies indicate the existence of a potential space between epithelium and capillary wall in normal animal lungs. My Fig. 82a, shows an electron micrograph from the alveolar wall of a rabbit's lung, published by Low (1953). I have added a tracing for the lettering of structures (Fig. 82b). In some places the basement membranes of alveolar epithelial and of capillary endothelial cells are in direct contact, but in other places they are separated by a distinct space. Low emphasized that these tissue spaces may be potential or apparent. I wish to add a comment on the importance of these spaces in pulmonary pathology. Traditionally, pathological thickening of alveolar walls was ascribed to two factors: (1) congestion and dilation of blood capillaries and (2) accumulation of exudate cells, such as leucocytes and macrophages, in the wall. As long as tissue spaces in the normal alveolar walls were hypothetical, the precise location of the exudate cells remained problematic. This question has been resolved through the electron microscopic demonstration of tissue spaces with their obvious potentiality for distention.

PART IV: REFERENCES

Altmann, H. W. (1955). *Klin. Wochschr.* **33**, 306.

Andresen, N., Holter, H., and Zeuthen, E. (1944). *Compt. rend. trav. lab. Carlsberg, Sér. chim.* **25**, 67.

Andrew, W. (1959). "Textbook of Comparative Histology." Oxford Univ. Press, London and New York.

Angevine, D. M. (1951). *Conf. on Connective Tissues 1st, Trans.* (C. Ragan, ed.), p. 13. Josiah Macy, Jr. Foundation, New York.

Bailey's Textbook of Histology (1953). (P. E. Smith and W. M. Copenhaver, eds.), 13th ed. Williams & Wilkins, Baltimore, Maryland.

Bakay, L. (1951). *A.M.A. Arch. Neurol. Psychiat.* **66**, 419.

Bakay, L. (1956). "The Blood-Brain Barrier: With Special Regard to the Use of Radioactive Isotopes." C. C Thomas, Springfield, Illinois.

Bakay, L., and Lindberg, O. (1949). *Acta. Physiol. Scand.* **17**, 179.

Bargmann, W. (1959). "Histologie und Mikroskopische Anatomie des Menschen," 3rd ed. Thieme, Stuttgart.

Bear, R. S. (1942). *J. Am. Chem. Soc.* **64**, 727.

Bear, R. S. (1952). *Advances in Protein Chem.* **7**, 69.

Beard, J. W., and Rous, P. (1934). *J. Exptl. med.* **59**, 593.

Becker, H., and Quadbeck, G. (1952). *Z. Naturforsch.* (**B**)**7**, 493.

Bennett, H. S. (1956a). *J. Biophys. Biochem. Cytol.* **2**, 99.

Bennett, H. S. (1956b). *J. Biophys. Biochem. Cytol.* **2**, 171.

Bessis, M. (1956). "Cytology of the Blood and Blood-forming Organs," (E. Ponder, transl. and ed.). Grune & Stratton, New York.

Best, C. H., and Taylor, N. B. (1955). "The Physiological Basis of Medical Practice," 6th ed. Williams & Wilkins, Baltimore, Maryland.

Bodenstein, D. (1955). *In* "Analysis of Development," (B. H. Willier, P. Weiss, and V. Hamburger, eds.), p. 337. Saunders, Philadelphia, Pennsylvania.

Bodian, D. (1942). *Physiol. Revs.* **22**, 146.

Boivin, A., Vendrely, R., and Vendrely, C. (1948). *Compt. rend. acad. sci.* **226**, 1061.

Bonner, J. T. (1947). *J. Exptl. Zool.* **106**, 1.

Bonner, J. T. (1959). *Sci. Am.* **201**(6), 152.

Bouin, M., and Bouin, P. (1898). *Bibliographie Anat.* **6**(1), 1.

Bouton, S. M., Jr. (1940). *A.M.A. Arch. Neurol. Psychiat.* **43**, 1151.

Brachet, J. (1957). "Biochemical Cytology." Academic Press, New York.

Bradfield, J. R. G. (1955). *Symposia Soc. Exptl. Biol.* **9**, 306.

Bremer, J. L., rewritten by Weatherford, H. L. (1944). "A Textbook of Histology," 6th ed. Blakiston, New York.

Briggs, R., and King, T. (1953). *J. Exptl. Zool.* **122**, 485.

Briggs, R., and King, T. J. (1955). *In* "Biological Specificity and Growth" (E. G. Butler, ed.), p. 207. Princeton Univ. Press, Princeton, New Jersey.

Broman, T. (1949). "The Permeability of the Cerebrospinal Vessels in Normal and Pathological Conditions." Munksgaard, Copenhagen.

Buchner, P. (1921). "Tier und Pflanze in intracellulärer Symbiose." Borntraeger, Berlin.

Chatton, E., and Lwoff, A. (1936). *Bull. soc. franç. microscop.* **5**, 25.

Clara, M. (1953). "Das Nervensystem des Menschen." Barth, Leipzig.

Clark, E. R., and Clark, E. L. (1933). *Am. J. Anat.* **52**, 273.

Cohn, H. M. (1926). *Arch. pathol. Anat. u. Physiol. Virchow's* **259**, 30.

Coman, D. R. (1944). *Cancer Research* **4**, 625.

Dandy, W. E., and Blackfan, K. D. (1913). *J. Am. Med. Assoc.* **61**, 2216.

Davson, H. (1956). "Physiology of the Ocular and Cerebrospinal Fluids." Churchill, London.
DeBruyn, P. P. H., Robertson, R. C., and Farr, R. S. (1950). *Anat. Record* **108**, 279.
Dempsey, E. W., and Wislocki, G. B. (1955). *J. Biophys. Biochem. Cytol.* **1**, 245.
DeRobertis, E. D. P. (1954). *In* "General Cytology," (E. D. P. DeRobertis, W. W. Nowinski, and F. A. Saez, eds.), p. 413, 2nd ed. Saunders, Philadelphia, Pennsylvania.
Di Stefano, H. S. (1948). *Chromosoma* **3**, 282.
Eames, A. J., and MacDaniels, L. H. (1947). "An Introduction to Plant Anatomy," 2nd ed. McGraw-Hill, New York.
Earle, W. R. (1943). *J. Natl. Cancer Inst.* **4**, 165.
Ehret, C. F., and Powers, E. L. (1959). *Intern. Rev. Cytol.* **8**, 97.
Elias, H. (1949). *Am. J. Anat.* **84**, 311.
Elson, D., and Chargaff, E. (1952). *Experientia* **8**, 143.
Engström, A., and Finean, J. B. (1958). "Biological Ultrastructure." Academic Press, New York.
Farr, R. S. (1951). *Anat. Record* **109**, 515.
Fauré-Frémiet, E., and Mugard, H. (1948). *Compt. rend. Acad. Sci.* **227**, 1409.
Fawcett, D. W. (1955). *J. Natl. Cancer. Inst.* **15**, 1475.
Fawcett, D. W., Ito, S., and Slautterback, D. (1959). *J. Biophys. Biochem. Cytol.* **5**, 453.
Fawcett, D. W., and Porter, K. R. (1954). *J. Morphol.* **94**, 221.
Fawcett, D. W., and Selby, C. C. (1958). *J. Biophys. Biochem. Cytol.* **4**, 63.
Fenn, W. O. (1922a). *J. Gen. Physiol.* **5**, 143.
Fenn, W. O. (1922b). *J. Gen. Physiol.* **5**, 169.
Finerty, J. C., and Cowdry, E. V. (1960). "A Textbook of Histology," 5th ed. Lea & Febiger, Philadelphia, Pennsylvania.
Fischer, A. (1946). "Biology of Tissue Cells." Gyldendal, Copenhagen.
Frankenhaeuser, B. (1952). *Cold Spring Harbor Symposia Quant. Biol.* **17**, 27.
Frey-Wyssling, A. (1955). *In* "Protoplasmatologia" (L. V. Heilbrunn and F. Weber, eds.), Vol. **2** (A,2), p. 1. Springer, Vienna.
Friedemann, U. (1942). *Physiol. Revs.* **22**, 125.
Friedemann, U., and Elkeles, A. (1932). *Deut. med. Wochschr.* **58**, 923.
Garnier, C. (1899). *Bibliographie anat.* **7**, 217.
Gerarde, H. W., and Jones, M. (1953). *J. Biol. Chem.* **201**, 553.
Goldmann, E. E. (1913). *Abhandl. Kgl. preuss. Akad. Wiss. Physikal.-math. Classe No.* **1**, 1.
Goldstein, L., and Plaut, W. (1955). *Proc. Natl. Acad. Sci. U. S.* **41**, 874.
Grafflin, A. L. (1947). *Am. J. Anat.* **81**, 63.
Grafflin, A. L., and Foote, J. J. (1938). *Anat. Record* **72**, 115.
Grafflin, A. L., and Foote, J. J. (1939). *Am. J. Anat.* **65**, 179.
Greep, R. O. (1954). "Histology." Blakiston, New York.
Gross, J. (1949). *J. Exptl. Med.* **89**, 699.
Ham, A. W. (1957). "Histology," 3rd ed. Lippincott, Philadelphia, Pennsylvania.
Hardesty, I. (1904). *Am. J. Anat.* **3**, 229.
Hardin, G. (1956). *Sci. Monthly* **82**, 112.
Harvey, E. N., and Marsland, D. A. (1932). *J. Cellular Comp. Physiol.* **148**, 75.
Hawn, C. V., and Porter, K. R. (1947). *J. Exptl. Med.* **86**, 285.
Helm, F. E. (1876). *Z. wiss. Zool.* **26**, 434.
Hess, A. (1953). *J. Comp. Neurol.* **98**, 69.
Hess, A. (1955). *A.M.A. Arch. Neurol. Psychiat.* **73**, 380.
Hevesy, G., and Hahn, L. (1940). *Kgl. Danske Videnskab. Selskab, Biol. Medd.* **15**, 1.

Holtfreter, J. (1948). *Ann. N. Y. Acad. Sci.* **49**, 709.
Hughes, A. (1952). "The Mitotic Cycle." Academic Press, New York.
Huxley, H. E. (1957). *J. Biophys. Biochem. Cytol.* **3**, 631.
Itano, H. A. (1957). *Advances in Protein Chem.* **12**, 215.
Iversen, S. (1960). *Nature* **187**, 86.
Jacobson, W. (1953). *In* "Nature and Structure of Collagen" (J. T. Randall, ed.), p. 6. Academic Press, New York.
Jacoby, F., and Marks, J. (1953). *J. Hyg.* **51**, 541.
Karpechenko, G. D. (1928). *Z. induktive Abstammungs—u. Vererbungslehre* **48**, 1.
Knisely, M. H., Bloch, E. H., and Warner, L. A. (1945). *In* "Liver Injury," 4th Conf., p. 22. Josiah T. Macy, Jr. Foundation, New York.
Korschelt, E. (1896). *Arch. mikroskop. Anat.* **47**, 500.
Kramer, H., and Little, K. (1953). *In* "Nature and Structure of Collagen" (J. T. Randall, ed.), p. 33. Academic Press, New York.
Laki, K. (1953). *Blood* **8**, 845.
Laki, K., Gladner, J. A., and Folk, J. E. (1960). *Nature* **187**, 758.
Lederberg, J. (1952). *Physiol. Revs.* **32**, 403.
Leuchtenberger, C. (1950). *Chromosoma* **3**, 449.
Lewandowsky, M. (1900). *Z. klin. Med.* **40**, 480.
Lewis, F. T. (1925). *Proc. Am. Acad. Arts Sci.* **61**, 1.
Lewis, W. H. (1922). *Anat. Record* **23**, 177.
Lewis, W. H. (1931). *Bull. Johns Hopkins Hosp.* **49**, 17.
Lloyd, D. P. C. (1949). *In* "Textbook of Physiology" (J. F. Fulton, ed.), 16th ed., p. 33. Saunders, Philadelphia, Pennsylvania.
Low, F. N. (1953). *Anat. Record* **117**, 241.
McClung Jones, Ruth, Editor (1950). "McClung's Handbook of Microscopical Technique," 3rd ed. Hoeber, New York.
Martin, A. V. W. (1953). *In* "Nature and Structure of Collagen" (J. T. Randall, ed.), p. 87. Academic Press, New York.
Marvin, J. W. (1939). *Am. J. Botany* **26**, 280.
Matzke, E. B. (1950). *Bull. Torrey Botan. Club* **77**, 222.
Maximow, A., and Bloom, W. (1952). "A Textbook of Histology," 6th ed. Saunders, Philadelphia, Pennsylvania.
Mayer, E. (1925). Verhandl. Deut. Pathol. Ges. 20th Meeting. *Zentr. allgem. Pathol.* **36**, Suppl., 327.
Mayer, E. (1930). *Arch. exptl. Zellforsch. Gewebezücht.* **10**, 221.
Mayer, E. (1935). *Compt. rend. soc. biol.* **119**, 422.
Mayer, E., and Schreiber, H. (1934). *Protoplasma* **21**, 34.
Mazia, D., and Prescott, D. M. (1954). *Science* **120**, 120.
Meeuse, A. D. J. (1957). *In* "Protoplasmatologia" (L. V. Heilbrunn and F. Weber, eds.), Vol. **2** (A,1,c) p. 1. Springer, Vienna.
Meves, Fr. (1897). *Arch. mikroskop. Anat.* **48**, 573.
Miller, W. S. (1947). "The Lung," 2nd ed. C. C Thomas, Springfield, Illinois.
Mommaerts, W. F. H. M. (1945). *J. Gen. Physiol.* **29**, 113.
Parker, R. C., Castor, L. N., and McCulloch, E. A. (1957). *In* "Cellular Biology, Nucleic Acids and Viruses," *Spec. Publ. N. Y. Acad. Sci. No.* **5**, 303.
Parker, T. J., and Haswell, W. A. (1940). "Textbook of Zoology," 6th ed., revised by O. Lowenstein, Macmillan, New York and London.
Patek, P. R. (1944). *Anat. Record* **88**, 1.
Pauling, L., Itano, H. A., Singer, S. J., and Wells, I. C. (1949). *Science* **110**, 543.

Pfeiffer, H. (1934). *Arch. exptl. Zellforsch. Gewebezücht.* **15**, 203.

Pollister, A. W., Swift, H. H., and Alfert, M. (1951). *J. Cellular Comp. Physiol.* **38**, 101 (Suppl. 1).

Popper, H., and Schaffner, F. (1957). "Liver: Structure and Function." McGraw-Hill (Blakiston), New York.

Porter, R. W., and Davenport, H. A. (1949). *Stain Technol.* **24**, 117.

Ranson, S. W. (1942). "The Anatomy of the Nervous System," 6th ed. Saunders, Philadelphia, Pennsylvania.

Robertson, E. B., and Duke-Elder, W. S. (1933). *Proc. Roy. Soc.* **B112**, 215.

Robinow, C. (1936). *Protoplasma* **27**, 86.

Robinow, C. F. (1945). *In* R. J. Dubos, "The Bacterial Cell," p. 351. Harvard Univ. Press, Cambridge, Massachusetts.

Rodriguez, L. A. (1955). *J. Comp. Neurol.* **102**, 27.

Rones, B. (1941). *Arch. Opthalmol.* **26**, 108.

Rous, P., and Beard, J. W. (1934). *J. Exptl. Med.* **59**, 577.

Saez, F. A. (1954). *In* "General Cytology" (E. D. P. DeRobertis, W. W. Nowinski, and F. A. Saez, eds.), 2nd ed., p. 215. Saunders, Philadelphia, Pennsylvania.

Samorajski, T., and Moody, R. A. (1957). *A.M.A. Arch. Neurol. Psychiat.* **78**, 369.

Schmitt, F. O., Hall, C. E., and Jakus, M. A. (1942). *J. Cellular Comp. Physiol.* **20**, 11.

Sharp, L. W. (1943). "Fundamentals of Cytology." McGraw-Hill, New York.

Sjöstrand, F. S. (1959). *In* "Symposium on Biological Organization, Cellular and Subcellular" (C. H. Waddington, ed.), p. 117. Pergamon Press, New York.

Sonneborn, T. M. (1959). *Advances in Virus Research* **6**, 229.

Stein, W. H., and Miller, E. G. (1938). *J. Biol. Chem.* **125**, 599.

Swift, H. H. (1950). *Physiol. Zoöl.* **23**, 169.

Sze, L. C. (1953). *J. Exptl. Zool.* **122**, 577.

Taft, A. E. (1938). *Arch. Neurol. Psychiat.* **40**, 313.

Tasaki, I. (1939). *Am. J. Physiol.* **127**, 211.

Tocus, E. C. (1959). The localization of I-131 labeled octoiodofluorescein in human brain tumors. Ph.D. Dissertation, Dept. of Pharmacology, University of Chicago, Chicago, Illinois.

Tyler, A. (1955). *In* "Analysis of Development" (B. H. Willier, P. Weiss, and V. Hamburger, eds.), p. 556. Saunders, Philadelphia, Pennsylvania.

Vendrely, C., and Vendrely, R. (1949). *Comp. rend. soc. biol.* **143**, 1386.

Walter, F. K. (1933). *Arch. Psychiat.* **101**, 195.

Weed, L. H. (1922). *Physiol. Revs.* **2**, 171.

Weed, L. H. (1923). *Am. J. Anat.* **31**, 191.

West, S. (1887). *Brit. Med. J.* **II**, 393.

Wilson, G. E. (1929). *Anat. Record* **42**, 243.

Windrum, G. M., Kent, P. W., and Eastoe, J. E. (1955). *Brit. J. Exptl. Pathol.* **36**, 49.

Winton, F. R., and Bayliss, L. E. (1955). "Human Physiology," 4th ed. Little, Brown, Boston, Massachusetts.

Wolbach, S. B., and Howe, P. R. (1925). *J. Exptl. Med.* **42**, 753.

Wolbach, S. B., and Howe, P. R. (1933). *J. Exptl. Med.* **57**, 511.

Woollam, D. H. M., and Millen, J. W. (1955). *J. Anat.* **89**, 193.

Yoffey, J. M., and Courtice, F. C. (1956). "Lymphatics, Lymph and Lymphoid Tissue." Harvard Univ. Press, Cambridge, Massachusetts.

Zetterqvist, H. (1956). The ultrastructural organization of the columnar absorbing cells of the mouse jejunum. M.D. Thesis, Karolinska Institute, Stockholm.

Part V

CLASSIFICATION AND IDENTIFICATION OF BIOLOGICAL STRUCTURES

Chapter 1

Principles of Classification in Different Areas of Science

A. Alternatives of Classification. Concepts of Relationship and Species

My teacher Heinrich Poll used the example of the telephone directory to introduce his students to the problems of classification. The white pages which list subscribers alphabetically supply telephone numbers and addresses, and sporadically indicate occupations, such as M.D. The yellow page listings are divided into occupational groups, and within each group the subscribers are arranged alphabetically. More explanations and cross references are found in the yellow pages than in the white pages. The white pages are a complete list of subscribers, whereas the yellow pages represent a selection. Both the white and the yellow pages are useful.

In natural science, classifications serve a wide range of purposes. On one end of the scale are systems of keys which allow identification of objects. On the other end of the scale are classifications which express natural laws. Classification may depend on observation or experimentation.

In inorganic chemistry procedures based on dichotomy and elimination permit the systematic determination of elements in unknown solutions. There are seven analytical groups for the identification of acid constituents or anions. Group I comprises acids whose silver salts are insoluble in water and in nitric acid, but whose barium salts are soluble in water. In group II are acids whose silver salts are soluble in nitric acid, but are insoluble, or difficultly soluble, in water, and whose barium salts are soluble in water. Groups III to VII follow a similar pattern. The seven analytical groups are based on certain characteristic solubilities and a definite sequence of procedures. The analytical grouping does not classify the general chemical behavior of the elements, as the periodic system of grouping does. Members of the same analytical group rarely belong to the same group in the periodic system (Treadwell, revised by Hall, 1932, p. 71).

The periodic system gives more insight into the general chemical behavior of elements than the analytical groups do, but it does not supply prescriptions for identification. In the modern form of the periodic table the elements are grouped by their atomic numbers, which indicate the number of positive charges on the nucleus of the atom and, at the same time, the number of electrons rotating about the nucleus. Classification by atomic numbers expresses the differences between elements in numerical terms, an achievement of the greatest theoretical value, but certainly not a substitute for analytical keys. However, methods of determination have been enriched by theoretical developments, and theoretical advances became possible by improved determinative techniques. In inorganic as well as organic chemistry, the interaction of the two approaches has been remarkably fruitful, as illustrated by the history of spectral analysis and chelating agents. Some physical procedures for establishing molecular and structural formulas of organic compounds were mentioned previously, such as melting point determinations (Part III, Chapter 3, C, 2) and infrared spectra (Part III, Chapter 8, Introduction). The use of empirical and structural formulas ("pictures") and of systematic, index, generic, and trade names was illustrated by the example of the insecticide dichlorodiphenyltrichloroethane (DDT) (Part III, Chapter 8, A, 1). The formidable task of classifying organic compounds has been handled with sustained success in Beilstein's Handbook.

Classification in botany and zoology is known as taxonomy. Like those of chemistry, biological classifications range from practical keys for identification to systematic arrangements expressing fundamental properties. Elaborate keys for the identification of animals, plants, and bacteria are available. All of them are based on the principle of dichotomy and elimination. Relationship between a number of individuals may refer to (1) characteristics suitable for identification by a key, (2) characteristics which are the expression of basic organization, (3) common origin, and (4) similar potentialities. Let us first discuss relationships which are derived from similarity in basic organization of two or more types of animals. In modern taxonomy similarities of embryologic development carry more weight than similarities of the adult organisms. This preference is based on the belief that, in some way, ontogeny reflects phylogeny. I give an example. Because of the presence of a notochord in their larval stages, tunicates have been classified in the phylum Chordata. The adult tunicates resemble coelenterates or sessile crustaceans, but not any vertebrates. The notochord is an elastic rod which develops ventrally from the central nervous system. The notochord is no longer present in the adult tunicate. It persists in the adult *Amphioxus,* and is the forerunner of cartilaginous or bony vertebral columns of fishes, amphibians, reptiles, birds, and mammals. Another feature common to all these vertebrates is the formation of the central nervous system by infolding of the surface with the subsequent production of a tubular structure (Fig. 3g–j). The third common trait is the appearance of pharyngeal clefts from which not only gills, but also endocrine organs in the neck region, of vertebrates originate. In terrestrial vertebrates these clefts disappear in later stages of the embryonic development.

Classification on the basis of embryonic development is not always a simple matter. This is illustrated by problems of insect taxonomy. In his 1957 review, van Emden stated that there are many insects in which the larval stages are incompletely known. Even in the well-explored beetle fauna of the British Isles, eggs and pupae of some species have not yet been discovered. On the other hand, insect larvae have been considered to be secondary developments, and therefore not indicative of phylogenetic relations. Van Emden discusses these problems comprehensively, with examples of agreement and disagreement between larval and adult groupings in insects.

Different principles of classification may or may not lead to different groupings. Biochemical and immunological relations between different species are of interest, both for theoretical and practical reasons. The degree of similarity of serum proteins can be determined by serological reactions. On this basis, sheep blood is closest to that of beef, most distant from that of dogs, and intermediary to that of pigs or horses (from Prosser, 1950, p. 107). Here the serological relationship parallels the traditional taxonomy.

Vertebrates can be divided into two large groups if one uses strong or weak control of body temperature as criterion (so-called warm-blooded and cold-blooded animals). Then birds belong with mammals since both classes maintain their body temperature irrespective of the environmental temperature. However, in other physiological and all anatomical characteristics birds are closer to reptiles whose body temperature, like that of amphibians and fishes, depends on the environmental temperature.

Faced with a choice of different classifications, one chooses that which offers the best correlation of important criteria. Herman Melville in "Moby Dick" (1851; p. 194 of the Modern Library edition, 1930) revolted against official zoology and insisted on classifying whales as fish because of their aquatic life, although he was aware of their anatomical and physiological characteristics. The zoologist prefers to classify whales as mammals, not only because they suckle their young, but because of similarity of respiratory, circulatory, urogenital, and central nervous systems, and a metabolism which keeps the body temperature constant. Irrespective of phylogenetic considerations (which could not enter into Melville's reasoning), I side with official zoology.

Up to this point taxonomy has been discussed with respect to the characteristics of adult and embryonic *individuals*. However, *ecological conditions* are also used for identification and classification. In differentiating between various species of wasps, the type of their colonies is important. In identifying caterpillars, the plants on which they feed serve as a guide. The ecological point of view is of obvious importance in the identification and classification of parasites, since the living host represents a peculiar type of environment. Tapeworms are a good example. The human body is the host for three types of tapeworms. They are morphologically different and develop in different intermediate hosts: *Taenia saginata* in cattle, *Taenia solium* in swine, and *Diphyllobothrium latum* in fish.

Among parasites the pathogenic bacteria have been classified with particularly efficient procedures, comparable to the methods of qualitative analysis in chemistry mentioned before. In "Bergey's Manual of Determinative Bacteriology" (1957) each type of bacterium is characterized by alternative criteria, for instance, rod-shaped or spherical (bacillus or coccus), motile or nonmotile, positive or negative with Gram stain, requiring or not requiring oxygen (aerobic or anaerobic). In addition to microscopy and cultivation, serological determinations (specific agglutination, etc.) and animal experiments are used for identification.

Enterobacteriaceae, a large family of Gram-negative bacilli, were studied extensively by Kauffmann. In a 1956 paper this author presented a lucid comparison between the use of biochemical criteria (and motility) on the one hand, and immunological criteria on the other hand. The salmonellas, which are the most important group of the Enterobacteriaceae, have been classified elaborately on the basis of their body and flagella antigens. The Kauffmann-White scheme makes provision for more than 3000 types of the salmonella group, but has been reduced considerably to make it practical. This scheme is intended as a *strictly descriptive classification,* irrespective of factors which might account for the immunological resemblance and difference between the various forms of Enterobacteriaceae (Kauffmann, 1957). However, interesting cytogenetic studies have been made in salmonellas (Demerec *et al.,* 1955), and recently antigenic polysaccharides of Enterobacteriaceae were prepared artificially (Lüderitz *et al.,* 1960). In the future it may be possible to express the immunological differences between salmonellas in terms of chemistry.

In identifying pathogenic bacteria, ecological criteria, such as types of hosts and distribution of bacteria in the host, are not entirely conclusive, but they are important enough to be mentioned here. Different types of bacilli share the property of acid fastness, i.e., a relatively high resistance to decolorization by acids after certain staining procedures. If acid-fast bacilli are found in the sputum of a patient, they will be interpreted as tubercle bacilli. Similar bacilli obtained from sebaceous material around the genitalia will be considered to be harmless smegma bacilli. For a long time meningococci and gonococci were separable only by the location from which the samples were obtained. Both types are Gram-negative diplococci characterized by great variability of the individual cocci, predominant occurrence inside of polymorphonuclear leucocytes, and failure to grow on ordinary media. No animal experiments were available to differentiate between the two types. If the diplococci were found in the meninges and cerebrospinal fluid, they were classified as meningococci, whereas they were identified as gonococci when obtained from the male urethra or the vagina and Fallopian tube of the female. Eventually cultivation methods and diagnostic sera have been developed which permit the separation of meningococci and gonococci.

At this point it seems appropriate to discuss *the unit of taxonomy, the species.* The Linnaean establishment of genus and species as basic levels of classification have proved extremely useful until this day. Sometimes it is controversial whether two forms should be considered subspecies (races, varieties, strains) within a species or two different species within a genus, but the vast majority of cases is undisputed. This is remarkable in view of the fact that the animal kingdom comprises approximately 1,000,000 species and the plant kingdom more than 250,000 species (Dobzhansky, 1951, p. 7). In spite of the practical usefulness of the *static* species concept of Linnaeus, it was not possible to agree on an operational definition. The modern *dynamic* concept, based on the variability of populations, is as follows. A species may show either continuous variability or discrete variability in the form of distinct races. The races within one species are defined as Mendelian populations with different frequencies of one or more genetic variants or chromosomal structures (Dobzhansky, p. 138). The term Mendelian populations means that fertile hybrids of two races show Mendelian segregation in their offspring.

In the new development of the species concept, genetic and ecological considerations are interwoven. "Genetic isolation" summarizes genetic and ecological factors which operate in favor of separateness, with the result that different species do not merge but remain distinct and may produce new species. Two species of the same genus may be isolated geographically in such a way that there is no opportunity of interbreeding. The question is how two species in adjacent habitats maintain their separateness. The genes of species hybrids are incompatible with each other, as a rule. The rare exceptions of successful interbreeding of species do not yield intermediary races with Mendelian segregation, but new species which breed true. Even races in adjacent habitats do not tend to random hybridization. If some inbreeding occurs in the border zone of the two habitats, the hybrids which result from Mendelian segregation have a slim chance of survival since they are not adapted to either of the two habitats. In this greatly condensed presentation I have attempted to recount the views of Dobzhansky in his "Genetics and the Origin of Species" (1951).

A clear presentation of the main differences between the old and new systematics is found in Mayr's "Systematics and the Origin of Species" (1942). The interest of taxonomists has moved from the species to the subspecies, and large samples of populations are the basis of modern taxonomic work. In the past, few or even single specimens appeared sufficient to establish a species.

Pioneering experiments on the interaction of genetic and ecological factors were included in the studies of Turesson (1922) with species of *Atripex, Hieracium,* and other plants which occur as large or dwarf forms, and with erect or prostrate stems. These differences appeared to be correlated with different habitats in Sweden. Therefore Turesson made two types of experiments: (1) cultivation of the different forms in rich and poor soil from different habitats, and (2) breeding for determination of genetic races within each species. It was possible to trace the phenotypic differences of the plants to genetic factors which determined their reactivity to poor and rich soil. Turesson suggested that these relations be indicated by the term *ecospecies.* Sewall Wright (1932) devised a system of symbolic pictures to express the relations between the gene combinations of species and genera and their environment. Some of his diagrams are reprinted in Dobzhansky's book (1951, pp. 9 and 279).

The problem of the origin of species became accessible to direct experimentation, for the first time, when it was demonstrated that a new species can originate by hybridization from clearly separated species. In the genus *Galeopsis* the species *G. pubescens* and *G. speciosa* usually give sterile hybrids. From the small minority of fertile hybrids a plant was obtained which was backcrossed to a pure *G. pubescens.* This backcross yielded a single seed which became the progenitor of a strain named artificial *G. tetrahit,* because it proved to be indistinguishable from an existing natural species known as *G. tetrahit* (Dobzhansky, 1951, p. 20). Most astonishing was Karpechenko's (1928) discovery that experimental *intergeneric* crossing could produce a new species which had no counterpart in nature. He named the new species *Raphanobrassica* since it was the result of crossing radish (*Raphanus sativus*) and cabbage (*Brassica oleracea*). All new species obtained by hybridization of two distinct species breed true and are characterized by their chromosomal assortment and macro- and microscopic properties of the plant's body. Some of Karpechenko's pictures of chromosome assortments and pods are reproduced in "Principles of Genetics" by Sinnott, Dunn, and Dobzhansky (1950, p. 370).

The new concept of species as defined by genetic isolation applies to *sexually reproducing organisms.* In any case, the concept of species cannot be the same for sexual and agamic organisms (Babcock and Stebbins, 1938). For instance, in bacteria the subdivision "of the mass of clones" into the species *Escherichia coli, Salmonella typhosa,* and *Salmonella enteritidis* is purely a matter of taste, according to Dobzhansky, as one may as well regard all of them as races within a single species (1951, p. 274). It seems to me that Dobzhansky's comment on the taxonomy of bacteria lost some of its weight when exchange of genic material became known. Studies of genic exchange in salmonellas by Demerec *et al.* (1955) were mentioned previously. However, according to Swanson (1957) the absence of haploid-diploid alternation in bacteria indicates that mutations in *Drosophila* and bacteria are not the same. Dr. W. M. Layton of these laboratories pointed out that it is hard to appraise this difference in view of the various types of mutations known (personal communication, 1961).

B. Specificity of a Single Characteristic

If two species can be distinguished by a large number of different characteristics, this is considered safe ground for classification. At the same time, there is a deeply rooted belief, or hope, that one single characteristic may permit identification of a species. A criterion which is supposed to achieve this is called *specific,* which means "making a species." An example from inorganic chemistry may illustrate the idea of *absolute specificity.* If the chelating agent dimethylglyoxime is added to a dilute ammonia solution containing nickel ions, the nickel will be precipitated completely. This does not occur with any of the other 97 elements tested. Here is a test which is absolutely specific for one metal, or, as Walton (1953, p. 72) puts it, "an analytical chemist's dream."

Are criteria with absolute specificity available in biological taxonomy? For some large taxonomic units one criterion may suffice to characterize all members. The possession of a vertebral column classifies any animal as a member of the subphylum Vertebrata (Craniata). For smaller groups, such as classes, it may be difficult to find one criterion which applies without exception. The great majority of starfish species have five arms, but *Pycnopedia helianthoides* starts with five arms and then changes to six arms and multiples of six as the animals grow older (Hyman, 1955, p. 247).

It was mentioned previously that various types of acid-fast bacilli can be distinguished and that the origin of the sample in which they are found is helpful in their identification. If discovered in the sputum of a patient with cavities in the lung, the acid-fast bacilli will be interpreted as tubercle bacilli. The acid-fast stain of bacilli is *specific under these circumstances,* in other words, it has *relative specificity.* This seemingly inconsistent term is useful since it covers the innumerable situations in which one criterion is to be interpreted in the light of other criteria.

Relative specificity of procedures and characteristics was discussed in various connections. The relative specificity of histological methods was illustrated by the examples of hematoxylin as a nuclear stain (Part III, Chapter 2, B, 3, d), and of silver nitrate solution as a stain for cell boundaries in a mosaic pattern (Part IV, Chapter 2, B). Relative specificity of enzyme reactions including a possible instance of absolute specificity, was discussed with reference to cytochemistry (Part III, Chapter 7, B, 3). The search for specific inductors (organizers) in embryology was recounted in the chapter on stimulus, response, and reactivity relations (Part II, Chapter 2, E).

The concept of a *specific titer* in immunology deserves a brief description here. Typhoid, paratyphoid A, and paratyphoid B bacilli are agglutinated by the serum from patients who recovered from an infection with any of the three types of bacilli. However, the titers or dilutions of sera which still agglutinate the different bacilli are different. Suppose at a titer of 1: 200 each serum agglutinates each type bacillus, and at a titer of 1 : 2000 the serum from an individual who was infected by typhoid bacilli agglutinates typhoid bacilli, but neither A nor B paratyphoid bacilli. On the basis of this quantitative difference reactions between bacilli and immune sera are termed specific. Evidently, the quantitative titer concept represents a useful extension of the originally qualitative concept of specificity. Immunologic methods of relative specificity have been used to characterize organs and cells. This subject will be discussed later (Part V, Chapter 2, C, 2).

C. Phenomena Crossing the Borders of Taxonomy

I mentioned above the six-armed starfish as a species differing from other starfish species which have five arms. It is quite frequent that a species differs from other species of the same genus or family with respect to a single conspicuous characteristic. This becomes even more puzzling if the peculiar characteristic is shared with a taxonomically remote species. A few morphological and biochemical examples from the plant and animal kingdoms may be given here.

The shapes of leaves of phanerograms cross taxonomic lines readily. The leaves of the chestnut oak are hardly distinguishable from chestnut leaves, and those of the willow oak resemble the leaves of willow trees; both oaks form typical acorns. As a rule, coniferous trees have needles or scale-like leaves, but the species *Agathis* (family Araucaraceae) has willow-like leaves. Phloroglucinol derivatives with similar anthelmintic activity are found in three taxonomically and geographically separate families: in a European fern (*Dryopteris Filix-mas*), in an Abyssinian plant of the rose family (*Hagenia abyssinica*), and in a species of the Euphorbia family (*Mallotus philippinensis*) (Zinner, 1955; Steinmetz, 1957). Interest in this subject developed when it became known that extracts of these plants are used as a tapeworm cure by the inhabitants of the respective countries. An astonishing example of crossing the lines of subclasses is the presence of only one cotyledon in the embryo of *Carum bulbocastanum*. This species is, in all other respects, a typical member of the carrot family (Umbelliferae) which belongs to the dicotyledonous plants (Strasburger, 1906, p. 539).

Patterns of cleavage in the development of eggs may serve as the first zoological example since this subject was discussed previously in connection with symmetry and asymmetry of animals (Part III, Chapter 6, A, 1). The fact was mentioned that spiral phases of cleavage occur in many forms of animals, irrespective of taxonomic grouping, and that mature animals with bilateral symmetry may result from embryonic spiral stages. The type of cleavage seems of little consequence. However, remarkable resemblances of cell lineages were observed in spirally cleaving eggs of widely separated taxonomic groups. In gastropods, lamellibranchs, and annelids, the four first blastomeres form three quartets of cells which produce the ectoderm. The fourth aggregate, which is derived from the left posterior macromere, produces the mesoblast (Costello, 1955, p. 214).

Similarity of adult forms in spite of taxonomic and geographic separation is well illustrated by the occurrence in North America and in Africa of blackbirds with "shoulder epaulets." The American bird (red-winged blackbird) belongs to the meadowlark family in the genus *Sturnella,* while the African bird is a member of the weaverbird genus (R. and M. Buchsbaum, 1957, p. 69).

The phenomena of bioluminescence are widely spread across the borders of taxonomy. According to the survey of Harvey (1952) luminous species occur in approximately 28 out of a total of 80 classes of the animal kingdom. The classes with luminous species are distributed in 13 out of a total of 25 phyla. This survey includes animals which produce light by the metabolism of their own tissue cells, as well as animals which contain luminous bacteria as symbionts. Luminous species are contained in some classes of protozoans, sponges, annelids, mollusks, arthropods, and chordates. It seems that all species of Ctenophora, but only some species of Cnidaria, are luminous. Among the vertebrates, fishes are the only class with luminous species. One species in a genus may emit light, and another closely allied species may not. Harvey emphasizes that erratic distribution of luminescence makes it impossible to relate the origin of this phenomenon to phylogeny (evolution). However, there is a certain correlation between luminescence and habitat. The majority of luminescent animals are marine, and most of these live in deep water. Only two species occur in fresh water. This ecological rule is independent of taxonomic relationships: within a genus of protozoa, the dinoflagellates, only the salt water species is luminous. A moderate number of luminous species is terrestrial, the fireflies being the best-known representative of light-emitting beetles. In the plant kingdom, luminescence is limited to bacteria and to some species of fungi (in the subkingdom of Thallophyta).

This discussion of the crossing of taxonomic borders by single characteristics may be concluded with examples from the biochemistry and pharmacology of mammals. A well-known example is the similarity of purine metabolism in human beings, apes, and Dalmatian dogs. The end product of their purine metabolism is mainly uric acid, while allantoin is the predominant end product in all other mammals, including monkeys and dogs other than Dalmatians. Sulfonamides are metabolized differently in different species of mammals. Acetylation of sulfonamides takes place in human beings and rats, but not in dogs. Some rabbits acetylate, but others do not. It is not known whether this difference in rabbits is based on genetic factors.

Chapter 2

Classification of Cells, Tissues, Organs, and Systems

Cells, tissues, organs, and systems can be classified on the basis of *static-morphological, developmental,* and *functional* criteria. Static-morphological criteria cover both the appearance of isolated structures at a given moment, and topographic relations. Developmental characteristics are related to questions of origin and potencies and to phases of life cycles. The term functional criteria conveniently summarizes all physiological activities, including biophysical, biochemical, and immunological aspects (see Table 1).

A. Static Morphological Criteria of Classification

The concept of species as used in sexually reproducing organisms cannot be applied to somatic cells (tissue cells). Therefore it is fortunate that in conventional terminology somatic cells are classified into different cell *types.* At this point it will be useful to coordinate the conventional classifications with the descriptive classifications presented in Part IV (Figs. 54, 65, and 70 to 74).

I used descriptive terms such as high-prismatic and low-prismatic mosaic, or cytoplasmic and fibrillar network. Conventional classifications into "epithelial cells" or "connective tissue cells" may give the impression that isolated cells can be identified, whereas it is their mode of aggregation or their location which supplies the cue to identification. A mosaic pattern derived from mesenchyme and forming the wall of capillaries and the inner lining of other vessels is known as endothelium or vascular epithelium. Some investigators claim that they can identify isolated endothelial cells, and even endothelial cells cultivated *in vitro.* However, neither nuclear nor cytoplasmic characteristics distinguish single endothelial cells from reticulum cells or fibroblasts. It is the vascular situation which characterizes endothelial cells (Figs. 65e, 75, and 87) in the same way in which cornerstones may differ from the other building stones by their location only.

Tissues are classified conventionally into four main groups: epithelial tissue, connective tissue with its derivatives blood and lymph, muscle tissue, and nerve tissue. This classification is based on a mixture of descriptive (static-morphological) criteria with developmental and functional characteristics. Therefore it is not surprising that members of some classes are very heterogeneous from the morphological point of view.

The group of *epithelial tissues* is relatively uniform, since these tissues usually consist of a closely packed mosaic of cells, either in a single layer or stacked in several layers. From a topographic and functional point of view epithelia are divided into surface epithelium and glandular epithelium, but surface epithelia which line mucous membranes may have glandular functions. Forms of single-layer surface epithelia are illustrated in Fig 71; stratified surface epithelia are shown in Figs. 63 and 66. In exocrine glands single-layered epithelium may form the acinus and the duct (Fig 87A). Some endocrine glands consist of follicles formed by one layer of epithelium (thyroid, Figs. 25, 26 and 37), others consist of a stacked mosaic (parathyroid, adrenal cortex, Fig. 87B). The peculiar pattern of the liver, with walls or trabeculae which are two cells thick, is illustrated in Fig. 87C.

The *connective tissue* group comprises a great variety of morphological structures: stellate cells such as the reticulum cells (Figs. 65f and 75); ramified cells of the bone (osteocytes); spindle-shaped cells associated with fibers (fibroblasts and fibrocytes, Fig. 73b); stacked vacuolized cells of the fat tissue; cells with the tendency to occur in pairs, such as the cells of mature cartilage; and finally the cells of the lymphatic and hemopoietic system, which aggregate in indistinct morphological patterns.

Muscle tissue is characterized by one functional criterion, the high degree of contractility. Morphologically, skeletal and cardiac muscle are somewhat similar since both form cross-striated fibers, whereas smooth muscle has an entirely different appearance. *Nerve tissue* consists of branching cells. The branches may form short or long fibers, and the fibers may be arranged as a network or in bundles. The functional unit of the nervous system is the neuron, consisting of a nerve cell with axon. Supporting cells and fibers are called neuroglia (Fig. 73c and d). The ependyma consists of a one-layer cell mosaic which forms the wall of the central canal in the spinal cord (Fig. 58) and the lining of cerebral ventricles (Fig. 79a, structures crossed by arrows *3, 4, 5,* and *6*). Ependyma is classified as nervous tissue for histogenetic reasons, since it is a derivative of the neural tube.

Identification and classification of cells and intercellular substances depends, to some extent, on the techniques used. As discussed earlier (Part IV; Fig. 72a–c), technical factors influence the decision whether certain cell aggregates are to be considered a mosaic texture or a mere crowding of isolated cells. I mentioned that nuclear territories may not appear separated by cytoplasmic partitions (cell boundaries) under the light microscope, but may show such partitions under the electron microscope. The classification of neuroglia cells depends on the techniques used for the presentation of their ramifications and fibers. The identification of blood cells is based largely on staining reactions which indicate either chemical properties, or empirical characteristics. Some characteristics of white blood cells are more distinct in smears and others are more distinct in sections. Fig. 83A–C illustrates neutrophile myelocytes from the human blood or bone marrow. With standard techniques (Romanowsky-Wright

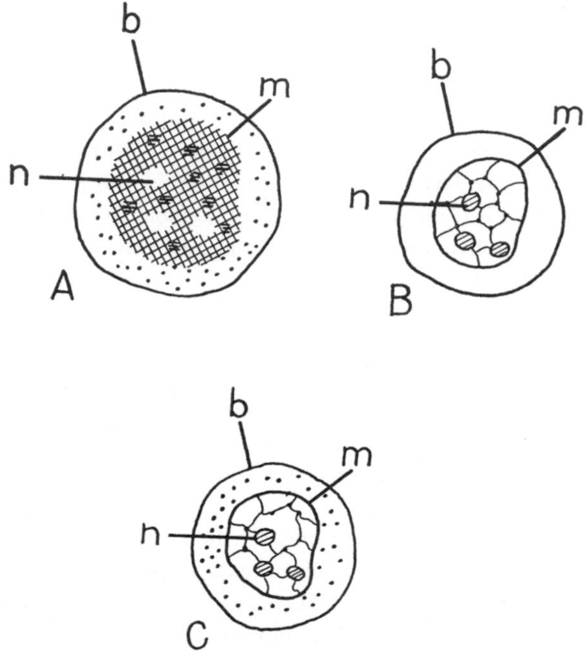

FIG. 83. Characteristics of human neutrophile myelocytes depending on treatment prior to staining. Diagrams based on Romanowsky-Wright stains. Results of different pretreatment: (A) appearance in smears, (B) usual appearance in sections, and (C) exceptional appearance in sections. *b*, Cell boundary, *m*, nuclear membrane, *n*, nucleolus. In the nucleus of the smear preparation (A) crosshatching and darker areas indicate variations in staining intensity. *Comparison:* Nuclear structures are most distinct in sections. Cytoplasmic neutrophile granules (fine dots) are distinct in smears, but rarely visible in sections.

stains) the delicate neutrophile granules are readily demonstrated in smears (A). The nuclear structures are cloudy and indistinct in smears, but distinct in sections (B). On the other hand, individual neutrophile granules are not visible, as a rule, in sections as indicated in diagram (B). For unknown reasons neutrophile granules may appear distinctly in a section and then the advantages of smear and section are combined, as illustrated in diagram (C). Although the combination (C) is exceptional, it is presented in textbooks without comments, as if it were the standard observation.

I recapitulate some *topographic aspects* in the classification of organs and systems (see Part III, Chapter 6, A, 2). The liver is a coherent-mass organ, in contrast to the bone marrow which is distributed through many regions of the body. The circulatory and nervous systems are coherent though widely rami-

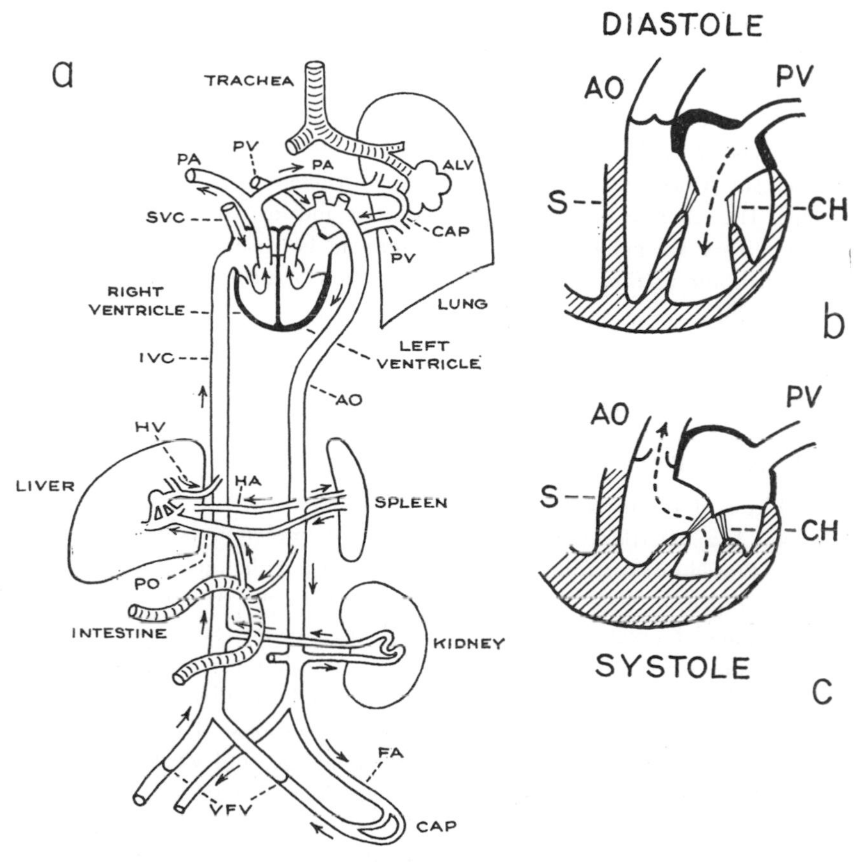

FIG. 84. Diagram of the circulation in thoracic and abdominal viscera, and in legs of man (a), and of two phases of contraction of left ventricle (b and c).

(a) Superior and inferior vena cava (*SVC* and *IVC*) carry blood to right atrium. Venous blood flows through funnel-like valve to right ventricle, from here through semilunar valves and pulmonary arteries (*PA*) to lungs (*ALV*, alveoli). After oxygenation in pulmonary capillaries (*CAP*) arterial blood flows through pulmonary veins (*PV*) to left atrium, left ventricle, and aorta (*AO*). *PO*, portal vein; *HA*, hepatic artery; *HV*, hepatic vein. *FA*, femoral artery leading to capillaries (*CAP*) in leg; *VFV*, valves of femoral veins.

(b) and (c) Solid black: wall of left atrium. Shaded: septum (*S*) and wall of left ventricle with projecting papillary muscles. *CH*, chordae tendineae connecting tips of papillary muscles with margin of cusps of atrioventricular valve; *AO*, aorta; *PV*, pulmonary vein. *Diastole:* atrium contracts, ventricle relaxes, atrioventricular valve opens, aortic valve closes. *Systole:* atrium relaxes, ventricle contracts, atrioventricular valve closes, aortic valve opens; contracting papillary muscles prevent atrioventricular valve from turning inside out.

fied, while the members of the endocrine and reticuloendothelial systems are separated. Diagrams of the various systems are found in textbooks of macroscopic anatomy, histology, and physiology. In the present book the circulatory system (Fig. 84) was selected to illustrate one of the systems.

The concepts of *organs and systems* are based on functional characteristics. In the kidney, for instance, epithelial tubules, blood capillaries, nerves, and connective tissue are integrated to perform the special excretory function of the whole organ. The two kidneys plus the system of ducts which control the discharge of urine are known as the urinary system (kidneys, renal pelves, ureters, urinary bladder, and urethra). Systems with different functions may have a common embryological history and therefore maintain close topographic associations. The relations between the reproductive and urinary systems are a good example. They are not only closely adjoining in both sexes, but in the mammalian male the urethra serves both the excretion of urine and the ejaculation of sperm.

B. Developmental Criteria of Classification

1. *Origin and Potencies*

Problems of origin and potencies of cells and tissues are not restricted to embryonic life, but continue through all phases of postembryonic life including old age. Procedures for following the fate of cell populations in the embryonic and adult organism were described earlier (see Figs. 30, 32, and 33). Experimental studies of potencies were mentioned in chapters dealing with stimulus, response, and reactivity (Fig. 4), explantation techniques (Fig. 19), living models, and the determination of polarity of whole animals and their parts. For the present discussion let us investigate two questions. To what extent can somatic cells be classified on the basis of their lineage, and what is the present state of the germ-layer doctrine? The two questions overlap to some extent.

a. *Classification of somatic cells on the basis of lineage.* In an earlier part I emphasized the difficulty of establishing cell lineages in vertebrates in contrast to the favorable situation in small invertebrate organisms with limited cell numbers, such as rotifers (Part III, Chapter 3, A). As far as human tissues are concerned, special efforts have been made to determine the development and potencies of blood cells and their forerunners in the bone marrow. Detailed pedigrees of these cells are found in textbooks of histology and hematology. My diagram, Fig. 29, illustrates possible developmental interpretations of transitional pictures of blood cells. The interpretation of three phases of a red blood corpuscle appears to be much safer than that of three phases of granulated white cells. My diagram, Fig. 85, covers the whole development of a red blood corpuscle and of an eosinophilic leucocyte. As soon as hemoglobin can be recognized, a cell can be identified as B, an early phase of red blood corpuscles; and as soon as eosinophilic granules become visible, the cell can be classified as B′, an early stage of eosinophilic leucocytes. It is reasonable to assume that the stages which precede B and B′ are so uncharacteristic that they are indistinguishable. This is illustrated by the two pictures A and A′. However, from their association with other cells, one may infer the series to which a cell of type A or A′ belongs. In an area of the bone marrow with numerous B and C cells, any A cells present would be interpreted as the earliest stage of red blood corpuscle formation. If blood smears of a patient with severe anemia show many B and C forms, one would be inclined to associate A types with these forms. Similarly A′ cells would be considered as potential eosinophilic leucocytes if associated

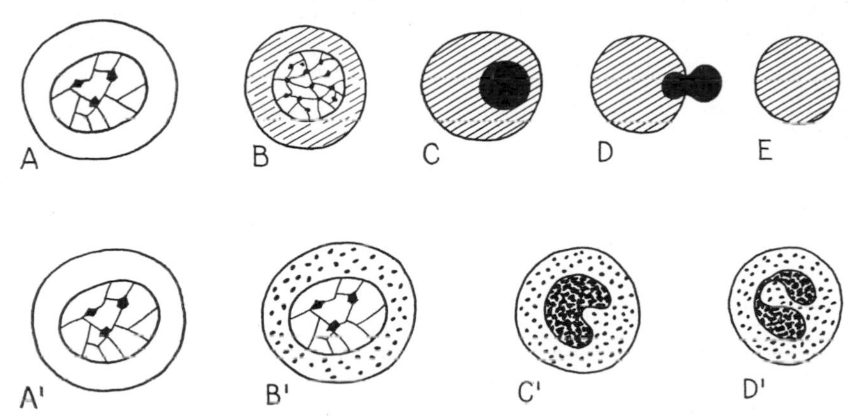

FIG. 85. Development of a red blood corpuscle (A to E), and of an eosinophilic leucocyte (A′ to D′). Maturation from left to right. B to E: shading indicates hemoglobin. B′ to D′: dots indicate eosinophilic granules. *Interpretation of the most immature stages:* A and A′ are morphologically indistinguishable but may be classified by inference from their spatial and chronological association with more mature forms (see also Fig. 29).

with B′ and C′ types in localized areas of the bone marrow or in smears from the blood of a case of eosinophilic leukemia.

In the examples above the potentiality of least mature cells was inferred from their associations with more mature forms on the basis of qualitative criteria. Potentialities which cannot be derived from the appearance of the individual blood cells may be suggested by numerical configurations. The normal composition of human blood cells is approximately 70% neutrophiles, 25% lymphocytes, and 5% other white cells. Suppose blood smears from two patients with leukemia show a number of very immature cells, which I will call X cells. Patient No. 1 has 20% neutrophiles, 40% myelocytes (immature neutrophiles), 10% lymphocytes, and 30% X cells. The X cells would be interpreted as forerunners of myelocytes. Patient No. 2 has 10% neutrophiles, 60% lymphocytes, and 30% X cells. In this case the X cells would be considered immature phases of lymphocytes.

Lineages of somatic cells have been investigated *in vitro,* as discussed earlier in connection with the appraisal of tissue culture techniques (Part III, Chapter 5, C, 2). The main points may be recapitulated here. Clones obtained by cultivation from single cells proved to be useful tools for physiological and immunological studies. However, a new complication was added to the problems of cell lineages, when single-cell clones dissociated into different strains although the conditions of cultivation had not changed. As a result of explantation, characteristics of tissue cells *in situ* may disappear, or remain unchanged, or be accentuated. Certain tumors whose cells cannot be identified in sections produce clearly recognizable nerve cells when cultivated *in vitro.* It is reasonable to classify such tumors on the basis of the *potency* of their cells to form nervous tissue. In other instances explantation produces cell forms and patterns of aggregation which cannot be identified as any known type of cells *in situ.* This difficulty has been emphasized by Parker *et al.* (1957), particularly with reference to cells which form undifferentiated networks or mosaics when settled on a surface. Cells of these two types are known as "fibroblasts" and "epithelial cells" in the tissue culture literature. Some tissue culturists prefer the terms "fibroblast-like" and "epithelial cell-like." I recounted earlier that alterations of cell types occurred either as a result of different culture media, or for unknown reasons (Figs. 72b, 76, and 77).

Many lineages of somatic cells are probable but not proved. Liver cell cords and small bile ducts are continuous, as shown in the diagram Fig. 87C. It has been controversial whether strands of undifferentiated cells in cirrhotic livers of mammals should be interpreted as atrophic liver cell cords or as regenerating bile ducts. Abercrombie and Harkness (1951) subjected rats to partial hepatectomy and subsequently explanted fragments of the regenerating livers. Circumstantial evidence made it seem probable that most of the epithelial cells which appeared in the cultures originated from bile ducts and only few from liver cords. However, direct demonstration of the lineages was not possible.

In an earlier chapter it was stated that different types of nuclei can be distinguished on the basis of shapes and chromatin distribution (Figs. 61, 64, and 65). Differences between nuclear types are helpful in identifying tissues, but the *groupings* of nuclei rather than the properties of individual nuclei characterize a tissue. Some unusual forms of nuclei occur only in one type of cells or tissues. Examples are the segmented nuclei of neutrophile leucocytes (Fig. 53), and the branched nuclei in spinning glands of caterpillars (Fig. 62).

In one stage of frog embryos Briggs and King (1955) observed marked morphological differences between interphase nuclei of blood cells, precursors of muscle cells, cells of the lens, the cornea, and the retina. Figure 1 of Briggs and King illustrates these differences, but the variability of each class of nuclei is not indicated. Differences between epithelial nuclei in the intestine, the kidney, and the ureter of Proteus were claimed by Heidenhain (1907); his Figs. 43 to

45 represent a few samples. If it could be demonstrated that nuclei are tissue- or organ-specific, many investigations of somatic cells would gain a solid foundation. However, three points need consideration: variations within the life cycle of nuclei (my Figs. 64 and 85), stability of characteristics under different environmental conditions, and statistical significance of differences between nuclear populations. Recent investigations do not indicate any biochemical differences between nuclei of different cells in the same organism. As Schechtman and Nishihara (1955) put it, "neither the chemical nor the serological data are sufficient to allow the conclusion that DNA or any other macromolecular constituent of the nucleus is organ- or tissue-specific." Immunological specificity of cells of different organs and tissues, which will be discussed later, seems to be restricted to the *cytoplasm.*

b. *Germ layers and their potencies.* Germ layers are transient structures which form at the end of the gastrula stage. Figure 3e′ and f′ and Fig. 32a–d show how, in amphibians, the original single layer of cells produces two layers by a process of migration and infolding of cell aggregates. The outer layer is called ectoderm (shaded in Fig. 32d) and the inner layer mesoderm (dotted in Fig. 32d). In a subsequent stage a third layer, known as entoderm, forms under the mesoderm. In some animals, such as *Amphioxus,* ectoderm and entoderm form first, and the mesoderm develops later between the two other layers as the result of a folding process of the entoderm (for illustrations of *Amphioxus* development see Huettner, 1949, Figs. 32–39). Mammals do not show a uniform behavior. In early stages of the pig embryo some cells of the blastodermic vesicle become detached and migrate toward the central space of the vesicle. These are the initial cells of the entoderm. The mesoderm forms last. In man the earliest stages are "telescoped" to the extent that, in the youngest known human embryo, the three germ layers are already established (Patten, 1953). In spite of these differences, there are two important features of the germ layers shared by all classes of animals: (1) the topographic separation into germ layers is a decisive step in the scaffolding of the embryo; (2) in the usual course of events certain organs and systems develop from each of the three germ layers.

The epithelial linings of the skin and of the openings of the digestive and respiratory systems are derivatives of the ectoderm. The bulk of the epithelial linings of the digestive and respiratory systems originate from the entoderm. Some epithelial structures of the genital-urinary system, such as the tubules of the kidneys and testes, are derivatives of the mesoderm. Evidently, mosaic patterns (epithelium) can be produced by each of the three germ layers. Similarly, each of the three layers can produce network and fiber patterns, as two examples will show. The supporting tissue of most organs (connective tissue) is derived from the mesoderm, but that of the central nervous system (neuroglia) is derived from the ectoderm. The cytoplasmic network in lymph nodes and bone marrow originates from the mesoderm, and a similar network in the thymus is produced by the entoderm.

As discussed earlier, most embryonic parts are endowed with more potencies than those which become apparent under ordinary conditions. Within the same germ layer distribution of certain potencies is much wider than one would expect. Figure 4 illustrates the unusual production of a lens from abdominal ectoderm, quite remote from the area of the head. Considerable overlapping of potencies between the three germ layers is revealed by creating conditions different from the normal development. This was done in the classic experiments on the regeneration of the lens (Colucci, 1891; G. Wolff, 1895). Normally the lens forms from the ectoderm (see Fig. 3k and l). However, when the lens of *Triton* larvae was extirpated, a new lens developed from the margin of the iris, which is a derivative of the mesoderm. Observations of this type invalidated the doctrine of "specificity of germ layers," which had been accepted for a long time. An excellent analysis of this subject is found in Oppenheimer's 1940 article "The Non-specificity of the Germ Layers." Mangold (1925) and Rugh (1948, p. 465) probably went too far by considering the germ layers to be merely temporary topographic patterns without any special biological properties. Experiments of Holtfreter (1939) showed differences in mutual adhesiveness between the three germ layers at the blastula and early gastrula stages of amphibians. In the first experiment loosely joined fragments of ectoderm and entoderm, without any admixture of mesoderm, were explanted. The two cell masses fused temporarily, but separated subsequently, though the cells of the same germ layer remained joined. In the second experiment fragments of ectoderm and entoderm were explanted with a thin layer of mesoderm between them. As a result the ectodermal and entodermal pieces remained joined indefinitely, forming a vesicle which enclosed the mesoderm cells. To some extent the mesoderm cells *in vitro* simulated their function *in situ,* which is to tie ectoderm and entoderm together.

Tissues and cells in postembryonic life are frequently classified according to their derivation from germ layers. It is customary to summarize reticulum cells, endothelial cells, fibroblasts, lymphocytes, plasma cells, and eosinophile leucocytes as mesenchymal cells which are derived from the mesoderm and endowed with immunological functions. There is no objection to this convenient label as long as it is understood that cell antigens are also produced by ectodermal and entodermal derivatives. This aspect is important in the reticuloendothelial system (Part V, Chapter 2, C, 1).

2. *Life Cycles of Tissue Cells*

Impressive qualitative changes occur during the life cycle of metamorphosing animals, such as the transformation from tadpole to frog, or from caterpillar through chrysalis to butterfly. The development of a bird or mammal from the fertilized ovum to the adult organism also comprises considerable changes though less abrupt than in metamorphosing animals. One would not expect changes of a similar magnitude during the life cycle of a single somatic cell, yet remarkable transformations do occur. There is quite a difference between a prismatic cell in the embryonic neural tube and its final stage which may be a ramified nerve cell with an axon of many feet in length.

In order to identify and classify a biological structure, it is necessary to know the variations of the structures within its life cycle. This is illustrated in Fig. 64, which shows that a mature nucleus of one type may be indistinguishable from an immature nucleus of another type. The main characteristic of mature tubercle bacilli, resistance to destaining by acid, is absent in young forms of bacilli soon after division. For some time, this difficulty in identifying young tubercle bacilli made it impossible to follow the fate of populations of tubercle bacilli in the host (see Part III, Chapter 3, A). Various aspects of cellular life cycles were mentioned previously in connection with mitotic and intermitotic duration (Fig. 60) and with mitotic and intermitotic metabolism (Part II, Chapter 2, H, 2). The contrast was stressed between meristematic cells which, in the postembryonic organisms, have kept the ability of dividing; and non-meristematic cells, which have lost this ability. The topographic distribution of meristematic cells was given in detail (Part III, Chapter 6, A, 3).

The basal cells of the epidermis are examples of immature meristematic cells, while liver cord cells represent mature meristematic cells. Sebaceous gland cells which are discarded in the process of secretion (holocrine cells), red blood corpuscles and polymorphonuclear leucocytes reach the climax of their functional activity and usefulness for the body when they approach the end of their life span. Artificial arrest of mitosis caused by colchicine was mentioned earlier. Natural arrest of mitosis produces giant chromosomes in larval organs of insects, particularly salivary glands. The giant chromosomes are approximately 100 times larger than ordinary chromosomes (Painter, 1934, Fig. 1). The giant chromosomes are fully developed in the late prophase, beyond which the mitosis does not proceed. When it turned out that the giant chromosomes resembled the normal chromosomes band for band, it became possible to tackle many problems of cytogenetics with the help of the enlarged units of the giant chromosomes. However, nothing seems to be known concerning the physiological role of the giant chromosomes and the giant cells in which they are located (Alfert, 1954, p. 162).

Cowdry distinguished between four "kinds of cell lives." He refers to mature cells which have lost their ability of dividing, as "fixed postmitotics" (nerve cells, leucocytes). Mature cells which are not found in division, as a rule, but are known to divide under special conditions, he calls "reverting postmitotics" (liver cord cells, cells of renal tubules). Immature cells which continuously reproduce are "vegetative intermitotics" (basal cells of the epidermis, stem cells in the bone marrow). Maturating cells with mitotic activity are called "differentiating intermitotics" (prickle cells of the epidermis, myelocytes of the bone marrow). For a detailed description see Cowdry's "Textbook of Histology" (1950, pp. 44ff., Fig. 7).

May I state explicitly there is *no antagonism between mitosis and cellular differentiation.* Dawson (1937) devoted a special article to the persistence of cytoplasmic differentiation during mitosis. Among the persisting structures he listed brush borders, cilia, myofibrillae and the granules of myelocytes. Similarly biochemical differentiation of cells does not remove their ability to divide. Dawson illustrates this point by hemoglobin-containing immature stages of red blood corpuscles, and I like to add the examples of cells of renal tubules and liver cords which can still divide. Not only completed differentiation, but also the process of differentiation is compatible with mitotic division. As a matter of fact, mitosis may be an instrument of differentiation. In a stratified squamous epithelium (stacked mosaic, Figs. 63 and 66) the basal layer contains the most immature cells. When a cell of this layer divides, either the two daughter cells remain in this layer, or one of them moves "upward," toward the free surface. Each cell which moves upward enters the process of differentiation. In the epidermis an original basal cell differentiates into a prickle cell, and the prickle cell into a keratinized cell representing the end of the life cycle. Mitoses leading to differentiation are properly termed *differential mitoses.* This concept was introduced by Koehler (1932) in studies of the development of wing patterns in butterflies.

From Cowdry's classification it becomes apparent that many or few life cycles of somatic cells may occur during the life cycle of the whole organism. Let us consider the life cycle of a mammal, starting with the fertilized ovum and concluding with the death of the individual. During this period the nerve cells of the central nervous system complete *one life cycle,* since these cells are not replaced between birth and death of the individual. In tissues which are subject to wear and tear new cells form and die continuously, which means that *many life cycles* of these cells are included in the life cycle of the individual. Examples of this type are the epidermis, the lining of the mucous membranes, and the blood-forming tissues. Finally there are tissues with periodic renewal, such as the inner layers of the uterus and vagina which are renewed at various intervals, depending on the type of animal. It is remarkable that in some animals, such as rotifers, the body cells do not multiply and are not replaced during a lifetime of the organism (Part III, Chapter 4, C).

Artificial life cycles are produced *in vitro* by transferring cultivated populations of bacteria or tissue cells from an old to a fresh medium. If tissue cells are cultivated in such a way that they form well-defined colonies, each transfer involves a reduction in colony size with subsequent regeneration of size and shape (see Fig. 21).

Procedures for determining the turnover of populations were discussed earlier (Part III, Chapter 3, A and B).

C. Physiological Criteria of Classification

An earlier chapter was devoted to the various relations between functions and visible structures in organisms (Part II, Chapter 2, G). Function-structure aspects pertaining to classification will be discussed here.

1. *Functional Structures Occurring in Several Organs*

a. *Glandular structures.* Blood vessels are functional structures present in most organs. Exceptions are retina, lens, cornea, and cartilage, which have no blood vessels (avascular tissues). Interfaces through which fluids or gases are exchanged may either serve the general metabolism of an organ or may be connected with specific functions. As part of the respiratory mechanism, pulmonary interfaces control the exchange of gases between alveolar airspace, alveolar lining, and blood capillaries of the alveolar wall (Fig. 82). Interfaces in the central nervous system and the potential sites of exchange were described earlier (Figs. 79, 80, and 81). In contrast to the blood capillaries of the brain, the capillaries of glands do not seem to possess selective permeability. Therefore, in glandular organs, the exchange between blood capillaries and glandular cells is controlled entirely by the glandular cells.

Let us consider the kidney as the representative of a gland. Figure 86 is a diagram of a nephron, the functional unit of the kidney. Exchange of material can occur at the following interfaces: between glomerular capillaries and the space of Bowman's capsule; between the bore (lumen) of first convoluted tubules and the cells of the tubules; and finally, between cells of the tubules and the peritubular capillaries. Potential spaces between the tubules and capillaries were mentioned previously. The substances contained in the blood reach the glomerulus through the afferent vessel. From the glomerular capillaries a portion of each substance leaks into the space of Bowman's capsule, thus contributing to the *glomerular urine.* The other portion travels along in the blood stream leaving the glomerulus through the vas efferens. By way of the peritubular capillaries each substance has the topographic opportunity to enter tubular cells from the outside. However, tubular cells have selective properties, both with respect to the substances in the glomerular urine and those in the peritubular capillaries. This is illustrated in Fig. 86 by following the fate of five substances. Water and sodium ions are reabsorbed in varying amounts, from the bore (lumen) of each tubular segment, and discharged into peritubular capillaries. Enough water and sodium ions remain in the tubular bore to form the final urine. Glucose, on the other hand, is almost entirely reabsorbed by the first convoluted tubules and is returned

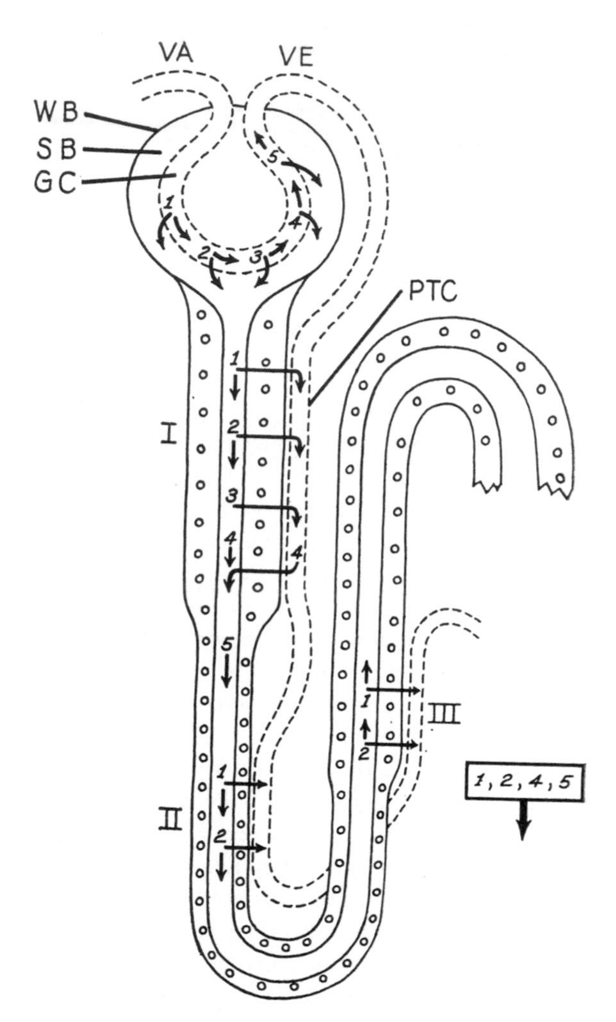

FIG. 86. Functional diagram of a human nephron showing the fate of five substances. *Glomerulus: VA,* afferent vessel (blood enters glomerulus); *VE,* efferent vessel (blood leaves glomerulus); *GC,* glomerular capillary (a single loop representing the capillary tuft); *WB,* wall, and *SB,* space of Bowman's capsule. *Tubular segments:* Open circles indicate nuclei of tubular cells. *I:* First convoluted segment (receives glomerular urine). *II:* Thin segment of Henle's loop. *III:* Thick segment of Henle's loop. *PTC,* peritubular capillary close to *I, II,* and *III;* potential spaces between capillary and tubule are disregarded. *Fate of five substances* indicated by arrows which represent net transfers (balance sheets) without other quantitative connotations: *1,* water; *2,* sodium ions; *3,* glucose; *4,* Diodrast; *5* inulin. Note: The last part of the nephron, the collecting tubule, is omitted. The final urine which is excreted into the renal pelvis contains four of the five substances mentioned here: *1, 2, 4* and *5* as indicated by the rectangle with arrow.

to the circulation. Diodrast, a radiopaque contrast medium, is completely eliminated from the circulation, since it is not reabsorbed from the glomerular urine and, in addition, is transported from the peritubular capillaries into the bore of the tubule. This an instance of tubular secretion (or excretion). None of the substances which are secreted by tubular cells occur in the body under natural conditions. Inulin, a polymer of fructose, is neither reabsorbed from the glomerular urine, nor added to the glomerular urine by tubular secretion. This is the basis for using "inulin clearance" as a measure of the glomerular filtration rate. If P, the concentration of a substance in the plasma, and U, the concentration in the urine, are expressed in the same units, and V is the flow of urine in milliliters per minute, the clearance in milliliters per minute is UV/P.

Two or more nephrons of the type shown in the diagram join to form a collecting tubule which possibly has some reabsorbing function also, but mainly collects the urine. The collecting tubules discharge the final urine through the papilla into the renal pelvis.

A glandular cell receives the same raw material from the blood as other cells receive, but utilizes the material in two different ways: part of the material serves the general metabolism of the particular cell, and part is handled in a specific way in the interest of the body as a whole. The specific functions of the kidney can be grouped as follows: (1) excretion of metabolic waste products, (2) maintenance of proper salt and water concentrations in the body, and (3) reabsorption of useful material, such as glucose, which leaks through the glomerular capillaries. Since the characteristic product of the kidney, the urine, is discharged from the body through a system of ducts, the kidney is an *excretory* exocrine gland. Most exocrine glands have a *secretory* function, which means that their end products are utilized within the body. The pancreas is an example of a secretory exocrine gland. Different structural levels of the pancreas are illustrated in Fig. 57A–G, with omission of the vascular relations. These relations are included in Fig. 87A, which shows the general pattern of an exocrine gland.

An endocrine function is also carried out by the kidney, namely the production of renin, probably localized in the juxtaglomerular body. In an earlier discussion of various relations between functions and visible structures (Part II, Chapter 2, G, 3), the juxtaglomerular body, the nervous part of the hypophysis, and the adrenal medulla illustrated hormone production by structures which differ from typical endocrine glands. Typical endocrine glands consist of a mosaic of cells having extensive contacts with blood capillaries. Solid masses of stacked cells are characteristic of the parathyroid and the adrenal cortex. This pattern is shown in Fig. 87B. In the thyroid, the glandular cells form the wall of alveoli which enclose colloid-containing spaces (Figs. 25, 26, and 37).

The raw material which reaches the endocrine cells through the blood capillaries is transformed into specific hormones. The iodides of the blood probably enter most cells of the body and are discharged back into the blood at a similar rate. The cells of the thyroid gland are distinguished by their capacity to store iodine by transforming the iodides into organic compounds (thyroxine, thyroglobulin). Similarly, the adrenal cortex makes and stores steroids from materials supplied by the blood stream. The thyroid is unique in that it possesses a special store house for the hormone other than the cells, namely, the colloid in the alveoli.

The liver is a mixture of an exocrine and endocrine pattern. At low magnification diagram Fig. 42 illustrates the extensive contact between trabeculae (liver

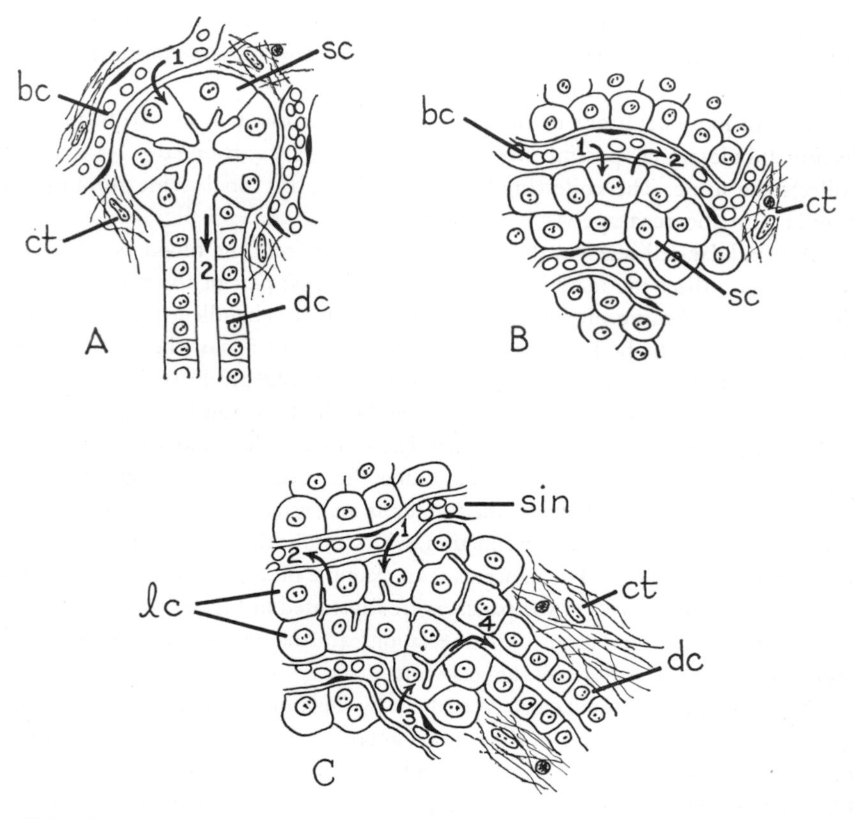

FIG. 87. Diagrams of basic patterns of glands. (A) Typical exocrine structure. Arrow *1*: Raw material moves from the blood capillary (*bc*) into a secretory cell (*sc*). Arrow (2): Transformed material is discharged through a duct (*dc*, duct cell). (*B*) Typical endocrine structure. Arrow *1*: As in (A). Arrow 2: Transformed material is discharged from secretory cells (sc) into the circulation (*bc*, blood capillary). (C) The liver, a mixed exocrine and endocrine structure. Arrows *1* and 3: Raw material moves from blood stream (*sin*, sinusoids) into liver cord cells (*lc*). Arrow 2: Transformed carbohydrates, fatty substances, and nitrogen-containing compounds are discharged into the circulation (nutritional material plus waste products, but no specific hormones). Arrow *4*: Breakdown products of hemoglobin (bile pigments) are combined with bile salts in the liver cord cells to form bile which is secreted from intra- and intercellular bile capillaries into bile ducts (*dc*, duct cells). Note: In all pictures potential spaces between blood capillaries and gland cells are shown; *ct* indicates connective tissue.

cell cords) and sinusoids and also the general topography of blood vessels and bile ducts. Figure 87C shows the transition from the trabeculae to the bile ducts, as seen at higher magnification. Materials present in the blood, such as glucose and amino acids, enter the liver cord cells (arrows *1* and *3*). The glucose is stored as glycogen, and the amino acids are synthesized to homologous proteins. On demand, glycogen is broken down into glucose by the liver cells and released into the blood for use in other places of the body (arrow *2*). Excessive amounts of amino acids are desaminated and the end product of this process, urea, is discharged into the blood (arrow *2*) for subsequent excretion by the kidney. The exchange of substances through the interface between liver cord cell and capillary wall resembles the transport mechanisms of endocrine glands. However, the specific products of the liver, such as glycogen and proteins, are fuel or construction material, whereas the specific products of endocrine glands are trigger substances. Concerning bile production the structure and function of the liver is that of a typical exocrine gland. The raw materials are bile pigments which are breakdown products of hemoglobin and enter the liver cord cells from the blood stream (arrow *3*). In the liver cord cells the bile pigments are combined with bile salts to form bile, which is secreted from intra- and intercellular bile capillaries into the bile ducts (arrow *4*). These ducts are continuous with the liver trabeculae, as shown in the diagram. The hemoglobin from worn-out red blood corpuscles is altered into bile pigment by metabolic activities of the reticulo-endothelial apparatus. Part of this apparatus are the endothelial cells which line the sinusoids of the liver. The fact that cells of blood capillaries prepare material for gland cells is exceptional. As a rule, the activities of the capillary wall are unspecific, while the gland cells have specific functions. This is well documented in the kidney (Fig. 86).

b. *Mucous membranes*. Mucous membranes are composite structures which form the inner layer of the digestive tube, the respiratory tract (including the small bronchi), parts of the reproductive and urinary systems, and finally the eyelids. Each mucous membrane consists of the epithelium, which is the surface layer, the tunica propria, and, in most instances, a muscularis mucosae. Figure 33 illustrates an intestinal mucous membrane in which the epithelium is formed by a mosaic of high-prismatic cells. In the mucous membrane of the oral cavity the surface layer consists of a stacked mosaic (Figs. 63 and 66); a similar type occurs in the esophagus and vagina. The tunica propria is a composite of connective tissue, blood capillaries, and varying amounts of migratory cells, especially lymphocytes; the connective tissue is mostly a cytoplasmic or reticular network. The components of the tunica propria are similar to those of embryonic mesenchyme (young connective tissue) and of granulation tissue which forms in wound healing and in chronic inflammation. The muscularis mucosae is a thin layer of smooth muscle fibers (not shown in Fig. 33). It separates the mucous membrane from the next layer which, in most instances, is loose connective tissue (submucosa).

All mucous membranes share the function of protecting against mechanical injuries the organ which they coat. The main protective factor is the mucus produced by cells of the lining epithelium or by glands which are located in the submucosa and whose ducts open onto the surfaces of the mucous membranes. Since the surfaces of mucous membranes are subject to continuous wear and tear, they show a high turnover of epithelial cells (Fig. 33). Besides their general protective function, the mucous membranes possess a variety of other activities with or without corresponding differences in structures. In the respiratory tract the epithelium is equipped with cilia which transport disturbing material away from the lungs. The mucous membranes of the stomach and intestines secrete not only mucous, but also digestive enzymes. In addition, the mucous membrane of the intestine controls absorption of digested food. In the tunica propria of some mucous membranes the concentration of lymphocytes can be very high. Lymph nodules are found in the tunica propria of the tonsils and various parts of the intestine. They are particularly conspicuous in the human appendix. Some mucous membranes can be classified as lymphatic tissues. The extent of lymphatic tissue in intestinal mucous membranes depends on environmental factors, such as food and bacteria. In the cavity of the uterus the mucous membrane (endometrium) is the site of implantation of the fertilized ovum. The placenta is a joint product of the transformed endometrium and the membranes of the embryo.

This survey may suffice to illustrate the unity and diversity of mucous membranes.

c. *The lymphohemopoietic system.* The topographic patterns of bone marrow and lymphatic tissue were described previously as discontinuous and scattered over many regions of the body (Part III, Chapter 6, A, 2). The lymphohemopoietic system is not the same at different periods of the life cycle and under different conditions of the internal and external environment. In the human embryo the formation of red blood corpuscles and granulocytes is concentrated, for some time, in the sinusoids of liver and spleen. After the fifth month hemopoiesis gradually shifts to the bone marrow, which becomes its only site by the time of birth (Wintrobe, 1951, Fig. 1). In the healthy adult active marrow is restricted to spongy bone, while fat tissue fills the cavities of long bones. In small animals most bones contain active marrow. In addition the spleen continues to be a blood-forming organ in adult rats and mice. Lymphatic tissue is found in the form of lymph nodes, or as nodular and diffuse aggregates of lymphocytes in mucous membranes and the spleen. The reason why hemopoietic and lymphatic tissues are mentioned here is that, under abnormal circumstances, they may occur in organs which normally do not contain such structures. Extramedullary hemopoiesis, which means formation of red blood corpuscles or granulocytes or both outside the bone marrow, may be a regenerative response to destruction of bone marrow by metastases of cancer. Lymphatic tissue frequently forms as a response to inflammatory stimuli. In the course of chronic inflammation the mucous membrane of the renal pelvis may eventually be studded with well-developed lymph nodules (M. Jacoby, 1927).

In myeloid leukemia the sinuses of liver and spleen are frequently filled with immature granulocytes, thus resembling the condition in the embryo. In practically any organ, accumulations of leukemic cells of the myeloid or lymphoid type can develop.

d. *Reticuloendothelial system.* Like the bone marrow and lymphatic tissue, the reticuloendothelial system belongs to the morphologically discontinuous organs. The peculiarity of the reticuloendothelial system is that its components become conspicuous only in response to special stimuli. If particles of a certain size come in contact with endothelial cells of blood or lymph vessels, or with reticular cells in lymphatic tissue, the particles will be phagocytosed. This process may be associated with a marked increase in volume of the phagocytosing cells. Sessile cells with the capacity of storing particulate material constitute the reticuloendothelial system. Some migratory cells which phagocytose particles are considered to be members of the reticuloendothelial system. Large mononuclear cells are usually included because of their probable transformation from and into sessile forms; polymorphonuclear leucocytes are generally excluded. Sessile cells with storage capacity are particularly prominent in spleen and liver. The endothelial cells of the liver sinusoids assume a stellate form when stuffed with particles (Fig. 75). Particulate material which stimulates the reticuloendothelial system may originate inside the body, such as red blood corpuscles and malaria pigment, or the material may have been injected artificially, such as colloidal suspensions of India ink, carmine, and trypan blue.

The reticuloendothelial system removes particulate material from the circulation and participates in the formation of antibodies. The formation of antibodies by this system was suggested by results of so-called blockade techniques. Large amounts of India ink or colloidal iron were injected intravenously. This was followed by injection of antigenic material. Animals treated in this way produced fewer antibodies than control animals which received antigenic material only. Indirect support of antibody production by the reticuloendothelial system came from experiments with Roentgen rays and nitrogen mustards. Both types of injury depress storage activity of the reticuloendothelial system and also antibody formation. Finally, a direct demonstration of localization of antigen-antibody reactions in reticuloendothelial cells became possible through the fluorescent antibody method which was described previously (Part III, Chapter 7, B, 2). A brief summary of immunological studies of the reticuloendothelial system is given by Wilson and Miles (1955, pp. 1254–1257).

Immunological changes after removal of the spleen or after irradiation of bone marrow are frequently ascribed to changed reactivity of the reticuloendothelial system. This seems to be justifiable in some, but not in all, instances.

Although the reticuloendothelial system plays a prominent role in production of antibodies and phagocytosis of particulate material, these capacities are not restricted to the members of the reticuloendothelial system. Phagocytosis of red blood corpuscles by liver cord cells has been observed in a man dying of liver cirrhosis (Rössle, 1907) and in dogs after experimental anaphylactic shock (Weatherford, 1935). Cell antigens and antibodies to cell antigens are probably produced by all cells of the body.

e. *Erectile structures.* One should differentiate between erectile organs and organs with variations in blood content without marked changes in organ volume. The blood content of all organs varies in response to changes of the external or internal environment (Part III, Chapter 6, A, 3). In a skeletal muscle the number of open (active) capillaries per unit of volume is 200 times greater after a period of work than after a period of rest (Krogh, 1922). Erectile organs are characterized by the fact that they consist mainly of blood vessels. Since these blood vessels have the capacity to change caliber, marked swelling of the organs can take place in a short time. The maintenance or cessation of the swollen state is controlled by special mechanisms. In the human body erectile structures are present in the large corpora cavernosa of the penis, the small corpora cavernosa of the clitoris, and in the conchae of the nose (see Körner, 1937, for human nose; Swindle, 1935, for other mammals). Particularly in dogs the spleen resembles an erectile organ. Methods for observing variations in the volume of spleens of dogs were described in detail, and it was stated that 20% of the total blood volume can be accommodated in the spleen of dogs.

f. *Parenchyma and stroma.* The traditional classification into parenchyma and stroma is based on the idea that some structures serve the *special* function of an organ (parenchyma), while other structures, such as blood vessels and connective tissue, carry out *general* functions and therefore are present in each organ (stroma). In the heart, the muscle fibers are considered the parenchyma, while the connective tissue between them is the stroma. In the liver, the trabeculae (liver cell cords) and bile ducts are the parenchyma, while the blood vessels and connective tissue form the stroma. However, the stroma in each of the two organs has its peculiarities. In rheumatic fever the stroma of the heart muscle forms special structures (Aschoff's nodules). The lining cells (endothelium) of the sinusoids in the liver are members of the reticuloendothelial system, as described before. Like other members of this system they phagocytose particles and produce bile pigment from the hemoglobin of worn-out blood corpuscles. The particular tendency of endothelial cells in the liver to transform into stellate cells (Kupffer cells) was mentioned (Fig. 75). The permeability of blood capillaries of the central nervous system differs markedly from that of other organs, as discussed previously (Part IV, Chapter 3, C, 2). The study of interfaces in the lung, kidney, and liver leads to the conclusion that the capillaries are integral parts of the special functional mechanisms of these organs. In the bone marrow the specific function, blood cell formation, rests entirely with cells of the connective tissue and blood vessels.

A distinguished hematologist and staunch supporter of the "dualistic doctrine" that granulocytes and lymphocytes have a different origin was disturbed when lymph nodules were discovered in the midst of myeloerythropoietic bone marrow of man. Therefore he declared that the lymph nodules in the bone marrow were *extra-parenchymatous* (discussed by Mayer and Furuta, 1924). Weaknesses of the traditional antithesis of parenchyma and stroma were presented clearly in Askanazy's 1923 article "Stromafunktionen." Of course, there is no objection to the use of the term stroma in a topographic sense, synonymous with interstitium, to indicate the location of blood vessels and connective tissue in an organ. In the cornea the term stroma denotes a layer of modified connective tissue, which lacks blood vessels but performs respiratory and nutritional functions by diffusion. Herrmann (1960) reviewed the metabolic interactions between the epithelial components and the stroma of organs, such as the choroid plexus, the ciliary body, and the cornea. One of the numerous interesting points of the review may be mentioned here. Epithelium and stroma showed opposite oxidation-reduction potentials. According to Herrmann, the maintenance of the different potentials controls the transportation of fluid and ions between epithelium and stroma, as necessary for proper functioning of these structures.

2. *Biochemical, Immunological, and Biophysical Criteria of Classification*

In a preceding discussion examples were given of functional classifications of whole organisms, either in agreement with morphological and genetical groups or crossing their borders (Part V, Chapter 1). It is not surprising that the same alternatives occur within the parts of individual organisms, from the cellular level to organs and systems.

The classification of functions of the digestive system can be complicated by activities of the intestinal bacterial flora. The gallbladder bile of rabbits contains deoxycholic acid conjugated with glycine. One might have ascribed the production of deoxycholic acid to the liver cells of the rabbit, but the compound is formed by intestinal bacteria (Bergström, 1959). Raising bacteria-free animals was described in connection with living models (Part III, Chapter 5, C, 3).

a. *Tissue antigens and antibodies.* Different organs possess antigenic properties which lead to the production of more or less specific antibodies. Uhlenhuth (1903) found that antibodies to lens antigen are highly organ specific and do not act on organs other than the lens; but the antigen is not species specific, since the antigen of any one species acts on the lens of mammals, birds, reptiles, and amphibians. It seems that evidence of organ antigens in the early chick embryo was reported first by Schechtman in 1948. Experiments by Ten Cate and Van Doorenmaalen (1950) with injections of chick lens into rabbits were described earlier in this book. The investigators found that the resulting lens antiserum of the rabbit produced specific precipitin reactions in saline extracts of embryonic chick lens, provided the lens extract came from chick embryos 60 or more hours old. Extended investigations on these lines by Langman (1958) were also recounted previously (Part II, Chapter 2, H, 2).

Antigens originating from organs, tissues, or cells were called cytotoxins in early publications, but the term tissue antigens is given preference now. A number of techniques are available to study immunological relations between organs, tissues, or cells. Morphological responses may indicate the action of organ-specific antibodies. Thyroids of rabbits were homogenized, and extracts of the homogenate were injected into intact rabbits whereupon histopathological changes occurred in the thyroids of the injected rabbits (Rose and Witebsky, 1956). I referred repeatedly to the fluorescent tagging of pneumococcal antibodies which permits localization of antigen-antibody reactions in histological sections (Part III, Chapter 7, B, 2). Radioactive tagging of antibodies to tissue antigens has been used extensively for localization studies (see Pressman and Sherman, 1951). The labeling of antibodies is done by introducing I^{131} into the globulin fraction of the antibody-containing serum. A study by Bale and Spar (1954) may be mentioned here in which rabbits were injected with preparations from different rat organs. Complement fixation was used to measure titers of rabbit serum antibodies against the injected tissue. Labeled antibodies prepared against rat kidneys localized in greater concentrations in kidney than spleen, whereas antibodies prepared against ovaries localized to a greater extent in spleen than in kidney. It should be emphasized that the specificity of tissue antigen-antibody reaction is always a relative one and, if possible, is expressed as a titer. The relations between organ specificity and species specificity may depend on the sensitivity of the immunological techniques, as Rose and Witebsky pointed out (1959). These authors studied the ability of thyroid autoantibodies from rabbits to react with thyroid extract of various species. If determined by the tanned cell hemagglutination test, positive reactions were observed with thyroid extracts of rabbits, dogs, guinea pigs, and human beings. If tested by complement fixation, the reaction was definitely positive only with thyroid extracts of rabbits, questionable with those from dogs and human beings, and negative with guinea pig material.

The concept of cell antigens and antibodies became remarkably fruitful in the application to problems of transplantation or tissue compatibility.

b. *Transplantation, immunological aspects.* Transplantation experiments for the determination of potencies in embryonic primordia were described earlier (Part II, Chapter 2, C). The question of compatibility arose when parabiosis was discussed as an illustration of artificial units larger than one organism (Part III, Chapter 4, B).

Transplantations are classified as *autoplastic* when made within the same organism; as *homoplastic* when made from one organism to another of the same species; and as *heteroplastic* when grafting from one organism to one of another species.

If a piece of skin is transplanted from one mouse to another, the transplant is vascularized readily. However, the subsequent fate of the grafted piece depends on immunological reactions between the host and the transplanted tissue. The development of antibodies in the host to the cell antigens of the transplant requires various periods of time, usually several weeks. No antibodies form if donor and recipient are uniovular twins or members of a closely inbred strain, and therefore, grafts are accepted. Grafts are rejected when donor and recipient are less closely related. A rejected piece undergoes necrosis (local death), disintegrates, and sloughs away. An accepted piece may gradually become indistinguishable from the host tissue. It is rarely possible to tell whether a permanently successful transplant still contains descendants of the transplanted cell population or whether the latter has been replaced by host cells. Transplanted avascular tissues, such as cornea and cartilage, can survive in hosts unrelated to the donor, provided the grafts are not vascularized by the host (Billingham and Hoswell, 1953; Peacock *et al.,* 1960).

It has been known for a long time that the final acceptance or rejection of grafts greatly depends on the maturity of the tissues and organisms. Transplantations between early stages of embryos of *different species* proved possible (Fig. 4). From the immunological point of view the compatibility of different cell types in the body rests on an antigen-antibody equilibrium. During embryonic differentiation some cell types multiply, while others regress (Fig. 30). Gradually the pattern stabilizes. As stated by Burnet (1949, pp. 60, 61), tolerance between the different cell types develops by elimination of reacting clones. After birth or hatching, maturation proceeds. In mature unrelated animals, Billingham, Brent, and Medawar (1953) observed *increase* in reactivity (sensitization) by repeated grafting. When an adult mouse of the CBA strain received a skin graft from an adult mouse of the A strain, the median survival time of the graft was 11 days. After a time interval a second graft of A strain skin was made. It survived for less than 6 days. Billingham, Brent, and Medawar discovered that the transplantation reactivity of a host can be *decreased* during embryonic life. Cell suspensions from various organs of strain A mice were injected into embryos (*in utero*) of strain CBA without interfering with their further development. The resulting adult CBA mice proved to tolerate skin transplants from the A mice. Injection into newborn mice was effective too. Conditioning of hosts for *heteroplastic* transplantation by Roentgen radiation or cortisone has been achieved mainly in transplantation of neoplasms (Part V, Chapter 4, C, 1, d).

With the use of immunological and other criteria Defendi, Billingham, Silvers, and Moorhead (1960) re-examined the identity and classification of eleven established lines of cultured tissue cells. The authors rightly assumed that identification is most reliable when various criteria confirm each other. The following five criteria were used: (1) morphological characteristics of fibroblasts or epithelium; (2) response to the cytotoxic action of rabbit antiserum which had been prepared against the nucleoprotein fractions of two human cell strains; (3) presence of detectable mouse transplantation antigens; (4) susceptibility to poliomyelitis virus; and (5) predominant chromosome numbers. In the tests for murine transplantation antigen, cell suspensions freshly prepared from known sources (tissues of various species) served as controls. An example may show the evaluation of results. Because of past experience, susceptibility to poliovirus supported the classification of a strain as human epithelium but cast doubt on a strain alleged to be human fibroblasts.

Of the eleven strains examined, eight were allegedly derived from human tissues, one from mouse tissue (Earle's L strain), one from rat and one from rabbit tissue. As a result of the five tests three lines of stated human origin and one line of stated rabbit origin showed the properties of mouse cells, similar to those of the L strain. Possible reasons of these discrepancies were discussed: biological alterations produced by cultivation *in vitro,* contaminations with the L strain cells, or errors in bookkeeping.

c. *Reactivity to infections and irradiation.* The *reactivity to harmful stimuli* can be used to characterize systems, organs, tissues, and cells. Differential sensitivity to infections plays a great role in the classification of pathological phenomena (Part V, Chapter 4). Tuberculous infection practically never involves the testis, although the epididymis is susceptible. The reverse is found in syphilis. In the spinal cord, the poliomyelitis virus destroys the motor nerve cells, but not the sensory nerve cells (Fig. 58). Differential susceptibility to poliovirus of cells cultivated *in vitro* was mentioned above.

The reactivity of organs and tissues to irradiation is important in practical medicine. Bone marrow, lymphatic tissue, and reproductive organs are most sensitive to Roentgen rays. Less sensitive are the skin, the inner lining (epithelium) of the gastrointestinal tract, and connective tissues. The remaining organs and tissues possess the lowest degree of radiosensitivity (Hempelmann *et al.,* 1952, p. 385). The relative sensitivity of an organ or a tissue to radiation depends mainly on the concentration of cells in the irradiated volume, and on their mitotic index. In other words, cells are more sensitive than intercellular substances, and dividing nuclei are more sensitive than nuclei in interphase.

3. *Merging of Anatomical, Physiological, and Pharmacological Criteria of Classification*

As pointed out repeatedly, classifications are most satisfactory if based on the correlation of several independent criteria. The merging of genetic and ecological characteristics in the new systematics of plants and animals was discussed previously (Part V, Chapter 1, A). Concerning classification of organs and systems it may be interesting to recall the two divisions of the autonomic nervous system which express a remarkable *parallelism of anatomical, physiological, and pharmacological criteria.* In contrast to the voluntary nervous system which controls skeletal muscles, the autonomic nervous system controls the heart muscle, the smooth muscles of the eye and viscera, and the glands of the skin and viscera. The autonomic nervous system is divided into the sympathetic and parasympathetic systems. *Anatomically* the sympathetic system forms a chain of ganglia right and left of the vertebral column, with connections to the thoracic and lumbar segments of the spinal cord. The cranial part of the parasympathetic system consists mainly of the vagus nerve, while the sacral part is formed by the pelvic nerve. *Physiologically* the two systems often supply the same organ, but act antagonistically. For instance, the rhythm of the heart is accelerated by sympathetic nerve fibers and retarded by parasympathetic fibers. *Pharmacologically* the two systems are distinguished as cholinergic and adrenergic nerve fibers or neurons: most sympathetic fibers release an adrenaline-like substance at the nerve endings, whereas most parasympathetic fibers release an acetylcholine-like substance. The classic two-color diagram of the sympathetic and parasympathetic systems by Meyer and Gottlieb (1911) has been reprinted in many textbooks (e.g., Winton and Bayliss, 1955, Fig. XV, 1). Instructive black-white diagrams are available, a very complete one by Woodburne (1957, Fig. 23) and a simplified one by Carlson and Johnson (1953, Fig. 148).

Chapter 3

Variations in Size, Volume, Weight, and Composition of Organisms and Organs

In an adult person the proportions of the skeleton, or the so-called build, are not likely to change under different environmental conditions, but his fat depots may increase or decrease. The pH of the plasma is very stable, but the concentrations of blood corpuscles and hemoglobin are relatively labile. If a person living at an altitude near sea level moved to an altitude of several thousand feet, a few days of dwelling in the new place would produce a marked increase in red blood corpuscles and hemoglobin. The reverse change occurs by moving from high to low altitude.

Assuming that the blood volume remains constant, differences in flow rate produce differences in blood content of regions and organs in the body (Part III, Chapter 6, A, 3). The concentration of blood cells is also subject to local variability. Under special conditions the volume of blood in the vascular system is altered. The three compartments between which body fluids can be exchanged are shown in Table 14. The present chapter deals both with fluctuating variations in chemical or morphological composition, and with directed, irreversible variations.

A. Morphological and Biochemical Concepts of Growth

Different morphological and biochemical criteria may be selected for studying increases in size, volume, and weight of biological structures. This is why the present chapter has been included in the discussion on classification and identification.

First, let us consider *morphological determinations* on the macroscopic level. A bear is measured from nose to tail, or standing upright. In hoofed animals usually the height at the shoulder is indicated. In birds the spread of the wings is an informative magnitude. One linear measurement will rarely give an adequate idea of the volume of an animal, whereas two linear measurements may.

Changes in young animals require consideration of various characteristics. By administration of moderate doses of propylthiouracil to young rats Goddard (1948) produced healthy dwarfs with normal proportions. The following determinations were made at autopsy: body weight, body length, volume of body (by immersion in water), skull length, skull width, femur length, and femur width. From these measurements specific gravity, surface area, and cephalic index were calculated. These data together supplied a reasonable basis for comparing the dwarfed animals with normal controls.

The *chemical composition* of the body changes with age and external environment. As the organism ages, the ratio of water to dry substance changes in favor of dry substance. Quantitative laws rule this change during embryological development, as pointed out by Needham in "Biochemistry and Morphogenesis" (1942). Within any group, such as mammals, animals become drier at the same absolute rate (Needham's Fig. 294 and Table 29). The different components of the dry substance vary in different ways in the developing embryo. In rats, pigs, and chicks the concentration of creatinine per whole embryo increases, while that of glutathione decreases. A large amount of such data is analyzed in Needham's chapter on "Heterauxesis," including the changes in composition of developing organs.

In postembryonic life, the process of aging, though connected with the genetic life span of each species, can hardly be separated from environmental factors. Suppose arteriosclerosis is a corollary of old age in man, yet diet and other factors may influence the deposition of cholesterol in the lining of arteries with respect to timing, localization, and intensity.

The ratio of water to dry substance is controlled in various ways in different species. In molting aquatic animals, such as lobsters, the concentration of dry substance in the tissues increases steadily until the time of molting. When the chitin integument is shed, the tissues swell by rapid intake of water, causing an increase in volume of the animal. As soon as the new chitin integument has formed and hardened, another period of increase in dry substance starts. Evidently this periodicity is based on an inner clockwork, although modifications by the environment are not excluded.

In mammals hibernation produces periodic changes in the biochemical composition of the body. In general, starvation makes stored fat decrease faster than proteins and minerals. Some diseases produce dehydration; others, retention of fluid.

The weights of organs are expressed as absolute figures in grams or as relative values, such as gram per organ per 100 grams of body weight. Some organs tend more than others to change in weight, parallel with the weight of the whole body. If body weights of experimental animals and contols are similar at autopsy, differences in relative weight of their organs can result from a number of factors. The weight and volume of the liver is increased by venous congestion, edema, excessive storage of fat, or infiltration with leukemic cells. Enlargement of the spleen is produced by many infectious conditions, by congestion, and by leukemic infiltration. If the body weight of the animal decreases, and the weight of the heart remains constant, the relative weight of the heart will be higher than usual. An increased weight of the heart cannot be interpreted unless it is stated whether the muscle walls of both ventricles were thicker than normal, or which ventricle was thicker if this condition existed in only one. Groups of normotensive as well as of hypertensive rats show variations of blood pressure. Benitz *et al.* (1961) found that the ratio of weights of left to right ventricles increased significantly with increasing systolic and diastolic blood pressures in hypertensive rats, but not in normotensive animals. The weight ratios of the two cardiac components showed no relation to body weight; in other words, relative weights did not give information in this instance.

In long-term studies on side effects of drugs in animals (chronic toxicity studies), increased relative weights of liver and kidneys are usually considered to indicate harmful effects. However, in such studies, relative weights of organs are meaningless without consideration of their absolute weights, of the body weight, and of the nutritional state of the animal. In drug experiments it is not very useful to classify some animals as lean and others as average or obese, unless their food intake has been recorded. If differences between organ weights of experimental animals and controls are observed, the determination of both wet and dry weights of the organs may give valuable cues to the interpretation of the data. The dry substance of organs may be analyzed for total nitrogen, fatty substances, or minerals. Determinations of glycogen in organs and tissues were described in detail as illustrations of quantitative histochemical procedures (Part III, Chapter 7, D, 1).

Since the composition of the whole body as well as of individual organs can vary in different ways and for different reasons, many *concepts of growth* are used. In practical medicine the term growth may be self-explanatory if applied to the whole organism. The statement that certain underfed children showed "inhibition of growth" usually means that the body weight was below that of well-fed children of the same race and age. I may also mean that the children with malnutrition had a shorter skeletal length than the other children. In individuals with definite signs of malnutrition, seemingly normal body weights indicate edema.

In biological studies the term growth should not be used without specification. Complex phenomena of growth should be presented as balance sheets of positive and negative factors. To state that the population of a city has grown may be interesting up to a point. In any serious study one would inquire into the ratio of birth rate over death rate and of immigrants over emigrants. In bacterial populations the counting of individuals supplies a first estimate. More information is obtained by counting the viable bacteria (Part III, Chapter 8, C, 1).

Tissue cell populations cultivated *in vitro* are measured by cell counts in suspensions without separating living and dead cells. If the populations form well-defined colonies on a solid or semisolid substrate, the mitotic rate and the rate of centrifugal migration can be determined microcinematographically. The balance sheet of growth of a cell colony *in vitro* involves six factors: centrifugal migration of cells, mitotic cell division, increase in cell size, centripetal migration, death of cells, decrease in cell size (Mayer, 1930). The net result of this balance sheet is reflected in an increase in area of the whole colony which can be measured. In the organism the morphological growth of organs and tissues can be expressed as a balance sheet composed of the same six factors plus production of intercellular substances contributing to the increase in volume. In dealing with cell populations of embryos or young animals, Weiss (1955, p. 7) considered growth to be "the surplus accruing in the balance sheet of a complex account."

Figures 30 and 32 illustrate how the migration of cell populations leads to changes in composition of embryonic structures. Single cells, too, may change their location. In the chick embryo primary reproductive cells which formed in the head area travel in the circulating blood until some of them settle in the primordia of the gonads. Relations between mitosis and cellular differentiation (differential mitoses) were mentioned above. In normal embryonic development, cell death plays an important role both in differentiation and in the balance sheet of growth. In the various primordia the distribution of dying cells follows a distinct spatial and chronological pattern (see Glücksmann's 1951 review). Scherbaum and Zeuthen (1954) were able to control several morphological and biochemical factors in populations of *Tetrahymena,* a ciliated protozoon. Different environmental temperatures produced separate effects on cell division and cell enlargement. Increase in cell volume was compared with changing protein content, and macronuclear volume was determined along with cytoplasmic volume. Some of these analyses were based on populations in which 85% of the cells divided simultaneously as a result of intermittent heat treatment.

The composition of the blood cell population changes during the life cycle of an individual under the influence of internal and external factors. Nucleated red blood corpuscles may be normal components of the circulating blood in young individuals, but may be indicative of abnormal regeneration in the adult. The proportion of lymphocytes to granulocytes changes with age, but also in pathological conditions. The concentration of red blood corpuscles and any one type of granulocytes in a blood sample is, at a given moment, the net result of the following factors: rate of production and maturation in the bone marrow; migration to the sinusoids and release into the sinusoids; distribution in the various areas of the circulatory bed; rate of destruction in the blood; escape from the blood capillaries into adjacent tissues. The sojourn of a particular cell type in the circulating blood is referred to as the life span of these cells. Methods of determining this life span were reported previously as examples of tagging procedures (Part III, Chapter 3, B).

B. Relations between Morphological and Biochemical Growth

Procedures for investigating relations between chemical and morphological *differentiation* were described in various connections (Part II, Chapter 2, E and H). Obviously, similar methods are needed for the study of relations between morphological and biochemical *growth* in embryonic and postembryonic life. The cleavage of the frog's egg leads to an increase in cell number in powers of two. The daughter cells get smaller and smaller since each daughter cell is half the size of its mother cell. Therefore the whole system does not increase in size, volume, or weight. This is illustrated in Fig. 3a–e, which shows that the blastula is not larger than the fertilized ovum before cleavage. The developing egg and embryo cannot increase in dry substance, since there is no source for such an increase, but the dry substance can change its composition. It is assumed that in the embryo nucleic acids increase at the expense of proteins.

A paper by Sze (1953) showed the following relations between cell number and deoxyribonucleic acid (DNA) in the frog embryo. From the fertilized egg to the gastrula stage the DNA content per egg remained practically constant, while the number of cells (estimated by nuclear counts) increased from one to more than 50,000. After the gastrula stage the DNA content of the embryo increased until it became six times that of the original content. This marked increase was determined at Shumway stage 19 when the estimated cell number was more than 400,000. W. M. Layton's comments on DNA estimates were given in the discussion of nuclear structures and nuclear material (Part IV, Chapter 1, A).

Some remarks on relations between morphological and chemical changes on the tissue level may be useful here. Increase in cell number and cell size cannot continue indefinitely without increase in dry substance. However, for limited periods, water intake may be the main chemical increase during morphological enlargement. Increase in collagenous fibers certainly involves increase in chemical collagen. In the growth of cartilage and bone, cell multiplication and production of ground substance are equally important. Under pathological conditions the proportions of components can change markedly, and can even be reversed. In the normal liver the mass of liver cord cells is much greater than that of connective tissue, but the ratio is reversed in advanced cirrhosis of the liver. The normal human liver contains much more protein than fat, but under abnormal conditions the fat may equal or exceed the proteins.

Two examples will illustrate how *changes in composition of organs can be compared with measurements of the whole organ.* The first example refers to the normal development of the eye and the second example to pathological developments in the kidney. Coulombre (1955) studied changes in size and composition of the retina of chick embryos with respect to morphological and chemical characteristics. Some of Coulombre's observations follow. The fourth and fifth days of incubation were a period of marked changes ("first critical period"). In the neural retina, intensive cell multiplication started which led to increase in thickness and area of this layer. In the pigmented layer of the retina, mitotic activity decreased at this time, but the cells increased in size so that the increase in area of the pigmented retina kept pace with that of the neural retina. The thickness of the pigmented retina remained unchanged, although cell shapes changed from prisms of medium height to a shallow type. (Using my diagrams, one may say that small cells with the proportions of Fig. 71b changed to large cells with the proportions of Fig. 71c). At the same time morphological and chemical differentiation proceeded. In the neural retina, axons developed and carbonic anhydrase became detectable (Clark, 1951), while in the pigmented retina the first pigment granules (melanin) appeared, together with tyrosinase activity. The appearance of red droplets in the cones on the 15th day of incubation was the starting point for determination of one of the carotenoids (astaxanthin), which increased in the retina until the 18th day after hatching. Changes in number of oil droplets per unit of area, and changes in oil droplet diameters were related to the age of the chick. Adenylpyrophosphatase (apyrase) concentrations fluctuated markedly between the 8th day of incubation and 18 days after hatching, but stabilized subsequently. As far as the whole eye was concerned, the wet weights increased rapidly from the 4th to the 8th day, but noticeably slowed from the 8th day until hatching (21 days). Axial diameters of the whole eye increased considerably from the 4th day to the 8th day, while pupillary diameters increased very little during this period. Equatorial and corneal diameters were intermediary, as shown in graphs (Coulombre's Fig. 5). This detailed account of Coulombre's study is given here to justify my previous comments on the concept of growth. Obviously it would be misleading to use the single word growth to summarize the quantitative morphological and chemical changes which take place in the developing eye of the chick.

The second example deals with pathological enlargement of an important component of the kidney without enlargement of the whole organ. It had been known for some time that kidneys of children who died of cyanotic heart disease contained abnormally large glomeruli. A quantitative study of this phenomenon was made by Bauer and Rosenberg (1960). Control kidneys were selected from individuals on the basis of the following criteria: presence of both kidneys; absence of chronic lung disease, of congestive heart failure, and of diseases of the blood or the hemopoietic system; age of individuals from newborn to 21 years of age. The abnormal kidneys were obtained from autopsies of patients with the tetralogy of Fallot, which is a congenital heart disease involving a hole in the ventricular septum, narrowing of the pulmonary artery, enlargement of the right ventricle, and abnormal position of the aorta. This condition is associated with abnormally high red blood corpuscle counts and hemoglobin concentrations. Measurements of the average glomerular size in 29 cases of

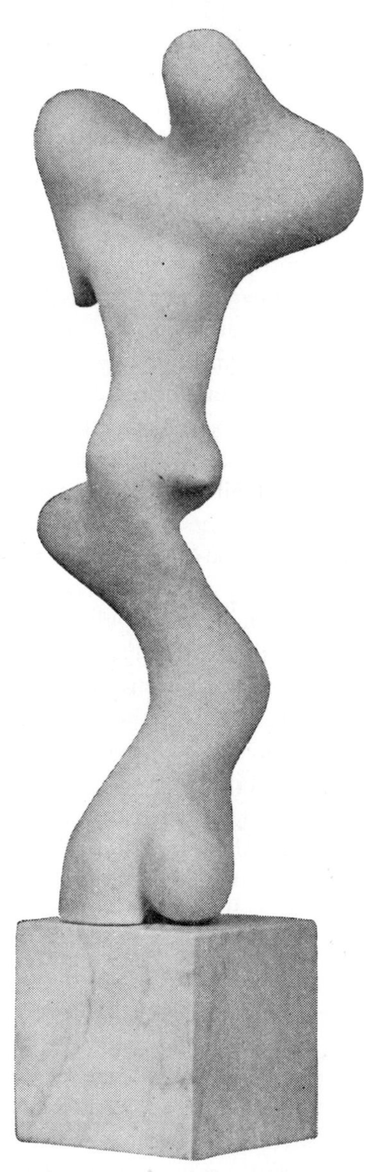

FIG. 88. "Growth." Marble statue by Jean Arp. The Solomon R. Guggenheim Museum, New York.

Fallot's tetralogy showed a significant enlargement of glomeruli, if compared to 35 control cases. Differences in size between normal glomeruli and glomeruli of children with tetralogy of Fallot were noticeable in infants, as documented by measurements and photomicrographs. The differences became more pronounced at the age of three years and older. The wet weights of the kidneys were not increased in the cases with Fallot's tetralogy. Obviously the renal tubules which constitute the bulk of the kidney were not enlarged in kidneys with large glomeruli. The enlargement of glomeruli was correlated with increase in hemoglobin concentration. This means high hematocrit values and a low concentration of plasma in the blood which reaches the glomeruli. The authors suggested that the increased filtration surface of the giant glomeruli may be an adaptation to the abnormal composition of the blood.

The two studies reported here dealt with clearly defined aspects of morphological or chemical growth. Confusing usages of the term growth occur in symposia or handbooks with multiple authors. One contributor takes it for granted that growth refers to increase in volume, while another contributor does not doubt that growth should be expressed as increase in nucleoproteins. Frequently growth has been defined as an increase in living substance. This is acceptable in metabolic studies which lead to a balance sheet of anabolism and catabolism. The usefulness of a balance sheet of morphological growth was discussed above. If properly defined, different concepts of growth can be equally operational, but it is unfortunate that some authors talk about "'true growth." Perhaps the least operational definition of growth I have found in the literature is "synthesis of biologically specific macromolecules." English-speaking biologists have marveled at the possibility of something "growing smaller," a problem that does not exist in French and German. One also encounters the opinion that all definitions of growth are futile at the present time, since more investigations are needed. A modern artist's idea of "Growth" is reproduced in Fig. 88.

False generalizations and analogies derived from S-shaped "curves of growth" were discussed in the chapter on models (Part III, Chapter 5, A). Berrill's (1955) useful article on curves of growth and related subjects was quoted in the chapter on the presentation of quantitative data (Part III, Chapter 8, C, 3).

C. Comments on Wound Healing and Regeneration

Repair in organisms and repair in man-made machines were compared in an earlier chapter (Part II, Chapter 2, H). Changes in the composition of a damaged organ are of interest here. Destroyed parts may be replaced by structures similar to the original ones. This is called regeneration. The alternative type of repair is the production of a scar consisting of connective tissue. The process of wound healing which terminates with scarring was described previously (Part III, Chapter 5, C, 2). In the skin and liver, regeneration as well as scarring is possible, while the functional elements of the central nervous system (neurons) cannot be regenerated. If the function of a destroyed part of the brain is taken over by another part, one speaks of functional compensation.

In the rat liver, which consists of several lobes with independent stalks, a special type of response occurs. After surgical removal of several lobes the remaining lobes enlarge until their total volume and weight equals that of the original liver. This is both morphological and functional compensation since the efficiency of the reconstructed liver is probably similar to that of the original liver. Conventionally this process is referred to as regeneration of the rat's liver. Mitotic activity and changes in chemical composition during this process were described in connection with parabiotic techniques (Part III, Chapter 4, B).

Regeneration is controlled by cellular, vascular, biochemical, hormonal, and nervous factors. For integrated studies of these factors, see Singer's publications (1959, 1960). Among other aspects the question of acetylcholine as a possible agent of regeneration in amphibians was analyzed by Singer. In the present book, the chapter on topographic morphology includes the sources of regeneration material, topographic distribution of meristematic cells, and cell migration in regeneration (Part III, Chapter 6, A, 3).

Chapter 4

Classification of Pathological Phenomena

Classifications used in present pathology stem from practical needs, tradition, and modern investigations in biology and medicine. Since pathology deals with abnormal conditions, the abnormalities may be deviations from the statistical or the teleological norm. As stated earlier (Part II, Chapter 1, D), most diseases are deviations from both norms. Pneumonia and diabetes affect a minority of the population and are, at the same time, undesirable. I mentioned instances in which the two norms lead to opposite classifications. Tooth decay (caries) is undesirable, but affects the majority of people in the United States. Arteriosclerosis is an undesirable condition, but statistically normal in old age groups. A statistically abnormal, yet innocuous, condition of human lungs was described in connection with radiological procedures (Fig. 16). In routine examinations, the lungs of some individuals showed radiopaque nodules resembling those seen in miliary tuberculosis or histoplasmosis, but without disturbance of the pulmonary function and general health of the great majority of these persons (Chaves and Abeles, 1952; see Part III, Chapter 2, A, 2).

In experimental studies on toxicity of drugs, any deviation of the drug-treated animals from the untreated controls is generally interpreted as *injury* induced by the drug. It is assumed that the statistical and teleological norm run parallel. A remarkable exception from this rule was discovered by Dessau and Sullivan (1961) in experiments with rats on the effect of two years' administration of chlortetracycline in the diet. At the end of the experimental period 70% to 90% of the rats which had received 0.05 to 1.0% of the drug in their diet were still alive, while only 50% of the untreated controls survived. The main factor responsible for the higher survival rate of the drug-treated animals was decrease in infections. Morphological examinations confirmed the beneficial effect of the drug administration on the suppurative lung disease common in laboratory rats. For descriptions of respiratory infections in laboratory rats see Innes *et al.* (1956).

Many phenomena that are classified as pathological take place in the course of normal embryonic and postembryonic development. Structural rearrangements necessarily involve death of many cells (Fig. 30B). This is particularly dramatic in metamorphosing animals. In mammals, the umbilical cord disappears after birth. This is a form of local death known as gangrene in pathology. Invasiveness and metastasis are characteristics of cancer, but can also occur in normal embryonic life (see Part V, Chapter 4, C, 1).

A. Biological Classification of Pathological Phenomena on the Basis of Abnormal Stimuli, Responses, and Reactivities

The triad of stimulus, response, and reactivity proved to be a useful scaffold for an orderly discussion of biological problems (Part II, Chapter 2, B). Let us now apply the same scaffold to phenomena and questions of abnormal biology, or pathology.

1. *Classification according to Abnormal Stimuli (Etiology)*

Emphasis on abnormal stimuli leads to the etiological or causative classification of pathological phenomena. The abnormal stimuli may be harmful physical or chemical agents, including the products of poisonous plants or animals. If relatively small living organisms settle on or in a larger organism, with undesirable effects for the latter, one says that the host has been infected by parasites. Harmful stimulation by parasitic viruses, bacteria, fungi, protozoa, helminths, and arthropods is the characteristic of the class known as infectious diseases.

Different abnormal stimuli can produce similar responses. Acute ulceration in the colon (large intestine) is produced by different bacteria, certain amebas, mercurial or arsenical poisoning, and uremia (failure of renal function). Microscopic nodules of the same type are the tissue response to tubercle bacilli, typhoid bacilli, spirochetes of syphilis, and other infectious agents. The *same abnormal stimulus* may cause different responses. Lung tissue responds to tubercle bacilli either by formation of small nodules (tubercles, Fig. 31d) or by the formation of an exudate in the alveoli (tuberculous pneumonia, Fig. 31f). The macro- and microscopic appearance of the two pathological phenomena is so different that, prior to the discovery of the tubercle bacillus, they were considered separate diseases.

2. *Classification according to Abnormal Responses*

a. *Systemic and local responses.* Most local conditions in the body are, in some way, subject to systemic controls. Similarly, local changes are likely to affect the whole body through the circulatory or nervous systems. Greater emphasis will be placed on either systemic or local aspects depending on the nature of the problem. The *systemic* aspect of arteriosclerosis is an abnormal cholesterol metabolism. The excessive cholesterol of the blood is deposited selectively in places of the arterial system which are subject to particular mechanical strain: this is the *local* aspect of arteriosclerosis. Diphtheria bacilli located in the throat produce a local inflammatory response in the mucous membrane which can lead to suffocation. In addition there can be a distant harmful effect on the heart muscle caused by toxin which is released from the diphtheria bacilli into the blood.

Various pathological conditions, such as hypertension and diabetes, produce abnormalities in the eye. For a comprehensive presentation see Sorsby's "Systemic Ophthalmology" (1951).

Acute inflammatory processes are characterized macroscopically by local swelling, redness, and heat, with pain as an additional indicator. Microscopically the acute inflammatory processes consist of changes in capillary caliber, redistribution of blood cells inside the capillaries, leakage of proteinic fluid from the capillaries resulting in edema, and emigration of leucocytes from the blood stream through the capillary wall. Massive accumulation of polymorphonuclear leucocytes in conjunction with fibrin exudation and local death of tissues is known as suppurative inflammation. Some of these phenomena can be seen only during life, whereas others are stable enough for study in fixed and stained preparations.

Inflammations are responses to a variety of stimuli: parasites including bacteria and viruses; inert foreign bodies such as splinters in the skin; physical agents such as heat (first and second degree burns); ultraviolet rays and Roentgen rays; and many chemical agents, croton oil representing the most active type. If the properties of the original stimulus or the reactivity of the tissues are altered, the acute phenomena may change to *chronic inflammation.*

The role of blood capillaries and humoral factors gradually diminishes in chronic inflammation. Fixed connective tissue cells become an important component. Polymorphonuclear leucocytes, the dominant migratory cells in acute inflammation, are replaced by eosinophile leucocytes, lymphocytes, or plasma cells in chronic inflammation.

More or less defined masses or nodes of chronic inflammatory tissue are known as *granulomas.* They occur frequently at the roots of dead teeth. The experimental production of granulomas was described previously (Part III, Chapter 5, C, 4). Except for its greater amount of capillaries, the granulation tissue of a healing wound is similar to early stages of chronic inflammation. A special class of tissue responses, which is seen in tuberculosis, syphilis, typhoid, and a few other infections, is known as "infectious granulomas" (tubercle in tuberculosis, gumma in syphilis). They are characterized by scarcity or absence of blood vessels and by aggregates of lightly staining, highly variable nuclei of the type illustrated in Figs. 64 and 65. The cytoplasm is either a network (Fig. 73a) or an unpartitioned multinuclear mass (Fig. 74). For historical reasons these structures are called "epitheloid cells," although they do not resemble any of the epithelial mosaics seen in Figs. 66 and 71. In the so-called giant cells of Langhans the nuclei are crowded as shown in Fig. 54b. Other important components of infectious granulomas are lymphocytes or plasma cells. A different type of giant cells (Reed-Sternberg cells) occurs in Hodgkin's granuloma, a tumor-like disease of the lymphohemopoietic tissues. The etiology of this disease is unknown.

Chronic inflammatory processes represent abnormal combinations of more or less normal elements. In appearance and function the fixed and migratory elements of inflammatory structures are not different from those of normal or regenerative tissues. The multinuclear unpartitioned or reticular cytoplasm of infectious granulomas (epithelioid and giant cells) are moderate deviations from normal structures. By contrast *neoplasms consist of biologically abnormal elements.* Neoplastic cells multiply and migrate in a way which is unrelated to the pattern and normal function of the organ from which they originate. Cancer cells infiltrate and destroy adjacent tissue, travel from their primary site to distant organs, either with the lymph or blood circulation, and settle in the new location to multiply again and invade the neighborhood. These characteristics of cancer cells are known as *invasiveness* and *metastasis.* Morphologically the neoplastic cells, or cancer cells, may or may not resemble normal cells. Associations of cancer cells vary from extreme irregularity to marked resemblance with normal structures, such as glandular or osseous tissue.

The transformation of a normal cell population into a cancer cell population may be the response to known carcinogenic stimuli such as irradiation, polycyclic hydrocarbons, and parasites, especially viruses and helminths. If cancer develops in the absence of external carcinogenic stimuli, the following internal factors are considered: tissue malformations, particularly in children; imbalance of hormones, particularly at the decline of sexuality; and finally genetic abnormal reactivity. As mentioned earlier some strains of mice have a high incidence of spontaneous tumors, whereas others have a very low incidence. This is a genetic difference in the reactivity of mice to normal laboratory conditions. Under abnormal conditions, such as restricted caloric intake the cancer rate decreases in strains of mice with otherwise high incidence. Since the cancer problems have been attacked from the biological as well as from the practical medical side, a special subchapter will be devoted to identification and classification of malignant tumors (Part V, Chapter 4, C).

Time relations between response and stimulus are different in inflammatory and neoplastic phenomena. Most inflammatory responses subside when the stimulating agent disappears. As soon as a splinter is removed from the skin, or the staphylococci in a boil have died, suppuration regresses and healing starts. By contrast, neoplastic responses not only continue after the stimulus has ceased, but in many cases a period of latency elapses between the time of stimulation and the appearance of the neoplasm. In experimental animals a relatively short application of chemical carcinogens is followed by irreversible neoplastic processes. In man skin cancer developing many years after a Roentgen ray burn is a well-known example. Neoplasms are not the only pathological phenomena without a definite time relation to the stimulating agent. In the late stages of syphilis the infectious agent is no longer present; this will be discussed subsequently.

3. *Classification according to Abnormal Reactivities*

After treatment with Roentgen rays or cortisone the organs of rats and mice are sometimes flooded by bacteria of unknown types. As stated earlier (Part III, Chapter 3, A), one must assume that these are bacteria which under ordinary conditions are hidden somewhere in the host tissues as a relatively small population in a steady state. It is not likely that Roentgen rays or cortisone increase the reproductive or invasive capacities of the bacteria. Therefore one ascribes the change in host-parasite relationships to a decreased resistance of the tissues, in other words, a change in reactivity of the host.

Pathological phenomena which are not readily classified on the basis of a characteristic stimulus or response are labeled reasonably as abnormal reactivities. Endocrine and metabolic deviations from the normal, which are rarely ascribable to a single factor, and rarely identified by a single criterion, can be classified as abnormal reactivities to normal environmental stimuli. Similarly, the various phenomena which occur in the course of infectious chronic diseases, such as tuberculosis and syphilis, are described best in terms of changed reactivity, since in each instance, the chain of events was started by one causative factor.

a. *Endocrine and metabolic diseases.* A healthy adult individual with a pulse rate of 60 per minute may show 120 pulses per minute after exercise, but reverts to his baseline after a short rest. A person with hyperthyroidism has a pulse rate of 120 or more at rest. The hyperthyroid patient responds to the condition of rest as the healthy person responds to exercise. The two individuals differ in reactivity. In this instance the rate of thyroxine production causes the difference. Yet, increased activity of the thyroid is not the whole story in so-called hyperthyroidism. One of the characteristic signs of this disturbance, exophthalmos, is definitely not produced by excessive thyroxine.

Obesity and diabetes are examples of abnormal metabolic reactivity. If a normal individual is exposed to the internal stimulus of appetite and the external stimulus of available food, he will eat enough to maintain his energy economy. The obese person eats more: his reactivity is different, but the factors responsible for this difference are unknown. The urine of a normal individual does not contain glucose unless a very large amount of glucose is consumed within a short period (alimentary glucosuria). The diabetic, on the other hand, excretes glucose in his urine when eating the usual food. In a number of patients, destruction of the islets of the pancreas is the cause of diabetes, but in certain forms of diabetes the islets are not involved. Therefore no single causative factor can be assigned to this disease. It seems that different conditions lead to a similar configuration of responses. The increased susceptibility of the diabetic to infection may be a consequence of his disturbed metabolism, or an additional independent indicator of his abnormal reactivity.

b. *Change in reactivity during the course of tuberculosis and syphilis.* Changes in reactivity were discussed in an earlier chapter (Part II, Chapter 2, F). Antibody formation producing change in reactivity was illustrated by the example of typhoid fever. In a person infected with typhoid bacilli the period of pathological responses (disease) comes to an end when enough antibodies have formed to neutralize the bacterial antigens. However, a steady state may develop known as the carrier stage. Without being ill, the host continuously harbors a moderate population of typhoid bacilli virulent for other people. While the different phases of host-parasite relationship are fairly understood in typhoid and many other infectious diseases, they are obscure in chronic tuberculosis and syphilis, because of periods of clinical latency and because of lack of definite antigen-antibody reactions in the latter two diseases. The tuberculin reaction possibly has shed some light on chemical relations between tubercle bacillus and host, but the serological tests used in syphilis have merely diagnostic value, as far as I know. Particularly inaccessible to investigation are the later phases of syphilis, in which the pathological tissues harbor very few spirochetes, if any. A summary of the course of tuberculous and syphilitic infections follows.

The so-called primary complex of tuberculous infection involves a small area in the lung and some bronchial lymph nodes. It seems that in many countries the majority of the population is infected with tubercle bacilli at any early age and responds with the primary complex without noticeable disturbance of health, resulting in scars of lung tissue and lymph nodes. If an adult individual develops pulmonary tuberculosis, two interpretations are possible. This person has been infected either by recent exposure to an outside source (reinfection), or by mobilization of bacilli which had survived in the scars of his primary complex (autoinfection). In either case the host experiences a second stimulation of his tissues by tubercle bacilli. The reactivity of the host to this second stimulus is determined by a number of factors which have been the subject of numerous studies. References are found mainly under the title of immunology in tuberculosis. Differences in reactivity of organs have been emphasized by Rich in his "Pathogenesis of Tuberculosis" (1951, p. 321, Table 17) from which the main points may be reported here. The tendency of circulating tubercle bacilli to lodge in particular organs is expressed in miliary tuberculosis, whereas the ability of the bacilli to survive and to proliferate in an organ is shown by the production of extensive tuberculous granulation tissue and necrosis (caseation). Lung and bone marrow are equally susceptible to both forms, but heart, skeletal muscle, pancreas, and thyroid are rarely involved in either type of tuberculosis. Spleen and liver are always involved in miliary tuberculosis, but are not frequently the site of large tuberculous masses. The opposite behavior is found in kidney, epididymis, adrenal, lymph nodes, brain, and skin.

Syphilis is distinguished by the relative importance of prenatal infection. Tuberculous infection of the fetus through the placenta has been observed, but is very rare, and the newborn with congenital tuberculosis is not viable. Syphilitic infection *in utero* is not unusual, and although many of the children with congenital syphilis die soon, survival for ten or more years does occur. Syphilis acquired in postnatal life starts as a local infection of the skin or a mucous membrane. The primary localized stage (primary chancre) does not produce noticeable disturbances of general health. Soon the spirochetes (*Treponema pallidum*) invade regional lymph nodes, lymphatics, and the circulating blood, leading to the secondary stage. This is manifested by various pathological changes of skin and mucous membranes. It has been said that the eruptions of syphilis can simulate practically any skin disease. The first two stages are summarized as "early syphilis." In the late (or visceral) stage of syphilis "infectious granulomas" of the type described above may develop in different parts of the body, such as the liver, the membranes of the brain, or the periosteum. These nodes, known as gummas, lead to local destruction of the organs affected. Other manifestations of late syphilis are chronic inflammatory processes in the wall of the thoracic aorta causing extensive disruption of the elastic fiber system of the aorta. Deformations of the aorta and aortic valves produce ruinous effects on the circulation. The brain may be affected in such a way that particular physical and mental disturbances result (general paresis of the insane), or special tracts in the dorsal part of the spinal cord may be destroyed, with characteristic neurological consequences (tabes dorsalis). As a rule, it is impossible to find spirochetes in the pathological tissues of late syphilis. Spirochetes can be impregnated with silver nitrate, but additional procedures may be needed to distinguish between spirochetes and spirochete-like structures in tissues (Sahyoun, 1939). As a rare exception spirochetes have been demonstrated in the brains of patients with general paresis of the insane. These findings allow two different interpretations. Either the demonstrated spirochetes were still virulent and therefore responsible for the brain injury; or these spirochetes were survivors of a population which had lost its virulence, at least for the particular host. Whichever interpretation is accepted, studies should be made to find the links in the long chain of events between the original infection with spirochetes and the final brain injury. The classification as "late syphilis" is based on history of early syphilis and persistence of positive serological tests in a certain proportion of the patients. A new test using treponemal immobilization by the serum of syphilitic patients (Nelson *et al.,* 1950) seems to reveal cases of late syphilis which gave negative reactions with older serological methods.

B. Medicopathological Classification of Diseases

1. *Morphological, Functional, Clinical, Etiologic and Epidemiologic Criteria*

A variety of criteria is needed for the identification and classification of most diseases. Influenza is a good example. There is no hard and fast line between a case of influenza and a case of severe cold with fever. During an influenza epidemic physicians will diagnose more cases as influenza and fewer as a common cold. In World War I soldiers exposed to phosgene poisoning developed so-called pseudomembranes on the surface of the respiratory tract. Similar pseudomembranes occur in influenza infections. Again, the interpretation of the pseudomembranes will depend on the circumstances. Bronchopneumonia which develops in the course of an influenza infection may be indistinguishable from ordinary bronchopneumonia. The identification of the influenza virus is rarely practical, although antigen-antibody reactions are available. The situation is complicated by simultaneous or secondary infection with bacteria such as *Hemophilus influenzae,* streptococci, staphylococci, and pneumococci. Death in influenza is produced by these bacteria rather than the virus infection.

Morphological characteristics of diseases consist of unspecific pathological phenomena, such as necrosis, which may occur in any organ, and of specific disturbances connected with the structure or location of an organ. Inflammatory exudates can form in any organ, but the accumulation of exudate in microscopic air spaces can occur only in the lung, a fact which places pneumonia in a class by itself. Circulatory disturbances of a moderate degree are tolerated by most organs, but not by the brain, because of its restriction to a rigid capsule, the skull. Inflammation, cellular death and scarring lead to the picture of cirrhosis in the liver because of the particular pattern of this organ, and to the different forms of nephritis and nephrosis because of the nephron architecture of the kidney.

Major destructions or alterations of organs are not compatible with continued function. Respiration is impossible in areas of the lung in which the air spaces of the pulmonary alveoli are filled by an inflammatory exudate. Therefore, the diagnosis pneumonia covers both the morphological and functional abnormality. Mechanical interferences with important functions are conveniently used in classifying pathological phenomena. Examples are the distortions and immobilizations of joints or cardiac valves, and obstructions of channels such as the digestive tube, the bile ducts, or the ureters. In the central nervous system, motor and sensory functions are related to special areas or tracts. Therefore combinations of morphological, topographic, and functional criteria have been used successfully in the classification of neuropathological phenomena. These points were discussed in connection with maps of the cerebral cortex (Figs. 48 and 49) and the spinal cord (Fig. 58).

In the liver and kidney, relations between morphological and functional disturbances are rather involved. Therefore attempts to classify liver and kidney diseases on a combined morphological and functional basis have led to confusion.

Morphological classifications have retained a leading role in clinical medicine in spite of increased emphasis on physiology and biochemistry. Because of their practical advantages, topographic criteria are used as the primary principle of classification in the "Standard Nomenclature of Diseases."

2. *The "Standard Nomenclature of Diseases"*

Uniform registry of diseases is desirable both in the management of hospitals, and in statistics of diseases. For these purposes "Standard Nomenclature of Diseases and Operations" was published first in 1932. The 5th edition (edited by Thompson and Hayden) appeared in 1961. The code consists of two groups of three digits, separated by a hyphen. The first three digits refer to "topographic classification," and the second three digits to "etiologic classification." An example will illustrate the system. The diagnosis "syphilitic aortic valvulitis" is coded 455-147. In the topographic classification 400 refers to the cardiovascular system in general, 451 to the valves of the heart, and 455 to the aortic valve. In the etiologic classification -100 indicates infections due to microorganisms, such as bacteria and viruses, -140 due to spirochetes, and -147 due to *Treponema pallidum* (*Spirochaeta pallida*), the infectious agent of syphilis.

Since more than two principles of classification would render the system too unwieldy to be practical, functional disturbances were included in the morphologic or etiologic groups. Thus diabetes mellitus is found under "Diseases of the Insular Tissue" in Chapter 8, "Diseases of the Endocrine System." The topographic code for the islets of the pancreas is 870. Decreased function of insular tissue is classified as the etiology of diabetes, with the number -785. Therefore, diabetes mellitus is coded 870-785. The decision to equate diabetes mellitus with decreased function of insular tissue represents a reasonable compromise although less than 80% of the cases of diabetes mellitus show pathological changes in the islets, and 10% of nondiabetic persons have similar changes (estimates based on data of Warren and LeCompte, 1952). Evidently factors outside the islets are essential in producing or preventing diabetes mellitus. If diabetes mellitus is clearly ascribable to a cause outside the pancreatic islets, a special etiologic number is provided, such as -781 for increased function of the adrenal cortex. The code number for this type of diabetes mellitus is 870-781. Paradoxically the topographic code number 870 stands for the islets which are not involved at all in this instance. However, no better solution could be found to maintain a system with no more than two principles.

It may be well to remember that the topographic and etiologic classification of diseases was established in 1761 by Morgagni's book "De sedibus et causis morborum." Probably Morgagni would be disappointed that *unknown etiology* is used as a criterion of classification in modern medicopathology. Diseases with unknown etiology are referred to as kryptogenetic or idiopathic. Thus, chronic ulcerative colitis without known etiology is distinguished from the dysenteric forms caused by bacteria or amebas. In Wintrobe's "Clinical Hematology" (1951, p. 707) pathological increases of red corpuscles in the circulating blood are defined as erythrocytosis if occurring in response to some known abnormal stimulus, and as erythremia when no abnormal stimulus is known. Similarly, leucocytosis refers to abnormally high white cell counts with a known cause, while in leukemia the cause is obscure. The abnormal conditions with unknown causes tend to be less reversible than their counterparts with known causes.

C. Identification and Classification of Malignant Neoplasms

Abnormal proportions of different tissues in an organ may be congenital or acquired in later life, as mentioned previously. Certain distorted tissue composites are known as benign neoplasms or benign tumors. There is no hard and fast line between benign tumors and some congenital or acquired tissue abnormalities. Local excesses of capillarization, known as hemangiomas, are probably always congenital. An excessive scar of the skin resulting from wound healing is called a keloid. A keloid may be indistinguishable from a cutaneous fibroma, which is a benign connective tissue tumor of unknown etiology. Benign tumors may be harmful and even fatal when their size and location interfere with the function of vital organs. In addition, injury to the tumor bearer can be produced by hemorrhage and infection of the tumor tissue.

In contrast to the merely *expansive* growth (increase in volume) of benign tumors, malignant neoplasms, or cancers, are characterized by their ability to *invade* adjacent normal tissue and to *metastasize* in places remote from the primary tumor. The formation of metastases is the result of invasion of blood or lymph vessels by the cancer cells, transportation within the circulatory system, and successful settling in different organs.

Cancer cells have been *compared to parasites,* and therefore the cancer-bearing organism is referred to as the host. Although cancer cells destroy host tissue, there is no evidence that they must live at the expense of host tissue. Some phases of animal parasites strikingly resemble the behavior of cancer cells. *Sacculina,* a parasitic decapod, attaches itself as a free-living larva to a young crab such as *Carcinus maenas.* In the course of its metamorphosis the larva penetrates into the interior of the crab. Inside the crab the parasite forms an undifferentiated cellular mass which migrates to a special area and, at the same time, develops a system of branches extending through the host (Caullery, 1952, pp. 87 ff.). This phase of the parasite shares with cancer the low level of differentiation and the invasion of the host. However, the rigid sequence of phases in the life cycle of *Sacculina* differs from the variable and erratic life cycles of cancerous formations. I do not know whether branches of *Sacculina* would be able to survive and grow when severed from the central body of the parasite. If so, there would be a remarkable resemblance with metastasis of neoplasms.

The traditional statement that cancer cells are *autonomous* means that they do not comply with the pattern of the organism as a whole. The *partial control* of cancer cells by hormones and other factors of the environment will be discussed later.

Attempts have been made to separate the biological approach from the medicopathological approach to neoplasms. When Eugen Albrecht founded the *Frankfurter Zeitschrift für Pathologie* in 1907, he suggested a program of cancer research along biological lines, as a branch of experimental embryology (*Entwicklungsmechanik*). It is fascinating to compare Albrecht's ideas with the state of cancer problems today. The *Annual Symposia on Fundamental Cancer Research* were established in 1946 and have been held regularly since that time (for introductory address, see R. L. Clark, 1947). Characteristic of this trend is a series of articles by F. M. Burnet "Cancer—a Biological Approach" (1957) and J. Huxley's book "Biological Aspects of Cancer" (1958). A sharp separation between biological and medicopathological cancer research proved to be impossible. However, approaches from the two different angles can be useful, particularly with respect to problems of classification of malignant neoplasms. The first part of the present discussion will deal with biological criteria and the second part with medicopathological criteria. A certain amount of overlapping will be unavoidable.

1. *Biological Classification of Neoplasms*

Morphological (descriptive), developmental, and functional criteria can be applied to cancer cells in the same way in which they were used for normal cells and tissues (Part V, Chapter 2).

a. *Morphological and developmental characteristics of cancer cells.* As stated previously, the appearance of cancer cells may or may not resemble that of normal cells. In the prostate carcinomatous cells may be indistinguishable from cells of the normal glands, although the glandular pattern is usually abnormal in the carcinoma (see Pessin, 1953, p. 625). Unusual variability in cell size, nuclear size, and nuclear shape is suggestive of malignant cells, though not specific. Similar morphological variations occur in regenerating tissue. Subcellular differences between malignant and normal cells in mice were described by Dalton (1959). He compared tumors of the thyroid, liver, and pigment-forming tissue with their normal cytological counterparts, limiting his study to the electron microscopic level.

At this point it may be useful to consider cancer first in relation to the cancer-bearing individual, and then the life cycle of cancer cells. Life histories of cancer cell *populations* will be discussed subsequently.

Malignant tumors of children are ascribed to *congenital tissue malformations.* A congenital tumor may be apparent at birth, or may be discovered during childhood, or at a later date. Congenital tumors of the kidney, known as Wilms' tumor, usually display their malignancy quite early, leading to death during childhood.

Malformations of the ovary containing skin, hair, bones, teeth, or other tissues, are known as teratomas. They may remain quiescent during the whole life of the individual, or the tissues may become malignant for unknown reasons.

Malignant tumors of adults are responses to a variety of known and unknown factors. *Chronic irritation* is an obvious carcinogenic factor in the skin and digestive system. Characteristic sites of such carcinomas are the hands of radiologists, the lower lip in pipe smokers, and the margin of the tongue opposite a jagged tooth. The role of cytogenetic, endocrine, and environmental factors in the transformation of normal cells into cancer cells will be analyzed in connection with the reactivity of cancer cells and precancerous cells. The *role of aging* in the production of cancer can be divided into three major aspects: cumulative effects of relatively weak irritating agents over a long period of time, changes in the internal environment of tissue cells, or changes in reactivity of tissue cells to the ordinary external environment.

The life cycle of cancer cells differs from that of normal tissue cells. Important points for comparison are: composition of chromosomes and their distribution during mitosis; potentialities and differentiation; cell multiplication and cell death.

After Hansemann's (1893) description of abnormal mitotic spindles in cancer cells, Boveri (1914) was the first to make chromosomal variations the basis of a theory of the origin of malignant tumors. The idea that the transformation of a normal tissue cell into a malignant tumor cell can be considered to be a *somatic mutation* gained acceptance mainly through the book of K. H. Bauer which appeared in 1928; Bauer credited several authors with this theory. Counts of chromosomes and analyses of their fate in cancer cells were greatly facilitated when tissue sections were replaced by squash preparations of ascites tumor cells. In contrast to solid tumors in which the cancer cells are either closely packed or enmeshed in connective tissue, ascites tumors consist of loosely aggregated or even isolated cells suspended in the fluid of the abdominal cavity. Squash preparations of such material show single cells displaying all structures in one plane with little danger of partial loss.

Representative of chromosome analysis in ascites tumor cells is the 1953 paper by Levan and Hauschka. The authors arrived at the conclusion that every ascites tumor of mice studied had a characteristic configuration of chromosome number variation so that the most frequent chromosome class could be used to characterize each "species" of ascites tumor. The total variability of chromosome numbers was much greater in ascites tumor cells than in normal mouse tissue cells. While the normal diploid range is 38 to 42 chromosomes, the total variation in cells of 13 ascites tumors ranged from 35 to over 1000 chromosomes. May I emphasize that, as yet, the chromosomal peculiarities of tumor cells have not shown any relation to their physiological characteristics — invasiveness and metastasis. A theoretical approach to this problem is found in the discussion of "Clonal Selection and Neoplastic Disease" which forms the last chapter of Burnet's 1959 book. According to this author, any somatic mutation theory of cancer rests on two assumptions: the development of abnormal clones from normal somatic cells and the persistence of such clones by selection. Then survival of the abnormal cell clones depends on their ability to invade adjacent normal tissues and to colonize in distant parts of the body.

Although chromosomal mechanisms appear to be of great importance in the origin of neoplastic cells, the role of mitotic activity in the increase of neoplastic cell populations should not be overrated. Contrary to widespread belief, mitotic indexes are not greater in neoplastic tissues than in normal tissues such as active bone marrow, spermatogenic cells in the testis, and the lining of intestinal crypts (Fig. 33). At the time of microscopic examination, some neoplasms may show a very low proportion of dividing cells, if any. Moreover, the rate of cell death is very high in many cancerous tissues. Therefore, changes in volume of neoplastic masses represent a balance sheet of growth like that of normal tissues: multiplication and destruction of cells, changes in volume of individual cells, and migration of cells. Intercellular substances which contribute to the volume of a tumor may be products of the tumor cells or of the invaded normal tissue (stroma).

The outstanding property of cancer cells is their failure to differentiate, or their differentiation in a wrong direction. Both abnormalities were termed *anaplasia* by Hansemann (1893). As Rous (1946, p. 335) remarked, the existence of neoplasms "thrust before the scientist the overlooked miracle of organization." The question of differentiation of neoplastic cells is complicated by the fact that their morphological and chemical differentiation are not always parallel. As a rule, multiplication and differentiation of normal cells are independent processes. A similar separation has been demonstrated in neoplastic cells. Lasnitzki (1951) treated mouse prostate cultivated *in vitro* with 20-methylcholanthrene. A lower dose of this carcinogenic hydrocarbon proved more effective in promoting cell division, whereas interference with normal differentiation was more pronounced with a higher dose. The author concluded that promotion of cell division and interference with normal differentiation are not related, but independent processes.

An earlier chapter dealt with relations between potentialities of cells and their *origin from the different germ layers* (Part V, Chapter 2, B, 1, b). I recapitulate the conclusions. In normal development each germ layer gives rise to a certain set of tissues. However, the potentialities of each germ layer are not restricted in this way. Under abnormal conditions, accidental or experimental, each germ layer may produce a much greater variety of structures than it does in the ordinary course of events. Any attempts to trace highly abnormal cells such as cancer cells back to one of the three germ layers are futile for two reasons. First of all, there is no method available to reconstruct the pedigree of cancer cells down to early stages of the embryo. Secondly, if such pedigree of cancer cells were available, this would hardly allow prediction of their reactivity in the post-embryonic organism. There is no objection to those cytogenetic classifications which refer to the *immediate precursors* of cancer cells without implying long pedigrees.

b. *Life histories of cancer cell populations.* The *life history of a cancer cell population* in the body can be divided into different phases. When a period of latency elapses between some carcinogenic stimulation and the development of a visible cancer, one looks for precancerous conditions. Cytologic studies of precancerous phases were possible in the skin of mice (Biesele and Cowdry, 1944), and the liver of rats (Jackson and Dessau, 1961) after treatment with chemical carcinogenics. Similar studies have been made in irradiated animals. Long periods of latency as occurring in man after Roentgen-ray burns are hardly accessible to investigation. Willis (1951) discussed long-delayed metastases without local recurrence of the extirpated primary tumor. He pointed out that in such instances emboli of tumor cells must have been dormant, and that the dormant cells might have been activated either by changes in their environment, or by changes inside the tumor cells. The longest interval observed by Willis was a period of fifteen years which elapsed between removal of a breast tumor (without local recurrence) and appearance of cerebral metastases.

Some case histories of human cancer suggest that tumors have regressed and finally disappeared. The *probability* of disappearance increases with the length of the tumor-free period after detection of the tumor. However, in view of possible dormant metastases no *absolute* proof is possible that a cancer has *disappeared completely.* Everson and Cole (1956) published a tabular analysis of 47 cases of malignant neoplasms with regression of tumors or disappearance of tumors over various periods of observation. Many of their cases received, after biopsy, some treatment which Everson and Cole (p. 366) considered "inadequate to exert a significant influence on neoplastic disease."

Regression of metastases has been observed mainly in clear-cell renal carcinomas ("hypernephromas"). Pulmonary metastases were discovered and followed roentgenologically by various investigators. In a case reported by Ljunggren *et al.* (1959), the Roentgen diagnosis was verified by biopsies from the lungs. These authors noticed disappearance of most pulmonary nodules while the kidney with the primary tumor was still in the patient. After removal of the diseased kidney the patient died from cerebral metastases, though the pulmonary metastases had remained minimal. In Jenkin's (1959) patient, pulmonary metastases first increased in size and number after removal of the primary tumor. However, from four to eight years after the operation the pulmonary metastases regressed steadily.

Observations of necrosis, encapsulation, and disappearance of metastases were well documented in transplanted rabbit tumors (Pearce and Brown, 1923; Brown, Pearce, and Van Allen, 1924). The tendency of metastases to regress varied in the course of transplantation generations. Changes in activity of cancer cell populations have developmental as well as physiological aspects. Studies with emphasis on the physiological side will be the subject of our next discussion.

c. *Reactivity of cancer cells and precancerous cells.* In his 1953 review J. Furth distinguished two classes of neoplasms: *conditioned and autonomous.* A tumor which is highly susceptible to normal host controls, such as hormones, is conditioned or dependent, while a tumor which is not visibly controlled by normal host factors is autonomous. The two terms are meant in a relative and quantitative sense. Irrespective of the type of primary stimulus, the neoplastic response of the target cells will be modified by the host environment. In the skin of rabbits and mice chemical carcinogens, ultraviolet radiation, and Roentgen radiation produce three phases of responses (Furth's Table 1): precancerous dermatitis, a conditioned tumor phase represented by warts, and an autonomous tumor phase in the form of squamous cell carcinoma. Furth emphasized that this is a diagrammatic simplification of the relation between cancer cells and their environment in the host organism.

Theoretically a population of cancer cells may originate in two different ways. The first alternative is the pre-existence of a minority of tissue cells with latent cancerous properties. As long as environmental conditions favor the multiplication and survival of the normal tissue cells, the potential cancer cells are restricted. When carcinogenic factors come into play the normal cells are killed while the potential cancer cells survive. The second alternative is transformation of individual normal tissue cells into cancer cells by somatic mutation, as the result of genic or environmental factors. The term "precancerous cells" covers latent pre-existing cancer cells as well as normal cells in the process of transformation into cancer cells.

A population of latent or potential cancer cells possesses a reactivity which is different from that of a population of fully developed cancer cells. This is well illustrated by studies on relations between *cancer and nutrition.* In numerous strains of mice restriction of caloric intake causes a marked decrease in the incidence of spontaneous cancers as well as of chemically induced neoplasms. Once a tumor has developed, distinct effects of caloric restrictions on the progress of the tumor cannot be demonstrated. These are some of the important results of the extensive work by Tannenbaum and Silverstone. A useful review by these authors (1953, pp. 451-501) includes records of Life Insurance Companies indicating greater incidence of human cancer in high body-weight classes, but the authors point out that reducing is not a cancer cure.

Similar to the effect of nutritional factors, one may distinguish between the *role of hormones* in the production of neoplasms and in the control of developed neoplasms. Circumstantial evidence suggests hormonal influences on the origin of human cancers, particularly in reproductive and endocrine organs. Well demonstrated is the effect of hormone treatment in carcinomas of the prostate and mammary gland. In experimental animals, hormones have been clearly established as carcinogenic factors. The first evidence in this direction was supplied by Lacassagne (1932), who induced breast cancer in male mice by the administration of estrogenic hormones. I mention Gardner's (1953) "Hormonal Aspects of Experimental Tumorigenesis" as a comprehensive review of this intricate subject. Dmochowski (1953) pointed out that the same hormonal factors which take part in the normal development of mammary glands also exert an influence on the development of neoplasms in these glands. The discovery of the milk factor by Bittner (1936, 1940) illustrated the way in which neoplasms result from the joint action of several factors. Certain spontaneous mammary carcinomas develop only in such strains of mice as have the proper genetic reactivity. In addition, the mice must be supplied with estrogen; in females from their ovaries, and in males by artificial devices. Finally, the mice must have received the milk factor when suckled either by their mothers or by foster mothers. This factor has many properties in common with viruses. The milk factor controls the incidence of spontaneous cancers in a way similar to cytoplasmic inheritance. Therefore, the factor has been compared to the kappa factor in *Paramecium aurelia* (Part IV, Chapter 1, E). Some neoplasms of mice, particularly those induced by carcinogenic hydrocarbons, seems to develop without the milk factor. Many environmental factors decide whether cancer develops in genetically suitable animals. Caloric intake was mentioned above. Different composition of the diet is another factor. The influence of environmental temperature deserves more study (Dmochowski, 1953).

It is important to recognize that, similar to the concept of poison, the *concept carcinogen should not be applied in an absolute sense.* A physical, chemical, or biological agent is a carcinogenic stimulus if the reactivity of the exposed tissue cells permits a neoplastic response. At a given moment the reactivity of the target cells depends on their past genetic and environmental history.

Quantitative relations between stimulus and reactivity are decisive. In the same species nitrogen mustards may be carcinogenic or carcinostatic depending on dose and duration of application (Steinhoff and Kuk, 1957). The polycyclic hydrocarbon 3-methylcholanthrene is known as a carcinogenic in rats and mice. However, Huggins and Pollice (1958) found that when this agent was applied to rats at a certain dose for a certain period of time, transplanted mammary tumors were retarded. The inhibitory effect was enhanced by simultaneous administration of androstan-17β-ol-3-one. As do chemical agents, ionizing radiation produces or suppresses cancer depending on dose, duration, and reactivity. Genetic and hormonal factors are prominent in determining this reactivity. An interesting illustration is found in studies by Bond *et al.* (1960) with rats of the Sprague-Dawley strain. In nonirradiated controls the incidence of breast tumors was 1 or 2%. Single exposure of the whole body to Roentgen rays produced breast tumors in 78% if the dose was 400 r, but in only 57% with a dose of 600 r. This can be explained by the known fact that an intact ovarian function is necessary for maximal breast tumor incidence. Evidently the dose of 400 r caused marked stimulation of the breast without damaging the ovaries. The authors emphasized that their data on the relation of dose and tumor incidence of the rat breast should not be generalized to other types of neoplasms or to other species. In a study on the modification of carcinogenic effects by dietary protein in rats, Griffin *et al.* (1949) emphasized that carcinogenic hydrocarbons should be considered in the same way as other toxic agents such as benzene, chloroform, arsphenamine, or selenium. Toxic effects are listed as the third item in my Table 2 which shows the general relations of stimulus, response, and reactivity in various areas of normal and pathological biology. Different chains of events may lead to a similar end phenomenon, cancer. Identical end effects of different factors are frequently seen in biology. As discussed earlier, melanin pigmentation of the skin occurs in a tanned white man, in a person with destruction of the adrenal cortex (Addison's disease), in a person of Mediterranean origin, and in a mulatto (Part II, Chapter 2, B).

A peculiar *energy metabolism of cancer cells* was discovered by Otto Warburg and his associates in surviving slices of tumor tissue (Warburg and Minami, 1923; Warburg, 1924). Whereas an oxidative (respiratory) type of metabolism prevails in normal tissue, tumor tissues predominantly ferment glucose. Both aerobic and anaerobic glycolysis are higher in neoplastic than in normal tissue. The energy metabolism of embryonic tissue and, surprisingly, of the adult retina resemble to some degree that of malignant tumors. Predominance of glycolysis over respiration was also observed in intact tumors *in situ* (Cori and Cori, 1925). Aisenberg's "The Glycolysis and Respiration of Tumors" (1961) covers the numerous aspects of this subject. The discussion in the last part of his book implies that the high glycolytic rate of energy metabolism is a constant and important characteristic of fully developed cancer cells, but does not shed any light on the way, or ways, in which normal cells are transformed into cancer cells. A negative experimental result had been obtained by Goldblatt and Cameron (1953). These investigators applied intermittent anaerobic conditions to normal rat fibroblasts cultivated *in vitro* to test possible transformation into cancer cells. One strain treated in this way developed conspicuous morphologic alterations, but implantation of such altered cells into the anterior chamber of the eye and various other places failed to produce tumors.

Little is known of the role of *blood vessels* in the biology of cancer cells. Some cancers form large avascular masses. As a rule, the center of such mass is necrotic, while the periphery shows an excess of cell multiplication over cell death.

There is no indication that, in general, cells in the wall of capillaries in a malignant tumor differ from those of other capillaries. Possibly the permeability of brain tumor capillaries for dyes is greater than that of capillaries of normal brain tissue (see Part IV, Chapter 3, C, 2). The pattern of capillaries in a malignant neoplasm is not adapted to hydrodynamic variations. Therefore, capillaries of tumors burst readily. Depending on frequency and location, the resulting hemorrhage may have serious clinical consequences. Direct nervous control of cancer cells is unlikely, although the presence of visible nerve fibers in cancer tissue has been claimed.

Neoplastic cells can be classified according to their reactivity to irradiation, nitrogen mustards, and other chemical agents, or to artificial virus infection. These factors are used as tools in biological cancer research and are studied, at the same time, for therapeutic effectiveness.

d. *Transplantability of neoplasms.* Probably, basic laws governing the acceptance and rejection of transplants are similar for normal and neoplastic tissues (Grüneberg, 1952, p. 467). Medawar (1943) stated that resistance to tumor homografting is directed against the foreign cells as such, irrespective of their malignant character. According to Billingham, Brent, and Medawar (1956) and Burnet (1959), cancer cells do not possess any antigens different from those of normal tissue cells and, therefore, are not "recognized" as aliens by the normal host tissues, if the genetic background of transplanted tumor tissue and host is similar. If the background is different, resistance to homoplastic tumor transplantation can be diminished by injection of finely dispersed tumor material into newborn animals before tumor transplants are made (Billingham, Brent, and Medawar, 1956).

Certain sites of the body, such as the anterior chamber of the eye, show low resistance to heteroplastic transplantation. This was observed first with tumor tissue (Greene, 1941). Subsequently, it was found that embryonic tissue, and even adult tissue, can persist in the anterior chamber provided the graft is not vascularized (Medawar, 1948). Similarly the cheek pouch of the Syrian hamster tolerates not only transplanted human tumor tissue but, for limited periods, also grafts of normal human skin (Resnick *et al.,* 1960). Tolerance to heteroplastic transplantation is enhanced if the prospective recipients are conditioned by Roentgen rays or cortisone. This permitted long-term study of human tumors in the hamster's cheek pouch (Toolan, 1951, 1954).

"Denaturation" of tumor tissue before grafting is another way to obtain successful transplants in hosts unrelated to the tumor donor. Apparently, explantation and cultivation *in vitro* causes denaturation in this sense. Colonies of Ehrlich's mouse carcinoma, cultivated in media free of mouse material for many years, produced tumors in most mice into which they were transplanted, although the mice were probably not genetically related (E. Mayer, F. Jacoby, and H. Mayer, 1933, unpublished data).

In their survey of transplantable and transmissible tumors Stewart *et al.* (1959) listed mouse, rat, and rabbit tumors which can be transplanted to unrelated hosts of the respective species. Although such tumors may produce a higher percentage of successful transplants within a given inbred strain, genetic relationship seems to be less important in the transplantation of cancer cells than it is in the transplantation of normal cells. In view of experiences with embryonic tissues, one might expect that undifferentiated (anaplastic) tumors can be transplanted with little difficulty to recipients not closely related to the tumor donor. The degree of correlation between anaplasia and malignancy of tumors will be discussed later. The usefulness and limitations of tumor transplantation in cancer research were reviewed by Klein in 1959. According to this author relations between malignancy and transplantability cannot be discussed in a general way. For studies of this problem he suggested replacing the concept of malignancy by definite criteria, such as rate of cell multiplication, degree of differentiation, invasiveness, ability to metastase, and dependence on hormonal and other controls.

e. *Mechanisms of invasiveness and metastasis.* Invasiveness is a frequent phenomenon in normal development. In the process of embryonic differentiation populations of cells invade new territories (Fig. 30). In chicks, two weeks after hatching the air sacks project from the lungs, invade the closed marrow cavity of the humerus, and finally replace the marrow (Bremer, 1940). This destruction of cortical bone and marrow is similar to destruction by a malignant tumor. Metastasis-like phenomena occur in chick embryos, such as the colonization of the gonad primordia by primary reproductive cells (gametes) which are transported in the blood stream from the place of origin. The choriocarcinoma of man presents a well-known transition from embryological to neoplastic invasiveness. Normally, chorionic villi of the placenta reach into the muscle wall of the uterus. Remainders of the coating layers of the villi (trophoblast) frequently survive in the uterine wall of women after termination of pregnancy. In some instances such portions of the trophoblast invade deeper blood vessels and are carried to the lungs, where they are arrested in the capillaries. In the lung the chorionic emboli may perish after some time, may persist without causing injury, or may form a malignant tumor (choriocarcinoma) with metastases in other organs. These tumors produce chorionic gonadotropin.

Metastases of neoplasms are frequently identified with the naked eye, while invasiveness can be seen only microscopically. The biochemical mechanisms of both phenomena are obscure. Although quantitative differences seem to exist between enzymes in cancer cells and those in normal tissue cells (Fenninger and Mider, 1954), no evidence has been found that cancer cells contain special enzymes by which they attack the normal living cells. Cancer cells *invade* adjacent structures by creeping into crevices, thus causing mechanical injury to the normal tissue and interfering with its nutrition. This can happen also among normal cells, as illustrated by the disintegration of tonsillar epithelium produced by invading lymphocytes (Fig 63). The remarkable fact that cartilage is not attacked by cancer cells may be due to the absence of crevices in cartilage. The *formation of metastases* in parts distant from the primary tumor depends on the following factors. First the cancer cells have to invade a blood or lymph vessel. Then they must be transported with the blood or lymph stream (retrograde transport also occurs). Traveling cancer cells arrested in the capillaries of an organ must be able to survive and multiply in this particular organ.

These problems have been studied in human autopsy material and in experimental animals. A classic paper by Schmidt (1903) showed circumstantial evidence that stomach carcinoma cells traveling through the pulmonary artery were arrested in the pulmonary capillaries, multiplied in this location, invaded the lymphatics and pulmonary veins, and subsequently were distributed by the systemic circulation. Zeidman *et al.* (1956) injected suspensions of rat tumor cells into the portal veins or renal arteries of rats. Cells of various rat neoplasms passed through the capillaries of the liver and kidney, except for cells of the Flexner-Jobling rat carcinoma. Subsequently, Zeidman (1961) used microcinematography to demonstrate the passage of cancer cells through capillaries. Suspensions of Brown-Pearce rabbit tumor cells were injected into a mesenteric artery of a rabbit, while simultaneously cinematographic pictures were taken in the region of the arteriocapillary junctions of the mesentery. By sacrificing the rabbits one month after microcinematography, the viability of tumor cells after passage through the capillaries was verified. At autopsy more than 95% of these rabbits showed tumor nodules.

f. *Tumors in various classes of animals, including invertebrates, and in plants.* Tumors have been described in larval stages of different classes of animals. Metamorphosis involves extensive destruction of old organs and tissues and the production of new ones. Errors in development are very likely to happen, and therefore the borderline between malformations and tumors in metamorphosing animals is even more vague than in mammals. Criteria of malignancy hardly seem applicable to larval stages. This point is frequently obscured by wrong emphasis on lethal effects of the abnormal structures.

Setting aside the questions of malignancy and etiology, one may define a tumor as a mass of tissue which is not organized or differentiated like the normal tissue of the particular species. In this broad sense, tumors occur not only in vertebrates, but also in many classes of invertebrates and plants. With respect to tumors in invertebrates the reader is referred to the 1950 review by Scharrer and Lochhead.

There is a vast difference between crown galls of sugar beets and human mammary carcinomas, but resemblances are not lacking. The production of new tissues, normal or abnormal, in the adult organism hinges on the presence of meristematic cells, both in plants and animals. Deviations from the normal diploid chromosome numbers were observed in cells of mouse tumors as well as in cells of sugar beet crown galls (Winge, 1927, 1930). Several generations of snapdragons (*Antirrhinum*) derived from one plant which had been irradiated as an embryo showed conspicuous alterations of leaves and stems with abnormal variability of cell bodies, intermitotic nuclei, and mitoses in the tissues (Stein, 1930). One of the remarkable cytologic abnormalities was the occurrence of more than 8000 chromosomes (estimated) in a single mass of unpartitioned cytoplasm (Stein, 1935, p. 310). As mentioned previously, a single cell of mouse ascites carcinoma may contain more than 1000 chromosomes (Levan and Hauschka, 1953).

From the physiological point of view it seems difficult to compare the so-called tumors of plants with neoplasms in vertebrate animals. In response to local injury the animal has the alternative of inflammatory responses or production of new tissue, regenerative or neoplastic. The plant responds to local injury only by producing new tissue in the form of calluses or galls. The reason for this fundamental difference is that the plant has neither a circulatory system nor migratory tissue cells and, therefore, is incapable of producing the complex phenomena of inflammation. These points have been stated clearly in a paper by Levin and Levine "The Role of Neoplasia in Parasitic Diseases of Plants" (1922).

g. *Causes of death and the question of metabolic cachexia in cancer.* Many experimental and clinical investigators consider the fatal effect of a neoplasm to be a decisive criterion of malignancy. The main *causes of death* are obstruction or destruction of vital organs by the primary tumor and the metastases. Fatal anemia can be produced by extensive metastases in the bone marrow or by hemorrhages from ulcerated tumor masses. Frequently necrosis and ulceration of neoplasms lead to fatal infections. Metastases of the peritoneum cause ascites, and increasing accumulation of fluid in the peritoneal cavity disturbs both nutrition and circulation. In experimental animals, especially mice, subcutaneous tumors may grow to such a size that they equal the weights of the animals, gradually interfering with most functions.

Are any toxic substances produced by the metabolism or decomposition of cancer cells? On the assumption that such toxic substances exist, the term *cancer cachexia* has been applied to the final condition in which many cancer patients die. The combination of emaciation, anemia, and a peculiar discoloration of the skin constitutes this cachexia. Willis (1953) rejects the idea of a special cancer cachexia, since practically all cancer deaths can be attributed to hemorrhages, infections, or interferences with vital functions, without invoking the production of special toxins. I agree with Willis on the basis of my own extensive experience with autopsies of cancer patients.

Fenninger and Mider (1954) expressed the opinion that in a large group of cancer patients death must be ascribed "to profound alterations in the metabolism of the host." They found the energy and nitrogen metabolism in human cancer patients as well as in experimental animals with tumors to be markedly different from the metabolism in normal organisms. Fenniger and Mider did not state the presence or absence of hemorrhages, infections, or mechanical disturbances of vital organs in the tumor bearers. Begg (1952) emphasized the fact that the fatal metabolic changes in the host produced by many neoplasms are *not specific* of neoplasms. Wright (1958, pp. 484-485) states that cancer cachexia is, at least in part, the result of a "metabolic drain" of the neoplastic tissues upon the host. Wright refers to studies by LePage *et al.* (1952) on protein metabolism in tumor-bearing rats. These authors showed that in starved animals the tumor tissue obtains an excessive share of amino acids, as judged by glycine-2-C^{14} utilization.

Although toxic substances or special enzymes produced by cancer cells have not been demonstrated, some circumstantial evidence seems to favor the idea that cancer cells interfere biochemically or immunologically with the functions of normal tissue cells (see Green's 1959 review). It has been claimed that the anemia in many cancer patients is "primarily of an auto-immune type" (Green *et al.*, 1957), not necessarily caused by hemorrhages or bone marrow destruction. The arguments of Green in favor of the antigen-antibody reactions between neoplastic and normal cells were not accepted by Burnet (1959, p. 195). Direct biochemical or immunological interactions between neoplastic and normal cells are hardly compatible with the views of Medawar (1958) and Burnet (1959) that cancer cells are *lacking* the antigens of other tissue cells and, therefore, are not recognized as foreigners by the host.

2. *Medicopathological Classification of Neoplasms*

For the purposes of practical medicine the following characteristics of neoplasms are important: primary and secondary sites; benign or malignant behavior; if malignant, degree of malignancy; reactivity to treatment; and finally the morphological label which is needed for communication and recording.

a. *Primary and secondary sites of neoplasms.* The *rules for deciding whether a neoplasm is primary or secondary* (metastatic) are based on past experience concerning the frequency of single tumors, on histological characteristics, and on possible routes of transportation. It is not easy to find these rules in the literature.

If an autopsy of a patient without history of previous tumor treatment reveals only one neoplasm, this is considered to be primary. If there are two or more neoplasms in the body, they may either be in relation of primary and secondary, or of multiple primary tumors. Observations in hospital pathology have shown that some tumors, such as carcinomas of the esophagus or rectum, frequently are the only detectable one at the time of autopsy. By contrast carcinomas of the liver are rarely found alone.[1] Therefore, if an esophagus and liver carcinoma (or a

[1] Primary liver carcinomas are frequent in some populations in South Africa and the Far East.

rectum and liver carcinoma) occur together, the probability of the liver carcinoma being metastatic is greater than the reverse. Neoplasms of similar size and composition may be found in paired organs, for instance, in both ovaries. Then it is impossible to guess whether both are primary or one is metastatic. The so-called Krukenberg tumor is a carcinoma which, as a rule, involves both ovaries, and one other organ, mainly the stomach. Since metastases are very rare in the stomach, it is reasonable to assume that the stomach is the primary site and the ovaries the secondary site of the neoplasm. Histological and circulatory criteria are used in distinguishing primary and metastatic neoplasms. Suppose a carcinoma with gland-like structures is found in the wall of the stomach, continuous with the mucous membrane, and tumor masses of similar structure are found in lymph nodes in the vicinity of the stomach. It is likely that this type of carcinoma originated from epithelium of the gastric mucosa, but it is unlikely that it originated from lymphatic tissue which does not contain epithelium. Secondly, it is known that the lymph vessels of the stomach drain to the neighboring lymph nodes. Both arguments support the interpretation that the stomach is the primary site, and the lymph nodes the secondary site, of the neoplasm.

Circulatory factors in the production of metastases may be recapitulated here. Cancer cells travel in the direction of the blood or lymph stream, or in a retrograde direction. Clumps of cancer cells, which reach the lungs through the pulmonary lymphatics or arteries are arrested in the capillaries, but may subsequently invade the pulmonary veins. This allows the cancer cells to spread through the systemic circulation. Single cancer cells of a diameter exceeding that of capillaries may temporarily assume an elongated and narrow shape permitting them to pass through the capillaries. Analysis of circulatory factors alone can hardly decide which of two tumors is primary and which metastatic. However, some circulatory distributions are more probable than others.

b. *Grading of malignancy.* Depending on the organ, the presence of invasiveness leads to different predictions concerning the future development of the neoplasm. If a tumor is removed surgically, microscopic evidence of invasion of lymphatics or blood vessels in the tissue adjacent to the tumor makes it probable that metastases have already formed in the vicinity and possibly also in distant places. This rule does not hold for thyroid tumors. Human thyroid adenomas with local invasiveness at the time of surgical removal showed a surprisingly weak tendency to metastasize (Warren, 1931). Of 34 cases with invasion of the thyroid capsule and local blood vessels, only two patients died of metastases whereas the 32 others were cured by the operation. Since one cannot assume that in 32 patients surgery prevented the spread of the tumor just in the nick of time, the conclusion is that thyroid tumor cells have a small chance to survive in the blood circulation or to colonize another organ. In other words, invasive adenomas or adenocarcinomas of the thyroid possess a relatively *low degree of malignancy.* As mentioned before, thyroid tumors are hormone controlled to a great extent. Under these circumstances the unspecified term cancer should not be applied to the thyroid gland of man or experimental animals. The low grade of malignancy of experimental thyroid tumors in rats was emphasized by Bielschowski *et al.* (1949).

Metastasis is an indicator of high-grade malignancy in most neoplasms. An exception is the behavior of clear-cell renal carcinomas ("hypernephromas") in man, since their pulmonary metastases are apt to disappear either after surgical removal of the primary kidney tumor or while the primary tumor is still present. This was mentioned previously in the discussion of spontaneous regression of neoplasms. Even the continuous presence of pulmonary metastases of these renal tumors seems to be compatible with a long relatively healthy life of the tumor bearer. Walter and Gillespie (1960) reported the case of a woman who died when 81 years old, having had pulmonary metastases for forty years after removal of a kidney with carcinoma ("hypernephroma"). During the last seven years of her life most pulmonary nodules increased in size, but some decreased. At autopsy, pulmonary as well as hepatic metastases were found.

Recurrence of neoplasms after surgical or other treatment is considered indicative of a high degree of malignancy in practical medicine. *Reactivity of neoplasms* to irradiation and chemotherapy was discussed previously from the biological point of view. Biological radiosensitivity of a particular tumor tissue does not entirely decide therapeutic success. The efficacy of radiotherapy is controlled by topographic factors, since normal structures surrounding the neoplasm should not be damaged by the irradiation; this was described in connection with procedures of regional anatomy (Part III, Chapter 6, A, 4). The success of surgical treatment depends on the completeness of removal of the neoplasm and on the absence of metastases at the time of operation. Since these are technical problems, local recurrence and metastasis after operation are not suitable criteria for estimation of biological reactivity of a neoplasm.

Abnormal variability of nuclei, frequency of cells in mitosis, and the occurrence of abnormal mitoses are valuable *cytological indicators of malignancy.* However, the absence of these cytological characteristics does not exclude malignancy. Exfoliative cytology has enriched tumor diagnosis during the last decades.

Attempts to correlate the histological appearance of neoplasms with the degrees of biological malignancy can be traced back to the introduction of the concept anaplasia by Hansemann in 1893, which was intended to characterize the peculiar immaturity of cancer cells as differing both from normal adult and normal embryologic cells. Hansemann stated that, as a rule, highly anaplastic forms of carcinomas and sarcomas are malignant, but he warned that the degrees of histological differentiation should not be overrated as indicators of the biological behavior of the tumor (pp. 92 and 93 of his 1893 book). Later investigators attempted to demonstrate *definite parallelism* between the histological pictures of cancers and their degrees of malignancy.

The most determined effort in this direction was made by Broders, starting in 1920. In his 1941 paper he divided carcinomas of the skin, lip, uterus, breast, and other organs into *four grades of histological differentiation,* with the claim that these grades corresponded with degrees of malignancy. The most differentiated grade 1 was supposed to have the lowest malignancy, and the completely undifferentiated grade 4 the highest degree of malignancy. The prognostic value of this scoring system was restated in subsequent publications by Broders and other members of the Mayo Clinic staff. Representative is a paper on breast carcinomas by Berkson *et al.* (1957) from which some data may be cited here. Survival for 15 years or longer after operation was observed in 76% of the patients whose carcinomas had been classified as grade 1; in 56% of those with grade 2; in 50% of those with grade 3; and in 46% of those with grade 4.

Most pathologists agree that histological grades are useful in recording morphological types of carcinomas. Photomicrographs of carcinomas graded according to Broders are found in various textbooks of pathology, for instance, in Boyd's "Surgical Pathology" (1942, Figs. 39–46). However, strong objections have been raised to the exclusive reliance on histological pictures in estimating the biological malignancy of tumors. Boyd (1942, p. 109) listed factors which are prognostically more important than microscopic properties of the tumor: age of the patient, duration and size of the tumor, its rate of growth (increase in volume) prior to operation, and the general appearance of the patient. Schiller (1953, pp. 1071–1072) classified the carcinomas of the cervix uteri into four groups according to the degree of extension of the tumor at the time examined. Of group 1, with carcinomas limited to the cervix, about 80% obtained five-year cures; in group 4, with carcinomas massively invading adjacent parts and extending to the wall of the pelvis, no five-year cures occurred. Groups 2 and 3 were intermediary. Schiller expressed the opinion that combined evaluation of the histological grade and the degree of extension yields the most accurate prognosis.

Is there any way of proving or disproving the claim that histological grades *alone,* without clinical data, are a reliable basis for prognosis? In order to test this claim both the histological and clinical procedures should be standardized. In a microscopic study of the uterine cervix carcinoma, Hueper (1928) rejected peripheral portions because of variable influences of adjacent normal tissue, and excluded central parts, "in which the original structure of the neoplasm is frequently altered." This shows how difficult it is to establish criteria for uniform, unbiased sampling of tumor tissue. On the clinical side there is little doubt that survival times are modified by postoperative treatment, particularly by different modes of radiotherapy. Therefore, patients should be separated into classes with different postoperative treatments. In consideration of all these factors it seems hardly possible to determine the prognostic value of histological grading alone. Probably there is no practical necessity to settle this question since clinical criteria are always available to supplement the histological data.

c. *Morphological identification: histogenetic versus descriptive classification.* In the "Standard Nomenclature of Diseases" (1961) tumors are classified first according to their sites and then according to their histological properties including histogenesis. Similarly, publications on experimental cancer research in animals use sites as well as histology and histogenesis as criteria for identification. Is there a general agreement on the rules of histological descriptions and records of neoplasms? Most pathologists give preference to histogenetic interpretations of tumors over merely descriptive statements. In descriptive terms, the elements of carcinomas *resemble* normal epithelial cells and those of sarcomas *resemble* normal connective tissue cells. In histogenetic terms, carcinoma cells and sarcoma cells are *derived* from epithelium and connective tissue, respectively. The two principles lead to similar classifications in all neoplasms which show a certain degree of histological differentiation. Thus the presence of stratified squamous epithelium in a malignant neoplasm leads to the diagnosis epidermoid carcinoma, whereas tubular or gland-like structures suggest the label adenocarcinoma. Malignant tumors containing cartilage or bone are identified as chondrosarcoma or osteosarcoma, respectively. It is customary to classify a neoplasm according to the *highest degree of differentiation* which can be detected. For instance, if most histological sections of a carcinoma show no differentiation, but one section contains a small area of gland-like structures, the diagnosis will be adenocarcinoma. Sometimes it is desirable to state the proportion of differentiated to undifferentiated areas in the sections examined.

The presence of connective tissue fibers in malignant tumors is interpreted in different ways depending on various conditions. If groups of carcinoma cells with typical mosaic (epithelial) arrangement are surrounded by connective tissue, the tumor cells either have invaded the pre-existing connective tissue or have stimulated the invaded tissue to produce fibers. In both cases the connective tissue is called the stroma of the tumor. Suppose neoplastic cells with elongated nuclei are not arranged in an epithelial pattern, but are associated with connective tissue fibers, similar to normal fibroblasts. One would assume that the fibers have been produced by the tumor cells, and the diagnosis would be fibrosarcoma.

Neoplasms without visible histological differentiation present problems of classification. Single cells cannot be identified as epithelial or connective tissue cells, and simple forms of cellular associations such as mosaic and network, are not reliable indicators of cytogenesis (Part IV, Chapter 2). How sure can one be that a neoplasm which did not form epidermis-like or gland-like structures originated from normal epithelium? What are the reasons for ascribing the origin of a tumor to normal connective tissue cells if neither fibers nor cartilage nor bone were produced by the tumor cells? The answer is that histogenetic interpretations are based not entirely on the appearance or pattern of aggregation of the neoplastic cells, but also on topographic and other *auxiliary criteria.* If an undifferentiated tumor is located in the deeper layers of an extremity, far from the skin, one would be inclined to assume that the neoplastic cells originated from connective tissue rather than from epithelium and, therefore, to classify the tumor as sarcoma. If a neoplasm is morphologically similar and continuous with a bronchus, one assumes that the tumor cells are derived from the bronchial epithelium, and the tumor is classified as carcinoma. Undifferentiated neoplastic cells seemingly attached to the wall of blood vessels, are interpreted as sarcoma. Similar cell masses distinctly separated from blood vessels are considered to be carcinoma cells. In undifferentiated tumors, parallel arrangement of oval nuclei, called palisading, is a criterion in favor of epithelial origin. Similar nuclei not arranged in a palisading pattern are hypothetically derived from connective tissue cells. Functional characteristics such as the production of specific hormones are valuable indicators of the origin of a neoplasm.

Because of the difficulties of histogenetic interpretation in undifferentiated neoplasms, a purely descriptive classification might be preferable. Such descriptive classification is possible on the basis of nucleocytoplasmic patterns, which are illustrated in Figs. 70 to 77. These descriptive patterns should be particularly useful in experimental cancer research. In human pathology conditions are not favorable to the best type of histological fixation. Therefore, diagnoses lean heavily on nuclei and their distribution, while cytoplasmic characteristics including cell boundaries do not carry too much weight. If the main nucleocytoplasmic patterns, namely, single cells, mosaic, cytoplasmic network, and unpartitioned multinuclear cytoplasm, cannot be identified in neoplastic tissue, the tumor is *not classifiable* in descriptive terms, much less in histogenetic terms.

Willis, in "Pathology of Tumors" (1953, p. 10) used four thyroid tumors to illustrate how the precision of identification decreases as the degrees of differentiation decrease. The first tumor was a highly differentiated thyroid carcinoma containing vesicles filled with colloid, very similar to normal thyroid. The second tumor showed definite glandular structures permitting the diagnosis adenocarcinoma, but without histological indication of thyroid origin. In the third tumor an epithelial pattern was recognizable, and therefore, the tumor could be classified as carcinoma without specification. Finally, a fourth tumor consisted of cells without any typical arrangement so that the histologist could identify it only as "a highly cellular anaplastic tumor of undetermined origin." In spite of undetermined origin and the absence of any epithelial characteristics, Willis decided to include the fourth tumor in the class "carcinomas of the thyroid." A tumor sample without signs of differentiation in many sections examined, may reveal differentiated structures after extended search. The so-called oat-cell tumor of the bronchi represents a well-known type of neoplasm which is undifferentiated throughout. The tumor tissue consists of variable oval nuclei in an ill-defined mass of cytoplasm. The descriptive name of oat-cell tumor is useful, since the tumor cannot be classified either as carcinoma or sarcoma.

Fischer-Wasels (1927) introduced the term meristomas for tumors too immature to be classified histogenetically. I prefer the longer term "unclassifiable malignant neoplasms" for two reasons: (1) both relatively mature and completely immature tumor cells must necessarily have originated from meristematic cells, and (2) cells which morphologically look undifferentiated may be differentiated biochemically. The existence of tumors of undetermined histogenesis and undifferentiated morphological pattern has been recognized in "Standard Nomenclature of Diseases" as well as in some publications on experimental cancer research such as that of Stewart *et al.* (1959). Most textbooks are reluctant to admit that the two categories carcinoma and sarcoma cannot cover all malignant neoplasms.

Cohn (1926) devoted a comparative study to the characteristics of carcinomas and sarcomas. Each neoplasm was classified in two ways. The first procedure was purely descriptive based on nucleocytoplasmic patterns, visible products of differentiation, and spatial relations between tumor cells and connective tissue fibers. The second procedure included all "auxiliary criteria" available, such as macroscopic location, attachment of tumor cells to capillary walls, and probable origin inferred from past experience. With the first procedure a large proportion of the tumors examined remained unclassifiable. The fact that so many could not be fitted into the categories of carcinoma or sarcoma was not surprising, since preference had been given to highly anaplastic types in this study. With the second procedure Cohn arrived at conventional diagnoses of all primary and metastatic neoplasms examined. For instance, tumors of the stomach, which consisted of small groups of uncharacteristic cells scattered between connective tissue fibers, were identified as the well-known diffuse (scirrhous) type of stomach carcinoma when auxiliary criteria were applied. In purely descriptive terms these tumors were unclassifiable for the following reasons: the nucleocytoplasmic pattern was indistinct, and there was no indication whether the connective tissue fibers were produced by the neoplastic cells or by the surrounding normal tissue. In other words, the neoplastic cells might as well have originated from normal epithelial cells as from normal connective tissue cells.

Evidently, alternative classifications, depending on the purpose, are as possible for neoplasms as they are for other diseases. Requirements for medicopathological and biological classifications are interdependent but not identical.

Part V: References

Abercrombie, M., and Harkness, R. D. (1951). *Proc. Roy. Soc.* **B138**, 544.

Aisenberg, A. C. (1961). "The Glycolysis and Respiration of Tumors." Academic Press, New York.

Albrecht, E. (1907). *Frankfurt. Z. Pathol.,* Vol. 1, pp. 221-247, 377-425.

Alfert, M. (1954). *Intern. Rev. Cytol.* **3**, 131.

Askanazy, M. (1923). *Münch. med. Wochschr.* **70**, 1107.

Babcock, E. B., and Stebbins, Jr., G. L. (1938). "The American Species of Crepis," *Carnegie Inst. Wash. Publ. No.* **504**, p. 1.
Bale, W. F., and Spar, I. L. (1954). *J. Immunol.* **73**, 125.
Bauer, K. H. (1928). "Mutationstheorie der Geschwulstentstehung." Springer, Berlin.
Bauer, W. C., and Rosenberg, B. F. (1960). *Am. J. Pathol.* **37**, 695.
Begg, R. W. (1952). *In* "Steroid Hormones and Tumor Growth." *Ciba Foundation Colloq. on Endocrinol.* **1**, 170.
Benitz, K.-F., Moraski, R. M., and Cummings, J. R. (1961). *Lab. Invest.* **10**, 934.
"Bergey's Manual of Determinative Bacteriology" (1957). (R. S. Breed, E. G. D. Murray, and N. R. Smith, eds.), 7th ed. Williams & Wilkins, Baltimore, Maryland.
Bergström, S. (1959). *Intern. Cong. Biochem. 4th Congr. Vienna 1958,* Vol. 4, Symp. 4, p. 160.
Berkson, J., Harrington, S. W., Clagett, O. T., Kirklin, J. W., Dockerty, M. B., and McDonald, J. R. (1957). *Proc. Staff Meetings Mayo Clinic* **32**, 645.
Berrill, N. J. (1955). *In* "Analysis of Development," (B. H. Willier, P. Weiss, and V. Hamburger, eds.), p. 620. Saunders, Philadelphia, Pennsylvania.
Bielschowsky, F., Griesbach, W. E., Hall, W. H., Kennedy, T. H., and Purves, H .D. (1949). *Brit. J. Cancer* **3**, 541.
Biesele, J. J., and Cowdry, E. V. (1944). *J. Natl. Cancer Inst.* **4**, 373.
Billingham, R. E., and Boswell, T. (1953). *Proc. Roy. Soc.* **B141**, 392.
Billingham, R. E., Brent, L., and Medawar, P. B. (1953). *Nature* **172**, 603.
Billingham, R. E., Brent, L., and Medawar, P. B. (1956). *Phil. Trans. Roy. Soc. London* **B239**, 357.
Bittner, J. J. (1936). *Science* **84**, 162.
Bittner, J. J. (1940). *Am. J. Cancer* **39**, 104.
Bond, V. P., Cronkite, E. P., Lippincott, S. W., and Shellabarger, C. J. (1960). *Radiation Research* **12**, 276.
Boveri, Th. (1914). "Zur Frage der Entstehung maligner Tumoren." Fischer, Jena.
Boyd, W. (1942). "Surgical Pathology," 5th ed. Saunders, Philadelphia, Pennsylvania.
Bremer, J. L. (1940). *Anat. Record* **77**, 197.
Briggs, R., and King, T. J. (1955). *In* "Biological Specificity and Growth," *12th Symposium Soc. Study Develop. Growth, Durham, New Hampshire, 1953* (E. G. Butler, ed.), p. 207. Princeton Univ. Press, Princeton, New Jersey.
Broders, A. C. (1941). *Surg. Clin. North Am.* **21**, 947.
Brown, W. H., Pearce, L., and Van Allen, C. (1924). *J. Exptl. Med.* **40**, 583.
Buchsbaum, R., and Buchsbaum, M. (1957). "Basic Ecology." Boxwood Press, Pittsburgh, Pennsylvania.
Burnet, Sir Francis MacFarlane (1959). "The Clonal Selection Theory of Acquired Immunity." Vanderbilt University Press, Nashville, Tennessee.
Carlson, A. J., and Johnson, V. (1953). "The Machinery of the Body," 4th ed. Univ. of Chicago Press, Chicago, Illinois.
Caullery, M. (1952). "Parasitism and Symbiosis" (translated by A. M. Lysaght). Sidgwick and Jackson, London.
Chaves, A. D., and Abeles, H. (1952). *Am. Rev. Tuberc.* **65**, 128.
Clark, A. M. (1951). *J. Exptl. Biol.* **28**, 332.
Clark, R. L., Jr. (1947). *Texas Repts. Biol. and Med.* **5**, 366.
Cohn, H. M. (1926). *Arch. pathol. Anat. u. Physiol. Virchow's* **259**, 30.
Colucci, V. S. (1891). *Mem. reale accad. sci. ist. Bologna* [5], **1**, 593.
Cori, C. F., and Cori, G. T. (1925). *J. Biol. Chem.* **64**, 11.
Costello, D. P. (1955). *In* "Analysis of Development," (B. H. Willier, P. Weiss, and V. Hamburger, eds.), p. 213. Saunders, Philadelphia, Pennsylvania.

Coulombre, A. J. (1955). *Am. J. Anat.* **96**, 153.
Cowdry, E. V. (1950). "A Textbook of Histology," 4th ed. Lea & Febiger, Philadelphia, Pennsylvania.
Dalton, A. J. (1959). *Lab. Invest.* **8**, 510.
Dawson, A. B. (1937). *Am. Naturalist* **71**, 605.
Defendi, V., Billingham, R. E., Silvers, W. K., and Moorhead, P. (1960). *J. Natl. Cancer Inst.* **25**, 359.
Demerec, M., Blomstand, I., and Demerec, Z. E. (1955). *Proc. Natl. Acad. Sci. U. S.* **41**, 359.
Dessau, F. I., and Sullivan, W. J. (1961). *Toxicol. Appl. Pharmacol.* **3**, 654.
Dmochowski, L. (1953). *Advances in Cancer Research* **1**, 103.
Dobzhansky, T. (1951). "Genetics and the Origin of Species," 3rd ed. Columbia Univ. Press, New York.
Everson, T. C., and Cole, W. H. (1956). *Ann. Surg.* **144**, 366.
Fenninger, L. D., and Mider, G. B. (1954). *Advances in Cancer Research* **2**, 229.
Fischer-Wasels, B. (1927). *In* "Handbuch der normalen und pathologischen Physiologie" (A. Bethe *et al.*, eds.), Vol. 14/II, p. 1341. Springer, Berlin.
Furth, J. (1953). *Cancer Research* **13**, 477.
Gardner, W. U. (1953). *Advances in Cancer Research* **1**, 173.
Glücksmann, A. (1951). *Biol. Revs. Cambridge Phil. Soc.* **26**, 59.
Goddard, R. F. (1948). *Anat. Record* **101**, 539.
Goldblatt, H., and Cameron, G. (1953). *J. Exptl. Med.* **97**, 525.
Green, H. N. (1959). *In* "Carcinogenesis: Mechanisms of Action" (G. E. W. Wolstenholme, M. O'Connor, eds.), *Ciba Foundation Symposium.* Little, Brown, New York
Green, H. N., Wakefield, J., and Littlewood, G. (1957). *Brit. Med. J.* **II**, 779.
Greene, H. S. N. (1941). *J. Exptl. Med.* **73**, 475.
Griffen, A. C., Clayton, C. C., and Baumann, C. A. (1949). *Cancer Research* **9**, 82.
Grüneberg, H. (1952). "Genetics of the Mouse," 2nd ed. Martinus Nijhoff, The Hague.
Hansemann, D. (1893). "Studien über die Spezificität, den Altruismus und die Anaplasie der Zellen." Hirschwald, Berlin.
Harvey, E. N. (1952). "Bioluminescence." Academic Press, New York.
Heidenhain, M. (1907). "Plasma and Zelle." Fischer, Jena.
Hempelmann, L. H., Lisco, H., and Hoffman, J. G. (1952). *Ann. Internal Med.* **36**, 279.
Herrmann, H. (1960). *Science* **132**, 529.
Holtfreter, J. (1939). *Arch. exptl. Zellforsch. Gewebezücht* **23**, 169.
Hueper, W. C. (1928). *Arch. Pathol.* **6**, 1064.
Huettner, A. F. (1949). "Fundamentals of Comparative Embryology of the Vertebrates," 2nd ed. Macmillan, New York.
Huggins, C., and Pollice, L. (1958). *J. Exptl. Med.* **107**, 13.
Huxley, J. (1958). "Biological Aspects of Cancer." Harcourt, Brace, New York.
Hyman, L. H. (1955). "The Invertebrates: Echinodermata." McGraw-Hill, New York.
Innes, J. R. M., McAdams, A. J., and Yevich, P. (1956). *Am. J. Pathol.* **32**, 141.
Jackson, B., and Dessau, F. I. (1961). *Lab. Invest.* **10**, 909.
Jacoby, M. (1927). *Z. Urol.* **21**, 241.
Jenkins, G. D. (1959). *J. Urol.* **82**, 37.
Karpechenko, G. D. (1928). *Z. induktive Abstammungs—u. Vererbungslehre* **48**, 1.
Kauffmann, F. (1956). *Zentr. Bakteriol.* **165**, 344.
Kauffmann, F. (1957). *In* "Ergebnisse der Microbiologie" (W. Kikuth *et al.*, eds.), p. 160. Springer, Berlin.
Klein, G. (1959). *Cancer Research* **19**, 343.
Koehler, W. (1932). *Z. Morphol. und Ökol. Tiere* **24**, 582.

Körner, F. (1937). *Z. mikroskop.-Anat Forsch.* **41**, 131.

Krogh, A. (1922). "Anatomy and Physiology of Capillaries." *Yale Univ. Press,* New Haven, Connecticut.

Lacassagne, A. (1932). *Compt. rend. acad. sci.* **195**, 630.

Langman, J. (1958). *Anat. Record* **130**, 329.

Lasnitzki, I. (1951). *Brit. J. Cancer* **5**, 345.

LePage, G. A., Potter, V. R., Busch, H., Heidelberger, C., and Hurlbert, R. B. (1952). *Cancer Research* **12**, 153.

Levan, A., and Hauschka, T. S. (1953). *J. Natl. Cancer Inst.* **14**, 1.

Levin, I., and Levine, M. (1922). *J. Cancer Research* **7**, 171.

Ljunggren, E., Holm, S., Karth, B., and Pompeius, R. (1959). *J. Urol.* **82**, 553.

Lüderitz, O., Westphal, O., Staub, A. M., and LeMinor, L. (1960). *Nature* **188**, 556.

Mangold, O. (1925). *Naturwissenschaften* **13**, 213, 231.

Mayer, E. (1930). *Arch. exptl. Zellforsch. Gewebezücht* **10**, 221.

Mayer, E., and Furuta, S. (1924). *Arch. pathol. Anat. u. Physiol. Virchow's* **253**, 574.

Mayr, E. (1942). "Systematics and the Origin of Species." Columbia Univ. Press, New York.

Medawar, P. B. (1943). *Bull. War Med.* **4**, 1.

Medawar, P. B. (1948). *Brit. J. Exptl. Pathol.* **29**, 58.

Medawar, P. B. (1958). *Harvey Lectures Ser.* **52**, 144.

Melville, H. (1851). "Moby Dick." Modern Library, New York, 1930.

Meyer, H. H., and Gottlieb, R. (1911). "Die experimentelle Pharmakologie als Grundlage der Arzneibehandlung," 2nd ed. Urban & Schwarzenberg, Berlin and Vienna.

Morgagni, G. B. (1761). "De sedibus et causis morborum," English translation by B. Alexander, London, 1769.

Needham, J. (1942). "Biochemistry and Morphogenesis." Cambridge University Press, London and New York.

Nelson, R. A., Jr., Zheutlin, H. E. C., Diesendruck, J. A., Austin, P. G., Jr., with the assistance of Stack, P. S. and Eagan, J. P. (1950). *Am. J. Syphilis Gonorrhea Venereal Diseases* **34**, 101.

Oppenheimer, J. M. (1940). *Quart. Rev. Biol.* **15**, 1.

Painter, T. S. (1934). *J. Heredity* **25**, 465.

Parker, R. C., Castor, L. N., and McCulloch, E. A. (1957). *In* "Cellular Biology, Nucleic Acids and Viruses" *Spec. Publ. N. Y. Acad. Sci.* No. **5**, p. 303.

Patten, B. M. (1953). "Human Embryology," 2nd ed. McGraw-Hill (Blakiston), New York.

Peacock, E. E., Weeks, P. M., and Petty, J. M. (1960). *Ann. N. Y. Acad. Sci.* **87**, 175.

Pearce, L., and Brown, W. H. (1923). *J. Exptl. Med.* **38**, 347.

Pessin, S. B. (1953). *In* "Pathology" (W. A. D. Anderson, ed.), 2nd ed., p. 605. Mosby, St. Louis, Missouri.

Pressman, D., and Sherman, B. (1951). *J. Immunol.* **67**, 21.

Prosser, C. L. (1950). *In* "Comparative Animal Physiology" (C. L. Prosser, ed.), p. 103. Saunders, Philadelphia, Pennsylvania.

Resnick, B., Farber, E. M., and Fulton, G. P. (1960). *A.M.A. Arch. Dermatol.* **81**, 394.

Rich, A. R. (1951). "The Pathogenesis of Tuberculosis," 2nd ed. C. C Thomas, Springfield, Illinois.

Rose, N. R., and Witebsky, E. (1956). *J. Immunol.* **76**, 417.

Rose, N. R., and Witebsky, E. (1959). *J. Immunol.* **83**, 34.

Rössle, R. (1907). *Beitr. pathol. Anat. allgem. Pathol.* **41**, 181.

Rous, P. (1946). *Am. Scientist* **34**, 329.

Rugh, R. (1948). "Experimental Embryology." Burgess, Minneapolis, Minnesota.
Sahyoun, P. F. (1939). *Am. J. Pathol.* **15**, 455.
Scharrer, B., and Lochhead, M. (1950). *Cancer Research* **10**, 403.
Schechtman, A. M. (1948). *Proc. Soc. Exptl. Biol. Med.* **68**, 263.
Schechtman, A. M., and Nishihari, T. (1955). *Ann. N. Y. Acad. Sci.* **60**, 1079.
Scherbaum, O., and Zeuthen, E. (1954). *Exptl. Cell Research* **6**, 221.
Schiller, W. (1953). *In* "Pathology" (W. A. D. Anderson, ed.), 2nd ed., p. 1044. Mosby, St. Louis, Missouri.
Schmidt, M. B. (1903). "Die Verbreitungswege der Karzinome und die Beziehung generalisieter Sarkome zu den leukämischen Neubildungen." Fischer, Jena.
Singer, M. (1959). *In* "Regeneration in Vertebrates" (C. S. Thornton, ed.), p. 59. Univ. of Chicago Press, Chicago, Illinois.
Singer, M. (1960). *In* "Developing Cell Systems and Their Control" (D. Rudnick, ed.), p. 115. Ronald, New York.
Sinnott, E. W., Dunn, L. C. and Dobzhansky, T. (1950). "Principles of Genetics," 4th ed. McGraw-Hill, New York.
Sorsby, A. (1951). "Systemic Ophthalmology," Butterworth, London.
"Standard Nomenclature of Diseases and Operations" (1961) (E. T. Thompson, A. C. Hayden, eds.), 5th ed. McGraw-Hill (Blakiston), New York.
Stein, E. (1930). *Strahlentherapie* **37**, 137.
Stein, E. (1935). *Z. induktive Abstammungs-u. Vererbungslehre* **69**, 303.
Steinhoff, D., and Kuk, B. T. (1957). *Z. Krebsforsch.* **62**, 112.
Steinmetz, E. F. (1957). "Codex Vegetabilis," 2nd ed. E. F. Steinmetz, Amsterdam.
Stewart, H. L., Snell, K. C., Dunham, L. J., and Schlyen, S. M. (1959). "Atlas of Tumor Pathology," Sect. XII, Fasc. 40. Armed Forces Institute of Pathology, Washington, D. C.
Strasburger, E. (1906). "Lehrbuch der Botanik," 8th. ed. Fischer, Jena.
Swanson, C. P. (1957). "Cytology and Cytogenetics." Prentice-Hall, Englewood Cliffs, New Jersey.
Swindle, P. E. (1935). *Ann. Otol., Rhinol. & Laryngol.* **44**, 913.
Sze, L. C. (1953). *J. Exptl. Zool.* **122**, 577.
Tannenbaum, A., and Silverstone, H. (1953). *Advances in Cancer Research* **1**, 451.
Ten Cate, G., and Van Doorenmaalen, W. (1950). *Proc. Koninkl. Ned. Akad. Wetenschap.* **53**, 894.
Toolan, H. W. (1951). *Proc. Soc. Exptl. Biol. Med.* **77**, 572.
Toolan, H. W. (1954). *Cancer Research* **14**, 660.
Treadwell, F. P. (1932). "Qualitative Analysis" (transl. and revised by W. T. Hall) 8th English ed., Vol. 1. Wiley, New York.
Turesson, G. (1922). *Hereditas* **3**, 211.
Uhlenhuth, E. (1903). *In* "Festschrift zum 60. Geburtstage von Robert Koch." Herausgegeben von seinen dankbaren Schülern, p. 49. Jena.
Van Emden, F. I. (1957). *Ann. Rev. Entomol.* **2**, 91.
Walter, C. W., and Gillespie, D. R. (1960). *Minnesota Med.* **43**, 123.
Walton, H. F. (1953). *Sci. Am.* **188**, 68.
Warburg, O. (1924). *Naturwissenschaften* **12**, 1131.
Warburg, O., and Minami, S. (1923). *Klin. Wochschr.* **2**, 776.
Warren, S. (1931). *Arch. Pathol.* **11**, 255.
Warren, S., and LeCompte, P. M. (1952). "The Pathology of Diabetes Mellitus," 3rd ed. Lea & Febiger, Philadelphia, Pennsylvania.
Weatherford, H. L. (1935). *Am. J. Pathol.* **11**, 611.

Weiss, P. (1955). *In* "The Hypophyseal Growth Hormone, Nature and Actions," (R. W. Smith, Jr., O. H. Gaebler, C. N. H. Long. eds.), p. 3. McGraw-Hill, New York.

Willis, R. A. (1952). "The Spread of Tumors in the Human Body." Butterworth, London.

Willis, R. A. (1953). "Pathology of Tumors," 2nd ed. Butterworth, London.

Wilson, G. S., and Miles, A. A., Editors. (1955). "Topley and Wilson's Principles of Bacteriology and Immunity," 4th ed. Williams & Wilkins, Baltimore, Maryland.

Winge, Ø. (1927). *Z. Zellforsch. u. mikroskop. Anat.* **6**, 397.

Winge, Ø. (1930). *Z. Zellforsch. u. mikroskop. Anat.* **10**, 683.

Winton, F. R., and Bayliss, L. E. (1955). "Human Physiology," 4th ed. Little, Brown, Boston, Massachusetts.

Wintrobe, M. M. (1951). "Clinical Hematology," 3rd ed. Lea & Febiger, Philadelphia, Pennsylvania.

Wolff, G. (1894-1895). *Arch. Entwicklungsmech. Organ.* **I**, 380.

Woodburne, R. T. (1957). "Essentials of Human Anatomy." Oxford Univ. Press, London and New York.

Wright, G. P. (1958). "An Introduction to Pathology," 3rd ed. Longmans, Green, New York.

Wright, S. (1932). *Proc. Intern. Congr. Genet. 6th Congr. Ithaca, New York, 1932, Vol.* 1, p. 356.

Zeidman, I. (1961). *Cancer Recearch* **21**, 38.

Zeidman, I., Gamble, W. J., and Clovis, W. L. (1956). *Cancer Research* **16**, 814.

Zinner, G. (1955). *Arzneimittel-Forsch.* **5**, 125.

Subject Index

C

D

E

N

O

P

R

S

U

V

W

X

Z